Proceedings of the
Michigan Morphometrics Workshop

Edited by

F. James Rohlf
Professor of Biological Sciences
State University of New York at Stony Brook

Fred L. Bookstein
Distinguished Research Scientist
The University of Michigan

Proceedings of the
Michigan Morphometrics Workshop
held at the University of Michigan,
Ann Arbor, Michigan from
May 16 through May 28, 1988.

Special Publication No. 2
The University of Michigan Museum of Zoology
Ann Arbor, Michigan
1990

SPECIAL PUBLICATION NO. 2
THE UNIVERSITY OF MICHIGAN MUSEUM OF ZOOLOGY

Sara V. Fink, Editor

The publications of the Museum of Zoology, The University of Michigan consist primarily of two series—the Occasional Papers and the Miscellaneous Publications. Both series were founded by Dr. Bryant Walker, Mr. Bradshaw Swales, and Dr. W.W. Newcomb. Occasionally the Museum publishes contributions outside of these series; beginning in 1990 these are titled Special Publications and are numbered. All submitted manuscripts receive external review.

The Occasional Papers, publication of which was begun in 1913, serve as a medium for original studies based principally upon the collections in the Museum. They are issued separately. When a sufficient number of pages has been printed to make a volume, a title page, table of contents, and an index are supplies to libraries and individuals on the mailing list for the series.

The Miscellaneous Publications, which include papers on field and museum techniques, monographic studies, and other contributions not within the scope of the Occasional Papers, are published separately. It is not intended that they be grouped into volumes. Each number has a title page and, when necessary, a table of contents.

A complete list of publications on Birds, Fishes, Insects, Mammals, Mollusks, Reptiles and Amphibians, and other topics is available. Address inquiries to the Director, Museum of Zoology, The University of Michigan, Ann Arbor, Michigan 48109-1079.

ISBN 0-9628499-0-1
ISSN 1053-6477

Contents

                                                                                                                    Page

Preface .... *v*

Part I. Introduction 1
  The Michigan Morphometrics Workshop 1
  Software 5

Part II. Data Acquisition 7
  1   Data Acquisition in Systematic Biology, *W. Fink* 9
  2   Digital Images and Automated Image Analysis Systems, *N. MacLeod* 21
  3   An overview of Image Processing and Analysis Techniques for Morphometrics, *F. J. Rohlf* 37

Part III. Analytic Methods
  Introduction and Overview, *F. L. Bookstein* 61

Section A. Multivariate Methods 75
  4   Traditional Morphometrics, *L. Marcus* 77
  5   Reification of Classical Multivariate Analysis in Morphometry, *R. Reyment,* 123

Section B. Methods for Outline Data 145
  6   On Eigenshape Analysis, *G. P. Lohmann* and *P. N. Schweitzer* 147
  7   Fitting Curves to Outlines, *F. J. Rohlf* 167
  8   Median Axis Methods in Morphometrics, *D. O. Straney* 179
  9   Application of Eigenshape Analysis to Second Order Leaf Shape Ontogeny in *Syngonium*
      podophyllum (Araceae), *T. Ray* 201

Section C. Methods for Landmark Data
  Introduction to Methods for Landmark Data, *F. L. Bookstein* 215
  10 Rotational fit (Procrustes) Methods, *F. J. Rohlf* 227
  11 Higher Order Features of Shape, *F. L. Bookstein* 237
  12 Conventional Procrustes Approaches, *R. E. Chapman* 251
  13 Resolving Factors of Landmark Deformation: Miocene *Globorotalia* DSDP site 593, *R.*
      Tabachnick and *F. L. Bookstein* 269
  14 Morphometrics and Evolutionary Inference: a Case Study Involving Ontogenetic and
      Developmental Aspects of Evolution, *N. MacLeod* and *J. Kitchell* 283
  15 Morphometrics and the Systematics of Marine Plant Limpets (Mollusca: Patellogastropoda),
      D. Lindberg 301
  16 Comparative Ontogeny of Cranial Shape in Salamanders Using Resistant fit Theta Rho
      Analysis, *S. Reilly* 311

Part IV. The Problem of Homology. 323
  17 Homology in Morphometrics and Phylogenetics, *G. R. Smith* 325

18 Using Growth Functions to Identify Homologous Landmarks on Mollusc Shells, *S. C. Ackerly* 339

19 Homology, Pattern, and Morphometric Framework in Cyrtid Radiolarians, *A. Sanfilippo and W. R. Riedel* 345

Combined References From All Chapters...363

Preface

The Workshop of which this volume is the somewhat augmented and supplemented Proceedings was mounted in Ann Arbor in May 1988. Its purpose was to determine the maturity of the consensus just then emerging regarding the interrelations among the best morphometric methods for evolutionary biology and systematics. At that Workshop it became clear to us that a careful new synthesis of old themes—ordination, transformation, superposition—was indeed worth mastering for most applications. Over the ensuing twenty-four months the new methods were reprogrammed and exemplified in the case studies collected in this volume. The whole system—methods lectures, programs, examples—was extensively tested at a second Workshop, on the campus of the State University of New York at Stony Brook, in June 1990. We found it to support an ever-greater variety of morphometric investigations, which will require a second volume for *their* collection, and so on. These Proceedings, and the associated system of programs on floppy disks, should be more valuable than is usual in this genre. Beyond the mere historical record of what was said in Ann Arbor, here are sharp and sturdy tools for you to exploit or develop further.

The synthesis put forward here combines three different thrusts in morphometrics that used to be separate: multivariate morphometrics (typically the analysis of distances and angles among landmarks, with an emphasis on ordination), the morphometrics of deformation ("Cartesian transformations," finite-element methods, and the like), and superposition-based morphometrics (the "Procrustes" analyses and their resistant variants). When carried out unwisely, even if in keeping with the example of the original publications, these methods may sometimes appear to lead to irretrievably contradictory analyses of a single data set. But when each is constrained and disciplined by integration with the others, they are remarkably consistent in statistics, in graphics, and in biological implications. Our purpose in this volume is to impart some of the tactics of each of these approaches and to show how by careful scrutiny, with modifications of published algorithms as necessary, they can be made to yield up the same interpretations. Using the associated program disks, you will be able to analyze your own data by most of these approaches; you will learn how to consider all the alternative displays for a single data set before choosing one, or more than one, for report and interpretation; you will learn some of the algebraic and geometric strategies by which one can prove that certain of these techniques must agree and others must omit information in predictable ways; and you will learn how to critique other methods by calibration against this shared framework.

A little history. Morphometrics grew naturally out of the tradition of accurate description in comparative anatomy. Before the 1960's, it was the concern of gifted amateurs, like D'Arcy Thompson and Julian Huxley. They and other pioneers explored the ways in which quantities can be reasonably attached to biological shape in light of understandings from developmental and evolutionary theory. The first branch of morphometrics to emerge as a praxis on its own, multivariate morphometrics, was almost wholly complete by the 1960's. Jolicoeur, Blackith, Burnaby, Hopkins and others contributed bits of logic and inference to the now-classical synthesis represented in Blackith and Reyment's *Multivariate Morphometrics* of 1971. This approach emphasizes a common core of frequently useful matrix manipulations involving variables the origin of which is never explored.

In the late 1970's and early 1980's, some methodologists turned their attention back to the more graphical methods of earlier years. Bookstein developed his implementation of Cartesian transformations during this period; Oxnard explored new applications of methods of mathematical optics, biomechanics, and computer vision; Blum invented and tested his "medial axis." Yet the variety of new graphical tools was not matched by any commensurate advances in the biometrics (narrowly construed) whereby data of this sort might be summarized and its covariation with causes or effects rigorously tested.

The third element in the synthesis appeared somewhat unexpectedly in the early 1980's. Working at first separately, Bookstein, Goodall, and Kendall developed the same statistical method to exploit *all* the information available in sets of landmark locations. The convergence was announced to statisticians only very recently—in 1986, in fact—and the first book-length treatment of the topic, Bookstein's *Morphometric Tools for Landmark Data*, will not appear until 1991. It is now clear that the old "multivariate morphometrics," which thought it was analyzing "distances" and "angles," was actually analyzing landmark location data instead. Likewise, the "outline" methods were summarizing relationships among "corresponding" points, somewhat akin to landmarks, and the superposition methods were graphical approximations to some of the same statistical themes.

For a few years now, morphometricians who emphasize the geometry of form have been exploring ways of explaining this convergence to our various (and very diverse) audiences. These Proceedings are the strongest evidence to date that, at least when supplemented by work at the blackboard, we think we've got it: we can teach mixed groups of biologists how to apply *all* the methods to a single data set, and make sense of all the computations and graphics.

The easiest way to approach this common core of morphometric method—by slogan, as it were—is to assert that the synthesis fills the lacunae in each of the component methods by carefully hybridizing it to another. There are thus three "methodological mosaics," corresponding to the three pairings we might invoke. (1) One interpretation of what we are preaching is the careful addition (via the deformation methods) of information about mean landmark locations to the multivariate analysis of covariances of those locations, or their distances, with group, size, or environment. (2) Another way of explaining exactly the same synthesis is as the careful addition of information about variance to the "Cartesian transformation" methods, so that when one draws a deformed grid, one has access to descriptors of the reliability of all its features. (3) Yet a third way of explaining the synthesis is as the overlayment of the superposition techniques by information about spatial coherence—autocorrelation of the little vectors it is encouraging our visual systems to intuit. But this requires that the distance functions usually reported there be replaced by others more central to the morphometric statistical literature.

When the original techniques are paired correctly, the resulting reports have all the same statistics. One may think of our synthesis, then, as a scheme of alternate *diagrams* for the same underlying information. For instance, you have been taught, correctly, that ratios aren't usually a very good way of analyzing distance data. But the multivariate analysis of landmark locations goes forward perfectly in terms of a particularly carefully crafted sort of ratio, the shape coordinate pair, which can, in turn, suggest new variables, themselves conventional ratios, which best summarize observed differences. As another example, it is well-known that the results of a multivariate ordination are in large part a function of the variables you choose. Yet once a set of landmarks has been fixed, a good argument can be made (and is made in one of the chapters here) that there are three "net morphometric distances" that underlie most biologically meaningful morphometric ordinations and explanations. These are size difference, uniform shape distance, and localized change ("bending energy"). Analysis goes best, and the relation of alternate techniques is clearest, when these pictures are kept carefully separate. As a final example, the features of deformation (and we're distributing to you a fairly nifty little package, TPSPLINE, for these graphics) can be usefully imagined as vectors in a few conventional multivariate analyses of geometrically distinctive "components," uniform and nonuniform. Thus features of deformation can be subject to visual inspection just like any other set of multivariate descriptors, checked for outliers and contemplated for hints of alignment with organismic form. For outline data without landmarks, several of the choices we are teaching coalesce; we will show you those, and also indicate which remain patent, to be weighed carefully in particular applications. These concerns apply especially to analyses of outlines with few, or any landmarks, such as typical plant leaves and certain microfossils.

Those of you who will read all the way through these Proceedings, and test the programs on the sample data sets supplied, will have been exposed to a great variety of good morphometric analyses and the appropriate reporting styles. Our subject is the translation of findings back and forth among these essentially equivalent hybrid methods. There is not one "best" method, but many; get to know them all. If you absorb the message of these chapters, then in your research and in your reading and reviewing you will be committed to at least three distinct methodological virtues:

1. You will apply every sort of graphical report to every data set. You will *always* look at mean differences as deformations; always compute and scatter features of these deformations; always scatter each landmark "separately" according to some convenient superposition; always test for spatial integration of patterns among such scatters. You will proceed in all this with confidence that the multivariate statistics of all these sheafs of reports are the same.

2. You will *always* consider how variously the common core of findings from these analyses may be expressed on the page: as vectors, tensors, deformations, "warps," or statistics. You

will be able to use all these tools to communicate the meaning of an analysis to all of your audiences.

3. You will understand how particular methods, from the "shear" through finite-elements, may be calibrated against this common core, both statistically and practically. You will be able to detect precisely where in a morphometric analysis information is lost, and you will be able to avoid that loss wherever you choose.

There is likely to be a second volume of these Proceedings incorporating new lectures from the 1990 workshop and further programs and worked examples. We are very eager to hear from you about your experiences with this new synthesis and your suggestions for its further dissemination. It is easiest to communicate with the editors by electronic mail. Write to both of us when you write, please.

We are very grateful for support from the Systematic Biology Program of the National Sciences Foundation for the subventions which enabled the 1988 and 1990 Workshops and the publication of these Proceedings at so low a price. David Schindel was an enthusiastic participant in most of the planning phases of both; perhaps he will accept a leather-bound copy of these Proceedings as a sufficient reward. We thank all the participants for their patience while we explored methods for teaching so interdisciplinary a synthesis. We especially wish to thank those who shared the podium with us—besides the authors of chapters in this volume, there were Julian Humphries, Barry Chernoff, Jim Cheverud, William Atchley, and Robert Ehrlich—and those who managed the computer systems on which sets of 30 participants would try out half a dozen methods in one frantic week within the confines of one or the other innocent evolutionary biology wing. At the 1988 meeting, the technological fixers included Julian Humphries, Norman MacLeod, and Leslie Marcus; at the 1990 meeting, Adrian Lema, Junhyong Kim, Steve Reilly, Dennis Slice, and, again, Leslie Marcus. In 1987-88, Bill Fink and Jennifer Kitchell wrote the grant proposal, overseeing the design of the curriculum and the selection of teachers, students, and vendors. They ably delegated the logistics of housing and feeding participants and faculty while they went about the business of mastering this stuff with the rest of us. In 1990, Bill and Sara Fink guided this volume through production, and Michael and Bonnie Bell managed logistics for the Stony Brook workshop at which the text and software were tested.

Special thanks go to Marie Josee Fortin for the tedious hours she spent formatting text, converting to yet another wordprocessor, and then formatting again for the production of camera-ready output. Thanks also to Norma Watson for copy editing the book so efficiently under a very strict deadline.

Your editors, while taking no responsibility for the bugs in the dormitories, would be interested in any bugs in the software or the text that you might find.

F. James Rohlf
Fred L. Bookstein
Stony Brook, New York, and Ann Arbor, Michigan

E-mail addresses, as of July 1, 1990, are:
`rohlf@sbbiovm.bitnet`
`Fred_L._Bookstein@um.cc.umich.edu`

July 1, 1990

Part I

Introduction

Introduction

This proceedings volume is based upon the workshop that took place from May 16 through May 28, 1988 at the University of Michigan. Most facilities were provided by the Museum of Paleontology and the Museum of Zoology. The workshop was sponsored by the grant "Workshop on morphometrics and systematics" (BSR 8801107) from the Systematic Biology program of the National Science Foundation. William L. Fink and Jennifer Kitchell were the Co-Principal Investigators. The summary that follows is based in part on their final report to the National Science Foundation.

The Michigan Morphometrics Workshop

The workshop had three goals. One of the goals was to bring experts in various morphometrics specialties together with highly motivated participants, with the expectation that the participants would be able to return to their institutions to teach the techniques and to integrate them into ongoing or new research programs. A second goal was to provide an environment where specialists in both morphometrics and systematics could interact and focus on conceptual and practical problems of integrating systematics and morphometrics methods. A third goal was to familiarize all concerned with advanced data acquisition systems for morphometric applications.

The organizers of the workshop (the Co-Principal Investigators plus F. L. Bookstein) set the agenda, chose the 16 lecturers, and selected 30 participants from the applicant pool. The agenda and choice of lecturers were designed to bring representatives of major schools of morphometrics to the Workshop. Participants were selected with the aim of getting a cross-section of the applicant pool (141 completed applications were received), with emphasis on scholars with a proven publication record (or one of great promise) and access to graduate participants. An effort was made to get reasonable representation from both small and large institutions, both public and private. University of Michigan faculty, staff, and graduate participants were provided with access to the workshop by participating as facilitators, usually helping participants with data acquisition.

Participants were expected to remain on campus for the entire two weeks of the workshop. Lecturers were asked to come for at least two days, the day of their presentation and the day following, for student interaction and consultation; and three off-campus Lecturers managed to attend the entire workshop (Humphries, Marcus, Rohlf). Rohlf and Bookstein were specifically charged with attending all lectures to provide commentary for the sake of continuity.

Lecturers were asked to provide drafts of their lectures for circulation among Lecturers and for dissemination to the participants in the form of a workbook. Lecturers were asked to address four issues in their lectures on particular methods: what biological assumptions were being made, when the technique was appropriate to a particular sort of data, when it was inappropriate, and how it related to systematic biology. The success of this charge was rather variable (but improved somewhat in the final versions of the chapters included in this volume).

Workshop Schedule

The workshop lecture schedule was organized into three parts: data acquisition and techniques, theory and applications of morphometric methods, and case studies of applications to specific problems in evolutionary and systematic biology. Laboratories were organized in three parts, as well: familiarization with data acquisition systems, applied data acquisition, and presentations of analyses done during the workshop.

The workshop began with introductory lectures to familiarize Lecturers and participants with what the workshop's goals were and what would be covered. The first afternoon laboratory session was assigned to vendors of data acquisition systems so that participants could become familiar with the equipment (see Chapter 1 by Fink). This also gave vendors a chance to demonstrate their systems. The second day began with lectures on data types and data gathering techniques, with emphasis on video-based systems (Fink and MacLeod, corresponding to Chapters 1 and 2, respectively), followed on the third day by a session on image processing (Rohlf, corresponding to Chapter 3). Richard Strauss lectured on PCA, shearing techniques, and truss analysis.

On the fourth day, the theoretical portion of the workshop was begun by Bookstein, who gave an overview of morphometrics (corresponding to the material in the Introduction to Part III of this volume). The next few sessions addressed the analysis of outline data by fitting functions (Rohlf, corresponding to Chapter 7), eigenshape analysis (Lohmann, Chapter 6), rotational fit methods (Chapman, Chapter 12), coordinate techniques (Bookstein, Chapter 11), Fourier methods (Ehrlich), and traditional multivariate methods (Marcus and Reyment, Chapters 4 and 5, respectively). A final lecture series on advanced uses of coordinate techniques for reconciling findings of diverse other methods was given by Bookstein (corresponding in part to the material in the Introduction to Part III, Section C of the present volume).

The systematics applications section of the workshop began with medial axes analyses (Straney, Chapter 8), followed on the next day with a presentation on morphometrics and analyses of development (Kitchell). Lectures on homology and morphometrics (Smith, Chapter 17), growth functions in Mollusc shells (Ackerly, Chapter 18), analysis of morphometric data as characters in phylogenetic analysis (Humphries and Chernoff) completed this third section of the lectures.

The final full day of the workshop was filled with student presentations of their research projects. A number of these reports lead to chapters in the present volume (Chapter 9 by Ray, Chapter 13 by Tabachnick and Bookstein, Chapter 15 by Lindberg, Chapter 16 by Reilly, and Chapter 19 by Sanfilippo and Riedel). The last half day involved discussions about the successes and failures of the workshop in meeting its goals.

Supplemental activities during the workshop included a tutorial on applications of SAS to the coordinate as well as traditional morphometric techniques (by Marcus). This was extremely valuable inasmuch as software for several of the techniques was unavailable or was difficult to run. Supplemental discussion sections addressed systematics issues in the context of the day's lectures on morphometrics. These informal, unscheduled discussions proved very fruitful in focussing participants and Lecturers attentions on what systematics issues were being adequately addressed by Lecturers and which were not. There were also supplemental presentations on use of the phylogenetics package PAUP and the UM mainframe (both by Fink).

Lecturers

Fred Bookstein
Center for Human Growth
University of Michigan
Ann Arbor, MI 48109

Ralph Chapman
National Museum of Natural History
Smithsonian Institution
Washington, D.C. 20560

Barry Chernoff
Field Museum of Natural History
Roosevelt Rd. at Lake Shore Dr.
Chicago, IL 60605

Robert Ehrlich
Department of Geology
University of South Carolina
Columbia, SC 29208

Douglas Erwin
Department of Geological Sciences
Michigan State University
East Lansing, MI 48824

William Fink
Museum of Zoology and
Department of Biological Sciences
University of Michigan
Ann Arbor, MI 48109-1079

Julian Humphries
Section of Ecology and Systematics
Cornell University
Ithaca, NY 14853-0239

Jennifer Kitchell
Museum of Paleontology and
Department of Geological Sciences
University of Michigan
Ann Arbor, MI 48109-1079

G. P. Lohmann
Woods Hole Oceanographic Institution
Woods Hole, MA 02543

Norman MacLeod
Museum of Paleontology and
Department of Geological Sciences
University of Michigan
Ann Arbor, MI 48109-1079

Leslie Marcus
American Museum of Natural History
Central Park West at 79th St.
New York, NY 10024

Richard Reyment
Paleontologiska Institutionen
Box 558
S-751 22 Uppsala, Sweden

F. James Rohlf
Department of Ecology & Evolution
State University of New York
Stony Brook, NY 11794-5245

Gerald Smith
Museums of Paleont. and Zoology
University of Michigan
Ann Arbor, MI 48109-1079

Donald Straney
Department of Zoology
Michigan State University
East Lansing, MI 48824

Richard Strauss
Dept of Ecology & Evol. Biology
University of Arizona
Tucson, AZ 85721

Facilitators

From the University of Michigan, Ann Arbor, MI 48109

Doug Eernisse
Museum of Zoology and
Department of Biological Sciences

Dan Fisher
Museum of Paleontology and
Department of Geological Sciences

Mary McKitrick
Museum of Zoology and
Department of Biological Sciences

William Stein
Museum of Paleontology

From other institutions

Tim Ehlinger
Kellogg Biological Station
Hickory Corners, MI 49060

Brian Bodenbender
Museum of Paleontology and
Department of Geological Sciences

Paulo Buckup
Museum of Zoology and
Department of Biological Sciences

Robyn Burnham
Department of Biological Sciences

Bernie Crespi
Museum of Zoology and
Department of Biological Sciences

Steve Dobson
Museum of Zoology and
Department of Biological Sciences

Mark Johnston
Museum of Paleontology and
Department of Geological Sciences

Tim Pearce
Museum of Zoology and
Department of Biological Sciences

Elena Tabachnick
Museum of Paleontology and
Department of Geological Sciences

Participants

Spafford Ackerly
Dept. of Geological Sciences
Cornell University
Ithaca, NY 14853

Jon Baskin
Department of Geosciences
Texas A & I University
Kingsville, TX 78363

Neil Blackstone
OML-Biology Department
Yale University
New Haven, CT 06511

Nancy Budd
Geology Department
University of Iowa
Iowa City, IA 52242

John Carr
Department of Zoology
Southern Illinois University
Carbondale, IL 62901-6501

James Collins
Department of Zoology
Arizona State University
Tempe, AZ 85287-1501

Peter Dodson
Laboratories of Anatomy
University of Pennsylvania
Philadelphia, PA 19104-6045

Michael Donoghue
Dept. of Ecology & Evol. Biology
University of Arizona
Tucson, AZ 85721

Marc Feldesman
Anthropology Department
Portland State University
Portland, OR 97207

Scott Hills
Geologisches Institut
ETH-Zentrum
CH-8092 Zurich, Switzerland

Craig Hood
Biological Sciences
Loyola University
New Orleans, LA 70118

Richard Jensen
Department of Biology
Saint Mary's College
Notre Dame, IN 46556-5001

Clarence Dan Johnson
Dept. of Biological Sciences
Northern Arizona University
Flagstaff, AZ 86011

David Johnson
New York Botanical Garden
Bronx, NY 10458

Gordon Kirkland
Vertebrate Museum
Shippensburg University
Shippensburg, PA 17257

Steven Leipertz
American Museum of Natural History
Central Park W. at 79th St.
New York, NY 10024

David Lindberg
Museum of Paleontology
University of California
Berkeley, CA 94720

Diana Lipscomb
Biological Sciences
George Washington University
Washington, D.C. 20052

Robert Owen
The Museum
Texas Tech University
Lubbock, TX 79409

Mary Rauchenberger
Smithsonian Institution
National Museum
Washington, D.C. 20560

Thomas Ray
SLHS
University of Delaware
Newark, DE 19716

Stephen Reilly
Dept. of Dev. and Cell Biology
University of California
Irvine, CA 92717

William Riedel
Scripps Institution of Oceanography
La Jolla, CA 92093

Roger Sanders
Fairchild Tropical Garden
Miami, FL 33156

Stephen Scheckler
Department of Biology
Virginia Polytechnic Institute
Blacksburg, VA 24061-0794

David Schindel
National Science Foundation
Washington, D.C. 20560

Michael Simpson
Department of Biology
San Diego State University
San Diego, CA 92182

Kuo-Yen Wei
Department of Geological Sciences
Yale University
New Haven, CT 06511

Stanley Williams
Department of Biology
San Francisco State University
San Francisco, CA 94132

Scott Wing
National Museum of Natural History
Smithsonian Institution
Washington, D.C. 20560

James Woolley
Department of Entomology
Texas A & M University
College Station, TX 77843

Software

As mentioned above, one of the important problems at the workshop was the availability and compatiblity of microcomputer based software for the various methods discussed. As a partial remedy, a set of floppy disks is being distributed with this volume. The programs are all designed to run on IBM PC compatible (MS-DOS compatible) microcomputers. While some of them should run on any compatible computer, most require the presence of a graphics adaptor and a graphics monitor. None of them require the presence of a math coprocessor chip but most will benefit greatly from its presence.

All programs are furnished in the form of executable programs. Some of the programs also include the original source code in whatever language the program was written in (FORTRAN, BASIC, or Pascal). Documentation is provided in the form of "readme" files. This documentation should provide enough information to install a program and to run a set of example data. In most cases it was not practical to include detailed information on exactly what operations the program performs. That information must be sought in the original literature (in some cases chapters in the present volume).

The following programs are included in the set of morphometrics software.

BURNABY and **SHEAR**   Programs to apply the Burnaby size correction procedure and the method of shearing (see Chapter 4). Written in FORTRAN (source code included) by N. MacLeod.

CANVAR and **PCA** General programs for canonical variates analysis and principal components analysis with various diagnostics and robust estimation procedures (see Chapter 5) included. Written in FORTRAN (source code included) by R. Reyment.

DS-DIGIT   A general program to capture x,y-coordinates from a digitizing pad. Written in Pascal by D. Slice.

EFA   A simple program to compute eliptic Fourier coefficients given an outline represented by a sequence of x,y-coordinates of points. Written in FORTRAN (source code included) by F. J. Rohlf.

EIGENS   A large set of programs to perform eigenshape analysis (Chapter 6) and related computations such as data conversion and plotting. General purpose 2 and 3-dimensional plotting programs are also included. Written by P. N. Schweitzer and G. P. Lohmann.

FC   A program to convert data from various input formats to the formats required by the various morphometrics software. Program written in Pascal by D. Slice and F. J. Rohlf.

GRF   Program for generalized rotational fitting (least-squares and resistant fit of two or more objects represented by x,y-coordinates, see Chapter 10). Written in Pascal by F. J. Rohlf and D. Slice.

IMAGE   A simple demo program to show the effects of various types of image enhancement operations. Written in Pascal by F. J. Rohlf and D. Slice.

LINESKEL A program to find the median axis. Written in Pascal by D. O. Straney. A hypertext user manual is included.

PROJECT   Written in FORTRAN (source code included) by F. L. Bookstein.

RELWARP This program computes relative warps (see Chapter 11) for a sample of specimens represented by coordinate data. The computational program was written in FORTRAN (source code included) by F. L. Bookstein and a driver program was written in Pascal by J. Kim.

RFTRA   A series of programs to perform least-squares and resistant fit superimpositions of two specimens represented by x,y-coordinate data. The methods are described in Chapter 12. The program was written in BASIC (source code is included) by R. Chapman.

SCALE3D A program to perform the numerical calculations of finite scaling analysis. The program was written in FORTRAN (source code is included) by James Cheverud.

TPSPLINE Program to compute the thin-plate spline transformation of one set of coordinates into another. It also computes its decompostion into partial warps (see Chapter 11). Written in Pascal by F. J. Rohlf.

In order to save space, the program files have all been compressed and grouped together in logical groups. A utility program, PKUNZIP, is included that will extract the various files. There is also a READ.ME file that lists the contents of the distribution disks and gives instructions on how to extract the files and install the programs. Each program has its own READ.ME file giving details about how a particular program is to be installed and used. Please contact the authors of the programs for more information about details of their operation and the existence of bugs (but also notify us so that the distribution copies of the software can be updated).

Part II

Data Acquisition

There has been increasing interest in the development of practical and low-cost methods to increase the accuracy and reduce the drudgery of acquiring quantitative data for morphometric analysis. This need has increased in recent years owing to the realization that taking a few simple linear distance measurements is not sufficient for most types of morphometric analyses. Many of the newer methods of analysis require the coordinates of landmarks (see Part III, Section C).

While one can construct coordinates from triangulations of linear distance measurements, it is much more direct to simply digitize the coordinates of the landmarks. But this results in additional effort being required to measure and record two— (or three)-dimensional coordinates rather than linear distances. Fortunately, the microcomputer revolution has brought with it the development of low-cost devices that can greatly reduce the effort required collect morphometric data.

Chapter 1 by Fink gives a general survey of microcomputer-based hardware and software appropriate for data acquisition in morphometrics. The devices range from computer-interfaced digital calipers to coordinate digitizers and image analysis systems. It, of course, covers only products available at the time of the Michigan Morphometrics Workshop. While this material will become dated very quickly due to the rate of progress in this field, this account should still be useful as an overview to the kinds of hardware and software available. It also provides a directory of vendors who can be contacted for current information.

The possibility of using microcomputer-based image analysis systems to capture morphometric data received particular attention at the workshop. Some systems simply allow a user to use a mouse to mark points on a image displayed on a video screen (simulating a coordinate digitizer). Other systems assist the user in various ways. For example, programs can automatically follow the outlines of structures. Systems can also perform various transformations on the image to make it easier to see particular features. Chapters 2 and 3 by MacLeod and by Rohlf, respectively, furnish overviews of image analysis. Their accounts include discussions of models of the image formation process itself and of methods to extract information from an image. There are also discussions of some of the limitation of what one should expect an image analysis system to do to automate data collection in a morphometric study.

Chapter 1

Data Acquisition for Morphometric Analysis in Systematic Biology

William L. Fink

Museum of Zoology and Department of Biology
University of Michigan
Ann Arbor, Michigan 48109

Abstract

There are several technologies now available for acquisition of shape information from biological specimens. The technology appropriate to any particular application depends on the nature of the specimens (their size and shape in three-dimensions, soft or hard-bodied, with or without identifiable landmarks) and the nature of the questions being asked. Two classes of measurement are typically used: distances and coordinates. A survey of data acquisition technologies for both of these kinds of data is included, with comments on the experiences of students with several systems used in the Workshop.

Introduction

Data acquisition has always been a major part of systematic biology, and a major bottleneck as well. Much effort has been spent to gather data on which to base systematic inferences, and only recently has there been improvement over methods common in the 19th century. Although my task is not to discuss the problem of *what* to measure of biological objects, but rather of *how* to measure, the latter really is dependent on the former, and that must be addressed, however briefly. My own experience in ichthyology, where a standard set of distance measurements has been in place for much of this century (see Hubbs and Lagler, 1941), shows that measurements are often made without proper consideration of the scientific questions being asked. As Strauss and Bookstein (1982) clearly showed, the standard ichthyological measures do not truly capture the shape of a fish, but rather redundantly and unevenly sample only certain aspects of its shape. Thus the first questions one must ask at the beginning of a systematic study, before any data are acquired, are "what do I want to measure?" and "why do I want to measure it?". Once those questions are answered, the technological issues are approachable.

The question "why do I want to take a measurement?" is one that each scientist must ponder before starting a morphometric study. It is likely that the study organisms exhibit some shape similarities and differences and that there is a need to quantify them to ask the standard systematic questions of how many taxa are present and how are they related, before the study can be extended to process levels. It is often true that the question of relationships is harder than the first, and one that needs to be addressed on a deeper theoretical level than the former.

The "what do I want to measure?" question needs thought as well. Systematists typically wish to measure those aspects of shape that contain information about group membership, and for most organisms we do not know what that is before an analysis is done. So the answer to this question is to try to capture as much information about shape as possible in the hopes of getting the information needed. The data acquisition technique should archive shape data in a format that allows use of various analytical approaches. Critical use of analytical methods is then needed to distinguish the systematically informative data from the uninformative noise.

There are two kinds of data useful in summarizing representations of form, distances and coordinates. Distances are the most common morphometric data, and have been in use for centuries; they can be taken using tools as primitive as string and a ruler. Distances are quantitative descriptions of the length or size of an object ("the dorsal fin is 10mm long"), or a measure of separation between two parts of an organism ("it is 25mm between the dorsal fin and the anal fin"). Coordinate data represent points in a grid, described by an x and a y. The grid can be the lines on graph paper, a digitizing tablet, or a digital video image. Coordinate data can be converted into distance data (as long as a reference standard measurement is provided), but distance data cannot always be converted into coordinate data. This makes the choice of data acquisition technology important since certain kinds of analyses can be performed only on coordinates. Outlines are summations of coordinate data around the periphery of a form.

In some studies, with some organisms, capturing but two dimensions of the form will be adequate to answer systematic questions. For others, three dimensions may be needed. This is an issue that will have to be addressed by each individual and will have a bearing on what kind of data acquisition system is appropriate.

Data Acquisition Devices

This section must begin with a warning that whatever device is chosen for data acquisition, rigorous accuracy and repeatability studies should be done before one commits to gathering research data. Never assume that the device is accurate. Periodic checks on accuracy should be made to insure that data are truly comparable over time. Some devices are more prone to problems than others, but *all* devices should be monitored for performance.

Data acquisition systems used in the Workshop are listed in the Appendix.

Instruments for Measuring Distance

These include the usual measuring devices used in many fields, such as ocular micrometers, dial micrometers, calibrated stages, and calipers. Devices for coordinate data, discussed below, can also be used for distances, with the proviso that there is an intermediate step of converting the coordinates to distances. Below is a list of some common traditional technologies, with comments on their strengths and weaknesses.

An ocular micrometer is a scale etched into the optics of a viewing device; depending on lenses used, the object being measured can be of virtually any size. However, most ocular micrometers are used for very small specimens that can be viewed through a microscope. To use them you focus on the object, move the specimen holder or microscope stage so that the initial point you wish to measure is under one of the lines of the micrometer, then count the number of lines to the second point. This distance is given in microns, in most cases. The advantage of the micrometer is that it is convenient to use and requires no other technology (except a tape recorder, or a pencil and paper for recording data). The disadvantages are that in most applications it is limited to specimens of small size, and its accuracy is particularly dependent on experience and training.

An improvement on the micrometer is the mechanized stage. The user moves the stage to measure the distance between parts of a specimen. This is obviously limited to specimens which fit under a microscope, usually a compound microscope.

Calipers are the standard tools for many standard measures in systematics that involve distances across a form. Vernier calipers have been replaced almost universally by dial calipers, in which the distance between two points is read on an analog dial or digital display, usually in millimeters and tenths of millimeters (although the calipers themselves allow measurements to the hundredths of millimeters). A relatively efficient way to use dial calipers is to read the measurements into a voice-activated tape recorder and then transcribe the data to a computer or calculator. The next step in efficiency is to purchase a caliper with a computer interface (e.g., Marcus, 1982). Several of these calipers are on the market, some with mechanical gears and analog dials, and some with magnetic strips and digital LCD's (Liquid Crystal Display). Depending on price, the calipers come with either a "dumb" serial connection (RS-232) which requires some programming to use, or a "smart" interface that minimizes programming needs. The "smart" interface usually is a good investment. In either case, a data acquisition program is needed in the computer to organize the incoming data. Several of these programs are available, most in BASIC, and some can be customized fairly easily. Calipers attached to a computer are a very efficient way to take data from macroscopic objects, but they are of limited value for small, near microscopic objects or for very large objects. Calipers can be purchased in several lengths. People planning to use calipers in the field should take note of the location and vulnerability of the gears and teeth on mechanical calipers and note that batteries are needed for the LCD instruments.

Instruments for Coordinate Data

There are several kinds of instruments available for gathering coordinate data, the most common of which include the pantograph, digitizing pads (also known as graphics tablets), video-based systems, sonic digitizers, and light-focusing microscopes. The last two devices can be used for three-dimensional objects, as discussed below.

The pantograph is the least expensive device for coordinate data acquisition. Its usual application is enlarging and reducing drawings. It consists of a series of metal bars connected together to form a series of levers. Adjustment of the positions of the connections between the levers determines the ratio of enlargement or reduction. The tracer, a sharp point attached to the underside of one bar, is moved about the specimen while a pencil causes the points or outline to be laid down (enlarged if that is desirable) on a piece of paper. The actual coordinates can be read by placing the marked paper over a piece of graph paper (enhanced by use of a light box) and then keyed into a computer. A more efficient strategy is to use a digitizing tablet to get the data into the computer. The greatest drawback of the pantograph is its inefficacy on specimens of a limited size, especially for small organisms. Specimens of less than 15 mm are approaching the lower limits of repeatability for this mechanism.

The digitizing pad has become a common device for getting positional information into a computer, and it is probably the most common tool used by morphometricians. A requirement for its use is that the object of interest must be essentially two-dimensional. For taking coordinate data from photographs, such as from a scanning electron microscope, or from a microscope and *camera lucida* (with the caveat noted below), a pad is an excellent choice. Another advantage of a digitizing pad is its resolution (480 lines/cm is common); currently, there is no other physical device that allows this degree of resolution with concomitant accuracy and repeatability. Whether such resolution is needed will depend on the size of the specimens, the analytical techniques used, and the degree of accuracy necessary to approach the problem. There are a number of digitizing pads on the market, from many manufacturers. To a user, the main differences in them include resolution, software support, and ease of installation and use. The major market for digitizing pads is the burgeoning Computer Aided Design (CAD) industry and new users may find that many of the commercial software data acquisition programs are far more complex than needed. There are several public domain acquisition programs available, some of which accompany this volume.

Most digitizing pads come with two pointing devices, a stylus and a cursor. The stylus looks much like a ball-point pen; its tip is touched to the pad at the desired coordinate to send the signal to the computer. The cursor resembles a mouse with a transparent plastic disc at its front. This disk usually has either a "bulls eye" or cross-hair built into it. A cursor is preferred to the stylus because the point being digitized is more visible and because the cursor is easier to keep steady.

Most digitizing pads are of two types, electrostatic or electromagnetic. Electrostatic digitizers use an electric field from the pad's surface and a cursor which is a capacitive pick-up, itself attached to a source of a sine wave signal equal to that of the tablet. As the cursor moves, it detects changes in the phase of the electrical field. While an electrostatic tablet is relatively unaffected by physical shocks and other environmental changes, it is sensitive to conductive materials and can even be affected by moisture. Electromagnetic digitizers are more common than electrostatic, and they are unaffected by conductive materials and environmental changes, making them more attractive. These pads have parallel copper wires embedded in them which are connected on one side to each other and on the other side by a multiplexer which scans them. The cursor has a small coil inside and acts as a transmitter. The cursor produces a current in each wire of the grid, proportional to its distance from the wire. The digitizer determines cursor location by scanning the output from the wires and analyzing the signals.

Most pads have areas of their surface dedicated to commands, so that a touch of the cursor in that area sends a message, such as "delete the last point taken" or "set the scale" to the computer. With a little programming this command area can be moved about, sized, and have its functions changed.

For systematic work, it is advisable to get a digitizing tablet with a translucent working area, so that it can be backlit. Digitizers usually plug into the serial port of the computer. Some have an optional display of the pad output. Their resolution and ease of use have led many manufacturers of video-based systems to incorporate digitizing pads as the main pointing device.

The main disadvantage of digitizing pads is the usual requirement of an intermediate step between the specimen and the pad. Three-dimensional objects must be reduced to two dimensions and objects too small or too large must be sized to fit on the tablet (usually about 300 mm or less). This intermediate step can be time-consuming and can introduce error into an analysis unless it is done carefully. For example, photographing specimens to digitize introduces the distortions of lenses at least twice: camera and photo enlarger. The photographs must be of sufficient quality that all the pertinent parts of the organism are clearly visible. Some workers project 35mm slides onto a digitizing pad, but usually this is not recommended because of the quality of most slide projector lenses. Many use digitizing pads in conjunction with a *camera lucida* (e.g., Jacobs and Claeys, 1987), a practice that can potentially introduce error into analysis due to the distortions in the optics of the *camera lucida* itself, and from the difficulty of getting the hand-held cursor as seen though the *camera lucida* properly and consistently aligned. Caveats about determining the accuracy of a data acquisition device apply especially to this kind of system and users should do repeatability studies with objects of known size before they commit to gathering data. It is not advisable that more than one person take data for a single study using this arrangement because of the distortions caused by differences in technique (usually unconscious) of different individuals.

The sonic digitizer is a device that overcomes some of the problems of the tablet digitizer, with some loss in resolution and repeatability, although the resolution lost will not affect many systematic applications. This device functions by measuring the time of arrival of sound, generated by the cursor, at small microphones placed along the frame of the digitizer. One small version consists of a microphone bar and a cursor, with the active area being about one square foot. Larger versions consist of two bars perpendicular to one another, and a cursor. Since there is no pad, these digitizers can be made in large sizes and are excellent for maps and other large, relatively two-dimensional objects. By extending a third microphone bar perpendicular to the other two, a three-dimensional digitizer is formed. The main drawbacks of sonic digitizers are inconsistency due to imprecise positioning of the sound source in the cursor, and the need for a relatively clear area between the cursor and the microphone bars. One should be aware that objects between the cursor and the microphones will deter consistently good results.

The Reflex^{TM} microscope is a light beam-based three-dimensional system. To use the system one focuses the microscope on the point to be measured, then focuses on that same point light beams generated by bright light emitting diodes (LED). When the light beam is sharply defined on a surface, the press of a foot pedal sends x, y, and z coordinates via a serial connection to the host computer for computation of distances, constructions of the image in three dimensions, and some statistical analysis. This is one of the few systems available which allows three-dimensional measurements on a specimen, directly in three dimensions. Its disadvantages include its limited range (rather small specimens are necessary) and the delicacy of the apparatus.

Video based systems are discussed in detail in the next chapter, but a general introduction to them now is appropriate. The flexibility and power of these systems make them the preferred data acquisition technology for many systematics problems. One strength of video data acquisition systems (VDAS) is that, because they are optically based, they can be used for specimens of virtually any size, from micron to meter range. Since the image of the specimen is being held in a computer's memory, many techniques are available to enhance the image, and in some systems the computer does image analysis and data acquisition. As with other optical systems, the VDAS is for data acquisition in two dimensions. Three-dimensional representations can be obtained by rotating specimens and redigitizing. When only a few measures in the third dimension are needed, it may be more economical of time to use calipers in combination with a VDAS (Fink, 1987).

Video based data acquisition systems usually consist of a minimum of a video camera (either analog or digital), frame grabber (or video digitizer), microcomputer, and usually, a video monitor. The major differences between individual systems involve resolution of the frame grabber and the capabilities of the software package. The largest investment commercial vendors make in these systems is the software, and its design and implementation will govern the system's market, almost regardless of the hardware being used.

Some of the important components of a VDAS are listed below, with some comments that should be kept in mind when using one or considering purchase of one. Perhaps the first choice one has to make is whether a black-and-white or color system is appropriate.

Most currently available VDAS are black-and-white systems, but some of these use "pseudocolor" to enhance the gray images. Black-and-white systems are less expensive to buy than color systems because color video cameras, color frame grabbers, and color monitors are all more expensive than their black-and-white equivalents. For many systematic applications, true color systems are not required, but for some, such as analysis of color patterns, a color system is necessary and completely justifiable. For most systematists, especially those working with preserved museum specimens, a black-and-white system is completely adequate, especially given its lesser price compared with true color VDAS.

"Pseudocolor" systems use a black-and-white camera with a color video monitor, and the image processor (located on the frame grabber) assigns colors to ranges of the usual gray scale for display of the image. The actual color of specimens is not shown. The computational overhead of some "pseudocolor" systems may cause them to have lower resolution than a black-and-white system.

Since VDAS's are optical systems, the first important component to consider is the lens. There are video attachments to virtually all dissecting and compound microscopes, although most are rather expensive. For macroscopic specimens, a good quality lens and bellows or extender ring setup is usually adequate. Since most video lenses are manufactured for much less demanding optical tasks than systematists require, it may often be a good choice to buy a high quality 35mm single lens reflex (SLR) camera macrolens and a video adaptor to attach it to the camera. For larger specimens, standard SLR lenses can be used. In all cases involving lenses, one should experiment thoroughly to determine the limits of the optics; for example, distortion increases near the edges of a lens. Nothing less than measuring good quality graph paper several times, over much of the area covered by the lens, is needed to insure that a lens is adequate for the task at hand.

Choice of a video camera is also important, and here there are two choices: analog vidicon tube cameras and digital (solid state, or so-called "chip") cameras. Vidicon cameras are larger, heavier, and prone to loss of picture quality over time. Because they are analogue cameras, however, compared to many less expensive chip cameras, they have higher resolution and are less expensive. Chip cameras use a photovoltaic silicon chip as the light detector; they can be very small, sometimes smaller than the lens attached to them, and they are not supposed to deteriorate over time. In both vidicon and chip cameras, the output is converted to an analogue signal. All of the systems used at the Workshop included chip cameras, and in all cases these cameras appeared to have adequate resolution. Color can be recorded either with a color camera or a black-and-white camera with several color filters.

The frame grabber itself is usually a printed circuit board that occupies a slot in the microcomputer. This board takes the incoming analogue signal from the camera and breaks it into small portions, each of which occupies a memory location in the random access memory (RAM) of the board, and each of which is represented by a pixel, a small rectangular or square area of the video screen. Resolution of the board is limited in part by size of the RAM. Most frame grabbers include enough onboard RAM to allow image manipulation and processing without depending on the RAM of the host computer. As RAM prices fluctuate, so do the prices of these boards, but the expectation is that prices will drop and thus board prices will drop as resolution increases over time.

Choice of the host microcomputer for a VDAS will determine which "family" of computers you join, how fast things will work and how much data analysis can be done without resorting to a mainframe computer. This workshop featured, almost exclusively, computers based on Intel processors and using Microsoft's MS-DOS operating system, the so-called IBM-PC and IBM-clone family. Apple systems are surprisingly rare in this part of the computer industry, in part because the Apple II and its permutations are not powerful enough to be used for serious image processing, and the Macintosh has until recently been a "closed" system requiring expensive and usually slow outboard add-ons. With the arrival of the Macintosh II, Apple has made a powerful and impressive computer that is a natural for image processing. We had hoped to have such a system for the workshop, but delays in getting these boards to market kept fully developed data analysis systems unavailable. Several Macintosh II frame grabbers are now on the market, and as software development moves forward, this family of computers is becoming an attractive (although generally more expensive) alternative to those discussed next.

Regarding the MS-DOS machines, as processor speeds have increased, competition has forced prices to drop, and the older XT and AT-style 8 and 16 bit machines have become very inexpensive (most commercial VDAS are now based on 80286, AT style machines). The newer 80386 32 bit machines now have clock speeds of 25mhz and over, and capacities for up to 16 megabytes of RAM. Because of the variety of processor speeds and proprietary schemes to make the machines faster, it is important to verify that any frame grabber considered for purchase is guaranteed to work in the computer of choice.

There are several methods for a user to interact with a VDAS, usually by moving a cursor around on the video monitor screen. The instruments that govern cursor movement include mouse, track-ball, joy stick, light pen, or digitizing pad. Most commercial systems use a digitizing pad or mouse, since track balls, light pens, and joy sticks are considered by many to be less accurate and/or more difficult to use. Mice come in two kinds, mechanical and optical, and most commercial systems which have a mouse use an optical one. The mechanical mouse has a small ball on its underside, and movement of the ball is interpreted by the mouse to indicate relative movement. Optical mice are placed on a pad with a grid imbedded in it, and a small sensor on the underside of the mouse totes up sightings of the grid to give relative movement. Mice also come with different numbers of buttons to push, from one to over twenty.

Finally, the video monitor is a very important component of a VDAS because data acquisition is an eye-tiring occupation. A monitor must have sufficient resolution for comfortable viewing of the image. Even the least expensive monitors have resolution greater than most frame grabbers, but that does not mean that a higher resolution,

higher quality monitor won't look better and be easier on the eyes. Color monitors in less expensive price ranges have lower resolution ratings than black-and-white monitors, so unless a system is going to use color, a black-and-white monitor is recommended.

Once the physical parts of a VDAS have been assembled, either by you or a manufacturer, the next crucial step is getting good software. Data acquisition programs should make it easy to start a session (e.g., have facilities for recording specimen number, catalogue number, etc., as appropriate), easy to actually record data, and easy to get data into a form that can be analyzed. Strategies for each of these portions of the data acquisition sequence vary depending on the experiences and expectations of the designers, and the greatest differences among data acquisition packages are in software. Some packages allow programming of the user interface, some allow the image to be captured to disk for archiving, some place data point information in spreadsheets; in short, there are many ways to approach data acquisition, and a potential buyer should become acquainted with all parts of a system before deciding whether to adopt it. Our experiences at the Workshop showed that many commercial systems, in trying to appeal to a wide variety of users with many applications, have software that is complicated and cumbersome. Some systems had software that was balky and made data acquisition awkward. One gets the impression that the software was designed to impress prospective customers rather than to be useful on a day-to-day basis. The software packages that include "programming" by use of macros written to do a particular task without forcing one to use the entire operating system were especially appreciated by Workshop students. Any prospective buyer of a VDAS should be very critical of the software offered, and use the software to do repetitious data acquisition of the kind that will be the normal application. Several vendors have noted our students' wishes and promise to try to make their software more flexible.

Rather than write software for a wide variety of applications, some vendors have taken a different approach and designed it for more restricted tasks. Two such systems shown at the Workshop were CODA (for coordinate data acquisition) and MorphoSys (primarily for outline acquisition). Both use the Imaging Technologies PCVISION-*plus*™ frame grabber, and both were developed in part by practicing systematists. These systems are "goal oriented" in getting to the task at hand right away. If the project you wish to do is one these packages are designed for, they are attractive and inexpensive alternatives to the other "turnkey" systems.

All commercial systems allow some degree of manipulation of the video image, and in some cases this is very sophisticated. Many have automated features for object counting, edge tracking, etc., but most of these are less useful to systematists than to some other users. Some systems also include statistical packages either bundled (included in the system purchase price) or as an option. Whether this will be attractive or cost effective depends on what kinds of analysis the user wishes to do. While the statistics packages offered usually do basic analyses, they do not provide many of the analyses highlighted in this volume. It is important that data be portable in ASCII form to other packages or to other computers, and, surprisingly, some rather expensive systems are unable to do this.

Another alternative is to buy the hardware and write software for it. Under most circumstances, this is not a trivial task, as much of the programming requires knowledge of assembler or higher level languages such as Fortran or C. Design of the user interface, data output standards, and many other things make this a complicated business that should not be done unless one has considered the commercial systems and found them truly wanting. An option that is not as difficult as starting from scratch is to use a commercial package of software like "Image-Pro" (Media Cybernetics, 8484 Georgia Avenue, Silver Spring, MD 20910), which runs on a wide variety of frame grabbers and comes with routines that can be integrated by the user for specific applications. Programming knowledge is still required, but very sophisticated image processing and data manipulation can be done with relatively less programming investment. One popular system shown at the workshop, that of D. Schindel, was based on software from Media Cybernetics.

Buyers of turnkey systems are usually also buying proprietary data acquisition software, and in some cases operating systems. Some vendors are willing to customize software for an extra charge. In any case, the stability of a company is a factor in a buying decision since program code is generally not available for modification or updating.

Comments from Workshop Attendees

A questionnaire sent to workshop students queried them on their experiences with the data acquisition systems and for suggestions to improve such systems. The respondents made it clear that they have specific needs and that most of the commercial systems don't meet those needs efficiently. The four top-rated systems included two which are large-scale commercial systems and two which are smaller scale, and which had systematists involved in their creation (see also the Appendix). The two large-scale systems are the OPRS from Biosonics and the Videometric 150 from American Innovision. Both of these are expensive (around $20,000), very sophisticated, and quite able to exceed the needs of most systematists. Both are flexible and allow user-programmed macros to simplify data acquisition. The OPRS system benefits from truly excellent support from the vendor; a new, much lower cost system was shown at the workshop and was met with great interest by workshop students. The Videometric 150 is the only true color system that was shown, and it appealed to those students whose study organisms have systematically useful color patterns; user support appears to be strong, as well. The two "smaller" systems rated highly are MorphoSys and CODA. Both are available as "turnkey" systems and both can be purchased as software for user-configured hardware. MorphoSys was designed primarily for outline acquisition, but has recently been modified to allow coordinate input; CODA is for coordinate input only. A fifth system, Jandel's JAVA, was also much commented on by respondents; some found the system awkward to use and overly complex, while others found it excellent and inexpensive.

One thing that most students wanted very much is a way to custom format data output. All of the systems shown place data in a proprietary format which had to be reformatted for input into the various morphometrics programs. In some systems, the data are placed into spreadsheets which can then be dumped to ASCII. In others, data conversion programs had to be written to get the data into manageable form. Many systems are dedicated to certain kinds of "standard" analyses (in some cases using proprietary statistical software) and some were not designed for ASCII output at all. Clearly, any purchase should be contingent upon demonstration by the vendor that the data can be output in a manner you require. It still may be necessary to do some programming to reformat data for different analysis programs.

Summary

The purpose of the devices discussed above is to aid the systematist in the collection of data to apply to the resolution of a scientific question. Do not let yourself be seduced by the technology, as this can become time consuming. By keeping the biological questions foremost, you can choose machines of appropriate resolution and sophistication. These machines can be great time savers, and allow approaches to problems that would have been difficult or impossible before their arrival. If you find that one kind of data acquisition system can't get the information you need, abandon it and go to something else.

The cost of most of these systems is dropping as their sophistication is increasing. Some of the systems are flexible enough that they can be purchased as multi-user facilities which can aid people in different fields, thus making their purchase more palatable to department chairs. New technologies are going to make this paper obsolete in short order, and every systematist should keep alert to new products and new opportunities in data acquisition.

References

Fink, W. L. 1987. Video digitizer: a system for systematic biologists. Curator, 30(1):63-72.

Hubbs, C. L. and K. F. Lagler. 1941. Guide to the Great Lakes and tributary waters. Cranbrook Institute of Science Bulletin, 18, 100 pp.

Jacobs, L. J. and H. Claeys. 1987. A digitizing tablet as an efficient and accurate tool in morphometric studies on nematodes. Ann. Soc. r. zool. Belg., 117:15-20.

Marcus, L. F. 1982. A portable, digital, printing caliper. Curator, 25(3):233-226.

Strauss, R. E., and F. L. Bookstein. 1982. The truss: body form reconstruction in morphometrics. Syst. Zool., 31:113-135.

Appendix. Data acquisition systems available to workshop participants.

Most of these systems are based on "AT-style" (80286) microcomputers of various brands. All run under DOS, 2.1 and above. Most systems will save images to disk and can output data as ASCII (some use their own spreadsheet programs). Manufacturers were asked to supply information about their products, including changes since the workshop; these comments have been included below.

* American Innovision,
Videometric 150
7750 Dagget St.
San Diego, CA 92111
(619) 560-9355

A true color video turnkey system (the only true color system at the workshop). Programmable user interface. Software package capable of sophisticated image manipulation. Since the workshop, both software and hardware have been changed to give higher resolution, an expanded macro command language, and several other features. Current cost is $21,500.

* BioSonics, *OPRS*
3670 Stone Way North
Seattle, WA 98103
(206) 634-0123

Flexible and programmable user interface. Image analysis and manipulation very powerful, some analytical software included in package, more is optional. Minimum system (frame grabber, software, digitizing pad and mouse) is $12,775; complete system including minimum system plus computer, microscope, video camera and monitor) is $25,500. A popular system with workshop attendees.

* Jandel Scientific, *JAVA*
65 Koch Road
Corte Madera,
CA 94925
(800) 874-1888

A flexible, relatively inexpensive, and attractive system of software for the TARGA-M8 and PCVision*plus* frame grabbers. Some students criticized the complexity of the system and its unintuitive help menus while others thought it the best system available at the workshop. Software alone costs $1,500; hardware accessories are available through Jandel.

* *MorphoSys*
C. A. Meacham and T. Duncan
University Herbarium
Univ. of California
Berkeley, CA 94720

Primarily designed for the acquisition of outline data, but has the capability to capture coordinate data, as well. Very popular with workshop students. Uses the PCVISION*plus* frame grabber. Software license fee for single user is $250; available from Exeter Software, 100 North Country Rd., Setauket, NY 11733, (516) 689-7838.

Olympus System, *Cue-2*
Available from

Olympus dealers

A powerful, but rather inflexible system that was not liked by participants. Designed for a market including histological and medical applications. Very awkward data output system. Documentation was poor and one workshop participant who liked the system rewrote parts of the manual for others to use. Approximate cost for basic system $18,000.

Pisces Microcomputer, *CODA*
19 Diana Drive
Scottsville, NY 14546

Specialized for coordinate data acquisition using a mouse and the PCVISION*plus* frame grabber. Image zooming is a useful feature. Cost of software is $300. Turnkey systems available, pricing depending on hardware chosen. Designed partly by working systematists, so the system works nicely for systematics oriented data acquisition.

* R&M Biometrics, *Bioquant*
5611 Ohio Ave.
Nashville, TN 37209
(615) 350-7866

Long a standard for video-based data acquisition in many biological disciplines. Students did not like the XT-based system shown at the workshop, as it was slow, of low resolution, and was awkward to use. A new true color system has been added to the line since the workshop. Current costs are approximately $19,000 for the Meg-M8 black-and-white system, $30,000 for the Meg-Vista color system.

* Reflex Measurement LTD., *Reflex Microscope*
9 Whitehall Park
London N19 3TS
England

Three-dimensional measurements on specimens through a modified compound stereomicroscope. A sophisticated optical system, but was used in the workshop with a computer user interface below the standards set by the video systems. Subsequent reworking of software has substantially improved the user interface. A new high precision model has also been recently added. For its application, a very attractive system that was enthusiastically received by some workshop students. Approximate cost is £11,400 for an entry level system.

Southern MicroInstruments, *Microcomp*
120 Interstate North Parkway East,
Suite 308

Atlanta, GA 30339

A rather inflexible system that was not used much by workshop participants. Clearly designed for specific markets which have needs different from those of systematists. Priced at about $10,400.

Chapter 2

Digital Images and Automated Image Analysis Systems

N. MacLeod

Department of Geological Sciences and
Museum of Paleontology, University of Michigan
Ann Arbor, Michigan 48109
Present Address: Department of Geological and
Geophysical Sciences, Princeton University
Princeton, New Jersey 08544

Abstract

The availability of powerful and relatively low-cost image analysis systems has substantially eased the time and labor intensive task of image acquisition and morphometric data collection. Recent technological improvements in biological image analysis are largely a consequence of the introduction of digital image formats. Digital images, while necessarily being of lower resolution than their analog antecedents, lend themselves to precise quantification and are readily transportable on a variety of electronic and magnetic media. Consequently, commercially available image analysis systems can fulfill several different roles in the systematic laboratory, from that of simple image or object documentation, to completely automated forms of object measurement and identification. This paper provides an overview of digital image formats, digitization procedures, spatial and brightness resolution in digital images, the general organization of typical digital image analysis systems and brief discussions of various hardware components used by systematists for digital image analysis.

Introduction

While the potential contribution of morphometrics to systematic biology and paleontology has been widely recognized, the routine implementation of these techniques has been impeded by 1) a general lack of emphasis on the development of quantitative data analytic skills that, until recently, has characterized the training of most young systematists, and 2) a specific lack of reasonably priced laboratory equipment designed to speed the laborious task of morphometric data acquisition. Fortunately, both of these barriers are rapidly being overcome; the first by the publication of a number of introductory textbooks in the field of biological morphometrics (e.g., Pimentel, 1979; Reyment, Blackith and Campbell, 1984; Bookstein et al. 1985; this volume), and the second by a recent increase in the availability of automated image analysis software packages and hardware/software systems for personal computers.

Ironically, now that image analysis technology is finally capable of providing the data acquisition and measurement tools that generations of systematists have hoped for, their relatively sudden appearance has left many systematists bewildered before the task of choosing among the many alternative computer systems and morphometric data analysis programs. In the case of the latter, it is sobering to realize that despite a clear need for the quantitative investigation of patterns of morphometric variation in organic form, there presently exists no commercially available, fully integrated software package that provides users access to a full range of specialized morphometric data analytic procedures. Of course, a number of more-or-less sophisticated statistical data analysis program packages are available. Although these programs can provide systematists access to techniques that may be usefully applied to certain types of morphometric data (e.g., distances between landmark points), it should be kept in mind that these programs were not specifically designed to deal with the general problem of morphometric data analysis, and, for the most part, fail to include adequate analytic procedures for many types of routinely collected morphometric data (e.g., outline coordinates, point coordinates). The software package that accompanies this volume represents an initial attempt at providing the systematic community with such a library of morphometric data analysis procedures. Owing to the diverse programming styles and research interests of its authors, however, this particular collection is perhaps best described as eclectic.

In addition to this very serious lack of adequate software for morphometric data analysis, few commercially available image acquisition/processing/analysis computer systems have been designed to meet the specific needs of research systematists. For example, even a brief survey of the current crop of computerized image analysis systems reveals large disparities among the features offered by various competitors (i.e., differences in the types of data that can be collected, spatial and brightness resolutions, degrees of automation of the various data collection operations, availability of image processing methods, techniques used to correct known sources of image distortion, compatibility with external data analysis program package formats, prices, etc.). Such diversity arises from the fact that many of these instruments were designed originally to fulfill a variety of different applied image analysis roles in various industrial and medical laboratories where methods of characterizing aspects of object size and shape have become somewhat standardized. This contrasts strongly with the rationale and goals of systematic research in which we rarely find ourselves in the enviable position of knowing which aspects of morphologic variation are most informative at the outset of an investigation. In order to provide systematists with a general overview of this heretofore unfamiliar technology, this paper is intended to introduce systematists to some of the basic concepts around which most automated digital image analysis systems are designed and to briefly discuss the implications of alternative hardware/software decisions that must ultimately be confronted by those who wish to employ morphometric data acquisition and analysis in their individual research programs.

Types of Images

The images we typically see recorded on photographic film (prints, negatives, slides, SEM micrographs, x-rays, etc.) are continuous tone images composed of a graded series of grey tones or colors that, at least in principle, blend smoothly into one another to reconstruct the original scene. Digital images, on the other hand, are composed of individual elements (termed picture elements or pixels) of discreetly quantitized brightness, with each pixel being separated from its neighbors by sharply defined boundaries. Close inspection of a digital image will almost always reveal the presence of small discontinuities due to the nature of the pixel boundaries (Figure 1). But, at normal viewing distances, distortions introduced by the conversion to many average and high resolution digital formats do not significantly alter our ability to recognize and measure the object or scene.

Keeping in mind the fact that digital images are wholly artificial constructs whose resolution is necessarily inferior to that of the original continuous tone image, one might wonder why it is necessary to go to the trouble of converting a continuous tone image to a digital format? As it turns out, however, there are several distinct advantages offered by digital images that offset this unavoidable, though in many cases slight, decrease in overall image quality. These include the speed with which digital images can be acquired and reproduced, their portability, and the ability to manipulate them on a pixel-by-pixel basis thereby enabling the implementation of a wide variety of automated image processing and enhancement techniques. In addition, digital images, due to their discrete nature, greatly facilitate the quantification of morphology by increasing the accuracy with which spatial relationships between objects or features can be measured both manually and through the use of automated boundary tracking and feature extraction techniques.

The Digitization Process

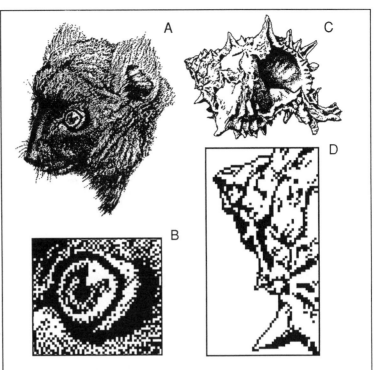

Figure 1. Examples of digitized continuous tone images. A) 72 dots per inch (DPI) scan of lemur head, B) 400× blowup revealing pixel-boundary discontinuities, C) 72 DPI scan of gastropod shell, D) 400× blowup of the spire.

The process of converting a continuous tone image to digital format is termed "digitizing," but the procedure is qualitatively identical to that of superimposing a cartesian grid over the image to be digitized and then sampling the brightness level of the image within each cell of the grid. In most types of automated image analysis systems, the original continuous tone image is initially "captured" by a black and white video camera. This is accomplished by focusing the image on the surface of a photosensitive tube or silicon chip. The camera scans along the surface of the exposed tube (or silicon chip) by rows and constructs a continuous electronic signal whose pattern of voltage variation is directly proportional to the pattern of brightness variation present in a horizontal transect through the original scene. This constitutes one line of video signal. Electronic markers are inserted into the signal to identify the beginning (or ending) of successive lines and of

successive frames, and the entire signal is continuously sequenced out of the camera.

Naturally, in order for any electronic device to sample (or for that matter to receive) an incoming video signal, the structure of that signal must be known. Standards for the timing and voltage level range of video signals have been established by the Electronic Industries Association (EIA), and the black and white (monochrome) television format most commonly used in image analysis is EIA RS-170.

An RS-170 video frame consists of 525 analog video lines. Within the active portion of the video line (Figure 2), voltages may range from +0.143v (black) to +0.714v (white) with intermediate voltages being assigned to various grey levels. The interval from 0.143v to 0.0v is termed the blanking level in which the corresponding video signal is considered "blacker than black". Each line

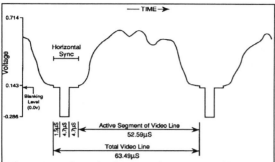

Figure 2. Schematic diagram of an RS-170 video line illustrating the relationship of the active portion of the video signal to the horizontal sync (after Baxes, 1984).

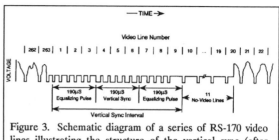

Figure 3. Schematic diagram of a series of RS-170 video lines illustrating the structure of the vertical sync (after Baxes, 1984).

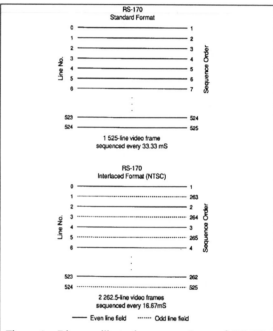

Figure 4. Diagram illustrating scan patterns of RS-170 standard and interlaced (NTSC) signal formats.

is separated by a horizontal sync that provides the signal used to reset the receiving device (e. g., video monitor, frame grabber) to the beginning of the next line. The RS-170 standard specifies that this horizontal sync be a signal sequence $10.9\mu S$ long during which the voltage drops from 0.0v to -0.286v. Upon completion of the 525 line video frame, a vertical sync signal sequence (Figure 3) is encountered that consists of a longer ($571.0\mu S$) series of 0.0v to -0.286v voltage drops. Reception of this vertical sync causes the system controller to reset the scan to the first video line of the next successive frame. Also, by convention, RS-170 requires that there be 22 "no-video" lines following the vertical sync. Thus, only 485 active video lines are present in any RS-170 video frame.

Thirty RS-170 video frames are either transmitted or received every second (one frame every 33.33mS). Unfortunately, this sequence rate is not quite fast enough to go unnoticed by the human eye and many people can detect a distinct image flicker that can be tiring when viewing standard format RS-170. To overcome this problem, most video systems employ a variation of standard RS-170 that tricks the eye into perceiving a sequence rate of 60 frames/sec. instead of the actual 30 frames/sec. This is accomplished by sequencing out all even-numbered lines first and then returning to the top of the frame to sequence the remaining odd-numbered lines (Figure 4). In other words, two virtually identical images, each 262.5 lines long, are alternated with one another at

16.67mS intervals; fast enough for the eye to blend the images together so that they seem to be continuous. This variation of the typical RS-170 signal is known as RS-170 interlaced format, or NTSC, and it has become the de facto standard for a wide variety of video applications, including image analysis.

Since the RS-170 standard dictates that only $52.59\mu S$ of active video signal exist per line, digitization consists of measuring the voltage of this continuous signal at a series of equal time intervals determined by the desired level of horizontal reso-

A   B

Figure 5.   Example of the type of image distortion produced by non-square pixels. A. Image produced via digitization using square pixels. B. Image produced via digitization using non-square pixels with a 4:3 aspect ratio. See text for discussion.

lution. This operation is carried out by the frame or line grabber (discussed below). For a horizontal resolution of 256 pixels per line, a voltage sample is taken once every 205nS, while for 512 pixels per line, the sampling rate is once every 103nS. In the same way, the vertical resolution of the resulting digital image is determined by how many of the 485 active video lines are sampled. For a vertical resolution of 256 lines, one of the two interlaced fields is used and a delay inserted such that 6.5 lines within the chosen field remain unsampled. Alternatively, for a vertical resolution of 512 lines, all of the 485 active video lines from the two interlaced fields are sampled, in addition to 27 lines from the two vertical sync and no-video fields (thereby introducing a certain falseness into to any manufacturer's claim of a resolution of more than 485 video lines from an RS-170 video signal for his frame or line grabber).

In addition to signal timing conventions, the RS-170 standard specifies that the horizontal length of the video frame be 4/3 the vertical height. Thus, all square grid (e. g., 256 x 256 pixels or 512 x 512 pixels) digitization schemes will produce pixels that have a similar 4:3 aspect ratio. This discrepancy is problematical in that any digital image composed of such "non-square" pixels will be systematically distorted (expanded) in the horizontal dimension, resulting in the physical distance between identical points on the same object appearing to change as the object is rotated within the frame (Figure 5).

Various methods are available for correcting measurements taken on digital images for the distortion brought about by non-square pixels. The most common of these adjusts the horizontal and vertical digitization rates to compensate for the 4:3 aspect ratio of the RS-170 frame. For example, this may be accomplished by digitizing 485 lines per frame but only 380 pixels per line, or by inserting a small delay ($6.58\mu S$) at the beginning and end of each video line so that only $39.44\mu S$ of the original $52.59\mu S$ signal is actually used for digitization. The former solution corrects for non-square pixels by sacrificing equivalence of spatial resolution in the horizontal and vertical dimensions, while the latter maintains this equivalence but truncates the frame from the left and right so that despite the fact that the image extends into these regions they are unavailable for digitization and analysis. Non-square pixel correction may also be accomplished by calibrating separate rulers for the horizontal and vertical dimensions and using this combination of rulers to scale all distance measurements and coordinate point locations.

All non-square pixel correction strategies described above sacrifice some aspect of image resolution in order to preserve the convenience of locating points and measuring distances on the image in terms of the pixel coordinate system. However, a few digital image analysis systems employ a more complex correction strategy that involves the use of an external coordinate system as the primary referent for the recording of positional information within the digital image. One way to implement this type of correction strategy is to map the non-square pixels onto a "virtual" coordinate system (e.g., the coordinate system of a digitizing pad or a rectangular grid constructed mathematically via appropriate transfer functions) that maintains a 1:1 aspect ratio. This alternative has the advantage of more faithfully preserving the equivalence of horizontal and vertical resolutions and utilizing the entire video frame as well as providing excellent spatial sensitivity for the purpose of object location and characterization. While the viability of a virtual coordinate mapping solution is ultimately dependent on the combined resolutions of the frame or line grabber, video monitor and digitizing pad (if present), this method appears to offer one of the more attractive general solutions to the vexing problem of non-square pixels. Users of RS-170 based image analysis equipment should familiarize themselves with the non-square pixel correction method (if any) employed by their system and periodically check their systems to be sure that they are producing data that are independent of this type of orientational bias as well as distortions arising from the use of various optical attachments to the camera (e.g., lenses, microscopes).

Adequacy of Resolution in Digital Images

One of the most obvious concerns of systematists who work with digital images is whether or not the digital image adequately represents the morphological complexity seen in the field, microscope, or photograph. Since video cameras can be mounted directly on optical microscopes or fitted with lens systems that enable them to image macroscopic objects (including photomicrographs), virtually anything that the systematist can observe can also be acquired by a digital image analysis system. But, because the digital image represents only a sample of the original continuous tone image, morphological detail present in the original at a scale finer than that of the digital sampling interval will not be consistently present in the resulting digital image.

What digitization rate is needed to adequately represent a given level of morphological detail within a particular image? The answer to this question must be given in two parts, one dealing with the required spatial resolution and the other addressing the matter of brightness resolution. In the spatial domain, fine scale morphological structure appears as high frequency variation in the voltage signal coming from the camera or other image input device. Images, or regions of images, that contain closely spaced brightness changes (= signal voltage) are said to have high spatial

frequencies, while those characterized by a more or less uniform range of brightness values are said to have low spatial frequencies. To represent the complete spectrum of spatial frequencies present within a continuous tone image, the image must be sampled or digitized at a rate at least twice as high as the highest spatial frequency (Baxes, 1984). This means that if a continuous tone image contains a morphological feature that is to be included in the digital representation of the image, it must be sampled in such a way as to allow at least two of the sample elements (pixels) to fall upon the feature itself. This relationship between the spatial frequency and the sampling rate is known within the signal processing literature as the Sampling Theorem or the so-called Nyquist Criterion.

This sampling principle also works in reverse. Thus, provided the minimum amount of detail present, or desired, within a particular image is known beforehand, it is unnecessary, even wasteful, to sample the image at a rate finer than twice that spatial frequency. Moreover, it should be kept in mind that one can never increase the level of resolution present in the original image regardless of the frequency of the digitizing or sampling rate. In cases where this has been attempted (e.g., demonstrations by ill-informed salespersons, technical reports by ill-informed image analysts) such spurious resolution over and above the level of the original is termed false or pseudo-resolution.

Since the physical size of the individual pixels is determined by the sampling rate, a digital image with low spatial resolution will contain artifacts of the digitization process that are a result of the outlines of the individual pixels being larger than a significant portion of the detail they are attempting to represent. The effect of undersampling an image's spatial frequency is to erroneously record high frequency brightness variation as lower frequency brightness transitions (Figure 6). Such under-representation of high frequency detail in a digital image is termed aliasing which, in extreme cases, may produce interference patterns that can obscure virtually all useful information in some

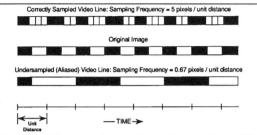

Figure 6. Diagrammatic example of aliasing. The middle line represents a series of alternating black and white fields within an original continuous-tone image. The upper line shows the result of sampling the image at a rate that will correctly represent the level of detail present in the original image, while the lower line illustrates the erroneous (lower frequency) pattern produced by a digitization scheme that undersamples the original image.

portions of the image. As spatial resolution increases, this aliasing effect will diminish until, at a given distance from the display, the viewer fails to notice it.

In most commercially available image analysis systems, spatial resolution is fixed by the frame or line grabber and cannot be increased or diminished without altering the hardware configuration. Low resolution systems tend to be either 128 x 128 pixels or 256 x 256 pixels along the horizontal and vertical axes. Normal resolution systems either adopt the TV standard of 485 x 380 pixels or are either 640 x 480 pixels or 512 x 512 pixels in overall dimension. Finally, high resolution systems are typically 1024 x 1024 pixels. The spatial resolution required by a particular study will vary with the complexity of the object and the type of data being collected. Manual data collection operations, however, are often able to tolerate (and in some cases may even have their repeatability improved by) lower spatial resolutions than more automated forms of data collection such as automatic boundary tracking and object location.

While spatial resolution may be an obvious resolution parameter of concern to systematists, brightness resolution should also be taken into consideration. During the digitization process each pixel is assigned a brightness or luminance value on the basis of the voltage present in that portion of the video signal. This value usually takes the form of an integer and the entire set of integers used for brightness quantification is termed the grey scale. By convention, a brightness of 0 represents pure black and the highest integer brightness value represents pure white. Between these two extremes, however, a wide variety of grey scales can exist. Brightness resolution is a function of how many levels are contained in the grey scale, and the Nyquist Criterion applies to brightness resolution in the same way that it does to spatial resolution. For programming convenience, grey scales are usually held internally by the image analysis system as binary numbers or bits. The number of bits assigned to the grey scale register determines the number of discrete levels in the digital grey scale. Thus, a 4-bit register has 2^4 or 16 grey levels, a 7-bit register has 2^7 or 128 grey levels, and an 8-bit register has 2^8 or 256 grey levels. Figure 7 illustrates the visual result of recording an image at different levels of brightness resolution.

Interestingly, from the standpoint of quantification of the digital grey scale, human perception of changes in brightness is logarithmic rather than linearly distributed (Figure 8). Under conditions of low overall illumination, any slight increase in the amount of light reaching the receptor cells of the human eye results in a relatively large increase in perceived brightness, whereas, under conditions of high overall illumination, the same slight increase in the amount of light reaching the receptor cells results in a much smaller perceived increase in brightness. This indicates that human visual sensitivity is greater in the darker regions of the grey level spectrum than in lighter regions. In acknowledgment of this fact, digital grey scales used in image analysis systems can be quantified on either arithmetic or logarithmic scales. Quantification on a logarithmic scale has definite advantages in terms of the naturalness of the images produced. However, this convention results in the brighter region of the grey scale being represented by fewer discrete levels than the darker region. Ideally, users should be able to switch between linear and logarithmic grey scales, though few currently available image analysis systems appear to possess this

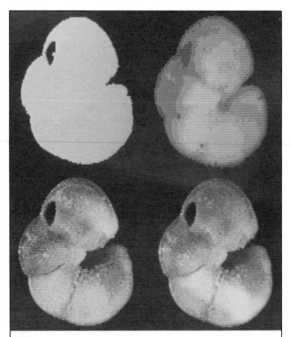

Figure 7. Digital image of a fossil planktic foraminifer recorded at different levels of grey scale resolution. Upper left - 2 levels (binary image); upper right - 8 levels (3-bit grey scale); lower left - 16 levels (4-bit greyscale); lower right - 256 levels (8-bit grey scale).

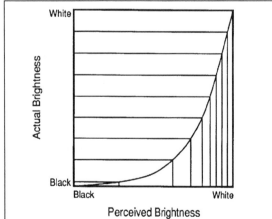

Figure 8. Generalized diagram illustrating differences in the human perception of brightness variation in different portions of the grey scale. Note that for equal increments of actual brightness change, human perception is most acute in the darker region of the spectrum.

capability.

It is often desirable to determine whether or not a digital image is utilizing the entire spectrum of available grey levels, or only a portion thereof. While there are many ways to summarize this information, one of the most useful is the image histogram (Figure 9). Image histograms are plots of the frequency with which the pixels comprising an image, or region thereof, have been assigned to the available spectrum of grey levels. Using this type of summary the viewer can immediately compare images, determine whether an image is utilizing the full range of the available grey level spectrum, and, in some cases, determine what type of image enhancements might be likely to improve the overall quality of the digital image (see F. J. Rohlf's chapter on image processing for a complete discussion of image enhancement techniques).

As with standard continuous tone images, the term "contrast" is used to describe the character of the distribution of grey values comprising a digital image. If a digital image histogram is bimodal, with most pixels being assigned either very light or very dark brightness values, the image is said to exhibit high contrast (Figure 10). Alternatively, if the distribution of pixels occupies only a narrow region in the middle of the image histogram the image is said to have low contrast characteristics (Figure 11). A well-balanced or "good contrast" image is generally composed of a more or less even distribution of grey levels. However, it should be remembered that terms like "good contrast" and "good image" tend to be subjective categorizations that are critically dependent on the type of information that the systematist is interested in extracting from an image.

Digital Image Analysis Hardware

A generalized diagram of an automated image analysis system is presented in Figure 12. Though images may be input to the system from a variety of media (e.g., video tape recorders, laser disk recorders, communications link through telephone lines), a video camera equipped with an appropriate lens system (consisting of either a research grade optical microscope and/or a high quality photographic lens system) provides maximum flexibility in terms of the range of images that can be acquired. Conversion of the analog image transmitted by the camera (or other image input device) to a digital format is accomplished by the frame or line grabber, which, in turn, is connected to an area of memory set aside for digital image storage. Once stored, the digital image may be reconverted to the analog format of the standard NTSC video signal and routed to a video display monitor for viewing. Since the incoming video signal from the camera is continuously digitized and written to memory by the frame or line grabber, the continuous reading of the stored image to the display monitor results in a "live" image being displayed. At some point, however, it is usually necessary to "freeze" the live image by suspending the writing procedure. Once frozen, the image stored in memory can be written to disk or tape for archival purposes, made available for image processing (which, due to the large number of

calculations involved, are usually handled by specialized image processing hardware) or subjected to various image measurement operations.

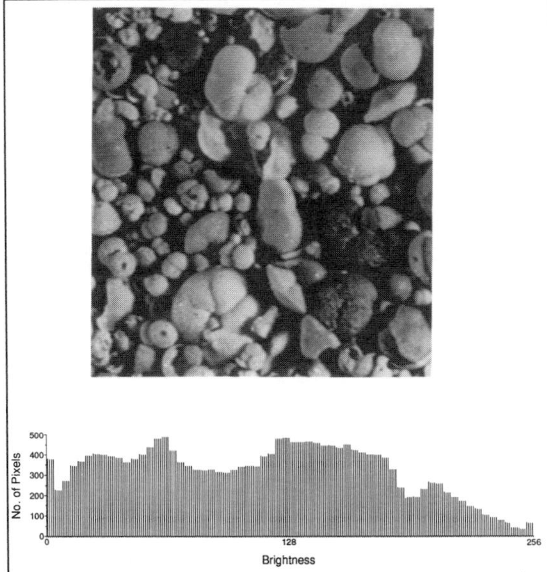

Host computer systems for image analysis vary widely in size, capability and expense from the small, stand-alone processors that drive so-called "smart" digitizing tablets up to mini- and mainframe computers. However, responding to the widespread presence of advanced microcomputers (IBM PC/XT/AT/386 line, IBM PS-2 series, Apple Macintosh™ line) in a large number of research laboratories, many manufacturers of image analysis hardware and software have recently begun to market systems designed to operate within the computing environment provided by these personal computers. [Note: At the time of the NSF-University of Michigan Morphometrics Workshop (May

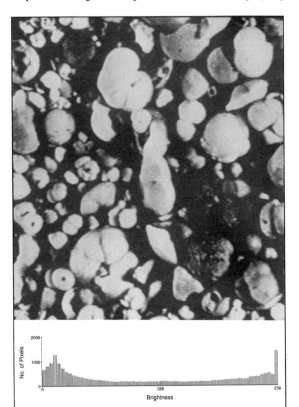

Figure 9. Example of an image exhibiting a large dynamic range and "good" contrast characteristics along with its digital image histogram. The high frequency of darker pixels is due to the large area occupied by the dark background in the image.

Figure 10. Examples of images exhibiting a bimodal (high contrast) distribution of grey tones (and a large dynamic range typical of high contrast images) along with its corresponding digital image histogram.

1988) almost all PC-based image analysis systems were designed to be hosted by the IBM PC/XT/AT/386 line of personal computers or their equivalents. During the intervening year, however, frame grabbers designed to operate with the IBM PS-2 series and Apple Macintosh™ SE and II personal computers have been released, and automated image analysis systems for all of these popular computer systems are currently available.] PC-based image analysis software packages typically range in price from $1,000 to $20,000 and offer cost-effective solutions to the problem of getting an advanced level of image analysis technology into the systematic laboratory. In addition, many of these systems allow some latitude in the selection of hardware configurations and have a modular software design, thus making it possible to extend the capabilities of computers originally purchased for different purposes (e.g., word processing, database management) to provide for immediate image analysis needs while retaining the option to increase a system's scope through the purchase of additional image processing/analysis modules at a later date. And, since image analysis hardware components may also be purchased individually, it has also become feasible for some users to design, assemble and program what are essentially "homemade" image processing/analysis systems to addresses their specific needs.

Most retailers of integrated PC-based image analysis systems offer users some degree of choice in the selection of video cameras, frame or line grabbers, and display monitors. The following sections provide general descriptions of these components along with a brief analysis of the significance of alternative selections.

Video Cameras

As the primary image input device, the video camera is a key component of any image analysis system. Most commercially available low and medium resolution systems will work well with any camera that supplies the RS-170 video signal, so the choice of camera type and model should be made on the basis of the requirements of the research to be undertaken. High resolution systems usually require the use of a high definition, non-RS-170 video signal. Currently, there are two alternative types of video cameras: "tube" cameras which use a photosensitive electronic tube to convert light

Figure 11. Examples of images exhibiting a bimodal (low contrast) distribution of grey tones along with its corresponding digital image histogram.

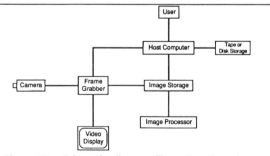

Figure 12. Schematic diagram illustrating the primary components and generalized configuration of a PC-based digital image processing and analysis system.

energy into an electrical signal, and the so-called "chip" cameras which utilize some type of photosensitive integrated circuit mounted on a small silicon chip for this purpose.

The most obvious distinction between these two types of video cameras lies in their respective resolutions, or their ability to distinguish between discrete objects in the field of view. Video camera resolution is measured as the number of individual points that can be resolved per line of video signal (often termed TV lines) and this information is usually provided in the manufacturer's specification summary. Typical tube camera resolutions fall within the range of 500 to 800 TV lines while typical silicon chip cameras may have resolutions as low as 300 TV lines. In addition, the output video signal from a photosensitive tube camera is a true analog signal while the output signal coming from a silicon chip camera is, in reality, a digital signal that has been converted to an analog format. This rather odd situation arises from the fact that the silicon chip used to convert light energy into the output signal is divided into pixels that "digitize" the scene during image capture.

Given the importance of good resolution to most systematic applications of image analysis, one might be led to reject silicon chip cameras as a group in favor of their photosensitive tube competitors. There are, however, several sound economic and scientific reasons for seriously considering a chip camera alternative. First, silicon chip cameras are far less expensive than their photosensitive tube counterparts and a top-of-the-line chip camera can have a resolution that equals or even exceeds that of an equivalently priced tube camera. Silicon chip cameras are more compact, weigh less, and are less susceptible to damage from rough handling than tube cameras due to the fragile nature of the photosensitive tube. In addition, chip cameras can be mounted vertically (the normal position for microscope and copy stand work) for long periods of time without damage to the photosensitive surface of the silicon chip. Most manufacturers of tube cameras expressly recommend that their products not be mounted in this position due to the possibility of metallic particles shed from the burning filament coming into contact with the light sensitive chemical films that coat the inside of the photosensitive tube. These light sensitive chemicals also deteriorate with time thus progressively losing their ability to become chemically excited by incoming light energy. This, in turn, leads to an inescapable reduction in image quality over the life of the tube. Furthermore, if a photosensitive tube is suddenly exposed to very intense light, the sensitivity of these chemicals may be temporarily (or permanently) altered leading to the production of residual or ghost images and burned spots. Chip cameras, on the other hand, are not affected by sudden exposure to bright light and their sensitivity should not diminish over time due to their solid state design. Finally, the light sensitive surface of the photosensitive tube must be electronically scanned in order to construct the video signal. In RS-170 cameras, this scan normally occurs at the RS-170 frame rate of 30 frames/sec. If an object within the field of view is moving during the course of this scan, the resultant image will be distorted to a greater or lesser extent depending on the object's speed of movement through the frame. Silicon chip cameras, however, can be designed in such a way as to allow all pixels to sample the scene at the same instant, thus freezing any motion within the frame far more effectively than tube camera designs.

Silicon chip cameras, nevertheless, have their own inherent problems which include susceptibility to aliasing effects that may lead to the production of artifactual interference (moiré) patterns when imaging objects or scenes containing regions of high spatial frequency and wide variability in the quality and reliability of various chip camera models. For instance, manufacturers of chip cameras who do not maintain strict quality control standards for their product (e.g., those who caution users to expect a certain number of "blemishes" [=bad pixels] on the light sensitive silicon chip) should be avoided.

Video cameras are sold without any type of lens system. Therefore, regardless of what type of object(s) one intends to image, a lens system for the video camera will have to be provided. The imaging of microscopic objects using light optics usually requires attachment of the camera to a research grade compound or stereoscopic microscope via photo-tube. For higher magnifications or improved depth of field, the output video signal from a scanning or transmission electron microscope may be directly input into the image analysis system *provided this signal conforms to a standard video format* (e.g., RS-170). Unfortunately, most scanning and transmission electron microscopes use non-standard video signal formats thereby necessitating the modification of these signals before transmission to an image analysis system. But, scanning or transmission photomicrographs, along with any other macroscopic photo, can be acquired by a video camera equipped with an appropriate macroscopic lens system. Video camera lenses are, for the most part, not of sufficiently high quality to be used for systematic research. Rather, single lens reflex (SLR) camera lenses are recommended, provided they can be fitted with a c-mount adaptor for attachment to the video camera. Optimal macroscopic lens configurations will vary from application to application, but a high quality SLR macro lens should suffice for most types of macroscopic imaging.

Lastly, most photosensitive tube and silicon chip cameras respond to changes in brightness levels within a scene in a non-linear manner, thus reflecting the non-linear brightness response curve of the human eye (discussed above). For each camera model, this deviation from linearity is quantified by the "gamma correction factor" which should be reported in the manufacturer's technical literature. A few image analysis systems allow the response of the frame grabber to be tuned to the response curve of the camera via specification of this gamma correction factor for particular hardware configurations.

Frame and Line Grabbers

For PC-based image analysis systems, frame and line grabbers are usually marketed as plug-in expansion boards that perform two functions. First, they accept an incoming analog video signal from the camera or other image input device (e.g., video tape recorder, laser disc) and convert this signal to a digital format. Second, they read a stored digital image from memory and convert it to a standard analog video format, supplying horizontal and vertical sync signals as necessary, for transmission to a video display monitor or other image output device.

Frame and line grabbers are rated by the number of pixels they divide the analog image into (128 x 128 and 256 x 256 = low resolution; 485 x 380, 640 x 480, and 512 x 512 = average resolution; 1024 x 1024 = high resolution) and by the number of discrete grey levels in the digital brightness or luminance scale (an 8-bit address or 256 grey levels is presently considered standard). Both types of boards are designed around a sync extractor, which sends a signal to the computer system controller whenever a horizontal or vertical sync is encountered in the incoming video signal, and a series of analog-to-digital (A/D) converters that sample the active video signal and send a binary number representing the sampled signal voltage to the appropriate image storage location.

The primary difference between line and frame grabbers lies in the amount of time needed to construct one digital frame from the incoming video signal. Frame grabbers employ high-speed A/D flash converters that enable them to construct a digital image at the frame rate of the video signal (30 frames/sec. for RS-170). Line grabbers, on the other hand, use less expensive successive approximation A/D converters to converge iteratively on a digital value representing the input voltage level. Since this convergence process may take up to $1\mu S$ to complete, most single converter line grabbers can digitize only one pixel per line per frame. Thus a 256 x 256 digital image would take a single converter line grabber 256 frames to complete the sampling process. Because of this limitation, most commercially available image analysis systems utilize the more time efficient frame grabbers for image digitization. But significant saving may be had for those whose research can tolerate the longer digitization times required by line grabbers.

Display Monitors

Video display monitors come in a wide variety of types, sizes and resolutions. Usually, a high (data) grade black-and-white monitor will be adequate for the majority of image analysis applications. But, despite the fact that the image is being recorded in black and white, systems that offer a pseudocolor image viewing option or employ color graphic overlays to indicate the position of the cursor during manual data acquisition operations and to identify previously located points and lines require a color (RGB) display monitor to take advantage of these features.

As with video cameras, resolution in video display monitors is rated in terms of TV lines and it is convenient (though not always possible) to achieve a reasonably close match between the resolutions of the camera, frame grabber, and display monitor. Reflecting the 4:3 aspect ratio of the standard video frame, display monitor picture tubes are rectangular in shape, and sized by the distance from the upper left to the lower right hand corners of the projected video frame. However, since physical size has little to do with resolution or image quality, it is often the case that a smaller, higher resolution monitor gives better results than a larger, lower resolution model. Though it is possible to project an image onto the video monitor used to communicate with the host processor, these monitors do not usually have sufficiently high spatial resolutions to accurately represent fine detail. In addition, the dialog menus and program icons used to control the image processing/measurement software take up space on the screen thus reducing the size of the frame that can be viewed with single monitor image analysis systems. Consequently, most PC-based image analysis system configurations require two video moni-

tors: one for system control and the other for viewing of the image.

Virtually all video display monitors marketed in the U.S., Canada, and Japan will accept the RS-170 interlaced video signal (NTSC). The European standard black-and-white video signal differs from RS-170 in consisting of 625 video lines/frame sequenced at a rate of 25 interlaced frames/sec. and is known as PAL/SECAM. Since these two signal formats are mutually incompatible, it is necessary to check the manufacturer's technical literature or specification sheets to be sure that any particular monitor has been designed to accept a particular frame or line grabber's output signal format.

The Role of Image Analysis in Systematic Research

Despite the novelty of being able to acquire, store, and manipulate images of living and fossil organisms, image analysis technology is beginning to make its importance felt in many different areas of systematic research. This can be seen as a reflection of the impact that the coming revolution in image handling capability will make on society as a whole. At present, there is little doubt that image analysis technology will have as large an effect on the ways in which future research in the biological sciences is conducted as did the development of optical microscopes (and more recently the development of transmission and scanning electron imaging instruments) in terms of providing answers to questions that were unanswerable before, not to mention fostering the consideration of new types of research questions. Broadly construed, most of the problems in systematic biology can be formulated in morphometric terms, e. g., predictions of the likelihood and order of chemical reactions taking place due to correspondences between the geometric arrangement of atoms in organic molecules, morphological descriptions of patterns of organismal development, or analyses of the character of morphological transitions taking place that result in the formation of new species. In each case it is easy to see that the quantification of patterns of geometric variation at a variety of scales can be used to test explicitly stated process-level hypotheses in ways that cannot be duplicated by more qualitative forms of analysis.

In particular, the availability of computerized image analysis systems has greatly stimulated both applied and theoretical research in the area of biological morphometrics. But, while morphometrics has played an increasingly important role in many different types of systematic research programs, its impact has yet to be appreciated in the area of phylogenetic reconstruction where, despite the logical formalisms of phylogenetic systematics, character analysis largely depends on qualitative assessments of structural similarity and difference. In my view, the proper role of morphometric analysis in phylogenetic systematics involves the precise, mathematical description of characters, including the identification of discontinuities in the distribution of variable characters that may be used to subdivide them into a number of discrete character states. This approach capitalizes on the power of morphometrics as a descriptive tool while at the same time allowing it to be integrated into a traditional program of phylogenetic analysis by avoiding the untested and largely unknown methodological complications that would inevitably arise from any attempt to use morphometric descriptors themselves (e.g., estimates of population means, variances, eigenvectors) within cladistic datasets (see Felsenstein, 1988).

In addition to enabling the quantitative investigation of organic size and shape changes, image analysis can serve the very important function of recording images for the purposes of archival documentation and publication. In the past, the only cost-efficient way of maintaining a large image database has been through the use of 35mm photography, which results in the production of large, single-copy sets of photographic negatives. These are cumbersome to store, difficult to use on a routine basis and usually require the expensive and time-consuming process of photographic printing on high quality paper for dissemination of the images to colleagues or editors. Digital image formats overcome many of these limitations by allowing images to be captured rapidly and stored on a variety of electromagnetic and optical media (e.g., hard disks, floppy disks, magnetic tape, laser disks). With appropriate equipment (see above), the quality of these images can be very high and in most larger cities publication-quality image printers are available for hard copy image reproduction at moderate cost. The real utility of digital images, however, lies in their inherent portability. Digital images can be reproduced on electromagnetic and optical media very rapidly and copies of the image can be sent over existing phone lines virtually anywhere in the world. [Note: though this process can be somewhat time consuming for large images sent over old telephone networks at present, these networks are currently in the process of being modified to handle digital information/image transfer at much more rapid rates.] This ease of transportation, coupled with the fact that individual images (e.g., images of holotypic or paratypic specimens) can be directly linked to sophisticated database management programs, will greatly increase access to up-to-date descriptive treatments of individual taxa and groups by the systematic community as a whole, as well as helping to facilitate communication between taxonomic specialists.

As for the future, it is clear that color digitization is desirable for a number of systematic applications as well as applications in the geological, metallurgical and materials sciences. Existing constraints on memory sizes in personal computers have thus far limited the availability of color digitization in the present generation of PC-based image analysis systems (since color video is produced by varying the hue and intensity of the three primary colors, color frame grabbers require at least three times the memory of their typical black-and-white counterparts). But, as more powerful personal computers and workstations make their way into industrial and research laboratories, future image analysis systems will be able to make direct use of color variations as an aid in object recognition and analysis. Unfortunately, this change to color capability will render much of the currently available image analysis hardware obsolete, necessitating the purchase of more expensive color video cameras, frame grabbers and perhaps color video display monitors. In addition, it can only be hoped that with time, digital image formats and numerical data storage conventions will become more standardized, thereby allowing images or data collected by one system to be used by other systems, including specialized morphometric data analysis program packages. At the moment, the unacceptably high level of incompatibility between images and data produced by image analysis software marketed by different manufacturers represents a formidable barrier for users who need to view, process, and analyze a wide range of images collected from a variety of sources but who do not possess the programming skills required to effect the necessary format modifications. Finally, recent advances in the field of computer vision have demonstrated that many of the problems involving inconsistent object orientation, uneven object illumination and partial object obscuration that have prevented all but the simplest objects to be automatically located and categorized within a field of view can, at least in principle, be overcome, thus suggesting that practical, fully automated object recognition may be a real possibility. Though automated computer vision at the level of detail required to make a contribution to systematic research remains a long way off, it is hard to think of a technological development in the biological sciences that holds as much potential for contributing to our understanding of the origin, development and present organization of life on earth.

Acknowledgments

I would like to thank the organizers of the NSF-University of Michigan Morphometrics Workshop (1988) for inviting me to participate in the program; F. J. Rohlf and J. A. Kitchell for reading and commenting on a previous draft of this manuscript; and Karen Klitz and Bonnie Miljour for providing the continuous tone illustrations used in Figure 1. The video images used in Figures 5, 7 and 10 were acquired, constructed and processed using the Biosonics Optical Pattern Recognition System (OPRS) and the BioScan Optical Image Measurement and Analysis System (OPTIMAS).

References

Baxes, G. A. 1984. Digital image processing: a practical primer. Prentice-Hall, Englewood Cliffs, 182 pp.

Blackith, R. E. and R. A. Reyment. 1971. Multivariate morphometrics. Academic Press, New York, 412 pp.

Bookstein, F. L., B. Chernoff, R. Elder, J. Humphries, G. Smith, and R. Strauss. 1985. Morphometrics in evolutionary biology. The Academy of Natural Sciences of Philadelphia, Spec. Publ. No. 15, 277 pp.

Felsenstein, J. 1988. Phylogenies and quantitative characters. Ann. Rev. of Ecol. and Syst., 19:455-71.

Pimentel, R. A. 1979. Morphometrics: the multivariate analysis of biological data. Kendall/Hunt Publishing Co., Dubuque, Iowa, 276 pp.

Reyment, R. A., R. E. Blackith, and N. A. Campbell. 1984. Multivariate morphometrics (2nd edition). Academic Press, New York, 233 pp.

Chapter 3

An Overview of Image Processing and Analysis Techniques for Morphometrics

F. James Rohlf

Department of Ecology and Evolution
State University of New York
Stony Brook, NY 11794

Abstract

This is a general introduction to methods for image processing and image analysis that are useful in morphometrics. Image processing consists of methods to enhance images, such as contrast enhancement, filtering, edge detection, etc. so that the desired details of the images are more evident, especially when viewed by a human. Image analysis is concerned with automatically isolating objects in the image and then obtaining descriptive information about the objects. Alternative sets of features may be mathematically equivalent in their ability to describe an object, but analyses based on different features may give different results. Some implications of this for morphometrics are also discussed.

Introduction

This paper cannot replace a detailed text on image analysis, but it should to serve as an introduction to those image processing and image analysis techniques that are useful (or are expected to become useful) in morphometrics. In addition, it considers some of the implications of the fact that large numbers of new kinds of morphometric characters are available once images of the organisms have been captured and manipulated by computers.

Models for the image-forming process itself are covered first. These are needed in order to understand some of the kinds of information present in an image as well as sources of distortion. The computer hardware involved in the scanning process is not discussed because it is covered elsewhere in this volume (chapters by Fink and by Macleod). The types of techniques available for enhancing an image to minimize the effects of known kinds of distortion are described as well as methods that transform the image to accentuate its desirable aspects. These operations, in which new images are created from old images, correspond to the field of image processing. The field of image analysis is concerned with methods for breaking a scene into its components (at least into object versus background), extracting useful descriptive information about the objects in the image, and interpreting this information (recognition of the objects and their relationships to one another).

The following texts are especially helpful general introductions to image processing and image analysis: Ballard and Brown (1982), Horn (1986), Pavlidis (1982), and Rosenfeld and Kak (1982). Journals that publish technical papers in this field include: Computer Vision; Graphics and Image Processing; and I.E.E.E. Transactions on

Pattern Analysis and Machine Intelligence. The former publishes Rosenfeld's extensive annual reviews of image processing and image analysis literature (the bibliographies usually have over 1,000 entries).

Image Geometry and Image Functions

A basic understanding of how an image is formed is important for an understanding of the methods used to obtain information about the geometrical form of the original object being studied. An image is treated as a two-dimensional pattern of brightness that is produced by an optical system such as a camera. An ideal pin-hole camera is the simplest model of the relationship between points on the object and points in the image (see Figure 1). Since light travels in straight lines, each point in the image corresponds to a particular ray of light projected back toward the scene containing the object. The direction is defined by the position of the point on the image and the location of the pin-hole. As a result of this geometry, the projection onto the image plane yields a perspective projection. The optical axis is the perpendicular vector from the pin-hole to the image plane (the length of this vector is f'). Consider a point P on the object. To compute its location, P', on the image plane, a coordinate system must be established. It is convenient to use the location of the pin-hole as the origin with the z-axis aligned with the optical axis and pointing toward the image (thus points in front of the camera will have negative z-coordinates). Let the x-axis extend to the right and the y-axis upwards. If the coordinates of P are given by the column vector $P = (x,y,z)^t$, then the vector of coordinates of P' can be found as follows:

$$p' = \frac{f}{p^t z} p \; , \tag{1}$$

where z is the unit vector along the optical axis. The elements of p' are

$$x' = f \frac{x}{z}$$
$$y' = f \frac{y}{z} \tag{2}$$
$$z' = f \; .$$

In order for the object scene to illuminate the image plane, the pin-hole must have a finite diameter to permit light to enter, but this leads to a blurring of the image. A solution is to use a lens rather than a pin-hole. When in focus, a perfect lens generates an image that obeys the same projection equations, as given above. The relationship between the focal length, f, of a lens and the distances to the object and the focal plane are shown in Figure 2 and in the equation:

$$\frac{1}{f} = \frac{1}{z'} + \frac{1}{-z}, \tag{3}$$

where z' is the distance from the lens to the image plane and $-z$ is the distance from the lens to the object (as above, z-coordinates are negative in front of the lens). If a point is actually at a distance $\bar{z}$, then it will be imaged as a blur circle of diameter

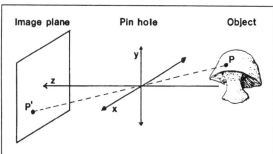

Figure 1. A model for a pin-hole camera. A point P on the object is projected onto a point P' on the image plane.

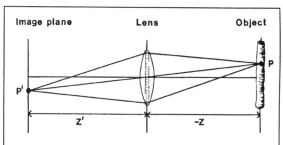

Figure 2. A model for the projection of a point onto the image plane in a camera with a lens. The z-axis is positive toward the image plane.

$$d \frac{\left| \overline{z}\text{-}z' \right|}{z'}, \tag{4}$$

where d is the diameter of the lens and $\overline{z}'$ is the z-coordinate of the point at which $\overline{z}$ is imaged in front or back of the image plane). Thus larger lenses have a smaller tolerance or depth of field.

The brightness, or image irradiance, at each point, (x,y), in the image can be represented by an image brightness function, $f(x,y)$. Irradiance is measured in watts per square meter of radiant energy falling on the image plane. The irradiance of a small area on the image plane, corresponding to a small surface patch at position P on the object, can be computed as

$$E = L \frac{\pi}{4} \left[\frac{d}{f} \right]^2 \cos^4 \alpha, \tag{5}$$

where L is the scene radiance of the object surface in the direction of the lens, d is the diameter of the lens (see below), f is the focal length of the lens, and α the angle between the vector p and the optical axis (see Figure 3). Therefore image irradiance is proportional to scene radiance.

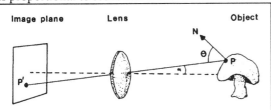

Figure 3. A model for the brightness of a point on the image plane as a function of the orientation of a patch on the surface at the corresponding point on the object.

Scene radiance, L, is power per unit foreshortened-area emitted into a unit of solid angle and is measured as watts per square meter per steradian. It is more complex to model accurately, but there are several important generalities. The foreshortening effect is proportional to $\cos \theta$, where θ is the angle between the surface normal and the vector p toward the lens. Thus, less light is directed toward the lens if the surface is directed *away* from the lens. Unless the surface is matte (an

ideal Lambertian surface that appears equally bright from all viewing directions and reflects all incident light), some light will also be reflected towards the lens. This will cause the object to appear glossy or mirror-like (specular). One usually wishes to minimize the effects of scene irradiance so that reflectance, which is a property of the object itself, can be measured. This can best be done by making sure that the objects of interest are evenly illuminated. If this is not possible, one can try applying various mathematical corrections to the resultant image. In the models shown above, image irradiance is a function of the product of a number of factors. Since most of the methods for image enhancement involve only linear operations, it is useful to use log-transformed brightness values as input for the methods described below.

Color

Color will not be considered in this review except to note that digitization of an image at more than one wave length captures more information about a scene. Color images require additional storage space and processing power in a computing system but having multivariate information at each picture point can enable more powerful techniques to be used to discriminate among different objects in a scene.

Blurring

Ideally, a camera's optics map a point in the scene into a point in the film, but in practice the point of light is spread out (blurred) as a result of the lens being slightly out of focus, diffraction rings, film grain, the camera not being perfectly steady, etc. The function that describes how a point of light is spread out is called a point-spread function (it is not imaged as a distinct small circle as implied by the equation given above). The spread of brightness values is often approximated by a normal curve. The effect of a point-spread function at a given point in the final image can be modeled by superimposing point-spread functions at each point in the image (with the height of each point-spread function being proportional to the brightness of the input point). The resulting brightness in the final image is the sum of the heights of the point-spread functions at each point. For a 1-dimensional image, this corresponds to

$$h(y) = \int_{-\infty}^{\infty} f(y\text{-}x)\, g(x)\, dx\,,\qquad(6)$$

where f corresponds to the point-spread function, g corresponds to the input function, and h is the resultant image function. This operation is called the convolution of the functions f and g and is symbolized as f⊗g. The function f is called the kernel of the convolution.

The 2-dimensional generalization is

$$h(x,y) = \int_{-\infty}^{\infty}\int_{-\infty}^{\infty} f(x\text{--}u, y\text{--}v)\, g(u,v)\, du\, dv.\qquad(7)$$

It can be shown that the convolution operation is both associative and commutative, f⊗(g⊗h) = (g⊗f)⊗h and f⊗g = g⊗f. The operation is well-behaved and easy to work with.

Spatial and Frequency Domains

The input function, f(x,y), can be modelled as the sum of an infinite number of sinusoidal curves. This allows the input function to be expressed as

$$f(x,y) = \frac{1}{4\pi^2} \int_{-\infty}^{\infty}\int_{-\infty}^{\infty} F(u,v)\, e^{i(ux+vy)}\, du\, dv,\qquad(8)$$

where

$$F(u,v) = \int_{-\infty}^{\infty}\int_{-\infty}^{\infty} f(x,y)\, e^{-i(ux+vy)}\, dx\, dy\qquad(9)$$

$F(u,v)$ is called the Fourier transform, $\mathcal{I}$, of $f(x,y)$. While $f(x,y)$ is always real, $F(x,y)$ is generally complex.

Certain operations are more easily performed on the Fourier transformation of a function than on the function itself. For example, it can be shown that the Fourier transform of the convolution of two functions is simply the product of the

Fourier transforms of each function considered separately.

$$\mathcal{I}(f\otimes g) = FG,\qquad(10)$$

where $\mathcal{I}(f) = F$ and $\mathcal{I}(g) = G$. Not all functions have a Fourier transform. Other difficulties are that the integrals are taken over the entire x,y-plane, whereas imaging devices produce images for only a finite part of the image plane: also digital computers must use discrete samples of these images. For an image with M,N rows and columns, the discrete version is

$$F_{mn} = \sum_{k=0}^{M\text{-}1} \sum_{l=0}^{N\text{-}1} f_{kl}\, e^{-\pi i(km/M + ln/N)},\qquad(11)$$

for $0 \le m \le M\text{-}1$ and $0 \le n \le N\text{-}1$. Its inverse transform is

$$f_{kl} = \sum_{m=1}^{M\text{-}1} \sum_{n=0}^{N\text{-}1} F_{mn}\, e^{\pi i(km/M + ln/N)},\qquad(12)$$

for $0 \le k \le M\text{-}1$ and $0 \le l \le N\text{-}1$. These expressions can also be given in terms of sine and cosines (which is more common in the morphometric literature), rather than as exponentials of complex numbers (which is more compact), using the Euler relation

$$e^{iu} = \cos u + i \sin u.\qquad(13)$$

The use of the Fourier transform assumes that the image is doubly periodic (replicates of the image repeat in both the x and y directions). Unless the image at the left edge happens to match that at the right edge (and the top also matches the bottom edge), there will be a discontinuity and some high-frequency components will be introduced. This problem can be avoided by making sure that there is a uniform background all around the object and that the entire object is within the image.

Digital Images

Of course, the actual images processed by digital computers must be represented as discrete samples of the image brightness surface over a finite range. The image is represented as a 2-dimensional array of measurements of brightness. This array usually has about 500 rows and columns (but devices are available that provide greater resolution). Each element of the array is an integer, usually recorded to 8 bits of accuracy, giving the average brightness of a small region in the image. This element is called a *pel* or *pixel* (short for "picture element"). Thus digital images can be treated as 2-dimensional tables of numbers. Figure 4 shows an image as an image surface; the brightness values for selected rows in the digitized image are plotted as a function of column position.

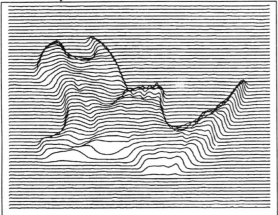

Figure 4. Digital image of a mouse mandible shown as an image surface. Brightness values for the selected rows plotted as a function of column position.

Problems of Sampling

Using a discrete, rather than a continuous, image introduces an effect called aliasing. It can be shown (e.g., Horn, 1986) that sampling an image function, $f(x)$, at intervals of Δx in the image (the spatial domain) is equivalent to replicating the Fourier transform of the image function, $F(x)$, at intervals of $1/\Delta x$. If there are frequencies in the original image greater than $1/\Delta x$, then components of F will interact to produce a composite image transform, F'. Basically, sampling causes information at high spatial frequencies to interfere with that at low frequencies (see Figure 5). This phenomenon is called aliasing, since a wave of frequency $\omega > A$ produces the same wave in the sample as a wave with frequency $2 A - \omega$. Therefore, the image should not contain frequencies smaller than half the sampling frequency if this problem is to be avoided (this lower threshold is called the Nyquist frequency). But some objects are better recognized at lower resolutions (where the effects of high frequency noise is averaged out).

One way to reduce the effects of aliasing is to use a pyramidal image data structure (see below), where the search for structure begins at low resolution and then resolution is increased as needed. Rather than redigitizing at lower resolutions (which would introduce aliasing), the lower-resolution images are computed as averages from the original high-resolution image. The consolidation that takes place as one creates lower-resolution images tends to offset the aliasing that would be introduced if one were to digitize at larger sampling intervals. The averaging attenuates the higher frequencies involved in aliasing. Algorithms have been developed to perform many types of image processing operations directly on data stored in a pyramid.

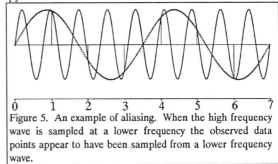

Figure 5. An example of aliasing. When the high frequency wave is sampled at a lower frequency the observed data points appear to have been sampled from a lower frequency wave.

Metrics

As described above, the usual scanning hardware produces a rectangular array of brightness values.

This rectangular spatial pattern is convenient for storage and indexing in digital computers, but it complicates the interpretation of the topological relationships among objects in the image. These considerations are important for the description of outlines using chain codes. The Jordan curve theorem states that a simple closed curve should separate an image into two simply-connected regions.

But consider the following binary image where the "0" state corresponds to the background:

0	1	0
1	0	1
0	1	0

If we adopt the principle of 4-connectedness (a point is considered adjacent only to its immediate neighboring points, left, right, above, and below it) then the four objects, "1", do not form a closed curve, yet the background cell in the center is not connected to the rest of the background. We thus have two background regions without a closed curve. On the other hand, if we adopt 8-connectedness (a point is also considered adjacent to its diagonal neighbors) then the four object cells form a closed curve but there is now only one background region because the center cell is now connected to the other background cells. One solution is to use the 4-connectedness principle for objects but the 8-connectedness rule for background (or vice versa). Horn (1986) suggests a type of 6-connectedness. In addition to up, down, left, and right, he considers cells to be neighborsif they are diagonally above and to the left or below and to the right. Of course, he could have arbitrarily chosen the diagonal directions above and to the right and below and to the left. If one had a hexagonal array, all six cells touching a particular cell would be considered neighbors. This would be much simpler, but standard hardware gives only rectangular arrays.

Data Structures for Digital Images

High-resolution images require considerable amounts of storage (a standard 480×512 8-bit image requires 245,760 bytes). That makes it important to use efficient methods for the storage and retrieval of images. The basic approach is to take advantage of the fact that the brightnesses of adjacent pixels are not independent of one another but are usually similar (the phenomenon of spatial coherence). There are two aspects to efficiency: compactness of storage and speed of retrieval of information. Both aspects are important. The techniques described below emphasize compactness of storage, since that is usually the limiting factor on the microcomputers most often used in morpho-metrics.

Run length encoding   This is a simple, and often very effective, technique for the efficient storage of images that have only a few levels of brightness (e.g., binary images). The image is stored as a continuous stream of bytes but with the start of each row marked with a special code (or else the byte offset of the start of each row stored in a separate array). The lengths of each of the runs of identical brightness values along each row are stored rather than the actual brightness values. For example, a row of brightness values might be:

0	1	1	1	1	0	0	0	0	0	0	1	0	1	1	0

The runs would then be (1, 4, 6, 1, 1, 2, 1). Note: in order to recover the brightness values each row must begin with the same brightness value, say 0. If a line happens to begin with a brightness value of 1 then an initial run of length 0 is inserted at the beginning of the row. Some image operations can be performed on the image in this compressed format. For example, the area of the object (the 1's in the image) is simply the sum of the even-numbered runs. Ballard and Brown (1982, pp. 58-61) show how to compute the horizontal and vertical projections of an image and its center of gravity from this storage representation.

Pyramids   In some applications it is useful to be able to process an image at varying degrees of resolution. One method is to partition the digitized image into non-overlapping regions of equal size and shape and then to replace each of these regions by the average pixel density in that region. This step is called consolidation. This is repeated recursively until there is only a single region with a brightness value equal to the average brightness in the original image. Using the average brightness value for a region rather than a value from the center of the region tends to reduce the aliasing effects that one would expect if one were just to redigitize the image at a lower resolution. Ballard and Brown (1982, pp. 109-111) show an example (from Tanimoto and Pavlidis, 1975) of an algorithm for edge detection using data stored in a pyramid.

Quad-trees Quad-trees (Samet, 1980) are an efficient method to store binary images. To convert data stored in a pyramidal data structure to a quad-tree one recursively searches the pyramid from top to bottom. If an element is "black" or "white" then form a terminal node in the quad-tree of the corresponding type. Otherwise, form an internal "gray" node with pointers to the results of the recursive examination of the 4 elements at the next level in the tree. Samet (1981a) gives an algorithm to directly convert a raster image (the usual storage is by rows) to a quad-tree. The method is efficient in that the image is read and processed a row at a time and the resulting quad-tree is of minimal size. Less space is needed by the algorithm than would be required if the entire image were read at once. There are many algorithms to perform image processing operations directly on images stored as quad-trees. The computation of the area of an object is easy. Samet (1981b) gives an algorithm for computing the perimeter of regions from a quad-tree representation. Samet (1983) shows how to perform a medial axis transformation from a quad-tree. A number of papers have been published describing the computation of various geometrical properties of objects from a quad-tree representation. Pavlidis (1982) is a convenient source of many of these algorithms.

One problem with quad-trees is that they are not translation independent—if an object is shifted in position by one pixel the quad-tree can be very different in structure. Scott and Iyengar (1986) developed a translation invariant version of the

quad-tree based upon the medial axis transformation (see below).

Filters

A filter is a function which produces new images that are transformations of an input image. The purpose is to produce images in which particular aspects of an image are accentuated or enhanced. The results can sometimes be quite impressive (such as revealing details hidden in shadow areas of a photograph, or the apparent sharpening of an out-of-focus image). Of course, the desired information must already be present in the image. What the enhancements do is transform the information so that the desired features of the image are more obvious to a human observer. It is useful to imagine an image as a surface where height represents the brightness at each point in the image. An example is shown in Figure 4 above, where each horizontal curve corresponds to a column in the digitized image. A transformation of such an image surface might, for example, smooth out the part of the surface corresponding to the background but steepen the sides of the hills corresponding to the object, so that it appears to have sharp vertical cliffs. The published literature on methods of enhancement is extensive and growing rapidly. Some standard methods are described below. These transformations cannot perform "magic." A great deal of effort can be saved by starting with simple images of well-illuminated scenes.

Contrast Enhancement

One of the first adjustments to consider is contrast enhancement. Brightness values are rescaled so that they cover the full dynamic range of the display device. For example, in a very low contrast image the brightness values may range from 150 to 200. Rescaling them to range from 0 to 255 will result in an image that humans find easier to interpret. If the digitizing hardware can perform this operation as it is acquiring the image, the researcher is able to make more efficient use of the resolution of the digitization of the brightness values. This is especially important if the digitizer cannot furnish at least 256 brightness levels. Contrast enhancement is useful even when working with an already digitized image. However, spreading out a range of 150-200 to 0-255 does not increase the number of distinct brightness values; they are simply spaced further apart. The operation is still useful, however; the image is more pleasant to look at even though it contains no more information than the original.

Histogram Transformations

Histogram enhancement techniques include contrast enhancement, as described above, but also make a non-linear transformation of the brightness values so that the brightness values not only cover the full dynamic range but have a frequency distribution of brightness values that take on a particular form. In theory, a uniform distribution can be obtained using the following transformation

$$g(q) = \frac{M}{N} \int_0^q h(p) \, dp, \qquad (14)$$

where M is the number of gray levels, N the number of pixels, and $h(p)$ the observed histogram of the number of pixels with each level of brightness p. The practical problem with this algorithm is that the input brightness values are discrete, so that the most we can do is to obtain a more even spacing of the values. The histogram need not be uniform since some classes may have many more entries than others. The results can be rather disappointing when the important information in the image is best represented by only a few distinct brightness values.

Hummel (1977) suggested a transformation called histogram flattening in which the histogram is made more uniform by randomly assigning pixels in the most abundant classes to other brightness classes. He suggested that the choice of pixels to be reassigned could be based upon the average gray-level in their local neighborhoods. On the other hand, Frei (1977) suggests that, for images to be interpreted by humans, the goal should be to produce a picture in which there is a uniform distri-

bution of perceived brightness levels. To do this, the distribution of displayed brightness levels should be hyperbolic rather than uniform. But this seems to apply only to images of objects that the viewer expects to see represented by continuous brightness levels. Simple high-contrast images (such as bone or shells laid on a sheet of black paper) are not improved by such transformations.

Adaptive contrast enhancement methods adjust the degree of contrast enhancement of a pixel depending upon the distribution of brightness of its neighboring pixels. Peli and Lim (1982) proposed a method in which the final image is a weighted combination of a smoothed image and the difference between the original image and the smoothed image. The weights can be a non-linear function of the brightness in the smoothed image so as to give greater increase in contrast to certain ranges of brightness values. An examination of the gray-level histogram can also be useful when trying to find a suitable threshold level to segment the image into regions. If the scene consists of just an object and background, then one would hope to find two peaks and the threshold would be placed in the valley between them. The results are not as clear-cut as one might expect. One problem is that pixels on the boundary of an object are expected to have intermediate gray values dependent upon their degree of overlap with the object versus the background. Other problems are shadows, uneven illumination, and noise.

Smoothing

Smoothing is a useful technique to eliminate unwanted fine detail in an image by averaging each pixel's brightness value with those of its neighbors. Such methods are sometimes referred to as low pass filters since they remove high-frequency details (fine undulations in the image surface) while preserving the low frequency information (large scale changes in the surface). Often the general form of an object is of interest rather than its fine details (texture). A simple analog method of smoothing is to digitize the image when it is slightly out-of-focus. Mathematically, smoothing corresponds to the convolution of the image function with a point-spread function; the brightness at each pixel is spread-out and averaged with adjacent pixels. This results in a linear, space-invariant (the same function is used over the entire image), moving average (each pixel is some type of average of its neighbors) filter. This can be implemented using the following equation:

$$g(x,y) = h \otimes f$$

$$= \sum_u \sum_v h(u\text{-}x, v\text{-}y)\, f(u,v), \qquad (15)$$

where h is the point-spread function (the kernel of the convolution), f is the input image function, and the summations are overall alignments at which the kernel overlaps the given pixel, at x,y.

There are many choices for h depending upon the type and degree of smoothing desired. A common choice is the values in the following array.

1/16	2/16	1/16
2/16	4/16	2/16
1/16	2/16	1/16

Usually this filter does not remove much noise from an image. To produce a stronger effect, one could either use a larger array of constants or else apply the filter repeatedly. Such simple local smoothing applied to the entire image often removes some important details in the image since the brightness values along the outline of an object will also be averaged with those of the background. This makes the boundaries of an object more difficult to detect by the human eye. A solution is to limit the smoothing to regions of relative homogeneous brightness levels. Nagao and Matsuyama (1979) proposed that one examine subregions around each pixel and then average its brightness value only with those in the most homogeneous subregion (an edge-preserving smoothing transformation). The procedure computes the variance in regions corresponding to the x's in the 4 possible 90° rotations of each of the first two patterns (within a

5×5 region centered on each pixel) and the one possible orientation of the third.

.	x	x	x	.		.	.	.	x	x		.	.	.	.	.
.	x	x	x	.		.	.	x	x	x		.	x	x	x	.
.	.	x	.	.		.	.	x	x	.		.	x	x	x	.
.	.	.	.	.		.	.	.	.	.		.	x	x	x	.
.	.	.	.	.		.	.	.	.	.		.	.	.	.	.

Nagao and Matsuyama (1979) suggested using the ordinary variance as a criterion of homogeneity. I have found slightly better results by weighting the center point when computing the mean and variance. One could also include a tolerance so that no averaging would take place if even the most homogeneous region was too heterogeneous. The method is time-consuming since nine variances must be computed at each pixel. If one knows that the gray-values in a particular region of an image should be uniform, one can limit the smoothing operation to that particular area and thus avoid the smoothing of edges.

Background Subtraction

In some applications it is possible to get rid of some of the complexity of the image (dirt spots on the lens, etc.) by subtracting out a constant background. If the images are correctly aligned, one can simply subtract the gray-values of the background image from that of the image under study.

Template Matching and Cross Correlation

A very common operation is that of matching a template pattern against an image. On a 1-dimensional image one might slide the template pattern, t = (-1, 0, 1), across an image, looking for the position of greatest match (which, in this example, is the position of the largest linear increase in image intensity). The distance between the image and a template (aligned at pixel y) is

$$d_y = \sqrt{\sum_x \left(f(x)\text{-}t(x\text{-}y)\right)^2}, \qquad (16)$$

where the summation is over all pixels in the image for which the template is defined in this alignment. Finding the location that minimizes d_y is equivalent to finding the location that maximizes the cross correlation function, R_{ft}, for f and t.

$$R_{ft} = t*f$$

$$= \sum_x f(x)\, t(x\text{-}y) \quad . \tag{17}$$

Note the similarity of this function to the convolution operation [in the convolution $t(x\text{-}y)$ is replaced by $t(y\text{-}x)$]. The same operation can be applied to a 2-dimensional image. The 2-dimensional template is "rubbed" over the entire image and a value is computed for each alignment tested. The continuous form of the 2-dimensional cross-correlation function is

$$t*f = \int_{-\infty}^{\infty} \int_{-\infty}^{\infty} t(u\text{-}x, v\text{-}y)\, f(u,v)\, du\, dv. \tag{18}$$

Edge Detection

The regions of rapid change in brightness values in an image, which often correspond to the boundaries or edges of objects, seem to convey much of the information about the shape and locations of objects in the image. Thus methods that enhance this aspect of an image are of particular interest.

Gradients The most obvious operation to consider is the computation of the magnitude of the gradient of the image surface at each pixel. If the result is viewed as an image, pixels located in regions of rapid change in brightness in the original image would appear as bright points. An object such as a dark leaf against a light background would thus appear as a set of bright pixels around the perimeter of the leaf. Such a transformed image may be simpler to process by computer program so as to obtain particular features of interest (e.g., the number of pixels brighter than a certain threshold could be used as an estimate of the perimeter of the

leaf). However a simple gradient computation may fail since it is very sensitive to noise in the image. Thus one may wish to use the gradient of a smoothed image or to use more complex algorithms such as those of Machuca and Gilbert (1981).

Horn (1986) shows a simple model for an edge in an image as a straight line separating two regions of different brightness.

$$E(x,y) = B_1 + (B_2\text{-}B_1)\, u(x \sin\theta - y \cos\theta + \rho), \tag{19}$$

where B_1 and B_2 are the brightness values in the two regions, $x \sin\theta - y \cos\theta = \rho$ is the equation of the separating line, and $u(z)$ is the unit step function:

$$u(z) = \begin{cases} 1 & \text{for } z > 0 \\ 1/2 & \text{for } z = 0 \\ 0 & \text{for } z < 0. \end{cases} \tag{20}$$

The gradient of this surface is the vector

$$\begin{bmatrix} \partial E/\partial x \\ \partial E/\partial y \end{bmatrix}, \tag{21}$$

where

$$\partial E/\partial x = \sin\theta\, (B_2\text{-}B_1)\, \delta\, (x \sin\theta - y \cos\theta + \rho)$$
$$\partial E/\partial y = \text{-}\cos\theta\, (B_2\text{-}B_1)\, \delta\, (x \sin\theta - y \cos\theta + \rho). \tag{22}$$

It is important to note that the gradient is coordinate system independent in that it maintains its magnitude and orientation relative to the underlying edge when the separating line is rotated or translated.

Laplacian The Laplacian of the surface defined above is

$$\nabla^2 E = \partial^2 E/\partial x^2 + \partial^2 E/\partial y^2$$
$$= (B_2\text{-}B_1)\, \delta'\, (x \sin\theta - y \cos\theta + \rho), \tag{23}$$

where δ' is the unit doublet, the derivative of the unit impulse $\delta(u)$. The Laplacian has the desirable properties of retaining the sign of the brightness difference across the edge, so we can determine which side is brighter and thus reconstruct the original edge, and it is a linear function of x and y.

Approximations for Digital Images

The simplest approximation is to estimate the derivatives of the surface at each point in the image by the differences in E_{ij} at adjacent pixels. Let the pixels around a point be represented by the following table.

$E_{i,j+1}$	$E_{i+1,j+1}$
E_{ij}	$E_{i+1,j}$

Note that the order of subscripts corresponds to the x and then the y dimension, not the usual row and then column convention with matrix algebra. The derivatives at the center of this 2×2 array can be estimated as

$$\partial E/\partial x = \frac{1}{2\varepsilon}((E_{i+1,j+1} - E_{i,j+1}) + (E_{i+1,j} - E_{ij}))$$
$$\partial E/\partial y = \frac{1}{2\varepsilon}((E_{i+1,j+1} - E_{i+1,j}) + (E_{i,j+1} - E_{ij})) \quad (24)$$

where ε is the spacing between the rows and columns. This formula (from Mero and Vassy, 1975) is the average of two finite-difference approximations and is an unbiased estimate of the slope for the point where the four pixels used in the above formula meet. The squared gradient can be used to produce a map of the location of high rates of change in brightness in the image. To get the actual direction of change one must refer back to the gradient itself. A more refined estimate of the slope of the surface can be obtained using the information from a 3×3 arrays of pixels.

$E_{i-1,j+1}$	$E_{i,j+1}$	$E_{i+1,j+1}$
$E_{i-1,j}$	E_{ij}	$E_{i+1,j}$
$E_{i-1,j-1}$	$E_{i,j-1}$	$E_{i+1,j-1}$

The Sobel estimate of the gradient (Shaw, 1979) is computed by cross correlating the above submatrix of E_{ij}'s with the weights in the following two tables (the first yields the change in the x-direction and the second gives the change in the y-direction).

-1	0	1
-2	0	2
-1	0	1

1	2	1
0	0	0
-1	-2	-1

From this 3 by 3 array of pixels we can also estimate the second partial derivatives as

$$\partial^2 E/\partial c^2 = \frac{1}{\varepsilon^2}(E_{i-1,j} - 2E_{ij} + E_{i+1,j})$$
$$\partial^2 E/\partial c^2 = \frac{1}{\varepsilon^2}(E_{i,j-1} - 2E_{ij} + E_{i,j+1}) \quad (25)$$

so that the Laplacian can be estimated as

$$\nabla^2 E = \frac{4}{\varepsilon^2}\left[\frac{1}{4}(E_{i-1,j} + E_{i,j-1} + E_{i+1,j} + E_{i,j+1}) - E_{ij}\right], \quad (26)$$

This function is zero both in areas of constant brightness and also in areas where brightness varies linearly. It represents subtracting the value of a central pixel from the average of its neighbors. This corresponds to the application of a template with weights $1/\varepsilon^2$ times the values in the following table:

0	1	0
1	-4	1
0	1	0

Note that one could just as logically rotate the coordinate system by 45° before approximating the derivatives. Linear combinations of this and the rotated template also produce estimates of the Laplacian. Horn (1986) states that the following template, times $1/6\varepsilon^2$, is popular and produces a particularly accurate estimate of the Laplacian.

1	4	1
4	-20	4
1	4	1

Kirsch Operators

The Kirsch operator (Kirsch, 1971) is a method for classifying the properties of small regions in an image. It can be used to detect whether a region gives evidence for an edge, a line, or is undifferentiated. For example, to determine the direction of a gradient at each pixel in the image one could compute

$$S(x) = \max [1, \max_{k} \sum_{k-1}^{k+1} f(x_k)], \qquad (27)$$

where the $f(x_k)$ are the 8 neighboring pixels to x and where subscripts are computed modulo 8. The value of k that yields the maximum indicates the direction of the gradient (with 3 bits of accuracy). This method can easily be implemented by matching (cross correlating) the following four templates with a span of $n = 1$ (similar templates can be used for larger values of n). The direction is then indicated by whichever template matches best.

-1	0	1
-1	0	1
-1	0	1

1	1	1
0	0	0
-1	-1	-1

0	1	1
-1	0	1
-1	-1	0

1	1	0
1	0	-1
0	-1	-1

Enhancement of Geometric Patterns

Methods have also been developed that accentuate particular geometric patterns in an image, rather than performing more general enhancements. Special attention has been given to line enhancement methods. This is both because linear features are often directly of interest (e.g., veins in insect wings or leaves) and also because linear features are useful as boundaries of objects. This does not mean that an object needs to have flat sides—it need only be relatively smooth. Small regions around the outline of a bone or a shell, for example, can be well approximated by a series of linear edges. Paton (1979) proposed several useful methods for finding linear features. In these methods templates are superimposed at various orientations over each pixel in the image. The templates are such that the presence of a line in the region of a pixel will result in a high cross-correlation. The maximal value obtained for templates centered on a given pixel is used as the output value for the given pixel.

Groch (1982) proposed a procedure for recognizing line-shaped objects in images by trying to follow a large number of lines starting from seed points found by searching along transects through the image. A regional operator is then applied to try to fill in gaps along more or less collinear line segments.

Three-dimensional Images

A detailed discussion of the recovery of 3-dimensional information from images is beyond the scope of the present review as it involves more complex techniques. Information on the 3-dimensional orientation of surfaces can be obtained from their pattern of reflectance (and thus requires information about their surface properties). For certain types of objects, one can gain 3-dimensional information by projecting stripes of light, perhaps from a laser, across the object at known angles. The apparent deflection from a straight line in the image can be related to the shape of the object.

Image Segmentation

It is usually desirable to break an image up into regions, or segments, corresponding to the logical subunits of the original scene. In morphometrics, one wishes to separate the image of an object from its background and perhaps to isolate different components of the image. The regions can then be analyzed separately. In some images the objects to be located are lines. One can adapt the line enhancement operators described above. Groch (1982), for example, used this approach to detect roads in aerial photographs. Biological images often contain linear features. Once an object has been separated from its background, contour-tracing algorithms can be used to trace its outline (see below).

Thresholding to Define Regions

While a human can usually easily recognize the component parts of a complex scene (such as background, outline of a wing, veins of a wing, cells between the veins, etc.) this is often a difficult task to perform automatically (Riseman and Arbib, 1977). With perfect, noise free, images one can isolate an object from a uniform background simply by finding a threshold level of brightness such that all pixels in the background are above or below the selected value. Castleman (1979, p. 311) gives a simple algorithm that can trace out the boundary between the object and the background for such images. He points out, however, that even small amounts of noise can send the tracking algorithm temporarily or hopelessly off the boundary. This is a problem with mosquito wings, for example, since the margin of the wing and the veins are covered with scales that can become dislodged and appear in unexpected locations when the wing is mounted on a slide. An additional problem is the fact that an image may not be illuminated evenly so that the background may differ in brightness in different parts of the image even after some of the standard image enhancement techniques have been applied. Thus the boundary tracking problem can become rather complex and usually must take into account a priori information about the geometrical properties of the particular class of objects being extracted, or else be supervised by someone who knows what the expected contour should be and thus can intervene and make corrections when necessary.

Region Growing

A complementary approach is that of "region growing" (Brice and Fennema, 1970) in which one first examines small regions in an image and then merges adjacent regions with similar properties (brightness, texture). This has been used, for example, to break aerial photographs into homogeneous blocks each representing a different type of forest, farmland, etc. Grainger (1981) reported an average accuracy of about 50% when this method was applied to 186 sample sites from New Forest, Southern England, for which both ground and densitometric data were available. This method is most often used with multispectral images (grey-scale images at each of several spectral bands). There is much more information for each pixel, and hence higher performance can be expected.

Contour Tracing

The most important technique for image segmentation in morphometrics is that of tracing the outline of a selected object. If the grey-levels of the object are distinct from the background, this is a relatively straight-forward task. Such images can be converted into binary images, where "1" corresponds to the selected object and "0" to the background pixels by thresholding. Pavlidis (1982) gives algorithms for finding the overall contour and also for finding the contours of any holes that may be contained within an object in such a binary image. In order to describe the algorithms, we need to adopt the following standard numbering system to refer to the 8 pixels adjacent to a given point. For example, the point above the given pixel, p_{ij}, is called the 2-neighbor.

3	2	1
4	p_{ij}	0
5	6	7

A contour is defined as the set of all pixels within the selected object that have at least one neighbor that is not part of the object. The strategy is to start with a point in the object whose 4-neighbor is not in the object, and then to trace the outline in a counter-clockwise direction.

1. Choose a point, A, in the contour such that its 4-neighbor is not in the object.

2. Set C = A, S = 6, and set the flag *first* = *true*.

3. While C ≠ A or *first* = *true*, do steps 4 to 10.

4. Set the flag *found* = *false*.

5. While *found* = *false*, do steps 6 to 9 at most 3 times (the purpose of this limit is to avoid looping on objects that consist of only a single pixel).

6. If B, the (S-1)-neighbor of C, is part of the object, then set C = B and *found* = *true*.

7. Else if B, the S-neighbor of C, is part of the object, then set C = B and *found* = *true*.

8. Else if B, the (S+1)-neighbor of C, is part of the object then set C = B and *found* = *true*.

9. Else set S = S + 2, modulo 8.

10. Set *first* = *false*.

11. End.

The algorithm must also be applied once for each hole in the object. When completed, one needs a description of the path traversed. One possibility is simply to list the coordinates of the points C. Another, more compact representation is to store the coordinates of only the first point and then store the neighbor code to indicate the direction taken when moving from one pixel to the next. When the contour is very smooth, further economy can be achieved by storing the derivative of the chain code. The change in direction will usually require fewer bits than the chain code (unless the contour often doubles back on itself). When the chain code sequence or its derivative contains sequences of identical codes, run-length coding can be used to reduce the amount of space needed to store a contour. In run-length coding one replaces a sequence of identical values with a special code and the length of the sequence.

Thinning

Thinning algorithms simplify the representation of the outline of an object by computing an internal skeleton that will contain useful information about the original outline. Straney (this volume) describes several different methods of defining what one means by a skeleton and different algorithms for their computation. They are usually applied to binary images (images that have already been thresholded). One method is to reduce the width of elongated objects in the image by "eating away" at the sides of objects while trying to avoid the deletion of pixels at the ends of the objects. There are several methods for carrying out such operations. Pavlidis (1982) gives the classical thinning algorithm. He also presents a simple approximate thinning algorithm that may be satisfactory for some purposes. Another approach is to compute the medial axis transformation, MAT. In a MAT skeleton, the pixels are located at centers of circles that touch the original outline at more than one place. For example, if the original object is a circle, then the MAT skeleton will be a single point at the center. If one codes each pixel in the MAT skeleton by the diameter of the circle that it repre-

sents, the MAT skeleton has the important property that it can be used to reproduce the original outline shape (i.e., the medial axis transform has an inverse). Blum (1973) describes the geometrical properties of MAT skeletons and some implications for shape description in biology. Bookstein et al. (1985) discuss and give examples of its potential application to morphometrics including that of Bookstein (1981). Thin figures seem to be represented well by MAT skeletons—the skeletons seems to have intuitively reasonable shapes. Bookstein et al. (1985) observe that, while it uses only information on the outline, the skeleton often has a structure that seems biologically appropriate. However, the skeleton seems less useful for wider objects. An important problem is that it is very sensitive to noise. Small changes in the outline (e.g., small bumps or indentations) can cause drastic changes in the form of the skeleton. The outline has to be quite smooth in order for a simple skeleton to be obtained.

There also has been some work generalizing the MAT to gray-level images. Dyer and Rosenfeld (1979) describe a simple algorithm to thin gray-scale images. For dark objects, their method changes each dark pixel to the minimum of its neighbors' levels provided this does not disconnect any pair of points in its neighborhood. This process can be repeated until the objects are sufficiently "thin". Wang et al. (1981) define a gray-scale generalization of the medial axis transformation (which they call a MMMAT, for min-max MAT). It allows one to reconstruct good approximations to the original image.

Texture

A general discussion of this topic is beyond the scope of this review as the complexity and diversity of types of surface textures possible in biological images is very large. However, the use of fractal curves and surfaces has attracted increased interest in many fields in the last few years. One type of application that seems useful in morphometrics is the use of fractal dimension as a description of the texture or complexity of an outline of an object.

For example, one can digitize the outline of a leaf at high resolution and then see how the apparent length of the outline changes as a function of the step-size (scale) used to measure the length of the outline. The fractal dimension of the outline curve can then be estimated using the relationship

$$D = \ln \frac{N}{\ln (1/S)}, \qquad (28)$$

where S is the step size and N is the number of steps. For a line in Euclidean geometry, a division of a line into segments of length $1/S$ results in S segments and hence a dimension of $D = 1$. When the outline is highly reticulate its length will be very long when one measures it with a small step size and as a result its fractal dimension will be greater than 1. Vlcek and Cheung (1986) describe the computation of the fractal dimension for several types of leaves. They show that fractal dimension is a useful descriptor of the irregularity of the leaf outline. The obtained values range from 1.02 for a rather smooth American basswood leaf to 1.28 for a white oak leaf. Long (1985) used fractal dimensionality to describe complex sutures in deer skulls and in ammonites. He found D-values from about 1.4 to 1.5. Morse et al. (1985) found D-values of about 1.5 for the outlines of a variety of plants during early spring. They point out that if insects and other arthropods living on these plants perceive the amount of space on the plant (for food and shelter) in relation to their body size, then small insects will perceive a much larger available habitat than larger insects when $D > 1$. They then show that the distribution of sizes of insects is in keeping with what one would expect if their abundance were proportional to the perceived amount of available habitat. Katz and George (1985) furnish a program in BASIC that estimates the fractal dimension of an outline represented by a set of x,y-coordinates. Slice and Gurevitch (in preparation) used this approach and found significant differences between species and trees of the genus Acer (Maples) with respect to leaf outline complexity. They found that the ordering of species with respect to mean fractal dimension was consistent with their subjective perception of outline complexity. D. Slice has developed a program, called FRACTAL-D, that performs these computations.

A similar idea holds for surfaces. In an Euclidean plane the subdivision into cells with a mesh size of $1/S$ will result in a surface area composed of $N = S^2$ equal sized cells, and hence a dimension of $D = 2$. As the surface becomes more complex, the surface area will increase and hence the fractal dimension will be larger than 2.

Boundary Representation

This is a very important topic. Most applications of image analysis to morphometrics have been concerned with the comparison and analysis of information that can be extracted from an outline (often supplemented by information on locations of morphological landmarks). But in order to use an outline in a quantitative analysis it is necessary to use an appropriate mathematical representation. Listed below are some of the most common approaches (some have been mentioned above). These approaches will not be described in detail since most of them will be covered elsewhere in this volume.

1. x,y-coordinates. One can simply save enough of the coordinates of enough points around the outline to capture its form with sufficient accuracy. Usually one will have more points in regions of higher curvature. Some methods of morphometric analysis use these coordinates directly. These raw coordinates can be used to derive other representations of the outline. For example, the elliptic Fourier method uses coordinates as input (rather than polar coordinates or tangent angles as in most Fourier studies).

2. Chain codes. The method of using chain codes (and differential chain codes) was discussed above.

3. Polar coordinates. If the outline is a simple convex shape, then it may be possible to describe the shape by giving the radius of

equally spaced vectors from some convenient origin to points along the outline contour. This method has been used in many Fourier applications.

4.   Tangent angle. While traversing the outline of an object, one can record the slope of a tangent to the outline at the current position and the distance traveled along the outline. This has the advantage that one can represent tangent angle as a function of arc length for any closed outline shape. This has also been used in many Fourier studies.

5.   Medial axis skeleton. Since the original outline can be recovered from a skeleton, the skeleton can be used as a method to encode an outline shape.

6.   Splines. Several studies have explored the usefulness of using splines rather than Fourier functions to describe the shapes of morphological structures. Some examples are Engles (1986) and Evans et al. (1985).

7.   Fractals. Barnsley et al. (1986) show that it is possible to determine the fractal curve that provides a close approximation to a given binary image. Remarkably, this representation required very few parameters to be estimated in order to fit the outline of complex objects with very complex outlines such as a black spleenwort fern frond.

Feature Extraction

Feature extraction is the task of obtaining the most important descriptive parameters from an image. These parameters represent not just an encoding of an image, but the isolation of particular parameters that can be used to distinguish an object in one image from another. These may consist of the usual distance measurements used in morphometrics (lengths, maximum widths, etc.), often the problem is more complex. Traditional measurements are often selected because they are easy to make using hand-held calipers. But with automation other types of measurements may be easier to

program in a computer. For example, once one has an outline contour it is easier to measure the area or the perimeter of an object than it is to measure its width. Thus with the availability of new technology one should not just duplicate conventional methods but explore other ways of describing differences among organisms. There are a large number of ways in which an object in an image can be described. Unless the different methods are linearly related, one does not expect them to give exactly the same results. Thus the choice of types of descriptors used in a morphometric analysis is expected to make a difference (see further discussion below). Unfortunately, it is unclear at this point how one should choose among the different systems. But in the important special case of systems of linear distance measurements, Strauss and Bookstein (1982) point out the advantages of taking measurements in the form of a "truss" rather than in the more conventional pattern that often has a lot of redundancy. One of the most popular approaches in morphometrics has been the use of Fourier coefficients to describe outlines of organisms. My comments on this topic is very short since it is covered elsewhere in this volume. The method of moment invariants has been used in a few morphometric studies. It is described in some detail below since different formulations of the method have been used and they raise some interesting issues.

Description of an Outline Contour

A common approach is the fitting of some mathematical function to the points sampled around the outline of an object. The parameters of the fitted function are then used in multivariate analyses as descriptors of the shape of the outlines. Various types of Fourier analysis are the most popular examples of this approach in morphometrics, but other functions have also been used. A brief outline is furnished below with references to more detailed accounts.

1.   *Fourier analysis of an outline expressed in polar coordinates.* In many morphometric studies points are sampled along the outline such that

vectors connecting them to some point of reference (or origin) are separated by equal angles. The lengths of these vectors (distances of each point to the origin) are then subjected to a Fourier decomposition (a 1-dimensional Fourier transformation). The resulting coefficients can be expressed in one of two ways: either as the coefficients of the sin and cosine terms in the Fourier series or in terms of their amplitude and phase angle. Kaesler and Waters (1972) provides an early example. When no landmarks are available it may not be possible to specify a unique starting point for the measurement of the angles (i.e., the vector that corresponds to an angle of 0). In such cases, only the amplitudes for each harmonic are used as descriptors. Younker and Ehrlich (1977) provide an example. A limitation of the use of this polar representation is that the outline of the object must be such that each vector crosses the outline only once. Thus the outline cannot be very complex.

2. *Fourier analysis of an outline expressed in terms of the change in tangent angle as a function of arc length.* Bookstein et al. (1982) refer to this as an intrinsic representation. Zahn and Roskies (1972) suggested that an outline be scaled so that its length is equal to 2 and then the following function computed for each point along the outline

$$\phi^*(t) = \theta(t) - \theta(0) - t, \qquad (29)$$

where t is the distance along the outline of a given point, $\theta(t)$ is the angle of a tangent line at that point, and $\theta(0)$ is the angle of a tangent line at the starting point of the outline. Thus $\phi^*(t)$ is the difference between the cumulative change in angle that one observes when moving along an outline and the change that one would expect if the outline were a perfect circle. The values are then subjected to a Fourier decomposition. This approach has the advantage that it can be used for any

closed contour (complete outline), regardless of its shape.

3. *Elliptic Fourier analysis.* Kuhl and Giardina (1982) proposed the separate Fourier decomposition of the differences in the x and y-coordinates as a function of arc length corresponding to the distance along the outline to each point (again scaled so the perimeter of the outline is 2π). Rohlf and Archie (1984) showed that this method has several advantages over the methods listed above.

4. *Splines and other functions.* Any function that can be made to pass through an observed set of points can be used as a description of an outline. Cubic splines and Bezier curves represent flexible families of curves that can be made to fit arbitrary configurations of points. Examples of the use of Bezier curves are given by Engles (1986) and examples of cubic splines are given by Evans, et al. (1985).

5. *Eigenshape analysis.* Lohmann (1983) showed that an outline, represented in his case by the $\phi^*(t)$ function, can be fitted by sets of empirical functions derived from the data. The advantages of this approach are that fewer functions are needed to describe the observed diversity among the objects under study, and it is not necessary to specify particular families of curves to be fit to the outlines (such as sums of sines and cosines or various types of polynomial functions). The Chapter 6 by Lohmann and Schweitzer in this volume is a general exposition of this method with examples.

6. *Fractals.* As mentioned above, Barnsley et al. (1986) have shown that it is possible to solve for the fractal curve that best approximates a given object outline. In their examples, very few parameters were needed to obtain a very close fit for objects with very complex outlines. In the case of a black spleenwort fern frond, the outline was described by a collage of four affine transformations which required 28 parameters, 8 of which were zero. So few

parameters were probably required because the frond does seem to be a case where the outline shows the property of self-similarity. Objects with less regularity may require many more parameters. The relevance of these functions for morphometrics needs further study. An important question is the extent to which objects with similar values of the parameters (i.e., those that are close together in the feature space) are similar morphologically.

In all of these cases, the results are a set of coefficients that can be used as measurements of descriptive variables for various types of multivariate analyses. Unfortunately, the results one obtains from multivariate analyses need not be the same for different types of shape descriptors. The descriptors obtained from different methods do not represent simple linear transformations of the same information. The relationships between some pairs of methods correspond to complex non-linear transformations of the original coordinate data. This implies that it is not sufficient for a method to be convenient computationally. In order to use a method one must be confident that it is appropriate for the description of the kinds of variation that ones expects to observe.

Moments of an Image Surface

One simple approach to the description of an image is to transform the image so that the brightness of the background is zero and the brightness values for the object are positive numbers. Then the brightness values can be treated as proportional to a 2-dimensional frequency distribution (a sample from a 2-dimensional probability density function). The 2-dimensional moments of this function can then be computed and used as the parameters of this distribution. For example, the mean in the x-direction is

$$\bar{x} = \int_{-\infty}^{\infty} x \, f(x,y) \, dx \, dy. \qquad (30)$$

The p, q central moment can be computed as

$$\mu_{pq} = \int_{-\infty}^{\infty} \int_{-\infty}^{\infty} (x-\bar{x})^p \, (y-\bar{y})^q \, f(x,y) \, dx \, dy. \qquad (31)$$

The order of a moment is the sum $p+q$.

A uniqueness theorem (Papoulis, 1965) guarantees that if $f(x,y)$ is piecewise continuous and has nonzero values in only a finite part of the x,y-plane (true by definition for brightness surfaces), then moments of all orders exist, the moment sequence is uniquely determined by $f(x,y)$, and the moments uniquely determine $f(x,y)$. Thus the moments can be considered descriptors of the image brightness surface and can be used to reconstruct an image brightness surface. Note that this method can describe the brightness surface, not just the outline of an object. However it is often applied to binary images to limit them to a description of the boundary of an object.

Moment invariants A problem with the use of raw moments as descriptors is that they are not invariant with respect to rotation, translation, and reflection of the object within the image. Hu (1962) and others have formulated functions called moment invariants which have this desired property. While useful as descriptors of an image, they have limitations. A practical problem is their sensitivity to rounding errors in the computation of the higher moments.

Average moments have been defined in two ways. Most workers suggest dividing the above moments by μ_{00}, the total density of the image (the volume under the surface). However, Dudani et al. (1977) suggest that one divide by n, the number of nonzero pixels in the image. These two methods are equivalent only for binary images (where $f(x,y) = 1$ corresponds to a point within the object and 0 otherwise). Yin and Mack (1981, p. 138) say that the latter method gives weak intensity invariance. An obvious property of central moments is that their values are invariant to translation of the object along the coordinate axes. In most studies the central moments are normalized in an effort to eliminate the effects of overall "size" of the image. As it is in morphometrics, this is not as simple as it might seem at first. The most common normalization (due to Hu, 1962) is

$$\eta_{pq} = \mu_{pq} / \mu_{00}^{(p+q)/2+1}.$$ (32)

for all p, q such that $p+q = 2, 3, \ldots$. This adjusts the moments to take into account the overall intensity of the image (i.e., the volume under the brightness surface). Other normalizations are described below. Because of a misprint in Hu (1962) the divisor is often, incorrectly, given as $\mu_{00}^{(p+q)/2} + 1$. Maitra (1979, p. 697) gives the correct formula (which is confirmed by Casasent et al., 1981, p. 127).

The above moments are not useful for most studies since the coefficients are still affected by such things as rotation of the image and its degree of contrast. Several methods have been proposed for obtaining functions that are invariant to such details about an image and thus are expected to describe just its form. Hu (1962) proposed a set of absolute orthogonal invariants, h_1, (based on the normalized moments, η_{pq}, above):

$$h_1 = \eta_{20} + \eta_{02}$$
$$h_2 = (\eta_{20}-\eta_{02})^2 + 4\eta_{11}$$
$$h_3 = (\eta_{30}-3\eta_{12})^2 + (3\eta_{21}-\eta_{03})^2$$
$$h_4 = (\eta_{30}+\eta_{12})^2(\eta_{21}+\eta_{03})^2$$ (33)
$$h_5 = (\eta_{30}-3\eta_{12})(\eta_{30}+\eta_{12})[(\eta_{30}+\eta_{12})^2-3(\eta_{21}+\eta_{03})^2]$$
$$\quad + (3\eta_{21}-\eta_{03})(\eta_{21}+\eta_{03})[3(\eta_{30}+\eta_{21})^2-(\eta_{21}+\eta_{03})^2]$$
$$h_6 = (\eta_{20}-\eta_{02})[(\eta_{30}+\eta_{12})^2-(\eta_{21}+\eta_{03})^2]$$
$$\quad + 4\eta_{11}(\eta_{30}+\eta_{12})(\eta_{21}+\eta_{03})$$

and a skew invariant:

$$h_7 = (3\eta_{21}-\eta_{03})(\eta_{30}+\eta_{12})[(\eta_{30}+\eta_{12})^2-3(\eta_{21}+\eta_{03})^2]$$
$$\quad -(\eta_{30}-3\eta_{12})(\eta_{21}+\eta_{03})[3(\eta_{30}+\eta_{12})^2-(\eta_{21}+\eta_{03})^2]$$

(34)

In addition to the position and mass invariance of the μ_{pq}, these functions are invariant to image rotation. They have been used in many applied studies. Hall (1979, p. 423) suggested the use of the logarithms of the h_i in order to reduce their "dynamic range." He does not state, however, what one should do when the $h_i \leq 0$ (which is often the case). As pointed out by Maitra (1979), absolute orthogonal invariants are sensitive to discretization and so are not computationally invariant, especially if one uses outline images. The example given by Hall (1979, p. 423) shows that they may vary over several orders of magnitude.

For practical computation, the formulas for the h_i can be simplified as follows:

$$h_1 = \eta_{20} + \eta_{02}$$
$$h_2 = A^2 + 4\eta_{11}^2$$
$$h_3 = B^2 + C^2$$
$$h_4 = D^2 + E^2$$ (35)
$$h_5 = BF + CG$$
$$h_6 = A(D^2-E^2) + 4\eta_{11}DE$$
$$h_7 = CF-BG,$$

where

$$A = \eta_{20}-\eta_{02}$$
$$B = \eta_{30}-3\eta_{12}$$
$$C = 3\eta_{21}-\eta_{03}$$
$$D = \eta_{30} + \eta_{12}$$ (36)
$$E = \eta_{21} + \eta_{03}$$
$$F = D(D^2-3E^2)$$
$$G = E(3D^2-E^2).$$

Dudani et al. (1977) proposed that Hu's (1962) coefficients should be normalized to correct for differences in the scale of an image. Since magnification (isotropic scale change in both x and y-coordinates) yields an equivalent image, descriptor functions should be insensitive to such a transformation. By dividing Hu's coefficients by various powers of h_1, the normalizations below achieve this scale invariance.

$$d_1 = h_2/h_1^2$$

$$d_2 = h_3/h_1^3$$

$$d_3 = h_4/h_1^3$$

$$d_4 = h_5/h_1^6 \qquad (37)$$

$$d_5 = h_6/h_1^4$$

$$d_6 = h_7/h_1^6.$$

Yin and Mack (1981, p.138) proposed a similar normalization but raised the moments to various fractional powers. The resulting moments are then in a more convenient numerical range.

$$y_1 = \begin{cases} 1 & \text{(intensity)} \\ h_1/\mu_{00} & \text{(silhouette)} \end{cases}$$

$$y_2 = \sqrt{d_1}$$

$$y_3 = \sqrt[3]{d_2}$$

$$y_4 = \sqrt[3]{d_3} \qquad (38)$$

$$y_5 = |d_4|^{1/4}$$

$$y_6 = |d_5|^{1/4}$$

$$y_7 = |d_6|^{1/6},$$

where an intensity image is one in which the $f(x,y)$ are equal to the actual image brightness values. In a silhouette image, all brightness equal to or larger than a specified threshold have been set to 1 and all values less than the threshold set to 0. Since the d_i and the y_i are scale invariant, the normalization of the μ_{pq} by division by $\mu_{00}^{(p+q)/2+1}$ has no effect.

Reddi (1981) proposed slightly different adjustments to h_2 to h_7 which were also intended to yield scale invariant functions.

$$r_2 = h_2/h_1^2$$

$$r_3 = h_3/h_1^{2.5}$$

$$r_4 = h_4/h_1^{2.5}$$

$$r_5 = h_5/h_1^5 \qquad (39)$$

$$r_6 = h_6/h_1^{3.5}$$

$$r_7 = h_7/h_1^5.$$

Contrast invariant moments, m_i, were proposed by Maitra (1979).

$$m_1 = \sqrt{h_2}/h_1$$

$$m_2 = h_3\mu_{00}/(h_2h_1)$$

$$m_3 = h_4/h_3$$

$$m_4 = \sqrt{h_5}/h_4 \qquad (40)$$

$$m_5 = h_6/(h_4h_1)$$

$$m_6 = h_7/h_5.$$

Maitra (1979) does not indicate what should be done when h_5 is negative (one could, arbitrarily, use $-\sqrt{|h_5|}$).

The problem of the normalization of the moment invariants is more complex than one might at first expect since the various adjustments described above can interact. In a discrete image, multiplication of the x and y-coordinates by a constant effects a scale change but no change in the numbers of rows and columns in an image, and hence no change in the "mass" of an image. On the other hand, a magnification of the original image implies that the digitized image is spread out over more pixels and thus the digitized image has a larger mass. Hu's (1962) normalization compensates by, in effect, reducing the image intensities so that the mass stays the same. But this lowers the contrast of the image. Radial and angular moment invariants were proposed by Reddi (1981). He expressed h_1 to h_7 in terms of angular and radial moments. While these are mathematically equivalent to the x,y-invariants described above, Reddi (1981) showed how the polar form allows one to generalize to higher order moment invariants more easily.

   White and Prentice (1987) compared the effectiveness of moment invariants, chain-code descriptors, Elliptic Fourier coefficients, and conventional measurements to discriminate between *a priori* defined groups. They found the chain-code descriptors to perform poorly and both the moments and the Fourier coefficients to perform well in their tests. However, Rohlf and Ferson (unpublished) found the method of moments to perform poorly, due in part to dependencies among some of the coefficients (they are not statistically independent) and sensitivity to rounding errors.

Use of moments to determine orientation   Since the ordinary moments are sensitive to the location and orientation of an object within an image, this information can be used to determine an object's location and orientation so one can move the object into a standard position for further processing. The first eigenvector of the variance-covariance matrix gives the direction of greatest variation. If one is working with an elongated object (such as a mosquito wing), then the vector is parallel to its long axis. Using the notation of Box 15.5 of Sokal and Rohlf (1981), the slope of this line is

$$b = \frac{s_{12}}{\lambda_1 - s_1^2}, \qquad (41)$$

where

$$\lambda_1 = \frac{1}{2}(s_1^2 + s_2^2 + D), \qquad (42)$$

and

$$D = \sqrt{(s_1^2 + s_2^2)^2 - 4(s_1^2 s_2^2 - s_{12}^2)}. \qquad (43)$$

In terms of the notation of the previous section, $s_{12} = \mu_{11}$, $s_1^2 = \mu_{20}$, and $s_2^2 = \mu_{02}$. Knowing the slope and position, the object can be rotated and translated into a standard position for subsequent analyses.

Reconstruction of Images

This topic is important for several reasons. First, if one can reconstruct the important features of an image from a set of measured parameters, that demonstrates that the parameters used are sufficient to describe the image. Of course, that does not prove that any of the parameters are directly interpretable biologically. One may have to perform various transformations on the parameters in order to put them into a form suitable for analysis and interpretation. The discussion on moment invariants, above, shows that different assumptions can suggest different transformations of the initial set of raw moments of an image surface. Reconstructed images may also be useful in themselves as convenient checks on whether the measurements are mutually consistent. If one measurement or more is inaccurate, the reconstructed image should look distorted. Strauss and Bookstein (1982) point this out as one of the advantages of the truss method.

   Summary statistics such as means, confidence regions, and principal component axes can be expressed in terms of the input variables. Fourier coefficients, for example, can be averaged to give a description of an average outline. These coefficients can then be used to construct a plot of the average outline. Points within a multivariate confidence region correspond to particular combinations of values of the input parameters. It is possible, for example, to show a confidence region for a set of morphometric shapes by constructing examples of various extreme images that still belong to the confidence region. Rohlf and Archie (1984) show examples of reconstructions of hypothetical mosquito wings representing extremes possible along each principal component axis. Thus the morphometrician is able to concentrate on the geometric aspects of the organisms under study without getting distracted by the large numbers of measurements, parameters, and various coefficients involved in the mathematical and statistical analyses being performed.

Acknowledgments

Dennis Slice assisted in getting software ready in time for the software demonstrations given as part of the presentation of this paper at the workshop (the IMAGE and FRACTAL-D programs for the IBM PC were shown to illustrate some of the points made in this manuscript). He and Karen Rohlf prepared the illustrations for this paper. Norma Watson provided many editorial suggestions. Their assistance is greatly appreciated. The discussion on the method of moment invariants is based on an unpublished study made in collaboration with Scott Ferson.

This work was supported in part by a grant (BSR 8306004) by the National Science Foundation. This paper is contribution number 761 from the Graduate Studies in Ecology and Evolution, State University of New York at Stony Brook.

References

Ashkar, G. P. and J. W. Modestino. 1978. The contour extraction problem with biomedical applications. Computer Graphics Image Processing, 7:331-355.

Ballard, D. H. and L. M. Brown. 1982. Machine vision. Prentice-Hall, New York. 523 pp.

Barnsley, M. F., V. Ervin, D. Hardin, and J. Lancaster. 1986. Solution of an inverse problem for fractals and other sets. Proc. Natl. Acad. Sci. USA, 83:1975-1977.

Blum, H. 1973. Biological shape and visual science (part I). J. Theor. Biol., 38:205-287.

Bookstein, F. L. 1981. Looking at mandibular growth: some new geometric methods. Pp. 83-103 in Craniofacial biology. Univ. Michigan Center for Human Growth and Development. (D. S. Carlson, ed.) Ann Arbor.

Bookstein, F. L., B. Chernoff, R. L Elder, J. M. Humphries, Jr., G. R. Smith, and R. E. Strauss. 1985. Morphometrics in evolutionary biology. The Academy of Natural Science Philadelphia. Special Publ. No. 15, 277 pp.

Bookstein, F. L., R. E. Strauss, J. M. Humphries, B. Chernoff, R. L. Elder, and G. R. Smith. 1982. A comment on the uses of Fourier methods in systematics Syst. Zool., 31:85-92.

Brice, C. R. and C. L. Fennema. 1970. Scene analysis using regions. Artificial Intell., 1: 205-226.

Casasent, D., J. Pauly, and D. Fetterly. 1981. Infrared ship classification using a new moment pattern recognition concept. SPIE, 302:126-133.

Castleman, R. R. 1979. Digital image processing. Prentice-Hall, Englewood Cliffs, New Jersey, 429 pp.

Cheung, E. and J. Vlcek. 1986. Fractal analysis of leaf shapes. Can. J. For. Res., 16:124-127.

Dudani, S. A., K. J. Breeding, and R. B. McGhee. 1977. Aircraft identification by moment invariants. IEEE Trans. Computers, C26:39-46.

Dyer, C. R. and A. Rosenfeld. 1979. Thinning algorithms for gray-scale pictures. IEEE Trans. Pattern Anal. Mach. Intell., PAMI-1:88-89.

Engles, H. 1986. A least squares method for estimation of Bezier curves and surface and its applicability to multivariate analysis. Math. Biosci., 79:155-170.

Evans, D. G., P. N. Schweitzer, and M. S. Hanna. 1985. Parametric cubic splines and geological shape descriptions. Math. Geology, 17:611-624

Ferson, S., F. J. Rohlf, and R. K. Koehn. 1985. Measuring shape variation of two-dimensional outlines. Syst. Zool., 34:59-68.

Frei, W. 1977. Image enhancement by histogram hyperbolization. Comp. Graphics and Image Processing, 6:286-294.

Grainger, J. E. 1981. A quantitative analysis of photometric data from aerial photographs for vegetation survey. Vegetatio, 48:71-82

Groch, W. D. 1982. Extraction of line shaped objects from aerial images using a special operator to analyze the profiles of functions. Computer Graphics Image Processing, 18:347-358

Hall, E. L. 1979. Computer Image Processing and Recognition Academic Press, New York, 584 pp.

Horn, B. K. P. 1986. Robot vision. M.I.T. Press, Cambridge, MA, 509 pp.

Hu, M. K. 1962. Visual pattern recognition by moment invariants. IRE Trans. Information Th., 8:179-187.

Hummel, R. 1977. Image enhancement by histogram transformation. Computer Graphics and Image Processing, 6:184-195.

Katz, M. J. and E. B. George. 1985. Fractals and the analysis of growth paths. Bull. Math. Biol., 47:273-286.

Kaesler, R. L. and J. A. Waters. 1972. Fourier analysis of the ostracode margin. Geol. Soc. Amer. Bull., 83:1169-1178.

Kirsch, R. A. 1971. Computer determination of the constituent structure of biological images. Computers and Biomedical Res., 4:315-328.

Kuhl, F. P. and C. R. Giardina. 1982. Elliptic Fourier features of a closed contour Computer Graphics and Image Processing, 18:236-258.

Lohmann, G. P. 1983. Eigenshape analysis of microfossils: a general morphometric procedure for describing changes in shape. Math. Geol., 15:659-672.

Long, C. A. 1985. Intricate sutures as fractal curves. J. Morph., 185:285-295.

Machuca, R. and A. L. Gilbert. 1981. Finding edges in noisy scenes. IEEE Trans. Pattern Anal. Mach. Intell., PAMI-3:103-111.

Maitra, S. 1979. Moment invariants. Proc. of the IEEE., 67:697-699.

Mero, L. and Z. Vassy. 1975. A simplified and fast version of the Hueckel operator for finding optimal edges in pictures. Proc. 4th. Intl. Conf. on Artificial Intelligence. Tbilisi, USSR. 650-655.

Morse, D. R., J. H' Lawton, M. M. Dodson, and M. H. Williamson. 1985. Fractal dimension of vegetation and the distribution of arthropod body lengths. Nature, 314:731-733.

Nagao, M. and T. Matsuyama. 1979. Edge preserving smoothing. Computer Graphics Image Processing, 9:394-407.

Papoulis, A. 1965. Probability, random variables, and stochastic variables. McGraw-Hill, New York.

Paton, K. 1979. Line detection by local methods. Computer Graphics Image Processing, 9: 316-332.

Pavlidis, T. 1982. Algorithms for graphics and image processing. Computer Science Press: Rockville, MD, 416 pp.

Peli, T. and J. S. Lim. 1982. Adaptive filtering for image enhancement. Optical Engineering, 21: 108-112.

Reddi, S. S. 1981. Radial and angular moment invariants for image identification. EEE Trans. Pattern Anal. Mach. Intell., PAMI-3:240-242.

Riseman, E. M. and M. A. Arbib. 1977. Computational techniques in visual systems, part II: segmenting static scenes. IEEE Computer Soc. Repository, R:77-87

Rohlf, F. J. and J. Archie. 1984. A comparison of Fourier methods for the description of wing shape in mosquitoes (Diptera: Culicidae). Syst. Zool., 33:302-317.

Rohlf, F. J. and S. Ferson. 1983. Image analysis. Pp. 583-599. in Numerical taxonomy. (J. Felsenstein, ed.) Springer-Verlag, Berlin

Rosenfield, A. and A. C. Kak. 1982. Digital picture processing, volume 1. Academic Press, New York, 435 pp.

Samet, H. 1981a. An algorithm for converting rasters to quadtrees. IEEE Trans. Pattern Anal. Mach. Intell., PAMI-3:93-95

Samet, H. 1981b. Computing perimeters of regions in images represented by quadtrees. IEEE Trans. Pattern Anal. Mach. Intell., PAMI-3: 683-687.

Samet, H. 1983. A quadtree medial axis transform. Communications ACM, 26:680-393.

Scott, D. S. and S. Iyengar. 1986. TID--A translation invariant data structure for storing images. Comm. ACM., 29:418-429.

Shaw, G. B. 1979. Local and regional edge detectors: some comparisons. Computer Graphics and Image Processing, 9:135-149

Sokal, R. R. and F. J. Rohlf. 1981. Biometry. Freeman, New York, 859 pp.

Strauss, R. E. and F. L. Bookstein. 1982. The truss: body form reconstructions in morphometrics. Syst. Zool., 31:113-135.

Tanimoto, S. and T. Pavlidis. 1975. A hierarchical data structure for picture processing. Computer Graphics Image Proc., 4:104-119.

Vlcek, J. and E. Cheng. 1986. Fractal analysis of leaf shape. Can. J. Forestry Res., 16:124-127.

Wang, S., A. Y. Wu, and A. Rosenfeld. 1981. Image approximation from gray scale "medial axes." IEEE Trans. Pattern Anal. Mach. Intell, PAMI-3:687-??.

White, R. J. and H. C. Prentice. 1987. Comparison of shape description methods for biological outlines. Pp. 395-402 *in* Classification and related methods of data analysis. (H. H. Bock, ed.). North-Holand, New York.

Yin, B. H. and H. Mack. 1981. Target classification algorithms for video and forward looking infrared (FLIR) imagery. SPIE, Infrared Technology for Target Detection and Classification, 302:134-140.

Younker, J. L. and R. Ehrlich. 1977. Fourier biometrics: harmonic amplitudes as multivariate shape descriptors. Syst. Zool., 26:336-342.

Zahn, C. T. and R. Z. Roskies. 1972. Fourier descriptors plane closed curves. IEEE Trans. on Computers, 21:269-281.

Part III
Analytic Methods

Introduction and Overview: geometry and biology

F. L. Bookstein

Center for Human Growth and Development
University of Michigan
Ann Arbor, Michigan 48109

Introduction

From definition of variables through publication of findings, a multivariate statistical analysis exploits our pre-existing notions of dissimilarities among objects. Morphometrics, in particular, is a formal treatment of our ideas about dissimilarity of geometrical form among biological objects; what makes it interesting is the interplay of statistical method with biological rules about procedures for discerning similarity.

"Distance" and Distance

To describe the diversity of morphometric tactics in the usual multivariate context, it suffices to discuss the variety of dissimilarity measures between objects — it is not necessary to discuss morphometric "variables" directly. (Just as well: there are far too many such variables.) Underlying all linear multivariate techniques is one single metaphor, the resemblance of these dissimilarities to ordinary (physical) distances in real physical space. Indeed, the findings of most linear multivariate strategies can be restated without error in the purely geometric language of vectors and angles. In this root metaphor, measured variables are considered to specify perpendicular axes of a Euclidean space of however many dimensions. "Distance" is computed by the ordinary extension of the Pythagorean Theorem: the squared distance between cases having measured values X_1, X_2, ..., X_k and Y_1, Y_2, ..., Y_k is $(X_1 - Y_1)^2 + ... + (X_k - Y_k)^2$, just as for real points on real paper, $k=2$, or in real space, $k=3$. In this circumstance, both variables and cases are described by orthonormal sets of vectors of "loadings" linked by one diagonal matrix of *singular values*. Distances between cases derive from the crossproducts of their loadings with respect to this diagonal matrix, and the covariances of the variables are the crossproducts of *their* loadings with respect to the same matrix.

In most applications of multivariate analysis, there is no other definition of "distance" at hand *except* this metaphorical one. In psychological research, for instance, one cannot observe the "distance" between two subjects' attitude profiles directly (not even in slides of brain tissue); the notion of the multivariate "distances" $\sum (X_i - Y_i)^2$ between profiles (along with the statistical machinery of whatever component analyses, cluster analyses, and the like are consequent upon them) is hence unambiguous. In nearly all applications of multivariate analysis outside morphometrics, there is no possibility of confusion between the statistical notion of "distance" and any physical distance in the real world.

The situation is different for the morphometric data which are the subject of the analyses in this book. Because the objects of our analysis co-exist with us in physical space, there is a prior notion of distance available which overrides that embedded in the fundamental multivariate metaphor. The physical distances involved in morphometrics are not those between pairs of whole organisms (although distances of that sort sometimes enter into other forms of analysis, such as the ecological, by way of map coordinates). Morphometric distances instead express the patterns of relative location among the parts of one organism in comparison to those of another.

In this introduction I will describe the variety of ways in which (real physical) distances can be brought to bear upon multivariate analyses of biological form that are otherwise conventional in their algebra and execution. Indeed, the survey of these implementations for the root metaphor of physical distance is *equivalent* to a complete survey of morphometric analysis as it applies to whole organisms: morphometrics above the tissue level.

Homology: Biological and Geometrical

All morphometric implementations of real physical distance within a multivariate statistical framework are governed by one crucial concept from biomathematics, the notion of *homology*. The ties between the morphometric and the biomathematical uses of this term are discussed in Section 1.2 of Bookstein et al., 1985. Morphometricians *qua* morphometricians have nothing much to say about "right" or "wrong" notions of homology, and the term will go undefined in these pages (see, however, Chapter 17); but it is necessary to say a few words about the *semantics* of this construct.

In theoretical biology, homology is a matter of correspondence between parts; thus, "the bones of the fish's jaw are homologous to the bones of the mammalian inner ear," or "the human arm is homologous to the chicken wing." This diction, unmodified, empowers only the most rudimentary

sort of morphometrics, the invocation of variables which represent "extents" of homologous parts without any additional geometrical content. Morphometrics based on this primitive utilization of the notion of physical distance is generally called "multivariate morphometrics" (cf. Reyment, Blackith, and Campbell, 1984). These variables are usually measured in cm (or cm^2 or cm^3), or log cm, or log ratios (differences of log cm), or various nonlinear transformations of these (such as degrees of angle). For instance, one common choice of variable in conventional morphometric analyses is the volume or weight of a well-delineated organ. For such variables, the physical notion of distance is present implicitly in the definition of volume as the integral of cross-sectional area and area, in turn, as the integral of lengths of parallel transects — for lengths are physical distances as described just above.

But the lengths, etc., that go into the integrals are not claimed separately to be homologous as extents upon the organism; they are simply conveniences in the computation of multiple integrals, which could be taken instead by surface integrals around the boundary (Green's Theorem). If, instead, the length of a linear structure, such as a long bone, is to be taken as a proper morphometric variable on its own, then the endpoints of the calipers which measure it must be themselves located upon homologous substructures: not, for instance, measured to the end of a bone spur on one form, a condyle on another. The primitive biological notion of homology gives us no further guidance regarding the precise measurement of "length," and, in particular, does not tell us what to do with structures that curve inconsistently between their "endpoints" over a sample of forms.

Beginning with D'Arcy Thompson, there has emerged an alternative extension of the notion of homology into biometrics, one in which the biological properties of the objects of study are considerably more richly articulated. In this replacement for extent-based morphometrics, the object of measurement is the relation *per se* between forms, not the single form: the same domain as for the root concept of homology driving this style of measurement. To pass from the biological to the biometrical context, homology must be considered as a *mapping* function, a correspondence relating *points to points* rather than parts to parts. This notion of homology can often be realized as a mathematical *deformation* (i.e., a smooth map) which can, in turn, often be described efficiently by way of its derivative (Bookstein, 1978). We will not use this further mathematical development here, but only the reduction to a point-to-point mapping.

In this revised version of homology, the ordinary integral extents of organs — areas, in two dimensions; volumes, in three — carry over without change. But the lower-dimensional reductions of extended data, such as "lengths" of three-dimensional objects, in general cease to be acceptable morphometric variables. They presume a prior knowledge of homology, that of the points at which the endpoints of the calipers are put down. If we would assert that prior correspondence directly, by recording sets of homologous locations of the same points over a sample of forms, then the higher-level concept ("length") is made quite unnecessary. We can found morphometrics purely upon a language of maps sampled by point-correspondences, without any mention of homologously measured "variables" at all. The variables aid in interpretation and publication of findings, but are not required for computation. On the contrary, much of the intellectual task of morphometrics is the explicit computation of the variables which best summarize observed findings, or best suggest an interpretation in terms of biological process, a posteriori, after all specifically statistical computations are completed.

To proceed in this wise, we represent the data base of a morphometric inquiry by samples of discrete points which correspond among all the forms of a data set. These points are called **landmarks**. That they are to be considered biologically homologous is, as it was for parts and

regions, a primitive concept, not the morphometrician's to argue for or against (although her experience may be helpful). Our survey of morphometrics then becomes a survey of reasonable measures of dissimilarity between corresponding sets of mathematical points *given that they have been previously assigned the same names*. In the notion of the naming ("bridge of the nose," "tip of the fin") is embodied the concept of biological homology represented by these maps. Different versions of homology lead to different computations of interspecimen distances. These do not necessarily share any common factor: in contradiction of the hopes of the early numerical taxonomists, there is no "true, underlying" morphometric distance between forms. There may, however, be factors underlying their covariance structure with other domains of measurement, such as the ecological or the biochemical. Techniques for the computation of such factors are available in the general statistical literature under the heading of "analysis of covariance structures," especially "Partial Least Squares": see Bookstein, 1991, Section 2.3.2.

A morphometric analysis of outline data is founded on the knowledge that to a particular curve in one form correspond particular curves on all other forms of a data set. For the analysis to be interpretable in terms of homology, it is required further that we know certain points which match from form to form upon those curves. The mathematical model of homology for curves must be founded on the correspondence of the labelled landmark points, and only then extended to the other points of a continuously curving form (and perhaps to the points inside the form as well) by one or another computational interpolation. When observed point-landmarks make no appearance in the interpolation rule, it is impossible to consider the computed correspondence as embodying any aspect of biological homology; the problems caused by this hiatus of meaning will be pointed out in due course.

I shall survey below the various forms of statistical "distance" as they are invoked in morphometric computations, and, for each, describe and critique the treatment of physical distances among homologues on which the statistical quantity is based. I will conclude that the root metaphor of a single "dissimilarity" between forms is inappropriate in morphometric applications. Instead, I shall show that morphometric dissimilarities need to be considered as a conceptual composite of *three distinct, incommensurate terms*: a net "size distance"; a classic non-Euclidean shape distance representing the logarithm of anisotropy of a uniform shape change; and a regionalizable residual that is usefully measured in units of inverse physical scale — a measure of "deformation per centimeter." At the close I will sketch the varieties of multivariate analysis to which the elements of this decomposition are suited.

In conventional multivariate morphometrics (cf. Reyment, Blackith, and Campbell, 1984), the concept of homology is present only implicitly. There is usually a tacit assumption that whatever scalar variables have been measured — diameters of forms or organs, net areas and volumes, angles of articulation between substructures, and diverse contrasts or ratios among quantities of this sort — are taken in a manner not too offensive to the biologist's sense of homology. These quantities are considered "homologous" from form to form by fiat of operational definition rather than according to any formal criterion by which the concept of homology may be transferred out of the biological domain into the statistical. Once the biological objects have been reduced to "variables" in this manner, the geometry of those objects is discarded. The only geometry remaining is that of the measurement space itself, a vector space of as many Euclidean dimensions as there were originally measured variables.

Such techniques will not be discussed further here. Any effectiveness they may have in systematic researches owes to the accidental alignment of conventional multivariate methodology with the themes of systematic exploration, such as group discrimination, rather than to any respect for the underlying properties of specifically morphometric data.

Outline Data

I begin this survey of distance-measures with the case of curving outlines of forms in two dimensions. (Certain extensions to three dimensions will be mentioned in the context of their two-dimensional equivalents.) These techniques all use landmark data relatively weakly: only a point or two is involved in specifying an interpolation rule extended to whole curves otherwise arbitrarily.

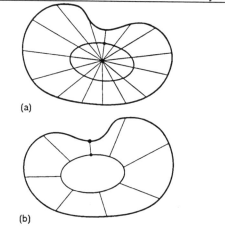

(a)

(b)

Figure 1. Variants of distance-measures for curving forms: distances. Once a computed homology is assigned relating a pair of outlines, their morphometric distance may be taken as the integral of the squared physical (geometrical) distance between homologous points around the outline. (a) Homology radial out of some center, ordinarily leading to radial Fourier analysis. (b) Homology linear in arc-length from some starting point, ordinarily leading to elliptic Fourier analysis.

Distances Between Homologous Points

The simplest versions of distance measures between curves rely on ordinary in-plane Euclidean distance between "corresponding" points on the forms. Figure 1 shows the two principal possibilities, as homology is computed out of a center or instead around the arc. The former possibility, but not the latter, can be directly generalized to three-dimensional data. The arbitrariness of these analyses is embedded in the unreality of the correspondence as it is actually presumed of the observed data. Whenever information is available, real organisms do *not* show correspondences which are uniform either in arc-length around an outline or in angle out of a center. These interpolations are defensible only in the absence of all other information about homology: in particular, only in the absence of all additional landmark information. It is no accident that they work wonderfully for sand grains.

There is a further problem with taking these distances to express specifically biological information. In most approaches the "homology" as executed is a function of the distance measure chosen, rather than vice-versa. Forms are rotated and translated (the radius method), or relabelled (the arc-length method), to minimize the integral squared distance computed between "homologues" over the family of all possible such maneuvers. Such a computation makes no sense when the purpose of morphometric analysis is to measure a *biological* homology that existed prior to the evolution of morphometricians. There is no theorem that the evolutionarily correct homology is that which minimizes apparent morphological distance. The error here is similar to the error underlying the thoughtless invocation of parsimony methods in phylogenetic reconstruction.

The caption of Figure 1 refers to "Fourier analysis." But, strictly speaking, the analysis of forms according to these distance measures is not any version of "Fourier analysis" (Chapter 6). Multivariate analysis of the integral squared distances between corresponding points of outline forms leads to ordinations which make no direct reference to Fourier coefficients or any other features of form. The Fourier analyses are, rather, a device for extracting features of a space of "variables" which may serve to describe the space(s) in which the forms described by these homology-free distances are ordinated. Then the Fourier

coefficients have no particular claim upon our attention: the "features" they supply have no direct translation into the language of biological form-correspondence. At best (Bookstein et al., 1982), Fourier coefficients may aid a numerical discrimination; they are no reliable guide to understanding homology or the biological processes that have modified form.

Derivatives

Two other variants of this general procedure are encountered in the literature of outline-processing. Both involve taking *derivatives* of the outline curves and measuring dissimilarity between forms in terms of squared differences of those derivatives rather than distances between the original paired point-loci. In one of these methods, Lohmann's *eigenshape analysis*, it is the first derivative of the outline the squared differences of which at "corresponding" points are integrated to form a net distance. (The distances are then normalized by a correction for "shape amplitude" which does not concern us here.) This method could be generalized to three-dimensional data, as the integral of squared angle between "corresponding" normals over a pair of surfaces. In another method, having historical priority by virtue of its usefulness in the early literature of computer vision, it is the curvature (essentially, the second derivative) of the outline whose differences are squared and integrated to generate a net distance between forms. In three dimensions, curvature is no longer a scalar, and this approach generalizes only with difficulty. Figure 2 illustrates these two possibilities by re-expressing each functional, the tangent-angle or the curvature, as a geometric distance all its own.

It continues to be the case that these methods rely on an arbitrary assignment of "homology" underlying the subtraction. In Lohmann's method, homology is taken linear in arc-length; in the curvature method, it may usefully be taken, depending on the application (handwriting analysis, recognition of enemy aircraft), either as uniform in angle along radii out of a center or as linear in arc-length.

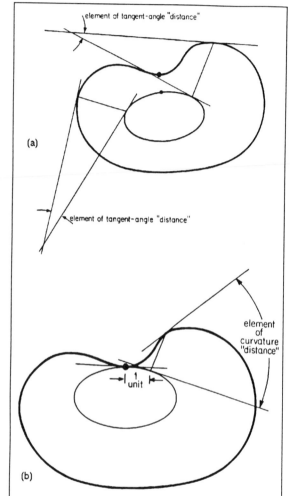

Figure 2. Variants of distance-measures for curving forms: differentials. Each has been visualized by a distance in the original physical space of the outline data, to a superposition upon one pair of "homologues" at a time. (a) Distances based on the tangent angle: eigenshape analysis. (b) Distances based on the curvature function: classic pattern recognition.

homology function cannot be dismissed (Bookstein et al., 1985, Section 2.3). Distances based on simple positional separation of homologous points, measured either radially or by arc-length, weight the largest-scale features most heavily, and so are less sensitive (though not insensitive) to confounding by incorrect homology. Distances based on the tangent angle, such as underlie the method of eigenshapes, are intermediate in their sensitivity. Nothing in the language of biological homology would indicate to us any resolution of these disagreements about weighting. I am aware of no justification of either of these choices — the nature of the computed homology, the order of the derivation before subtracting and squaring — in any biological terms. Rather, all methods of this class seem badly in need of a device for taking homology into account in setting the correspondence of curves prior to computation of squared distances. In practice, such an accounting consists in the explicit *observation* of homology at a selection of discrete mathematical points. Then any computation of distances between curves must *begin* with the computation of distances between landmark data sets, the topic to which I now turn.

Landmark Point Data

The tie between geometric distance and statistical distance is much richer for point data than for unlabelled curve data. There are more ways to express prior biological knowledge, and more possibilities for interpreting findings as features of evolutionary interest.

Methods Ignoring the Spatial Ordering of Landmarks

A first category of distance-measures for landmark data accept the fact of labelled points as the subject of analysis but ignore everything about the points except their pairing. The distance between forms is taken to be the sum of squared distances $\sum |P_i - Q_i|^2$ between all pairs of identically named points, regardless of the spacing of those points in the forms separately. This statistical

In general, the distance that is based on curvature weights smaller-scale irregularities of perimeter much more heavily than large-scale features. The mismatching of these features is given a great deal of credence in the computation of distance, but the possibility that the small-scale features are merely misaligned under the arbitrary distance underlies the family of techniques called *Procrustes analysis*, originally developed by Gower to describe the distance between alternate multivariate analyses of the "same" data. In that original application, indeed, corresponding "points" shared' nothing but their names (case numbers, variable numbers), and had no further meaning vis-a-vis each other — there is no equivalent of "homology" in general multivariate analysis.

Procrustes distances are usually taken as a minimum over all possible superpositions of the forms or according to various arbitrary rules which do not concern us here. As already remarked in connection with homology of outlines, the concept of a "superposition rule" has no meaning in the context of comparisons via biological homology.

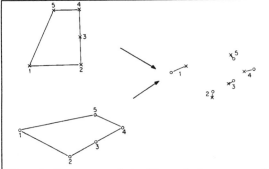

Figure 3. Procrustes analysis. The method sums squared distances between corresponding landmark points at the "optimal" superposition without attending to their spatial configuration.

The "truss" An early attempt by the Michigan group at tying physical geometry to multivariate geometry was the *truss* method (Bookstein et al., 1985). In this approach, a set of landmark locations is divided into neighborhoods of four points each, and each neighborhood is measured exhaustively by all six interpoint distances. For small size variation, the ordinary multivariate distance-measure $\sum (X_i - Y_i)^2$ based on truss data is an oddly weighted approximation to the Procrustes distance $\sum |P_i - Q_i|^2$ relating the same set of forms. For larger size ranges, the conventional multivariate distance $\sum (X_i - Y_i)^2$ applied to the *logarithms* of the truss length measurements approximates a standardization of Procrustes distance to forms of varying scale. But this distance may be computed more expeditiously by the formula of Kendall (1984) making no reference at all to trusses. Except as a device for collection of data when landmark locations cannot be digitized, the truss analysis gains one nothing in a multivariate context, and should be superseded by the scheme of three distance functions to be described presently. When the "landmarks" of the truss are instead intersections of arbitrary radial lines with an outline, multivariate analysis of the truss reduces to that of the ordinary integral radial distance for the outlines (again, oddly weighted); in the absence of true landmarks, there is no advantage to computing a truss at all.

Spatially Ordered Landmarks: Triangles

The remainder of this discussion treats methods which take into account the starting configuration of landmark locations more explicitly. We begin with the simplest case, a triangle of landmarks.

In various papers over the last few years, I have shown the utility for subsequent multivariate statistical analysis of a particular concept of distance between triangles which explicitly separates out size and shape components. By now this formalism is perhaps familiar to the student of morphometrics (see, for instance, Bookstein, 1986, or Appendix 4 of Bookstein et al., 1985); in any case, it will only be sketched here. Suppose we have three landmark locations for each of a pair of organisms. We can imagine these locations to describe a *triangle* for each form. (The triangle has no biological reality but is a convenient way to speak of the set of three landmarks as a unit.) Compute the sum of the squares of the edge-lengths of each triangle (the sum of the squares of the distances between the landmarks in pairs), and set it aside: it will be our **net size measure**.

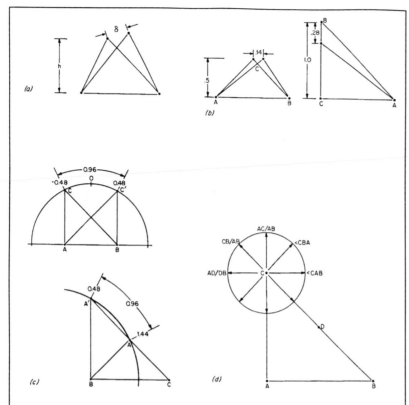

Figure 4. Analysis of a triangle of landmarks. (a) The two-point registration and the distance function δ/h. (b) Shape distance between the forms is invariant to choice of baseline for the construction. (c) The same is true for large changes. (d) Shape variables are made commensurate by this distance function.

For small amounts of shape variation, the appropriate measure of shape distance between the two triangles is the scaled distance between the two images of the moving point in this registration, the quantity δ/h in Figure 4a. It can be shown that this scaled quantity is the ratio of principal strains of the change between triangles treated as a uniform deformation: that is, the ratio of the axes of the ellipse into which this deformation takes a circle inside one of the triangles (Bookstein, 1991, Section 6.2). In this manner the distance δ/h provides a tight link between multivariate statistics and the study of deformations by their derivatives (the method of biorthogonal grids, which we shall not treat here). As a side-benefit, it follows immediately that this shape distance is independent of the edge of the triangle chosen for registration (Figure 4b), since the shape of the aforementioned ellipse is not a function of any choice of edge. For larger changes of shape (Figure 4c), the appropriate distance between forms is really the *logarithm* of the ratio of the principal strains of the transformation; the quantity δ/h is an approximation for small distances. The finite ratio is the integral of the *infinitesimal ds/h* over "straight lines" in this triangle space when the lines are construed in the appropriate non-Euclidean geometry as circles perpendicular to the baseline (cf. Bookstein, 1991, Appendix 2).

Now choose, arbitrarily, one side of the first triangle, and its homologous edge in the second triangle. Rescale the triangles independently so that this edge has length 1.0 in both forms. (In this step, size information is lost; but we have already set it aside for later retrieval.) Now place the second triangle down atop the first triangle so that the two edges of unit length lie precisely upon one another. All the information about the shapes of the triangles then inheres in the positions of the third landmark (Figure 4a), the one still free to move. We call the coordinates of the third vertex to this registration upon the other two the *shape coordinates* of the triangle *to the specified baseline*.

When distances are measured between triangles in this way, all conventional shape measurements of the same triangle — ratios of sides, angles, and the like — can be shown to reduce to directions in this space in the vicinity of a "mean form." The shape metric here thus provides a means of rendering all these alternate shape measures commensurate: they are represented by their gradients in shape space, with equal lengths corresponding to equal variances under a null model of circular noise of landmark locations (Bookstein, 1986) and angles corresponding to correlations under the same model. For instance, for the triangle in Figure 4d, a change of 2.7° in the angle at vertex C represents the same amount of shape distance as a change by 0.05 in the ratio of sides through that vertex, and these two shape variables have gradients at 90° in shape coordinate space (Bookstein, 1991, Section 5.1.3), and so may be expected to be uncorrelated under the null model specified.

Already, in this simplest instance of three landmarks, we have a way of modelling *allometry* that supplants the usual investigation in the space of "extents" (multivariate *n*-vectors in units of log centimeters). Allometry, the correlation of size measures with shape measures, may be seen in that space as the inconstancy of coefficients of the first principal component of the logarithms of measured distances (Jolicoeur, Teissier) or as the angles between ratio variables and the manifold of constant size however defined (Mosimann); see Bookstein 1989c. In the present context, this same phenomenon is viewed as the correlation of shape, measured by position of the moving point C in Figure 4, with the a size variable sequestered at the outset. That variable exists in another space (in this case, a one-dimensional space, a line). For a triangle of landmarks, allometry is observed quite simply in the multiple correlation between size (summed squared edge-length), on its line, and shape, in its plane. That is, we have a size distance between objects, and, separately, a shape distance; allometry is the correlation between these two distances when shape distance is projected along the direction of maximum correlation.

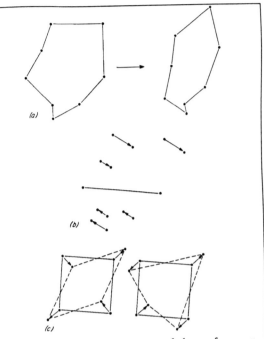

Figure 5. The uniform component of change for many landmarks. (a) Picture to an arbitrary baseline. (b) Two changes of a square of landmarks showing the same Procrustes distance. (left) Uniform; (right) Purely inhomogeneous. (c) For more landmarks. (left) Uniform; (right) Growth-gradient.

Spatially Ordered Landmarks: The General Case

These two aspects of landmark change each generalize directly to the case of more than three landmarks. Size is again best taken as the summed squared distances among all the landmarks in pairs, or, equivalently, the summed squared distances from all the landmarks to their common centroid case by case (Bookstein, 1991, Section 4.1). This single quantity will be referred to below as Centroid Size. The equivalent of circle-to-ellipse distance for triangles is now a bit more complicated; it is the length of the appropriate average (taken at unit height) of all the little vectors V in Figure 5a when two of the landmarks are fixed and all the others imagined to move "with respect to them." The complexity arises from the requirement that this "average" must be independent of the choice of those fixed landmarks. One appropriate computation, which involves generalized least squares, is explained in Bookstein, 1991, Section 7.2; for another, see Rohlf's affince Procrustes algorithm in this volume; for a third, try out the formula corresponding to Figure 5a, in Chapter 11.

The existence of this uniform component as a separately meaningful aspect of the general shape

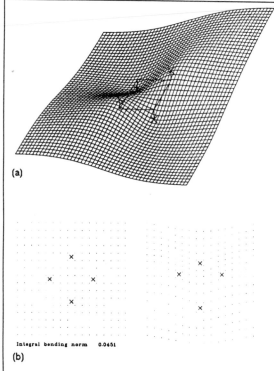

(a)

(b)

Integral bending norm   0.0451

Figure 6. Bending-energy distance between a pair of forms. (a) A thin metal plate over a non-planar armature. (b) The same viewed as the deformation of the square into a kite. The metal plate adopts the position which minimizes the net bending energy.

change indicates a major problem with the Procrustes distance measure (Bookstein, 1989a). *Procrustes distance conflates two sorts of shape change that should be kept separate.* In each of Figures 5b and 5c, we show a starting shape and two others at the same Procrustes distance. But these changes correspond to completely different biological descriptions. In one, the changes cannot

be localized: they are the same at every landmark — a uniform change of vertical:horizontal proportion by 1.2:1. In the other, different changes are observed in different parts of the form, and a regional description is not only possible but required. The Procrustes distance formula sums these two components in a relative weighting which is a highly artificial function of the landmark spacing. The two components need instead to be ordinated and interpreted separately.

After size distance is estimated, as the log ratio of summed squared interlandmark distances, and after the uniform component of shape distance δ/h is estimated by the construction of Figure 5a, the information that remains is entirely *local*. It is best formalized by an algebraic approach that I have recently borrowed from the mathematics of interpolation. It is explained in greater (but still inadequate) detail in Chapter 11. There is a family of landmark-based morphometric distance functions which are *identically zero* for all uniform transformations. The most suitable of these distance functions, the **bending energy of the thin-plate spline**, corresponds to a certain physical energy associated with the landmarks according to the metaphor shown in Figures 6 and 7, the change of shape between two landmark configurations interpreted as a "four-dimensional bending" of one onto the other. Just like any other statistically useful distance measure, the bending energy from a standard (or mean) form is a quadratic form in the landmark coordinates of the variable form. It differs from other formulas, such as the Procrustes, in that the formulation is a highly nonlinear function of the landmarks' mean configuration. Shifts of a landmark inconsistent with shifts at its neighbors are weighted heavily in this distance measure; the nearer these neighbors, the more heavily the discrepancy of shifts is weighted (the smaller the armature of Figure 6a for a given vertical discrepancy, the greater the bending energy required to accommodate the metal to its arms). By virtue of this incorporation of mean landmark configurations *in extenso*, the bending-energy matrix has a spectrum of great biological interest. The eigenvectors of the matrix of bending energy, which

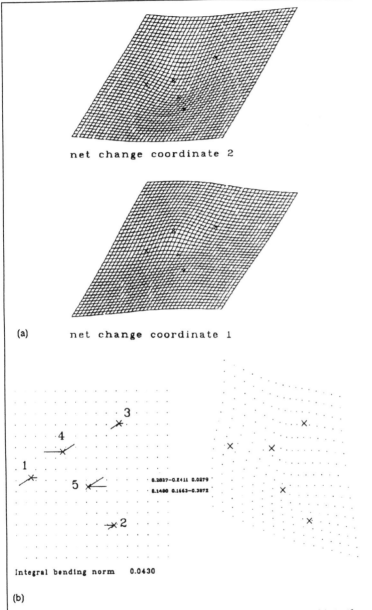

net change coordinate 2

net change coordinate 1

(a)

3

4

1

5

0.2837−0.2411 0.0279
0.1400 0.1663−0.3872

2

Integral bending norm   0.0430

(b)

Figure 7.   Bending-energy distance between a pair of forms.   (a) A "four-dimensional" plate: arbitrary changes of landmark location separately in *x* and *y* directions.   (b) The equivalent deformation.   Bending energy is independent of translation, rotation, change of scale, and any uniform component of shape change as previously defined.

I call *principal warps*, are orthogonal both with

respect to Procrustes distance and with respect to bending energy.   That is, the cross-product of landmark shifts in any fixed Cartesian direction is zero between any two of these features, but also the energy of the sum of two is the sum of their energies separately, with no cross-spectrum. This situation is exactly analogous to the familiar fact that principal components are orthogonal in Euclidean *n*-space and also with respect to the sample covariance matrix, so that the variance of a sum of principal components is the sum of their variances separately.   In fact, the principal warps are computed by the same subroutines which compute ordinary principal components outside morphometrics. Each eigenvector may be assigned a (real physical) scale, in cm^{-2}, and observed changes have components — "features" — on each scale.   In this way the general approach to morphometric distance I am suggesting continues to incorporate our emphasis on properties based in real physical distance; for the nonlinear part of shape change, the bending energy incorporates that emphasis in a manner that is both mathematically elegant and statistically quite unprecedented.   I know of no analogue to so natural a distance function anywhere else in applied multivariate analysis.

In most applications with substantial numbers of landmarks, the bending energy is seen to be a superposition of diverse processes like those in Figures 6 and 7 at different physical scales.   Each process is identified with a vector coefficient — so much of it in the *x*-

coordinate, so much in the *y*-coordinate; the rest of the observed change is a single uniform transformation, and this decomposition is unique. The basic component of nonlinear distance, then, is something like rigid motion of one (geometrically arbitrary) subset of landmarks with respect to another; the different components correspond to the diverse spacings characterizing such pairs of subsets. Again, to fill out this very brief sketch, the reader should examine the examples in Chapter 11 below and in Bookstein, 1991, Section 7.5.

Information about curving of outlines does not add any additional channel of integrated physical distance to this scheme. Rather, any remaining discrepancies between curves after interpolation of scenes via the thin-plate spline formula that matches up the available landmarks are to be resolved, up to a predetermined tolerance, by specification of additional non-landmark point pairs upon the curves. These new "landmarks" are *deficient*, lacking one coordinate each; but the expression of morphometric distance remains the same. See Appendix 1 of Bookstein, 1991, or Bookstein, 1989b.

Thus the proper multivariate analysis of a homology map, whether the data be of landmark points only or involve outlines as well, very nearly ignores the sums or integrals of physical distance between corresponding points underlying the outline methods or the Procrustes landmark analyses. These sums of squared distances in real space are relevant to landmark morphometrics only at one single step, for the computation of Centroid Size. Otherwise, good morphometrics is more subtle than that. The multivariate distance which ought to underlie shape analysis in accordance with a homology function is not at all a simple invocation of the Euclidean distances to which we are accustomed in the other, metaphorical applications of multivariate statistics.

Yet another overview of the method of bending energy, emphasizing a critique of its divergence from Procrustes distance, may be found in Bookstein, 1989b.

Features and Multivariate Analysis

Landmark data support a joint language for talking about geometry, homology, and multivariate statistics all at once. In my view, "multivariate morphometrics" should comprise not the syllabus of books presently going by that phrase in their titles but instead the following few themes:

(1) **Size change** is described by a single quantity, usually most usefully taken as the logarithm of the ratio of net Centroid Size scores.

One often encounters, in practice, shape changes which can be described with only a very few parameters which apply to the entire form or to large regions of the form. These most-global transformations include (2) the **uniform component of shape change** introduced above, with distance measured as log-anisotropy. They also include (3) the **rigid motion** of one subset of landmarks with respect to another (translation and rotation without change of shape of either separately). The best description of rigid motion is by the Euler parameters of kinematics; we will not go into these here.

A special case of rigid motion is the displacement of a single landmark point over a background of a configuration of many others not changing their shape. In this special case, morphometric distance is equivalent to the scaled distance moved by that point, and Procrustes programs like RFTRA arrive at the same ordination as the methods recommended here. (For more on this condition of agreement, see Chapter 10.) Otherwise, distance measures supplied by Procrustes-type programs are unusable (Bookstein, 1991, Section 7.1), even though their graphics may be suggestive.

The description of other aspects of shape change, those which are not identical over half or the whole of the form, generally expresses the homology map by expansion in an orthonormal series of functions defined over the form. These functions may apply only to outline points, in which case they are best viewed as varieties of simple *displacement*, the counterexample of eigenshape analysis notwithstanding; or else they apply to landmark points "dragging along" entire bounded regions of organisms in two dimensions or three. These functions may be considered in three large categories.

(4) *Some functions underlying a decomposition of homology maps incorporate no information from the actual form under study or from its observed variation.* These include, for outline data, Fourier expansions (in three dimensions, spherical harmonics); and, for landmark data, the *growth-gradient* models fitted by projection methods as in Bookstein 1991, Section 7.4.

Each of the methods (1)–(4) mentioned thus far is a projection into a feature space defined a-priori, before the onset of computation. Each should be computed as a statistical least-squares or robust least-squares fit in a manner independent of all arbitrary assumptions beyond the list of landmarks involved. Notice that the "feature space" of no shape change, the basis for analysis by the Procrustes methods, is the one-dimensional space of size changes; the residuals from a Procrustes analysis are not projections, i.e., measurements.

(5) Other functions into which homology maps may be decomposed incorporate information about the "typical" form under study, but not about its variability. The only example of this sort of decomposition known to me is that using my principal warps (eigenfunctions of the bending-energy matrix introduced here and in Chapter 11).

(6) A final class of functions, presenting a different analogy to ordinary principal components, are orthogonal statistically as well as geometrically, and so depend upon the entire matrix of shape differences under study as well as upon the mean landmark configuration. This class of decompositions includes eigenshapes (for outlines), the (computationally trivial) factor analysis of a scatter of shape coordinates for triangles or of the uniform component derived from more numerous configurations (Bookstein, 1991, Section 5.3.3 for

an example, see Chapter 13 of this volume), and my *relative warps* nonlinear (scale-specific) principal components for the variation remaining after partialling out all linear (uniform) changes. The nonlinear principal components (Bookstein, 1991, Section 7.6) are features orthogonal both according to covariance, i.e., statistically according to bending energy — just as ordinary principal components are orthogonal both statistically and geometrically — and are computed by the same sort of algorithm, one involving simultaneous diagonalization. Whereas the principal warps (feature style 5, no reference to variances) are gotten from a joint diagonalization of the bending-energy matrix and the identity (Procrustes) matrix, the relative warps are extracted via simultaneous diagonalization of bending energy and the observed covariances of the shape coordinates. I intend this technique to wholly supplant principal component analysis of arbitrary variables of extent as applied to landmark data. (Recall that the use of truss lengths for studies of allometry has already been superseded by direct inspection of correlations of shape coordinates with size.) Each relative warp expresses the summation of correlated distributed effects of larger and larger geometrical import, all over the form. For example, the first relative warp of a set of landmarks with respect to bending energy is usually found to have the appearance of some sort of growth-gradient or other systematic disproportion graded one-dimensionally across some transect of the form; the last few are highly local. Demonstrations are presented in Chapters 13 and 14 of this volume.

When all these features have been computed and inspected, they empower the usual two grand multivariate strategies, the study of patterns of covariation *within* and *between* domains of measurement.

(7) *Analysis of the correlations among the separate morphometric feature spaces.* There is no single distance measure for morphometric analysis, and there is no single statistical analysis of any morphometric data set; as we explained in the Preface, all good analyses of landmark data are

"hybrids." After a set of landmarks is analyzed or ordinated according to each of the methods (1–6) separately, one combines the analyses as one would any other collection of incommensurable information: by the study of two-dimensional and higher-dimensional scatterplots of the features with respect to each other and in the light of biological understandings. For instance, allometry is the examination of correlations of size change with components of uniform and nonuniform shape change; growth-gradients may be aligned with known biomechanical constraints or with properties of cell division; and so on.

(8) Analysis of the structure of covariation of all of these morphometric descriptors with other measurement schemes exogenous to morphometrics, including descriptions of group differences, ecophenotypy, changes over evolutionary time, changes over ontogenetic time, apparent selective value, and so on. The principal import of morphometric analyses in the larger context of the biological sciences is borne by the covariances of morphometric descriptors with measurements outside the morphometric domain (Bookstein, 1991, Section 6.5.3 and Appendix 3). The technology I have surveyed in this short introduction incorporates my best current understanding of the role which geometric information can play in the understanding of biological processes.

References

Bookstein, F. L. 1978. The Measurement of biological shape and shape change. Lecture Notes in Biomathematics, v. 24. Berlin: Springer, 191 pp.

Bookstein, F. L. 1986. Size and shape spaces for landmark data in two dimensions. (With Discussion and Rejoinder). Statistical Science, 1:181-242.

Bookstein, F. L. 1989a. Comment on D. G. Kendall, "A survey of the statistical theory of shape." Statistical Science, 4:99-105.

Bookstein, F. L. 1989b. Principal warps: Thin-plate splines and the decomposition of deformations. I.E.E.E. Transactions on Pattern Analysis and Machine Intelligence, 11:567-585.

Bookstein, F. L. 1989c. "Size and shape": a comment on semantics. Systematic Zool., 38:173-180.

Bookstein, Fred L. 1991. Morphometric tools for landmark data). Book manuscript, accepted for publication, Cambridge University Press.

Bookstein, F. L., B. Chernoff, R. Elder, J. Humphries, G. Smith, and R. Strauss. 1982. A comment on the uses of Fourier analysis in systematics. Systematic Zool., 31:85-92.

Bookstein, F. L., B. Chernoff, R. Elder, J. Humphries, G. Smith, and R. Strauss. 1985. Morphometrics in evolutionary biology. The geometry of size and shape change, with examples from fishes. Academy of Natural Sciences of Philadelphia, 277 pp.

Kendall, D. G. 1984. Shape-manifolds, Procrustean metrics and complex projective spaces. Bulletin of the London Mathematical Society, 16:81-121.

Reyment, R. A., R. Blackith, and N. Campbell. 1984. Multivariate morphometrics, 2nd ed. London: Academic Press, 232 pp.

Part III
Section A

Multivariate Methods

Multivariate statistical methods have played different roles in morphometrics. A number of years ago canonical variates analysis of a suite of linear distance measurements for two or more groups of samples was the basis for the field of "multivariate morphometrics" (Blackith and Reyment, 1971). This method of analysis has the desirable property that it "takes into account" levels of variation and covariation found within the groups being studied when it makes comparisons among groups and it does this for a number of measurements simultaneously. Since the largest component of variation found within a sample is usually due to size, the method has the fortunate effect of greatly reducing the effect of size differences among the specimens in the different groups being compared.

Important early examples of this approach are Jolicoeur (1959) and Blackith and Reyment (1971). In Chapter 4 Marcus gives a broad survey of the application of multivariate techniques to the analysis of linear distance measurements in morphometrics. Chapter 5 by Reyment covers similar ground but with the emphasis on what to do "when things go askew" — when there are bad data points (outliers) and/or the assumption of a multivariate normal distribution does not hold.

The problem with this "conventional" approach is that suites of linear distance measurements usually do not capture much information about the overall shape of the organism. This is true even when a very complete set of measurements is used. For example when the pattern of measurements corresponds to a truss (Strauss and Bookstein, 1982) only the lengths of the edges in the truss are entered into the analysis —

not the information about pattern of connections among the edges. Thus multivariate statistical methods do not take into account the geometrical pattern of the measurements on the organism. They are only able to take into account the empirical covariances between pairs of variables.

But multivariate methods are not limited to the analysis of such data. In recent years the approach has been to record x, y (and sometimes z) coordinates of landmarks and then use various techniques to derive variables that capture important aspects of the differences among two or more organisms. These methods (the subjects of Chapters 6 through 16) take the geometrical configuration of the sample data points into account. The multivariate statistical analyses described in Chapters 4 and 5 can then be performed on these derived variables. This approach is expected to result in more powerful tests since fewer variables will be used and they will focus more sharply on specific aspects of shape differences, rather than representing the use of a miscellaneous collection of somewhat redundant measurements.

But the use of such derived variables means that the results of the analyses will seem a bit more abstract and harder to interpret directly from the numerical results. For this reason the effective use of graphics becomes even more important as a tool in data analysis. It is especially important to take advantage of the fact that most of the suites of derived variables emphasized in this proceedings allow one to reconstruct the outline or the configuration of landmarks. In such cases the geometrical meaning of principal component axes or a discriminant functions, for example, can be seen from plots of hypothetical organisms representing various positions along an axis. Rohlf and Archie (1984) and Ferson et al. (1985) are examples using multivariate analyses of elliptic Fourier coefficients as variables. The clear differences in shape were difficult to appreciate from an examination of the numerical results by themselves.

References

Blackith, R. E. and R. Reyment. 1971 Multivariate morphometrics. Academic Press: London, 71 pp.

Ferson, S., F. J. Rohlf, and R. Koehn. 1985. Measuring shape variation of two-dimensional outlines. Systematic Zool., 34:59-68.

Jolicoeur, P. 1959. Multivariate geographical variation in the wolf, *Canis lupus* L. Evolution, 13:283-299.

Rohlf, F. J. and J. Archie. 1984. A comparative study of wing shape in mosquitoes. Systematic Zool., 33:302-317.

Strauss, R. E. and F. L. Bookstein. 1982. The truss: body form reconstruction in morphometrics. Systematic Zool., 31:113-135.

Chapter 4

Traditional Morphometrics

Leslie F. Marcus

Department of Biology, Queens College
City College of New York, Flushing, NY 11367

"It is no more use trying to be traditional than trying to be original."

T. S. Eliot

Abstract

Classical multivariate statistics applied to morphometrics is reviewed in terms of exploratory and confirmatory analysis. Brief explanations are given of principal component analysis (PCA), principal coordinate analysis, factor analysis, canonical variate analysis, use of Mahalanobis D^2, and discriminant analysis. Statistical assumptions are given for each method and some of the difficulties in application and interpretation are discussed. The methods are illustrated using two data sets: 1) twelve skull measurements for 574 mice of the genus *Zygodontomys* distributed over fifteen collecting localities, with special emphasis on one locality, Dividive, with 68 specimens; 2) seven external measurements of skins for 129 species of birds from Mediterranean climates. The complete data for the latter example are published in Blondel et al. (1984). The mouse example serves to illustrate applications of multivariate methods to geographic variation and low-level taxonomic studies; the bird example, based on species means, serves to illustrate a larger-scale comparison of faunal realms and ecomorphological variation. The Dividive data are given in Appendix 1 and the complete *Zygodontomys* data set is on a supplied disk.

Path models are illustrated with Wright's classical fowl example and several alternative models given. A separate analysis of a *Brizalina* data set, is given in Appendix 3.

The Jackknife and Bootstrap re-sampling techniques for estimating standard errors in principal component analysis are applied to the fowl data (Wright, 1968) and mouse data from Dividive. These results are compared with the asymptotic standard errors for the two data sets, and an experiment on the effect of sample size in re-sampling, using the fowl data set, is discussed for the asymptotic and re-sampling methods.

Most of the computations discussed were done using the Statistical Analysis System (SAS) software package on an IBM PC AT or clone. Some of the re-sampling experiments were done on a main frame computer using main frame SAS. When a SAS procedure was unavailable for doing an analysis, programs were written in the SAS matrix language called Interactive Matrix Language (IML). These programs are included on an accompanying disk and include: Bootstrap Analysis for PCA; Jackknife Analysis for PCA; asymptotic standard errors for PCA; confidence limits, F-test and correction for bias for Mahalanobis D^2; and minimum spanning tree. A short SAS program for reading in the complete mouse data set and producing the Dividive subset is given as well.

Introduction

The word "traditional" is used here to mean a body of statistical techniques available for morphometric analysis which have been widely applied in the past 20 or 30 years. They include among others, principal component analysis, principal coordinate analysis, factor analysis, discriminant analysis, canonical variate analysis, and multivariate analysis of variance. Most of these will be reviewed from the viewpoint of descriptive and inferential statistics. Tests of hypotheses and confidence intervals traditionally have been based on methods derived from assumptions of multivariate normality—and this is the classical multivariate statistics discussed in many texts. The users of these classical techniques, however, did not pay close attention to the shape or geometry of the biological objects being studied. Instead, measurements of form usually were analyzed as distances. The multivariate techniques used were those that could be applied to any assemblage of continuous variates. The recent emphasis on landmark data and statistical methods that take the geometry of the biological objects into consideration has lead to a new morphometric perspective discussed in many contributions in this volume (Bookstein, 1982, and this volume).

I will review the more traditional or classical techniques that have been applied largely to measurement or distance data. Exploratory and confirmatory analysis will be contrasted in this discussion. Exploratory analysis includes the search for pattern, use of descriptive statistics, and graphs; confirmatory analysis involves models based on biological concepts and uses classical inferential statistics for testing the fitted models. Inferential statistics depends on probability models for data and observed variation.

The single most important requirement in statistical inference is random sampling, though this may be extremely difficult to achieve in practice, and more often difficult to evaluate or verify for samples commonly used in morphometrics. A recent attempt to model data analysis without the requirement of random sampling is discussed in Diaconis (1985).

Confirmation of biological models, formulated before or after we collect data, requires that biological hypotheses be translated into statistical hypotheses. Most often descriptive tools are used to present a simplification or summary of data without specification of a model. For example principal component analysis is used for ordination or pattern description, and sometimes for confirmation of prior notions, but without rigorous hypothesis testing. However, inferential statistics for realistic probability models are frequently not available because the probability model is unspecified or we don't have enough a priori information or data to construct or fit an adequate model.

In my discussion of traditional multivariate techniques used in biological research, I will point out, where I can, how inference and confirmation may or may not be possible or appropriate.

Morphometrics

Morphometric questions come from a variety of studies including: analysis of form related to growth-both as summarized by carefully measured ontogenies and cross-sectional data; the nature and origin of polymorphism—sexual dimorphism, life stages, and other within-population polymorphisms; taxonomy—geographic variation in centroids and covariance structure, variation within and differences among taxa, assignment of individuals to taxa; adaptation and origins of adaptation—relation of form to environment and habitat, ecomorphological convergence, evolutionary sequences; functional questions—size in relation to physiological limits, allometric relations at different levels in the taxonomic hierarchy and their meaning, and so on. I will include indirectly some discussion of species recognition and boundaries, but I will not discuss descriptive and inferential phylogenetic analysis. For two good discussions of morphometrics see

Reyment (1985) for a more traditional view and Bookstein (1982) for the newer geometry-based morphometrics.

Most of my discussion will be about measurements taken on a continuous scale, including distances among well-defined landmarks (the "distance" of Bookstein et al., 1985), or minimum and maximum diameters, distances between tangents and points, tangents and tangents, as well as angles, areas and volumes. All of these will, it is hoped, be repeatable and carefully described in the methods part of a particular study. I also include appropriate common transformations such as logarithms, sums of distances, for example over parts, or ratios of distance measures between parts. The recently proposed use of homologous landmarks, sometimes summarized in the form of a truss, for recovering information about form for many evolutionary taxonomic studies is attractive for rigid two-dimensional or three-dimensional objects. The truss, however, can only be applied to individual parts of a jointed skeleton or to whole organisms such as fish treated as static (or dead). "Dynamic morphometrics" dealing with articulated or mobile parts has yet to be developed. The need for strict homology depends on the purpose of the study. For example, in functional problems, such as the study of weight bearing in bipedal organisms, measures of minimum supporting long bone shaft diameter are appropriate, or, for studies of flight, measurements of wing area and aspect ratios are the necessary ingredients for physical modelling.

Sampling Requirements

In a higher level taxonomic study we may include relatively few specimens per taxon, perhaps even only one as an exemplar for each taxon, if within-taxon variability is very small relative to among-taxa differences. For example, within-sample skeletal coefficients of variation are rarely more than 3 in birds, and this variation may be insignificant in the context of comparisons among higher taxa. If we are interested in detailed geographic variation or covariance structures, we need relatively large samples from single localities, from one

sex if there is sexual dimorphism, controlled for age of the individuals and time of collection. All of these potential sources of variation and others need to be carefully examined and explained, and then taken into consideration in the context of the specific problem or analysis being undertaken.

Many rules of thumb have been offered for determining sample size. For descriptive statistics the biological variability realized as individual or sampling variability and the required size of confidence intervals or standard errors should determine sample size. One can use as a rough guide the fact that the standard error of most commonly used statistics is inversely proportional to the square root of the sample size (Sokal and Rohlf, 1981). Note that for measures of variability (e.g., standard deviations and coefficients of variation), the constant of proportionality is greater than for measures of location (e.g., means and medians), and these statistics will require larger sample sizes to get acceptable standard errors. For statistical inference one must consider power of tests, i.e., the level of difference that one is trying to detect and the probabilities of Type I and Type II errors (op. cit.).

Types of Data

I will not discuss frequency type categorical data, or continuous data viewed as categorical data, whose study requires huge numbers of individuals if the number of categories is large even though the data might be relatively easy to score and collect. Some ordered discrete data such as fin ray counts in fish are often dealt with as continuous data.

I will mostly emphasize measurements in terms of linear dimensions and distances or their transforms. Proportions are popularly used (for example all measures divided by a standard length) in some fields, but have come into disfavor more recently (Atchley et al., 1976). However, a simple ratio or proportion might suffice to answer a simple question. Univariate and bivariate graphs or plots are descriptive tools that should be included in any analysis of data, but not necessarily published.

Always examine bivariate plots or scattergrams of the relations of the two variables to see if ratios may be appropriate. Furthermore, variables that satisfy statistical assumptions for inferences will not satisfy those same assumptions in the form of ratios.

A logarithmic transformation is often used. Natural logarithms using a base e or base 10 are most popular, and the choice makes no difference in most analyses. Logarithms have been justified partially from the multivariate generalization of the bivariate allometric equation (Jolicoeur, 1963) and the frequently observed greater homogeneity of coefficients of variation compared to variances for different sized characters in the same taxon, and over taxa for the same character. When coefficients of variation are similar for different sized variables, then a log transformation will usually make their variances more similar. Variances and covariances of logged data are unit free. Again, if the probability distribution assumptions are satisfied for the original data, they will not be satisfied for non-linear transformations such as logs; and if the logged data satisfy assumptions, the original data will not satisfy those same assumptions. For example, data whose logarithms have a normal distribution are distributed according to a log-normal distribution before the log transformation. However, either the raw data or logged data or both may satisfy the not very powerful tests of assumptions for sample sizes frequently used in systematic studies.

Size and Shape

The search for "size" adjustments has been a long one, especially in studies of organisms with different "shapes" where one is interested in size and size-related shape differences. Mass is one natural size measure which has some physiological and functional importance; but in the typical systematic study of linear dimensions, where a useful value of mass is unknown or unobtainable in preserved material, it may be difficult to compute a widely useful size measure and more than one may be relevant (Mosimann, 1970, and Mosimann and James, 1979). Bookstein (1989) has recently

published a valuable essay on the semantics of "size and shape", and Rohlf and Bookstein (1987) discuss the "shear" and other methods of size adjustment. Thorpe (1988) has developed Multiple Group Principal Components Analysis, or MGPCA, which is equivalent to one form of the Burnaby size adjustment in Rohlf and Bookstein (1987). See also Reyment, Blackith and Campbell (1984).

Data Analysis, Distribution Assumptions and Exemplar Data

When variables are well defined, in the sense of being accurately determined and verified, and measured with some required precision on similar biological objects, their distributions are most often summarized in terms of means, variances, and covariances or correlations. We will ignore statistics involving higher moments such as univariate skewness and kurtosis in this review of multivariate analysis. This results from the pervasiveness of normal theory in traditional morphometrics, since sample means, variances, and covariances are sufficient statistics for estimating the parameters of a multivariate normal probability distribution. Multivariate normality is sometimes justified empirically (though samples are rarely large enough to be very sure) or sometimes we depend on the simplicity and availability of the theory and statistical methods developed from that theory. Frequently there aren't enough data, and the alternative non-parametric and re-sampling methodology is as yet not widely available in software packages.

There have been some good applications of distribution-free or empirical discriminant functions (Howarth, 1971 in geology; see SAS Statistics, Version 6.03 for the PC). This author feels, however, that, in many circumstances, normality assumptions are appropriate for biological samples. Reyment (this volume) discusses some robust techniques which control for non-normality, especially in terms of influential and outlying observations. I will discuss later some recently popular distribution-free inferential techniques, the Jackknife and the Bootstrap, based on re-sampling, that do not depend on multivariate normal assumptions.

In studies where one is using measures on an exemplar to represent a taxon—or on centroids to particularize taxa—there may be no probability theory available, and the results must be descriptive. A probabilistic point of view may take the form: "what we see is one of a number of potentially possible outcomes of the historical processes leading to our observations." Beyond this statement, probability models are difficult to apply and inferential statistics in these circumstances should be used with great care.

Vectors of Means and Variance-Covariance Matrices

The minimal basic statistics required for classical multivariate analyses are means and sums of squares and cross-products. Vectors of means and variance-covariance matrices or distance matrices are derived from them and are easy to represent in geometrical terms for two and three dimensions; and the algebra extends to any number of dimensions.

Individual organisms representing one sex and one age class randomly drawn from a locally panmictic population over a short period of time I will call a "homogeneous sample." These will be the basis of many studies, though frequently with less homogeneity than desired. A set of measurements on many variables for such a homogeneous sample may be conceptually described as a cloud of points, one point for each individual, plotted in a space of as many dimensions as there are numbers of variables. Figure 1 is a bivariate example for a large data set. The centroid (vector of means), and variances and covariances (in the form of a symmetric matrix) are sufficient to model this cloud as a multivariate normal distribution. The cloud is most dense at the centroid and thins out in ellipsoids of decreasing concentration as one moves away from the centroid. Small data sets even if actually from normal distributions may not show this pattern clearly.

If the data do not in fact come from a multivariate normal distribution, then that summary will not be adequate. For example, the data may not be symmetrically distributed about the centroid, but rather be strung out or skewed more in one direction or another, or may not be represented by a linearly arranged concentration of points, implied by normal theory. They may be better represented instead by curved lineaments of concentration and thus not be representable by ellipsoids of concentration. Again, these patterns may be difficult to discern without lots of data. If polymorphisms are present there may be several regions of dense concentrations of points. Always plot your data in the form of scattergrams and histograms to see if univariate and bivariate normality are plausible if you are going to use techniques based on that model. These are necessary, but not sufficient conditions for multivariate normality, however. Since linear combinations of normally distributed data are also normal, all plots of derived linear

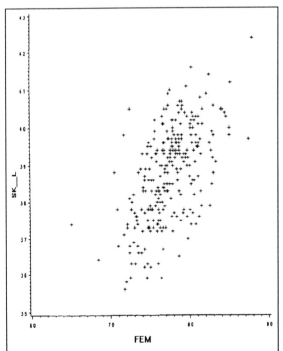

Figure 1.   Plot of skull length, ordinate, against femur length, abscissa, for 276 female fowl skeletons (Dunn, 1922). Both were measured in millimeters.

combinations such as principal components and canonical variates should produce normal-looking plots as a further check. However, linear combinations may tend to be more normal than the original data because of a central limit kind of effect, recalling that the sum or mean of independent observations from any distribution will tend to normality as the number of observations increases. Additional plotting techniques based on distances for individual specimens from the centroids are also useful for looking for outliers (see Reyment, this volume).

Exemplar data may represent each taxon as a point; or the entire data matrix of individual specimens combined over several taxa may be thought of as clouds of points of varying density and separation; or perhaps as a scatter of more or less clustered points (Figure 2). In this case normal theory is irrelevant.

In either case, when we summarize data in terms of centroids each surrounded by ellipsoids of concentration, most of the classical statistical techniques may be described algebraically and geometrically in terms of linear transformations of cartesian coordinate axes, even if the data do not justify the multivariate normal model. In that case, many of the methods whose derivation depends on that model will not have optimal properties; or if inferential statistics are used, reported probabilities will be incorrect. However, modest departures from normality and other violation of assumptions may not be too important, as in the analysis of variance, which is quite robust to non-normality and moderate inequality of variances for balanced designs (Sokal and Rohlf, 1981). Transformations, such as to logarithms, sometimes make the data more nearly normally distributed.

All of the classical methods find new axes which are determined by rigidly rotating the old axes about an origin (usually put at the centroid of all the data). Some, in addition, then place the axes at angles different from 90 degrees with respect to each other. Rigid rotations do not distort the distances between points or specimens, while those, such as canonical variates analysis, that allow the

axes to be at angles other than 90 degrees do distort distances.

These simple operations may seem to form a narrow framework within which to recast data and they sound deceptively simple. However, they have provided a powerful methodology for examining many biologically interesting questions. Even three dimensions are difficult to depict using two-dimensional graphs on paper and screens. More than three are impossible to display except as projections in two dimensions. Even in two dimensions there is no natural ordering of individual observations, as there is in univariate analysis where we can order our original measurements uniquely on each axis. We may find that these conceptually simple methods are as much as we may safely apply given the small amount of data frequently available.

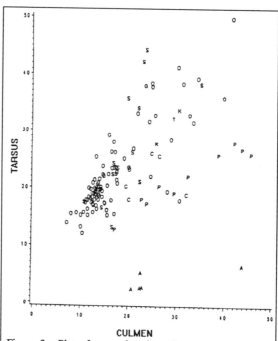

Figure 2. Plot of tarsus length, ordinate, against culmen length, abscissa, for means of 129 species of Mediterranean birds in eight "Orders". Both measured in millimeters.

Recasting Data Through Rigid Rotation (Principal Component Analysis and Related Techniques)

Principal component analysis (PCA) has been the most popular way of examining biological multivariate data arrays. I will follow a convention of representing each specimen's data as a row in the data array or matrix; and each column will represent all of the specimen or taxon values for each variable. PCA is basically a model-free and distribution-free technique for rigidly transforming data according to a simple maximization principle. We can, however, represent the data in the form of equations which will become useful when we contrast PCA to factor analysis.

In order to make inferences about the populations from which we are sampling, dis-tribution assumptions are necessary. I reemphasize that if we are going to make inferences, no matter what the technique, our sample must be a random sample from the population or populations of interest.

Two quite different examples are discussed: one where there is some hope of drawing statistical inferences; and another where biological hypotheses can be formulated, but where it is difficult or perhaps impossible to draw statistical inferences. The *Brizalina* data are analyzed separately in an Appendix. In a study of the rodents belonging to the genus *Zygodontomys* (their species assignment is undergoing revision), fifteen samples are available over a major part of its geographic range (Figure 3). From 29 to 68 suitable specimens are available from each locality. Robert Voss, Department of Mammalogy, American Museum of Natural History, measured 12 skull characters (Figure 4 and Appendix 1) to the nearest 0.05 mm. using hand-held dial calipers. All of the data for the 574 specimens are supplied on the accompanying disk. The mea-surements for the largest sample from the locality Dividive is given in Appendix 1. The specimens represent both males and females collected from an area no more than 10 kilometers in diameter and during less than six months' time. The rodent skulls grow throughout life. Voss defined five tooth-wear classes, of which only stages 2, 3 and 4 are included here (see Voss et al., in press for further details). The investigator would like to see how the 12 cranial measures vary and co-vary and if the variation and covariation structures can be simplified and are similar over the fifteen localities. We will use principal component analysis to do this.

In another study, Blondel measured 8 traditional characters, weight and 7 standard skin measurements as distances, for 100 species of birds from Mediterranean climates that breed in Chile, Provence or California (Blondel et al., 1984). Data for an additional 29 species that breed in the temperate bio-climate of Burgundy were also included in the study. Each observation published in Blondel et al (1984) represents the mean of five male birds, except in rare cases where fewer than five were available. Blondel et al. wanted to present the pattern of differences among families, habitats, food categories, foraging strategies and geographic regions in a simple graphical form. Principal components were used and are described below and in greater detail in Blondel et al (1984).

Bivariate scattergrams for both examples are

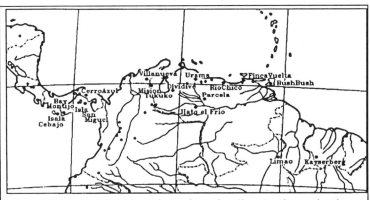

Figure 3. Map of southern Central America and northern South America showing sample localities for *Zygodontomys* included in the data discussed.

given in Figures 5 and 2. We see in the *Zygodontomys* data, presented here for Dividive, that there is high correlation between growing parts (Figure 5a: Condylo-incisive length [CIL] and Length of the Diastema [LD]); less for some parts than others (Figure 5b: CIL and Brain-case breadth [BB]); while tooth dimensions do not change after eruption and show weak or no relation to the growing parts (Figure 5c: CIL and Breadth of Molar One [BM1]). On the other hand, there is a great deal of variation among the taxa in the bird study as the species breeding in these areas represent many orders of birds included in the study.

I will first describe principal components for bivariate data and then generalize to the 12 variable rodent and 8 variable bird examples.

In Figure 6a, I have plotted a line through the scatter of points, the same data shown in Figure 5a, and have dropped perpendiculars, some labeled h, from the data points to the line and marked them with an *x* on the line. This line has the property that the amount of variance (or standard deviation) along the line among the *x*'s is the greatest for any line we can pass through the scatter of points, no matter how the data are distributed. At the same time, the sums of squares of the lengths of the subtended lines labeled h from the points perpendicular to the fitted line will be smaller than for any other line we can draw. This line and a line perpendicular to it through the centroid (joint means of the two variables) and parallel to the lines labeled h form new axes called principal axes. They are also the major and minor axes of the family of concentric concentration

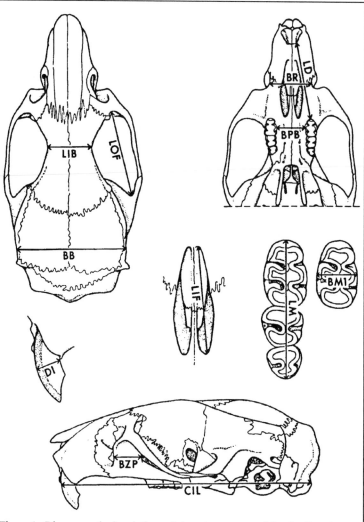

Figure 4. Diagrammatic descriptions of characters measured for the *Zygodontomys* data.

ellipses for a bivariate normal model fit to the data (whether the data follow a normal distribution or not).

In terms of the original axes, now moved to the centroid of the data, we have merely found a new set of axes (it can be thought of as a rigid rotation about the centroid) on which to project our data so that for the major axis, or first principal axis, the data have maximum variance, and for the minor axis, or second principal axis, perpendicular to the first, they have least variance. The variances along these new axes and along the old axes still add up to the same total variance (Table 1) and this is easily shown using the Pythagorean theorem. When we plot the data using these new axes (Figure 6b), we have defined new variables, called principal components, that are uncorrelated. Principal components are variables whose values, called principal component scores, represent linear or weighted combinations of the original variables. The weights or coefficients, the f values in the equations below, represent the cosines of the angles by which the axes are rotated (see Jolliffe, 1986, or Neff and Marcus, 1980, for a fuller explanation).

$$\text{Principal Component Score}_1 = f_{11}x_1 + f_{21}x_2,$$

$$\text{Principal Component Score}_2 = f_{12}x_1 + f_{22}x_2.$$

We summarize this in vector and matrix form more compactly as:

$$s_i = \mathbf{F'}x_i,$$

where s is the vector of scores for an individual specimen, x is the data vector for that specimen, and $\mathbf{F'}$ is a matrix of coefficients.

At the same time we may represent each vector of observations x as:

$$x_i = \mathbf{F}s_i.$$

Either set of equations is like multiple regression equations where the coefficients are regression coefficients, and the principal components are regressed on the original data, x_1 and x_2 for each specimen or taxon; or vice versa. However, there is no error term in these principal component equations, and the principal component scores are not directly observable. The lack of an error term emphasizes the non-statistical nature of this principal component analysis in two dimensions. The method of finding the f's and the variances for the scores is that of finding eigenvectors and eigenvalues, respectively, of a square symmetrical matrix (the matrix of variances and covariances, or for standardized data with means 0 and standard deviations 1, the correlation matrix). The eigenvectors as rows of $\mathbf{F'}$ or columns of $\mathbf{F}$ are constrained to have length 1, i.e., the sums of squares of their elements must add to 1. In this formulation, the eigenvalues represent the variances of the scores, and the columns

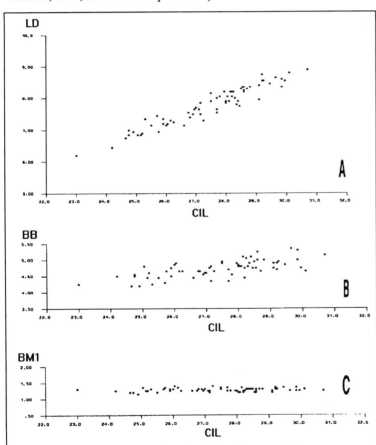

Figure 5. Bivariate scattergram for pairs of *Zygodontomys* skull measurements for the locality Dividive with 68 individuals. CIL = condylo-incisive length is plotted on the abscissa for all three plots and measurements are in millimeters. A) LD = length of the diastema against CIL; B) BB = braincase breadth against CIL; and C) LM = length of the molar tooth row against CIL.

of the eigenvector matrix **F** are the partial regression weights or coefficients for a specific principal component and at the same time are the cosines of the angles between the principal component axes and the original variable axes.

"Loading" is a popular term for the coefficients scaled in one or another way, a term borrowed from factor analysis. I like to think of the loadings as the correlations between the original variables and the scores. Using this definition, loadings are obtained by a scaling which multiplies the coefficients by the standard deviation of the scores (the square root of the eigenvalue = l_j) and divides them by the standard deviation (= sd_i) for their respective variable (like stand-ardized partial regression coeffi-cients):

$$loading_{ij} = f_{ij} \sqrt{l_j} / sd_i .$$

I have avoided using the term "loading." Others, however, have found the term convenient when not as rigidly defined and may use it for different scalings of the eigenvectors. It is important to note that the principal components for standardized data obtained from the matrix of correlation coefficients are not the same as principal components for the variance-covariance matrix even when both are presented in terms of loadings as defined here. For the correlation matrix, since the data are standardized to have standard deviation, $sd_i = 1$, then the formula above for the loading simplifies.

It behooves every researcher who publishes a PCA to specify the form of the data analyzed. Any transformation or standardization should be specified in a discussion of methods along with the form in which the loadings are given.

When we compute the eigenvalues and eigenvectors of the full 12×12 and 7×7 variance-covariance matrices of the logged data for the two data sets, we define principal components (as many as there are variables), as linear combinations of the variables, the first having the largest variance of any linear combination on the data, the second having the next largest variance of what

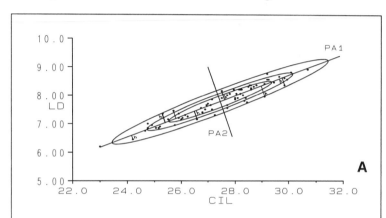

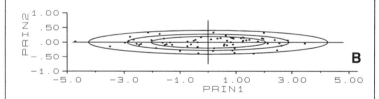

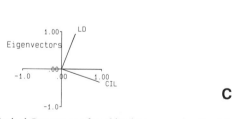

Figure 6. Illustration of Principal Components for a bivariate case, using the data plotted in Figure 5a: A) The sums of squares of the verticals, h, are minimum. Ellipses are best fit bivariate normal ellipses as 50%, 75% and 95% contours, and major and minor axes are shown; B) same data plotted on new principal component axes–the figure in a), is just rotated to make the major and minor axes the principal axes and the origin is at the joint means (centroid) of the data; C) eigenvector coefficients plotted to show variable loadings as part of a biplot.

variance is left in the data, and so on down to the last which is the most invariant combination of our data, that is, it has the least variance.

We may then represent a data vector as:

$$x_i = \mathbf{F}\, s_i + e_i$$

where we have introduced a vector of residuals e. Unlike our 2 variable model above, we may retain only 2, 3 or r of the p (the total number of characters measured) principal components in our summarization of the data. Then the matrix $\mathbf{F}$ will have p rows and r columns. There will only be r scores for each individual, and the portion of x not summarized by the components retained will reside in the e vector. The e_i can similarly be expressed in terms of the last p-r principal components not used in the presentation. If p components are used, then of course e will be 0.

A two-dimensional plot or three-dimensional projection of the principal component scores in the plane or volume defined by their respective principal axes will then provide displays summarizing the most variability (in terms of variance) we can display using new axes in two or three dimensions. The principal component scores are uncorrelated and their respective eigenvectors are at right angles to each other or orthogonal, that is, their inner product is 0.

The coefficients, as the elements of the eigenvectors (columns of the matrix $\mathbf{F}$), will contain information on the relative contribution of the original variables to the principal components derived from them; and also, as rows in the matrix of column eigenvectors, the coefficients will tell us how the variance of each variable is distributed over the components.

It has sometimes been found useful to name a component in terms of its dominant set of coefficients or the pattern of coefficients. For example, if all the coefficients of the first component have the same sign it is sometimes called a "size" component. It is size only in the sense that the correlation of this principal component with all of the variables has the same sign, i.e., if an individual is larger in any variable on the average its score on the first principal component is larger. Then the others are sometimes called "shape" components as they reflect contrasts in measurements or differences of weighted sums of variables for non- logged data, and products of weighted ratios for logged data. However, one must be careful in making too facile an interpretation, because if the first component is "size" as described above, then all other components must be "shape". This comes from the requirements that the scores be uncorrelated over components and that the eigenvectors are orthogonal. This means that their inner products, the sum of products of corresponding elements, are 0. So if one eigenvector has all signs the same, then all others must have plus and minus signs, and be "shape" components.

A useful way of graphically showing the eigenvector coefficients is to plot them on the PC scores plot, in the form of a "biplot". A vector

Table 1. Principal component analysis for two variables from Dividive data.					
Untransformed data			Log transformed data		
Simple statistics					
	CIL	LD		LOG_CIL	LOG_CIL
Mean	27.44	7.77	Mean	3.31026	2.04739
s	1.65	0.61	s	0.06092	0.08039
Covariance matrices					
	CIL	LD		LOG_CIL	LOG_LD
CIL	2.7071	0.9652	LOG_CIL	0.0037114	0.0046956
LD	0.9652	0.3757	LOG_LD	0.0046956	0.0064636
Total variance = 3.0828			Total variance = 0.0101750		
Correlation matrices					
	CIL	LD		LOG_CIL	LOG_LD
CIL	1.0000	0.9570	LOG_CIL	1.0000	0.9587
LD	0.9570	1.0000	LOG_LD	0.9587	1.0000
	Eigen-value	Cumu-lative %		Eigen-value	Cumu-lative %
PRIN1	3.0549	99.1	PRIN1	0.009981	98.1
PRIN2	0.0280	100.0	PRIN2	0.000194	100.0
Total	3.0828		Total	0.010175	
Eigenvectors					
	PRIN1	PRIN2		PRIN1	PRIN2
CIL	0.9408	-.3389	LOG_CIL	0.5995	0.8004
LD	0.3389	0.9408	LOG_LD	0.8004	-.5995

corresponding to each original variable may be projected onto the principal component scores display (Krzanowski, 1988; Gabriel, 1971), using the eigenvectors for coordinates. The biplot will be a best two-dimensional representation of the mean centered data in PC terms. The plot of the first two eigenvectors for the *Zygodontomys* data from Dividive is given as an inset on Figure 6c. When there are a lot of data points, plotting the scores and eigenvectors on the same graph may be messy. For data well summarized by two principal components, the data can be nearly reconstructed from a biplot; as each specimen's observed value for a variable will be nearly the product of the vector to the point representing the specimen and the vector for the variable, both vectors originating at the origin.

Other display techniques in low-dimensional space are discussed in Reyment, Blackith and Campbell (1984), Neff and Marcus (1980) and Gnanadesikan (1977).

For logged data, the first component of the covariance matrix may sometimes be interpretable as a vector of allometric coefficients (Jolicoeur, 1963), which also reflect shape differences. Isometry is represented by each coefficient being equal to $(1/p)^{0.5}$. Coefficients less than this amount reflect negative allometry and those greater, positive allometry. Bivariate allometric coefficients may be determined from ratios of eigenvector coefficients (see Shea, 1985, for an up-to-date discussion in the context of growth allometry). Frequently, however, beyond the first few components, the coefficients are near 1 for only one variable on a component and near 0 for all of the rest. This implies that the component is highly correlated with that variable, or just summarizes information for that single variable and tells us nothing about association of variables for interpreting variability and covariability.

It is important to emphasize that the pattern of coefficients represents the peculiarities of the sample of data at hand and there can be considerable sampling variability for small samples. At worst, it may represent introduced variability due to measurement error by the researcher. One wants to see biologically interpretable patterns which

Table 2. Principal component analysis of logarithms of Dividive data.

Variance-covariance matrix × 100

	CIL	LD	LM	BM1	LIF	BR	BPB	BZP	LIB	BB	DI	LOF
CIL	.3711	.4696	.0575	.0255	.3636	.4254	.4928	.4695	.2241	.1185	.4994	.3447
LD	.4696	.6464	.0292	.0217	.4979	.5381	.6486	.5877	.2795	.1599	.6016	.4360
LM	.0575	.0292	.1575	.0831	.0174	.0951	.0445	.1031	.0826	-.004	.1066	.0444
BM1	.0255	.0217	.0831	.1762	.0196	.0992	.0280	.0504	.0491	-.001	.0618	.0167
LIF	.3636	.4979	.0174	.0196	.5453	.4044	.4974	.4836	.2195	.1449	.4663	.3486
BR	.4254	.5381	.0951	.0992	.4044	.7536	.6705	.5822	.3343	.1396	.6263	.3995
BPB	.4928	.6486	.0445	.0280	.4974	.6705	.9362	.7023	.3226	.1832	.7267	.4462
BZP	.4695	.5877	.1031	.0504	.4836	.5822	.7023	.9718	.2893	.1298	.6402	.4282
LIB	.2241	.2795	.0826	.0491	.2195	.3343	.3226	.2893	.3093	.0800	.2869	.1952
BB	.1185	.1599	-.004	-.001	.1449	.1396	.1832	.1298	.0800	.1021	.1601	.1239
DI	.4994	.6016	.1066	.0618	.4663	.6263	.7267	.6402	.2869	.1601	.9293	.4712
LOF	.3447	.4360	.0444	.0167	.3486	.3995	.4462	.4282	.1952	.1239	.4712	.3686

Eigenvectors of variance-covariance matrix of logged data

	PC1	PC2	PC3	PC4	PC5	PC6	PC7	PC8	PC9	PC10	PC11	PC12
CIL	.2742	-.074	.0614	-.002	.1585	-.136	-.067	.3045	.0203	-.113	.0784	.8692
LD	.3519	-.271	.1210	.1585	.1400	-.030	-.098	.4712	-.262	-.313	.4132	-.417
LM	.0441	.4482	-.141	-.036	.2932	-.117	.3031	.1952	.6726	-.132	.2404	-.137
BM1	.0292	.4537	-.054	.1187	.2676	.6202	.2189	.2574	-.403	.1480	-.139	.0497
LIF	.2845	-.421	.0476	.2708	.4323	.3376	.2201	-.500	.2091	-.129	-.077	.0232
BR	.3564	.4344	.1527	.3291	-.215	.1682	-.613	-.250	.1618	-.043	.1076	.0019
BPB	.4135	-.060	.1161	.0193	-.690	.1741	.4867	.1406	.1580	-.030	-.152	.0034
BZP	.3914	-.040	-.885	-.138	-.042	-.044	-.089	-.086	-.109	.1071	.0257	-.027
LIB	.1798	.2704	.0490	.4691	.0912	-.622	.3172	-.193	-.332	.0387	-.164	-.031
BB	.0932	-.121	.1332	.0859	.0166	-.010	.0842	-.008	.0584	.8398	.4847	.0195
DI	.3987	.2038	.3247	-.728	.1641	-.082	.0590	-.284	-.196	-.015	.0445	-.073
LOF	.2570	-.119	.1084	-.029	.2271	-.119	-.257	.3509	.2395	.3348	-.664	-.204

Table 3. Principal component analysis of logarithms of Dividive data.

PC	Eivenvalue	Cumulative%
1	0.456573	72.85
2	0.036284	78.64
3	0.033649	84.00
4	0.025552	88.08
5	0.023436	91.82
6	0.013521	93.98
7	0.012521	95.98
8	0.009079	97.42
9	0.006374	98.44
10	0.005768	99.36
11	0.003131	99.86
12	0.000833	100.00

are relatively stable over repeated sampling. Since no biological model is used to generate the components, there is no special reason to search for biological sources of variation and covariation explained in a principal component analysis. Gibson et al. (1984) have recently provided some data to indicate that, in their systematic morphometric study of a series of homogeneous data sets, little past the first principal component was interpretable for the birds they studied due to sampling error alone. I will discuss this point later in detail and present some new results.

If the first few principal components usefully summarize and display your data, always look further and see what is not summarized by the components retained. Every principal components analysis should include an analysis of the residual variances and covariances of the data not so summarized, and also the residual values represented by the vector e for the data itself. This is seldom done in actual applications. While the standard statistical packages may present us with residual covariance or correlation matrices, none that I am aware of partition the data itself into a part contained in the components retained or displayed, and a residual part. It is not difficult to obtain such residuals in a package like SAS (using matrix routines such as PROC IML in version 6). The residual covariance matrix may be computed by forming that part of the covariance matrix explained by the principal components retained and subtracting this from the original covariance matrix:

$$\text{VARCOV}_{\text{residual}} = \text{VARCOV}_{\text{original}} - \mathbf{F L F'} ,$$

where $\mathbf{F}$ contains the r eigenvectors retained as columns, and $\mathbf{L}$ is a diagonal matrix of the r eigenvalues. Similarly a residual data matrix may

be constructed as:

$$\mathbf{X}_{\text{residual}} = \mathbf{X} (\mathbf{I} \text{-} \mathbf{F} \mathbf{F'}) ,$$

where $\mathbf{X}$ is the mean centered data matrix, and $\mathbf{I}$ a diagonal matrix of ones. This is just an extension of formula 23 for more than 1 vector in Rohlf and Bookstein (1987). $\text{VARCOV}_{\text{residual}}$ can also then be formed as:

$$\text{VARCOV}_{\text{residual}} = \mathbf{X'}_{\text{residual}} \ \mathbf{X}_{\text{residual}} / (n\text{-}1) .$$

As another way of looking at the residuals, BMDP provides a useful partition of the distance from each observation to the centroid of the data into two parts, one based on the components retained, and a residual part (Dixon, 1983). The residual part can be used for detecting outliers in homogeneous data sets (Hawkins, 1974 and 1980).

The variance-covariance matrix, the principal component variances (eigenvalues) and coeffi-

Table 4. "Loadings" = correlation coefficients between original variables and PC scores.

	PC1	PC2	PC3	PC4	PC5	PC6	PC7	PC8	PC9	PC10	PC11	PC12
CIL	.9617	-.073	.0584	-.002	.1260	-.082	-.039	.1506	.0084	-.044	.0228	.1303
LD	.9354	-.203	.0873	.0997	.0843	-.014	-.043	.1766	-.082	-.093	.0910	-.047
LM	.2375	.6802	-.206	-.045	.3576	-.108	.2702	.1482	.4278	-.080	.1072	-.032
BM1	.1486	.6510	-.075	.1429	.3086	.5433	.1845	.1848	-.242	.0847	-.058	.0108
LIF	.8233	-.343	.0374	.1854	.2834	.1681	.1055	-.204	.0715	-.042	-.019	.0029
BR	.8773	.3014	.1020	.1916	-.120	.0712	-.250	-.087	.0471	-.012	.0219	.0002
BPB	.9133	-.037	.0696	.0101	-.345	.0662	.1780	.0438	.0412	-.008	-.028	.0003
BZP	.8483	-.024	-.520	-.071	-.020	-.016	-.032	-.026	-.028	.0261	.0046	-.002
LIB	.6907	.2929	.0512	.4264	.0794	-.411	.2018	-.104	-.151	.0167	-.052	-.005
BB	.6235	-.228	.2419	.1360	.0252	-.011	.0933	-.008	.0461	.6313	.2685	.0056
DI	.8837	.1274	.1954	-.382	.0824	-.031	.0217	-.089	-.051	-.004	.0082	-.007
LOF	.9045	-.118	.1036	-.024	.1811	-.072	-.150	.1742	.0996	.1324	-.194	-.031

Residual covariance matrix × 100 after removal of PC1-3.

	CIL	LD	LM	BM1	LIF	BR	BPB	BZP	LIB	BB	DI	LOF
CIL	.0246	.0192	.0173	.0023	-.005	-.012	-.029	-.003	.0053	-.004	-.001	.0175
LD	.0192	.0493	.0080	.0215	-.003	.0018	-.027	-.009	.0152	-.007	-.032	.0070
LM	.0173	.0080	.0691	.0009	.0308	-.040	-.023	-.011	.0047	.0033	.0085	.0172
BM1	.0023	.0215	.0009	.0966	.0519	-.017	-.015	-.011	-.018	.0085	-.019	.0040
LIF	-.005	-.003	.0308	.0519	.1107	.0053	-.051	-.017	.0265	.0032	-.026	-.005
BR	-.012	.0018	-.040	-.017	.0053	.0972	.0010	-.003	-.003	.0000	-.071	-.005
BPB	-.029	-.027	-.023	-.015	-.051	.0010	.1495	-.003	-.013	-.001	-.034	-.046
BZP	-.003	-.009	-.011	-.011	-.017	-.003	-.003	.0087	-.013	.0012	.0274	-.000
LIB	.0053	.0152	.0047	-.018	.0265	-.003	-.013	-.013	.1344	.0131	-.066	-.006
BB	-.004	-.007	.0033	.0085	.0032	.0000	-.001	.0012	.0131	.0511	-.015	.0044
DI	-.001	-.032	.0085	-.019	-.026	-.071	-.034	.0274	-.066	-.015	.1530	.0004
LOF	.0175	.0070	.0172	.0040	-.005	-.005	-.046	-.000	-.006	.0044	.0004	.0580

cients (eigenvectors) for the logarithms of the *Zygodontomys* data are given in Tables 2 and 3. The correlations between the principal components and original variables (loadings) together with the residual covariance (after the first three components) are given in Table 4. The original correlation matrix and residuals to the correlation matrix (both after the first and first three components) are given in Table 5. The residual analysis for *Zygodontomys* does not point to any exceptional data. While the variance-covariance matrix was used in the original data analysis, it is very difficult to examine residuals in that form. It is far easier to look at the residuals in terms of the correlation matrix.

Stem and leaf diagrams of the residuals before and after extraction of the first principal component is instructive for the Divide data (Figure 7). Note that for the residual data, the mean has been added back in so that we are looking at similar sized variables as raw data. For CIL, condylo-incisive length of the skull, most of the variation is in the direction of the first principal component and the histogram of the residuals is more symmetrical than for the raw data. Apparently much of the age-related growth variation has been removed from the data (Voss et al., in press). Width of the first molar, BM1, however, retains essentially all of its variability as would be predicted from the value of the coefficient, 0.029, for BM1 in the first eigenvector. Note in the residual matrix from correlations after removal of PC1 (Table 5) that most of the molar variance and much of the variance for some other characters such as braincase breadth is retained. However, the only sizeable portion of covariance retained is for the relation between tooth row length and first molar width, again apparently because these characters do not change with growth. The residual matrix after removal of PC1-3 (Tables 4 and 5) still shows half of the variance for the tooth characters and still some for the braincase measures but little remaining covariance. The second and third eigenvalues are very similar, so there is not any clear pattern of

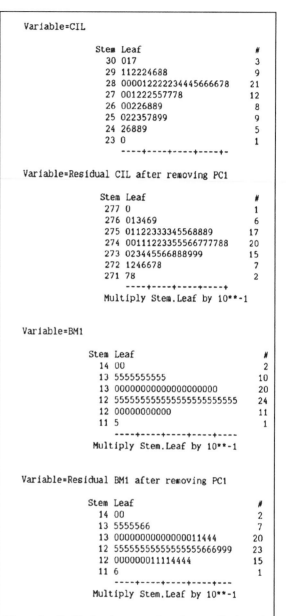

Figure 7. Residuals after removing the contribution of Principal Component one from the data for Divide using all 12 variables. Note: for CIL the range of the data has been reduced from 23.0-30.7 to 27.2-27.7; while for BM1 with a small coefficient on principal component one the residuals have essentially the same distribution as for the raw data.

relation among the variables past the first principal component.

Blondel et al. (1984) discuss the important individual species' data which are not well summarized by the two-dimensional principal component plot presented in their paper (reproduced here as Figure 8). In that analysis, some individual species were clearly distinct as determined for principal components past the second, and that result is not expressed in the plane of the first two principal components.

Principal Coordinate Analysis

If our data are in the form of distances or similarities among individuals or taxa we may wish to use the method of principal coordinates, one of a series of methods called multidimensional scaling (Kruskal and Wish, 1978). Principal coordinates can provide the same displays and the method has the same summary characteristics as principal components for data as a matrix of Euclidean distances squared between individuals or taxa. The distances among objects are maximally summarized by the first, then the second, down to the last principal coordinate as in principal components analysis. The method is sometimes called "dual" to principal components analysis (Gower, 1966a), in that principal components analysis is based on the character-by-character sums of square and cross-products matrix, or **R** matrix, of the mean-centered data; while principal coordinate analysis is based on the individual-by-individual distance squared matrix, which can be transformed to a sums of squares and cross-products or **Q** matrix (Pielou, 1984). The **R** and **Q** matrices have the same eigenvalues, and the eigenvectors of one can easily be obtained from the eigenvectors of the other (op. cit.).

A generally useful result, not taken advantage of much in applications (however, see Reyment, this volume) is that any data matrix, whether mean centered or not, can

Table 5. Correlation matrix.

	CIL	LD	LM	BM1	LIF	BR	BPB	BZP	LIB	BB	DI	LOF
CIL	1.000	.9587	.2379	.0998	.8082	.8043	.8360	.7817	.6614	.6088	.8504	.9319
LD	.9587	1.000	.0914	.0642	.8387	.7711	.8338	.7415	.6252	.6224	.7762	.8933
LM	.2379	.0914	1.000	.4989	.0592	.2761	.1159	.2635	.3741	-.031	.2785	.1843
BM1	.0998	.0642	.4989	1.000	.0633	.2722	.0690	.1219	.2104	-.010	.1527	.0654
LIF	.8082	.8387	.0592	.0633	1.000	.6309	.6962	.6644	.5345	.6142	.6551	.7775
BR	.8043	.7711	.2761	.2722	.6309	1.000	.7982	.6803	.6926	.5032	.7485	.7580
BPB	.8360	.8338	.1159	.0690	.6962	.7982	1.000	.7363	.5995	.5928	.7792	.7596
BZP	.7817	.7415	.2635	.1219	.6644	.6803	.7363	1.000	.5277	.4123	.6737	.7155
LIB	.6614	.6252	.3741	.2104	.5345	.6926	.5995	.5277	1.000	.4501	.5352	.5782
BB	.6088	.6224	-.031	-.010	.6142	.5032	.5928	.4123	.4501	1.000	.5199	.6388
DI	.8504	.7762	.2785	.1527	.6551	.7485	.7792	.6737	.5352	.5199	1.000	.8052
LOF	.9319	.8933	.1843	.0654	.7775	.7580	.7596	.7155	.5782	.6388	.8052	1.000

Residual matrix from correlations after removal of PC1. Note high residual variance and covariance for LM and BM1.

	CIL	LD	LM	BM1	LIF	BR	BPB	BZP	LIB	BB	DI	LOF
CIL	.0751	.0591	.0095	-.043	.0165	-.039	-.042	-.034	-.003	.0091	.0005	.0621
LD	.0591	.1250	-.131	-.075	.0686	-.050	-.020	-.052	-.021	.0392	-.050	.0473
LM	.0095	-.131	.9436	.4636	-.136	.0678	-.101	.0621	.2101	-.179	.0687	-.030
BM1	-.043	-.075	.4636	.9779	-.059	.1418	-.067	-.004	.1077	-.103	.0213	-.069
LIF	.0165	.0686	-.136	-.059	.3222	-.091	-.056	-.034	-.034	.1009	-.072	.0329
BR	-.039	-.050	.0678	.1418	-.091	.2303	-.003	-.064	.0866	-.044	-.027	-.035
BPB	-.042	-.020	-.101	-.067	-.056	-.003	.1659	-.038	-.031	.0233	-.028	-.066
BZP	-.034	-.052	.0621	-.004	-.034	-.064	-.038	.2804	-.058	-.117	-.076	-.052
LIB	-.003	-.021	.2101	.1077	-.034	.0866	-.031	-.058	.5230	.0195	-.075	-.046
BB	.0091	.0392	-.179	-.103	.1009	-.044	.0233	-.117	.0195	.6112	-.031	.0748
DI	.0005	-.050	.0687	.0213	-.072	-.027	-.028	-.076	-.075	-.031	.2191	.0059
LOF	.0621	.0473	-.030	-.069	.0329	-.035	-.066	-.052	-.046	.0748	.0059	.1820

Residual matrix from correlations after removal of PC1-3. Note still high residual variance proportions for LM, BM1, LIB, and BB.

	CIL	LD	LM	BM1	LIF	BR	BP	BZP	LIB	BB	DI	LOF
CIL	.0663	.0392	.0714	.0089	-.011	-.023	-.049	-.005	.0157	-.022	-.002	.0474
LD	.0392	.0763	.0252	.0638	-.004	.0026	-.034	-.011	.0340	-.028	-.042	.0143
LM	.0714	.0252	.4386	.0053	.1049	-.116	-.061	-.028	.0214	.0259	.0222	.0713
BM1	.0089	.0638	.0053	.5484	.1673	-.047	-.037	-.027	-.079	.0635	-.047	.0158
LIF	-.011	-.004	.1049	.1673	.2030	.0083	-.071	-.023	.0645	.0136	-.036	-.012
BR	-.023	.0026	-.116	-.047	.0083	.1290	.0012	-.003	-.007	.0001	-.085	-.010
BPB	-.049	-.034	-.061	-.037	-.071	.0012	.1597	-.003	-.024	-.002	-.037	-.078
BZP	-.005	-.011	-.028	-.027	-.023	-.003	-.003	.0089	-.024	.0037	.0289	-.001
LIB	.0157	.0340	.0214	-.079	.0645	-.007	-.024	-.024	.4346	.0738	-.122	-.017
BB	-.022	-.028	.0259	.0635	.0136	.0001	-.002	.0037	.0738	.5007	-.049	.0228
DI	-.002	-.042	.0222	-.047	-.036	-.085	-.037	.0289	-.122	-.049	.1647	.0007
LOF	.0474	.0143	.0713	.0158	-.012	-.010	-.078	-.001	-.017	.0228	.0007	.1573

be represented as the matrix product of the eigenvectors of the among-variables sums of squares and cross-products matrix (R analysis), the square root of the diagonal matrix of eigenvalues, and the eigenvectors of the sums of squares and cross-products matrix among individuals (Q analysis). This most powerful result in matrix theory is called the singular value decomposition theorem (Joreskog et al, 1976, Krzanowski, 1988, and Reyment, this volume). However, if one mean centers the data, say by variables or columns as one would do in an R analysis, then this duality is only true for this mean centering; and not true if one mean centers by columns as is commonly done in a Q analysis by rows.

Principal coordinate analysis is especially useful when one has only association data or distance data, as in DNA annealing or immunology studies. Then principal coordinate analysis provides the best reduced-dimension display of the data in the least squares sense described above. However, to faithfully display useful relations among the data points, the data must have properties of a metric. Some data are not well behaved, and negative eigenvalues may occur. Still, useful plots may be produced from the principal coordinates corresponding to the positive eigenvalues.

Non-Metric Multidimensional Scaling

Non-metric multidimensional scaling is a further generalization, with principal coordinate analysis as a special metric case. Among the best-known

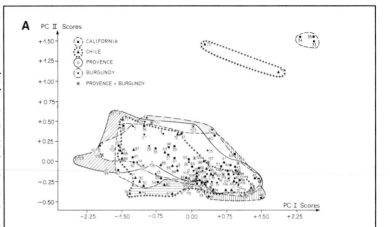
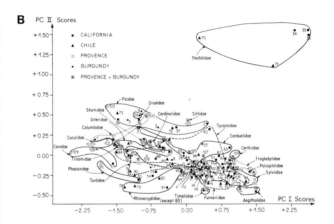
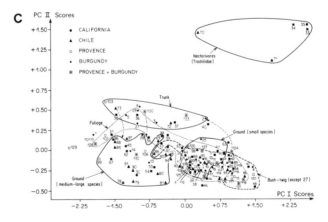

Figure 8. Figures of Principal Component Scores for PC1 and PC2 from Blondel et al. (1984; original Figure 7). "Envelopes" have been placed around subsets of the data on a priori grounds: A) Regional envelopes; B) Taxonomic envelopes-some families are outlined; and C) habitat stratum envelopes.

approaches is that of Joseph Kruskal (Kruskal and Wish, 1978). A variety of programs and algorithms is available in a number of statistical packages (for example, ALSCAL as an addendum to mainframe SAS, and NTSYS by Rohlf (1988). Rohlf (1972) compares this technique to principal components and principal coordinates as ordination techniques. Non-metrical multidimensional scaling provides "best" low-dimensional displays or coordinates in terms of distances in the display that are as nearly as possible monotone functions of the distances among individuals or taxa. As the name implies, data do not have to have metrical properties, and the method provides plots of relations among entities, which may prove useful. This technique has not been used much in ordinations in systematics.

Correspondence Analysis

Another form of eigenvalue-eigenvector analysis is correspondence analysis (Greenacre, 1984), widely known to ecologists as reciprocal averaging (Pielou, 1984). As used in ecology, its attraction has been that counts or frequencies for both quadrats and species are scaled in the same way. One finds the eigenvalues and eigenvectors for the sums of squares matrix formed from this doubly scaled matrix. It is claimed by its proponents (see Reyment, this volume, for further references) that one low-dimensional plot shows structure relating to both quadrats and species, and the relation of one to the other. This is another form of principal components analysis on a specially scaled matrix displayed as a biplot (discussed above).

The scaling or data transformation is motivated from analyzing frequency data and is akin to contingency table analysis. This row and column scaling is not very intuitive when applied to continuous measurements. Applications of correspondence analysis using measurement data have been made in systematics by Petit-Maire and Ponge (1979) and more recently by Werdelin (1983, 1988). The column and row scalings are a kind of "size" adjustment. However, this is a peculiar scaling for measure-

ment data, and other means of controlling for size need to be explored with these data.

In summary, principal components analysis and allied procedures are powerful methods for ordination and exploration of homogeneous systematic data sets, as well as those for which a maximum variance ordination is informative as in the Mediterranean bird study. It summarizes variances, covariances (or correlations) and distances, and may be used to motivate insightful questions in analysis of data prior to model construction. Other multi-sample applications will be discussed later.

I have emphasized some of the inadequacies of principal components analysis related to biological hypotheses. The fact that principal components are uncorrelated may lead to data summaries which may or may not be biologically relevant. If one is trying to find a causal or functional basis of varia-

Table 6. Wright fowl analysis - 276 female crossbred chickens. Published results (Wright, 1968).

	Mean mm.	SD mm.	CV %	Correlation L	B	H	U	F	T
L	38.77	1.26	3.25	1	0.584	0.615	0.601	0.570	0.600
B	29.81	0.93	3.13		1	0.576	0.530	0.526	0.555
H	74.64	2.84	3.80			1	0.940	0.875	0.878
U	68.74	2.73	3.97				1	0.877	0.886
F	77.34	3.20	4.14					1	0.924
T	114.84	5.00	4.35						1

Components of variables in White Leghorn population by Hotelling's method with complete apportionment of self-correlations to six factors.

	px1	px2	px3	px4	px5	px6	
L	0.7426	0.4536	0.4922	-0.0205	0.0073	-0.0007	
B	0.6975	0.5886	-0.4085	-0.0008	0.0021	-0.0144	
H	0.9477	-0.1583	-0.0259	0.2182	0.0472	0.1623	
U	0.9403	-0.2125	0.0071	0.2033	-0.0423	-0.1654	
F	0.9287	-0.2351	-0.0382	-0.2139	0.1841	-0.0323	
T	0.9407	-0.1908	-0.0296	-0.1950	-0.1945	0.0446	
Σ	4.5677	0.7141	0.4122	0.1731	0.0758	0.0569	5.99998*
%	76.1	11.9	6.9	2.9	1.3	0.09	100.0

* Note this sum should be as given here, but is 0.59998 in Wright.

Above as eigenvectors.

	e1	e2	e3	e4	e5	e6
L	0.3745	0.5368	0.7666	-0.0493	0.0265	-0.0029
B	0.3624	0.6954	-0.6363	-0.0019	0.0076	0.0604
H	0.4434	-0.1873	-0.0403	0.5244	0.1714	0.6804
U	0.4400	-0.2515	0.0111	0.4886	-0.1516	-0.6914
F	0.4345	-0.2782	-0.0595	-0.5141	0.6687	-0.1154
T	0.4402	-0.2258	-0.0461	-0.4687	-0.7065	0.1870

tion and covariation, principal component analysis may be inadequate. If on the other hand one is looking for clusters, cluster analysis would be more appropriate. PCA should be recognized for what it is, a data projection and rotation technique summarizing most of the variability in the data, where one may search for patterns and clusters in displays and get some idea of influential and associated variables giving rise to these displays.

Factor Analysis, Path Analysis and LISREL

If we want to summarize morphometric data as weighted combinations of variables, i.e., latent variables, motivated by our biological understanding of the morphology, the context of the study should determine whether these latent variables are to be correlated or not. Factor analysis provides a richer modeling framework for doing this. Factor analysis incorporates variance and covariance summarizing features found desirable in principal components, but it also requires additional assumptions. We may fit models to data using factor analysis or path analysis. Sewall Wright invented path models and path analysis, a part of which is closely related to the factor analysis methodology independently developed in the social sciences.

Wright (1921, 1954, and 1968) was thinking in terms of causal models which could be quantified for explaining morphometric relationships of variables and could be fit to data. Factor analysis, too, depends on a model that expresses observed variables in terms of underlying "causal" factors or latent variables. Cause is used here both in its more formal sense of "cause and effect," or in a weaker structural sense to relate the correlations we see among measured

variables in terms of a path diagram of interrelations among variables. We may or may not know in advance how many factors (causes) or latent variables there are, or how they relate to the measured variables. In other words, we cannot draw a correct path diagram in advance of our analysis. This corresponds to an exploratory factor analysis and the intention is to get some idea of the factors or latent variables, or to test weak conceptions of what is going on causally or structurally in our data. A preliminary principal

Table 7. Results computed from Dunn's data.

	Mean	SD	Correlation					
	mm.	mm.	L	B	H	U	F	T
L	38.78	1.25	1	0.583	0.621	0.603	0.569	0.602
B	29.80	0.93		1	0.584	0.526	0.515	0.548
H	74.68	2.83			1	0.937	0.877	0.874
U	68.89	2.73				1	0.878	0.894
F	77.36	3.21					1	0.926
T	114.95	4.99						1

Eigenvectors and eigenvalues for correlation matrix above with asymptotic standard errors.

	e1	e2	e3	e4	e5	e6
L	0.348+.021	0.525+.083	0.774+.057	0.046+.054	0.028+.028	-0.012+.022
B	0.324+.024	0.704+.068	-0.626+.076	0.028+.049	-0.016+.028	-0.071+.020
H	0.443+.009	-0.166+.028	-0.057+.023	-0.546+.041	0.350+.070	0.593+.095
U	0.440+.010	-0.250+.026	0.004+.046	-0.474+.043	-0.358+.102	-0.625+.059
F	0.434+.012	-0.289+.028	-0.057+.051	0.502+.052	0.613+.062	-0.311+.096
T	0.440+.010	-0.232+.027	-0.040+.046	0.471+.052	-0.610+.070	0.395+.095
Vals	4.568+.389	0.716+.061	0.412+.035	0.168+.014	0.079+.007	0.053+.005
%	76.1	12.0	6.9	2.8	1.3	0.09

Results based on logarithms of Dunn's data.

	Mean	SD	Correlation					
			L	B	H	U	F	T
L	3.657	0.032	1	0.586	0.621	0.602	0.569	0.603
B	3.394	0.031		1	0.585	0.528	0.518	0.551
H	4.312	0.037			1	0.937	0.878	0.873
U	4.231	0.039				1	0.879	0.894
F	4.347	0.041					1	0.925
T	4.743	0.043						1

Eigenvectors and eigenvalues for correlation matrix above.

	e1	e2	e3	e4	e5	e6	
L	0.3482	0.5303	0.7711	0.0438	0.0294	-0.0131	
B	0.3257	0.7001	-0.6307	0.0261	-0.0135	-0.0711	
H	0.4437	-0.1690	-0.0558	-0.5459	0.3372	0.5998	
U	0.4401	-0.2533	0.0050	-0.4724	-0.3503	-0.6294	
F	0.4341	-0.2885	-0.0548	0.4999	0.6202	-0.3012	
T	0.4401	-0.2294	-0.0379	0.4757	-0.6147	0.3848	
Eig. Vals.	4.571	0.717	0.410	0.168	0.080	0.053	5.999
%	76.2	12.0	6.8	2.8	1.3	0.09	

components analysis would accomplish some of this intent, but without implying a more formal factor analysis or path model. We may then be able to produce a path model for further testing (Voss et al., in press).

If we have stronger ideas about relationships among our measured variables, for example related to ontogenetic or genetic causal mechanisms (Zelditch, 1987 and 1988), then we can fit an a priori factor model, which is relatively easy to visualize using Wright's path modelling. We can estimate the relative influence of the factors on the measured variables and their mutual inter-correlations. In addition we can partition out the uncorrelated residual part of the variables, due to unspecified causes and perhaps to measurement error, which are unique to those variables and not shared through the factors with other variables.

I use Wright's (1921, 1954, and 1968) fowl example, which has been analyzed using PCA and factor analysis in many textbooks and manuals including Bookstein et al. (1985). In the early 1920's the poultry laboratory of the University of Connecticut at Storrs, raised large numbers of individuals of inbred and cross-bred lines of poultry. Dunn (1922) published measurements on several hundred skeletons of these birds. The morphometric variables were length and breadth of the skull, length of the femur, length of the humerus, length of the tibia and length of the ulna. The chickens varied in age from 159 to 2513 days from hatching, but since chickens complete their growth early, there are no detectable age trends in any of the variables.

Wright analyzed the correlation matrix of these 6 measurements for a sample of 276 females from a cross-bred line. I have reanalyzed the raw published data, but have not been able to obtain quite the same correlation matrix as produced in Wright's papers. I attribute this perhaps to different round-off rules in our computations. There are some obvious errors in the published data. Most of these can be found because Dunn also included the ratio percentage,

100 (Skull breadth/Skull length) ,

and a sum of all the length measurements for each bird in his table of measurements. These serve as checks (like parity checks) for the data, and inconsistencies between these indices and those computed from the raw data point to obvious typographic errors. This is a good reason, almost always forgotten in recent publications, to present ratios or redundant information in tables of data. The results of Wright's original analysis are given in Table 6. I give the correlation matrix of the raw data and the covariance matrix of the logged data in Tables 7 and 8 and plotted skull length against femur length in Figure 1.

The variances for the six principal components (eigenvalues) and the eigenvectors or regression coefficients for the variables are given in Tables 7 and 8, for the correlation matrices, and covariance matrices respectively. Note that the number of cases (276) is 46 times the number of variables; a much larger sample size than is usually available in most taxonomic studies.

Wright (1968) comments on the lack of economy of this presentation. Starting with 15 correlation coefficients among the variables and remembering that the variances have been re-

Table 8. Covariance matrix of logs.						
	L	B	H	U	F	T
L	0.001052	0.000594	0.000764	0.000776	0.000767	0.000852
B		0.000978	0.000694	0.000656	0.000674	0.000751
H			0.001440	0.001413	0.001385	0.001444
U				0.001580	0.001452	0.001547
F					0.001728	0.001676
T						0.001898
Eigenvectors and eigenvalues for covariance matrix above.						
	e1	e2	e3	e4	e5	e6
L	0.2733	0.6220	0.7276	-0.0827	0.0448	-0.0114
B	0.2430	0.6837	-0.6806	-0.0307	-0.0029	-0.0969
H	0.4334	-0.0745	-0.0413	0.5454	0.2369	0.6718
U	0.4537	-0.1813	0.0496	0.5255	-0.2499	-0.6483
F	0.4717	-0.2631	-0.0477	-0.4401	0.6813	-0.2194
T	0.5001	-0.1951	-0.0312	-0.4742	-0.6444	0.2661
Eig. Vals.	0.00697	0.00080	0.00042	0.00027	0.00013	0.00008
%	80.3	9.2	4.8	3.2	1.6	1.0

scaled to one in the correlation matrix, there are 6 eigenvalues or variances for principal components, and 36 coefficients (of course, with similar scaling constraints). We have replaced the six 1's on the diagonal and 15 correlation coefficients with 41 numbers, albeit presenting different and interesting aspects of the data. "The analysis does, however, bring out clearly the pattern of relations on focusing on the components with absolute values greater than 0.05. " (from Wright, 1968; note "components" should read coefficients in the terminology used here).

Wright found several more economical and satisfactory ways to express the relationships among the variables in terms of biological models. He was aware that many models were possible, but presented two and preferred the relatively simple one shown in the path diagram below (Figure 9a) which relates to overall growth and relations of parts.

All six variables regress with the same sign on a factor for "general size" which explains most of the variability and covariability. Growth toward a larger size is reflected in the linear measurements. The relative values of the coefficients do measure how closely they are associated with this tendency and the limb bones have larger coefficients than the skull measures. Wright then provided three other uncorrelated factors—a "head" factor, a "wing" factor, and a "leg" factor—and six special factors, one for each original variable. The factor coefficients not included in the path model are set to 0, a priori. The variance not explained by the four factors is left as residual variance for each variable in the special factors. A small amount of residual correlation also remains as a residual correlation matrix. Wright was not fully satisfied by the fit, although he thought this was the

best that could be done for this kind of measurement data. Most of the correlation among the original variables is explained by these factors, and all of the variance. A variation of Wright's ad hoc fitting technique is available in a FORTRAN program in the Appendix of Bookstein et al, 1985.

In terms of a modern factor model we would express the vector of measurements x_i for the ith individual on the 6 variables as:

$$x_{1i} = f_{11}s_{i1} + f_{12}s_{i2} + \mathbf{f_{13}s_{i3}} + \mathbf{f_{14}s_{i4}} + e_{1i}$$

$$x_{2i} = f_{21}s_{i1} + f_{22}s_{i2} + \mathbf{f_{23}s_{i3}} + \mathbf{f_{24}s_{i4}} + e_{2i}$$

$$x_{3i} = f_{31}s_{i1} + \mathbf{f_{32}s_{i2}} + f_{33}s_{i3} + \mathbf{f_{34}s_{i4}} + e_{3i}$$

$$x_{4i} = f_{41}s_{i1} + \mathbf{f_{42}s_{i2}} + f_{43}s_{i3} + \mathbf{f_{44}s_{i4}} + e_{4i}$$

$$x_{5i} = f_{51}s_{i1} + \mathbf{f_{52}s_{i2}} + \mathbf{f_{53}s_{i3}} + f_{54}s_{i4} + e_{5i}$$

$$x_{6i} = f_{61}s_{i1} + \mathbf{f_{62}s_{i2}} + \mathbf{f_{63}s_{i3}} + f_{64}s_{i4} + e_{6i}$$

Note that all of the terms in bold are 0 in Wright's model. The model, like the PCA equations above, is written more compactly in matrix form as:

Table 9. Jackknife results for eigenvectors and eigenvalues of correlation matrix based on original Dunn data.

	e1	e2	e3	e4	e5	e6
L	0.349+.012	0.526+.071	0.773+.044	0.045+.053	0.029+.027	-0.012+.022
B	0.325+.015	0.703+.059	-0.627+.061	0.029+.043	-0.014+.027	-0.071+.019
H	0.444+.004	-0.166+.023	-0.057+.041	-0.541+.043	0.329+.139	0.608+.089
U	0.440+.004	-0.252+.021	0.004+.044	-0.470+.045	-0.331+.124	-0.636+.065
F	0.434+.004	-0.290+.027	-0.057+.054	0.499+.065	0.623+.086	-0.289+.115
T	0.440+.005	-0.232+.022	-0.039+.041	0.466+.066	-0.627+.080	0.370+.136
Cosine of angles between all original and jackknife eigenvectors < 0.9992						
Vals.	4.570+.101	0.715+.066	0.413+.041	0.168+.023	0.079+.011	0.055+.010
%	76.1	12.0	6.9	2.8	1.3	0.09

Bootstrap results for eigenvectors and eigenvalues of correlation matrix based on original Dunn data (based on 1000 samples).

	e1	e2	e3	e4	e5	e6
L	0.347+.012	0.522+.075	0.770+.047	0.046+.053	0.027+.028	-0.014+.021
B	0.323+.015	0.703+.061	-0.620+.065	0.029+.045	-0.015+.028	-0.069+.020
H	0.444+.004	-0.164+.022	-0.057+.043	-0.541+.043	0.353+.134	0.568+.101
U	0.441+.004	-0.248+.021	0.002+.046	-0.470+.045	-0.363+.129	-0.603+.086
F	0.434+.004	-0.286+.028	-0.057+.054	0.499+.065	0.589+.095	-0.314+.119
T	0.441+.005	-0.229+.021	-0.039+.040	0.466+.066	-0.583+.103	0.401+.130
Cosine of angles between all original and bootstrap eigenvectors < 0.9997						
Vals.	4.562+.102	0.727+.066	0.412+.041	0.169+.023	0.079+.011	0.052+.009
%	76.1	12.0	6.9	2.8	1.3	0.09

$$x_i = \mathbf{F}s_i + e_i \quad ,$$

where x_i is the vector of measurements for chicken i, $\mathbf{F}$ is the 6×4 matrix of coefficients relating the factors to the s_i vector of scores on the four factors for chicken i, and e_i is the vector of residuals not explained by the factors for chicken i. The factors, which are usually scaled to have variance one in factor analysis, are uncorrelated with each other and with the residuals, and the residuals are further not correlated with each other (if they were we might want to describe more factors).

Contrast these constraints to PCA, where the residuals are not constrained to be uncorrelated. In factor analysis the goal is to reproduce the correlations as closely as possible with few factors. The factors are not ordered in terms of variance explained as in PCA. Rather the goal is interpretation of "causal" relations among the variables, and additional constraints must be added to arrive at a unique solution. The stringent requirement that the factors be uncorrelated may even be relaxed as the latent variables or "causes" may be reasonably correlated.

Wright's factor model provides both a powerful explanatory model for data and a compact representation. Wright's path model was refit using the program LISREL (LInear Structural RELations; Joreskog and Sorbom, 1985), which allows one to specify models in a path diagram and then translate the path diagram into systems of equations which can be fit to the data in the form of covariance or correlation matrices. Using the Maximum Likelihood method of fitting the model, which makes an assumption of a multivariate normality for the data and which I feel is appropriate here, a chi-square goodness of fit statistic is 11.88 with 3 degrees of freedom and the model fits fairly well. LISREL essentially produced Wright's original results. A fit of the same model to the correlation matrix computed from the published raw data, and to the covariance matrix of the log data, is also presented (Tables 7 and 8). The model is based on a biological hypothesis and fits the data

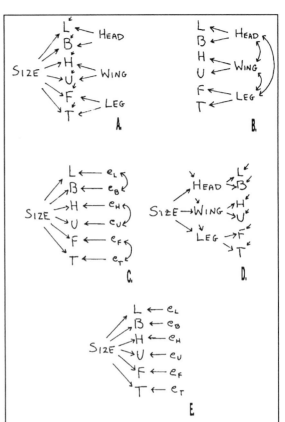

Figure 9. Path diagram fitted by Wright and LISREL. Chi-squared is 11.88 for models a through d, but the degrees of freedom (d.f) differ. All unlabeled terms are for errors: A) Wright's original path diagram with four factors and 3 d.f. freedom; B) an alternative model suggested by Wright and fitted by him, with three correlated factors and 6 d.f.; C) a one factor model with correlated errors and 6 d.f.; D) a two level factor model with 6 d.f.; E) an inadequate model using only one factor with a chi-squared of 131.37 and 9 degrees of freedom.

with an economy of parameters. An alternative model presented in Wright (1968) that fits three correlated factors gives exactly the same chi-square value, but requires three fewer parameters (Figure 9b for path diagram). Two other models, one with correlated residuals, and a two-level factor model (Rindskopf and Rose, 1988; see Figure 9c and d) also fit with the same chi-squared value with 6 degrees of freedom.

Rindskopf and Rose introduce the concept of discriminability to deal with examples like this where several models fit equally well, and point out that this situation can be avoided by careful design. Unfortunately data like the fowl data with only two variables per factor produce models that are not discriminable. More variables would have to be measured per factor to avoid this problem. The choice among the different models is a choice based on one's biological understanding of the problem and data.

Zelditch (1987, 1988) has recently applied these techniques to data on a series of measurements of rat skeletons taken from Olson and Miller (1958). A careful consideration of growth and ontogeny in relation to the measured variables is required to apply such models.

The LISREL program includes least-squares techniques as well as maximum likelihood methods for fitting models. The maximum likelihood technique provides two goodness-of-fit measures and an asymptotic chi-squared goodness-of-fit test which provides an approximately correct probability level for large data sets from a multivariate normal distribution. A hierarchical set of models shows how much better a less specified model fits relative to a more specified model. For example, a single factor fit to the fowl data gives a chi-squared value of 131.37 with 9 degrees of freedom. The improvement of fit can then be estimated by the change in chi-square, which is also a chi-square with degrees of freedom equal to the difference in degrees of freedom between the two models. In this case, adding Wright's three skull, femur and tibia factors reduces chi-squared value to 11.88 with 3 degrees of freedom, a highly significant change of chi-squared equal to 119.49 with 6 degrees of freedom.

Confirmatory factor analysis through LISREL is a tool for modeling relationships between variables based on a developmental or function model. Attempts to fit models to the *Zygodontomys* data have not been successful. We had hoped to see if a single model applies over the range of the species. We anticipate that this technique, richer than principal components or exploratory factor analysis, may allow us to identify more than one factor even though our sample sizes are far smaller than those used in the social sciences. The LISREL program is powerful, but is not easy to use. Very large data sets, of the order of the number of fowl specimens, are required to get stable solutions.

It would be inappropriate to look at the Mediterranean bird data using factor analysis as we cannot provide a probability model for the data, and all relevant species are in the data set.

Multivariate Analysis of More Than One Sample

The complete *Zygodontomys* data set consists of samples from fifteen geographic localities. The bird data set has species from different geographic realms and habitats and may be divided into subsets according to a number of criteria (Blondel et al. 1984 and Figure 8). For *Zygodontomys* we want to compare samples of organisms from the localities. For the birds we want to compare different subgroups of taxa.

In the bird example, since the data analyzed is based on means or centroids for species the problem is inherently a multiple sample problem. However, data for variation within species were unavailable, and the questions being considered were at a higher level—the comparison of different faunal realms. We wanted to "test" whether the collective morphologies of the birds that live in a similar climatic regime were similar when collected from widely separated geographic regions.

In the *Zygodontomys* data we have data from a large part of the range of the taxon (Figure 3). Some are mainland and some are insular. We could summarize the data as was done with the birds, e.g., compute centroids for each locality and then do principal components on these centroids to summarize and display our data in reduced dimensions. Or we could combine all of our data over localities in one grand matrix including all specimens and do a principal components on this array "to see what happens" (Figure 10). There are major problems with both of these approaches. For the rodents, the variability among individuals within samples is a large part of the total variation among samples. In addition, the precision in estimating the centroids is a function of locality sample size and covariance structure within each population. If samples sizes are different, precision differs. We would also ignore patterns of variation and covariation within samples in our evaluation of the relations among samples if we had only done a principal component analysis on the centroids. Both within- and among-sample variation and covariation patterns must be considered.

An analysis of the total variance-covariance matrix confounds within- and among-sample variability, especially in the interpretation of coefficients in principal components, and we have no idea of the relative contributions of among-sample and within-sample variability to the ordinations we see when we plot principal component scores. I have seen studies where this approach was taken and the plots display no interpretable structure. There is too much overlap of samples and too many points in the plots of the hundreds of cases. If the groups are very different, far apart in the sense of gaps between samples or groups in ordinations, we may then get "useful?" displays. However, the loadings may only give us a clear indication of the variables contributing to the separations if the differences are parallel to one or more of the PC axes. This type of approach too often leads to the following statement: "The pattern I thought (or knew) was there, is!", though we

may not see it expressed quite that way in print. This is sometimes a marvelous way to learn that the methodology "does work" in capturing and summarizing distances among specimens, and this does have pedagogical value. We need to do somewhat better in research.

I frequently hear the point raised that principal component analyses are more objective as we

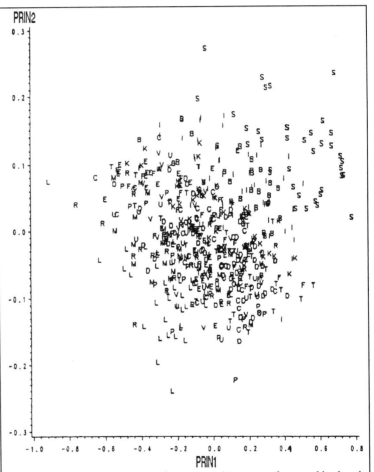

Figure 10. A plot of principal component one and two scores for a combined analysis of all 574 specimens from 15 localities. This type of analysis is not recommended. Each plotting symbol is the first letter of a locality name. Note that some of the bigger separations, such as for the island population from San Miguel are indicated, and one may see other distinctions, summarized better by canonical variates and Mahalanobis D^2, by careful examination of this plot and additional principal component scores plots.

have not prejudiced our analyses with the subjective hypotheses that use of a priori groups implies. In one way or another, we always use what we know and I am advocating using such a priori knowledge about group or locality membership when we can. There are genuine circumstances when we cannot separate mixed samples and do not have a clear understanding of a priori groups. In that case principal component analysis as an ordination may be informative and other techniques such as cluster analysis should be explored.

The bird data consisting of species centroids include a wide array of sizes and types of birds from several orders (Figures 2 and 8). The variation is presumed small within-species relative to that among higher taxa. We are ignoring the within-species variance and covariance. Geographic variation is also likely to be small relative to among-species differences. Sexual dimorphism is not relevant for the bird data as only males were measured.

Mahalanobis Distance and Canonical Variates Analysis

Multivariate distance measures are available to compare centroids that take into consideration the variance and covariance of variables within a priori designated subsets of the data, such as samples from different localities as in the *Zygodontomys* data. Mahalanobis distance or Mahalanobis distance squared, D or D^2, use the within scatter for calibration. This can be visualized in two dimensions and generalized to more. Differences among centroids are weighed more heavily in the directions along minor PC axes within samples than differences among centroids in the direction along the major PC axes within samples. Correlation within samples is taken into account.

Graphical demonstrations of the required transformations are given in two useful papers by Rempe and Weber (1972) and Campbell and Atchley (1981). A brief description of the analysis is given in Reyment (this volume) and here.

The data are pooled within samples to estimate a common within or pooled sample covariance matrix and its principal components. All of the data are rotated and represented relative to the within principal axes, with the coordinates arbitrarily centered at the centroid of all of the data. Each principal axis is then put on the same scale by dividing through by the within standard deviation along its axis. This standard deviation is the square root of the eigenvalue of the pooled-within principal component corresponding to that axis. This re-scaling of axes moves the original coordinate axes so they are no longer at right angles, which distorts distances between data points. The pooled sample covariance matrix can be used to estimate the contours of the average concentration ellipse or ellipsoid for each sample. This rotation and re-scaling transforms the ellipse or ellipsoid to a circle or spheroid.

Mahalanobis distance is then the Euclidean

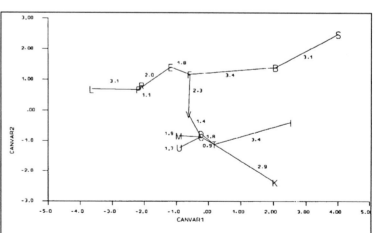

Figure 11. Plot of canonical variate one and two means for each locality for the *Zygodontomys* data. A minimum spanning tree connects the closest means in 12 space and Mahalanobis D^2 is given for each link of the tree. Note the crossing and distortions in this two-dimensional figure. Slight distortion is also introduced by not making the scales quite the same.

distance between sample centroids in this transformed data space. It may be measured with a ruler, when there are only two variables, or calculated using the Pythagorean theorem for any number of variables or dimensions. Visualization is not possible past three variables. What is usually published or displayed is that two- or three-dimensional projection which collectively maximizes the distances among the centroids (Figure 11 for *Zygodontomys*). This corresponds to doing a principal components analysis of the centroids (weighted by their sample sizes), which have been rotated and rescaled as described above. The new variables are called canonical variates which are plotted on canonical axes.

Many program packages provide canonical variate analysis as a multi-sample or multi-group ordination technique. It is important to notice that canonical variate analysis involves the among-centroids sums of squares and cross-products matrix as well as the pooled-within matrix. The among matrix is like the among sums of squares in an analysis of variance, where the contribution of each mean to the sum of squares is multiplied by its sample size. Gower (1966b) has suggested that, in many biological analyses, sample size differences are not due to design but rather to sampling variability alone and that an ordination of un-weighted centroids is more informative. When the sample sizes are equal the results are the same. The ordination of centroids can be done through a principal components analysis of the variance-covariance matrix of the unweighted canonical variate centroids, or from the matrix of Mahalanobis D^2 using a principal coordinates analysis. The results are the same.

I have introduced canonical variate analysis from the viewpoint of inter-centroid distances, but it can also be discussed from the goal of displaying data in a reduced space of uncorrelated variables which maximally separate the samples or taxa, relative to the within variation and covariation. These variates are the canonical variates and the Mahalanobis distances are Euclidean distances in the space of these variates.

Canonical variates, like principal components, are linear functions of the original variables weighted by coefficients. There is a strong temptation to interpret these coefficients as one interprets principal component coefficients or loadings. This is more difficult, more akin to, but more complicated than interpreting principal component coefficients from a covariance matrix where the different variables may have markedly different variances. While the vectors of coefficients in principal components analysis are orthogonal, this is no longer true for canonical variate coefficients, even though the canonical variates themselves are uncorrelated.

In order to make the canonical variate coefficients more interpretable, they are usually standardized by multiplying them by the pooled-within standard deviation of their respective variable. This renders them unit free. Another transformation frequently used for interpretation is the correlation coefficient between the canonical variate and the original variate. This is analogous to a "loading" in principal component or factor analysis. Rencher (1988) has recently shown that such correlations are merely functions of the univariate F's for the separate variables, and therefore the correlations add no additional multivariate information for interpretation of the coefficients. Rencher's result is only given for one of the three possible ways of computing these correlations, that based on the pooled-within sample variability. The second is based on the total variability, and a third uses the among variability to transform the canonical variate coefficient into a correlation coefficient. I suspect that Rencher's result holds for all three choices. Nevertheless, SAS presents the user with all three of these uninformative scalings in its canonical variate procedure CANDISC.

Bargmann (1970) has gone so far as to say that we should not attempt to interpret the canonical coefficients. The purpose of canonical variate analysis is separation of populations in terms of within-group variability, not interpretation of coefficients in a causal, structural or path framework. This argument harkens to my earlier criticism of principal components relative to path modeling or confirmatory factor analysis. In other words, if our goal is to find interpretable models that explain group differences as a function of the within- and among- group covariance patterns, then we should construct our analyses with this in mind. Bookstein et al. (1985) and Rohlf and Bookstein (1987) have discussed this problem and offer the ad hoc method of shear, and some others, as ways of taking into consideration within structure when a major part of the pooled-within variance-covariance structure is summarized by a "size" factor. Computational schemes to estimate multiple group path or factor models are not currently available, except in some special forms as discussed in Zelditch (1988) and Joreskog and Sorbom (1985).

When the desired result is data reduction, e.g., finding a reduced set of variables for expressing the relationships among samples, or discrimination (see below) then stepwise procedures may be more appropriate. However, stepwise procedures do not guarantee an optimal subset of the variables, and too many researchers interpret the order of the variables entering the analysis in terms of their relative importance. This is to be avoided, and alternative variable selection methods are to be recommended (Hocking, 1983).

Discriminant Analysis

When the purpose is identification or assignment of specimens of unknown affinity to a priori groups, then classical discriminant analysis corresponds to computing the Mahalanobis distance from our unknown to each of the a priori centroids and assigning the unknown to the group with the smallest distance. If we feel justified in pooling the variance-covariance matrices over samples, we can use the pooled data within to compute our distances. Even if there is heterogeneity in variance-covariance structure within groups we can

Table 10. Number of observations classified into LOC: based on pooled variance-covariance matrices.

	Bus	Cer	Div	ElF	Fvu	IsC	Kay	Lim	MoB	Par	RiC	SMi	Tuk	Ura	Val	Tot
BushBush	28	0	0	0	2	0	0	0	0	0	0	0	0	0	0	30
CerroAzu	2	8	2	1	3	0	1	0	5	1	1	0	5	5	4	38
Dividive	0	2	39	3	5	1	4	0	2	2	0	0	3	7	0	68
ElFrio	0	1	2	31	4	0	0	1	3	3	1	0	0	0	2	48
Fvuelta	1	0	1	5	18	0	0	0	1	2	1	0	0	0	1	30
IslaCeba	0	0	0	0	0	50	0	0	0	0	0	0	1	0	0	51
Kayserbe	1	0	1	0	0	0	40	0	0	0	0	0	1	0	0	43
Limao	0	0	0	2	1	0	0	30	1	0	2	0	0	0	1	37
MontijoB	0	5	1	0	0	0	0	0	25	0	1	0	1	0	0	33
Parcela2	0	0	3	3	2	0	0	1	0	12	4	0	0	2	2	29
RioChico	0	0	2	2	0	0	0	2	0	6	16	0	0	2	3	33
SanMigue	5	0	0	0	0	0	0	0	0	0	0	34	0	0	0	39
Tukuko	1	3	3	0	1	0	3	0	6	0	0	1	13	4	1	36
Urama	0	3	3	0	0	0	0	1	2	2	1	0	2	13	3	30
Valledup	0	2	0	2	3	0	0	0	1	1	1	0	1	7	11	29
Total	38	24	57	49	39	51	48	35	46	29	28	35	27	40	28	574

Number of observations classified into LOC: based on locality variance-covariance matrices.

	Bus	Cer	Div	ElF	Fvu	IsC	Kay	Lim	MoB	Par	RiC	SMi	Tuk	Ura	Val	Tot
BushBush	29	0	0	0	1	0	0	0	0	0	0	0	0	0	0	30
CerroAzu	1	24	2	0	2	0	0	0	2	0	0	0	1	4	2	38
Dividive	0	4	49	1	1	1	1	0	1	0	0	0	4	3	3	68
ElFrio	0	0	0	42	2	0	0	0	1	0	2	0	0	0	1	48
FvueltaL	0	1	0	0	26	0	0	0	0	2	0	0	0	0	1	30
IslaCeba	0	0	0	0	0	50	0	0	0	0	0	0	1	0	0	51
Kayserbe	0	1	0	0	0	0	42	0	0	0	0	0	0	0	0	43
Limao	0	0	0	1	1	0	0	32	0	2	0	0	0	0	1	37
MontijoB	0	0	0	0	0	0	0	0	31	0	0	0	1	0	1	33
Parcela2	0	0	1	0	2	0	0	1	0	20	2	0	0	1	2	29
RioChico	0	0	2	1	0	0	0	0	0	1	26	0	0	1	2	33
SanMigue	0	0	0	0	0	0	0	0	0	0	0	39	0	0	0	39
Tukuko	0	2	2	0	1	0	0	0	2	0	0	0	25	3	1	36
Urama	0	1	0	0	1	0	0	0	1	2	1	0	3	21	0	30
Valledup	0	2	0	1	3	0	0	0	1	0	0	0	1	0	21	29
Total	30	35	56	46	40	51	43	33	39	27	31	39	36	33	35	574

compute the Mahalanobis distance of the unknown to each centroid, using the variance-covariance matrix for that group, and assign the unknown to the closest group, or make no assignment at all if it is too far from any one of them. These rules derived from a multivariate normal model also involve the determinant of the variance-covariance matrix when these are presumed different over samples (Hand, 1981 and SAS Statistics Manual).

The procedure which pools the within-sample variance-covariance matrices, assumed or judged to be homogeneous, may be summarized in the form of the few canonical variates as displays which depict the best low-dimensional plots of the overall discriminating power. However, the less important canonical variates may have considerable discriminating power for some groups, and multiple discriminant analysis is to be preferred for identification. Sometimes hybrid terminology is used for these methods, e.g., "canonical discriminant analysis" in PROC CANDISC of SAS, but I feel that this jargon adds confusion to an already complicated subject (Neff and Marcus, 1980). Too often canonical variates analysis, really an ordination procedure, is treated as a discrimination procedure. The confusion arises because the mathematical treatment is essentially the same.

The use of Mahalanobis distances as descriptive statistics, canonical variates for ordination, Multivariate Analysis of Variance for hypothesis testing, and classical discriminant analysis for assignment of unknowns, involves a major constraint and a difficult to validate assumption, namely, the homogeneity of the within-covariance matrices used to estimate the common ellipsoid of concentration. If there are sizable differences among samples in covariance structure, I know of no way to display data in reduced dimensions taking into consideration simultaneously the within and among variability and co-variability structure. However, different orientation of the ellipsoids, that is different correlation structure, is worse than differential inflation (Dempster, 1969).

The appropriateness of pooling can be determined by comparing the covariance matrices. For our *Zygodontomys* data, for example, the first principal axes all of the samples point in directions that differ by no more than 15 degrees from each other (Voss et al., in press). The major component of variation is pretty much in the same direction for all of our samples. However, Bartlett's test for equality of covariance matrices indicates that the localities are heterogeneous, apparently differing largely in inflation. Furthermore, discrimination is improved for some localities by using the separate covariance matrices for assignment of the original samples to their respective populations (Table 10). However, we obtain nearly as good discrimination, and useful ordinations as well, using Mahalanobis distance statistics and ignoring the heterogeneity.

Acceptable distance statistics have not been defined for the case of heterogeneous covariance matrices, except for some special cases when two samples are compared, or when the covariance matrices differ mainly in terms of inflation (Dempster, op. cit.). Many suggestions have been made over the years for such measures, but there is no good solution to this problem that I am aware of, and a matrix of distance statistics, one for each between group difference, does not suffice to describe the distances among samples or taxa.

There are good alternatives to multivariate-normal-based discrimination in the identification or assignment problem. If it is unreasonable to accept normality assumptions, or when covariance matrices may not be sufficient statistics to represent the probability distributions, then there is a rich set of available procedures. More-or-less ellipsoidal data with not too dissimilar covariance structures are already taken care of by looking at distances from individuals to all group centroids as mentioned above. If the differences are large enough, the classical methods may be found to be relatively robust. In any case, if a large enough set of test unknowns (samples whose identity is known, but is submitted to the procedures as unknown) gives correct identification, don't worry too much about the underlying assumptions.

On the other hand, where the data structures are non-linear, or populations exist in several interspersed clumps, then the methods derived from normal theory cannot work well. A sensible nonparametric solution has been offered based on relative frequency of nearest neighbors in the measurement space. It is implemented in SAS, but I am unaware of any published morphometric applications.

Another more powerful method is that of empirical discriminant functions. This amounts to smoothing the data in multivariate space to estimate empirically the multivariate probability density functions for each a priori group. An unknown is assigned to the population with highest density at its location, or to none at all if all densities are too small. While these methods have been applied to geological data (Howarth, 1971), I am unaware of published morphometric applications. See Hand (1981) for a discussion of this methodology and further references. The implementation of SAS (Version 6.03 for the PC) and accompanying Statistics manual give an extensive discussion of these techniques and apply them to data for three samples of iris specimens using four variables. There are the data which R. A. Fisher (1936) used when he invented classical discriminant analysis. These data are widely reproduced in many texts. They are not as non-normal nor heterogeneous as the data described in Howarth (1971); and not a good example to test this methodology.

Some Useful Graphical Techniques

A minimum spanning tree provides an especially useful supplement to canonical variates and distance displays as well as to displays of principal component scores in two or three space. It provides a way of depicting distortions in the low-dimension display relative to the entire set of canonical variates, distances or principal components. The minimum spanning tree is the non-rooted connection among the centroids which has minimum length. If the minimum spanning tree is superimposed on a two- or three-dimensional display, then each centroid is connected to its nearest neighbor in the space of all of the variables (12 space for the *Zygodontomys* example, and 8 for the birds). Centroids that are far apart in the high-dimensional space may appear close together in the low-dimensional graph as the display is a projection, and the minimum spanning tree will demonstrate some of these discrepancies (Figure 11 for *Zygodontomys*). If the canonical variates are all scaled the same in a plot, and the links of the tree labeled with the numerical D values in the full space, then comparing the actual distances on the two-dimensional graph to the multidimensional D values can give an idea of the magnitude of the distortions. One can also compute and tabulate residual distances not displayed and these will be informative as to what is not summarized by the graphs. This type of residual analysis should be routine, but is seldom done, though published displays including minimum spanning trees are quite common.

Inference and Confirmation

Multivariate Analysis of Variance

When the data reasonably can be assumed to come from multivariate normal distributions with homogeneous variance-covariance matrices, there are tests of hypotheses for equality of group centroids, and tests of linear contrasts among groups. Also confidence ellipsoids may be found for centroids and functions of centroids. In other words, analysis of variance can be generalized as multivariate analysis of variance or MANOVA. The multivariate generalization of the one and two sample t-tests are Hotelling's T^2 tests of the hypotheses that the sample centroid equals a known centroid and equality of centroids respectively. T^2, multiplied by a function of sample size(s) and number of variables, has Fisher's F-distribution when the Null Hypothesis is true. However, there is no universally accepted MANOVA test statistic for more than 2 groups corresponding to F in ANOVA. The several alternative tests proposed (for example SAS GLM offers four different test statistics) are powerful against different kinds of alternatives (see for example Harris, 1975, and Morrison, 1976). All of these statistics may be stated in terms of canonical variates and Mahalanobis D^2, which like, tests in ANOVA, may be related to various biological hypothesis of interest.

Multiple comparisons and contrasts available for ANOVA are more complicated for MANOVA since, in addition to comparison of centroids two or more at a time, one may also consider the many possible combinations of variables for finding homogeneous subsets of populations for different subsets of variables. This problem is addressed in some of the general multivariate texts (for example Morrison, 1976; Harris, 1975, and Johnson and Wichern, 1982).

The test that the population Mahalanobis Δ^2 (the population parameter estimated by D^2) equals 0 is equivalent to testing equality of centroids for two groups using the Hotelling T^2 test. Note that Mahalanobis D^2 is a biased estimate of Δ^2 (Sjovold, 1975). An approximate confidence interval algorithm for Δ^2 is given by Bargmann (1970) and is given here as a

Table 11. Mahalanobis D^2, correction for bias, F test and confidence bounds for *Zygodontomys* data.

Values above diagonal are D^2, those below unbiased D^2.

	Bus	Cer	Div	ElF	Fvu	Is C	Kay	Lim	MoB	Par	RiC	Smi	Tuk	Ura	Val
Bus	0.00	13.7	16.1	17.7	11.3	16.6	17.4	38.5	22.7	22.1	24.8	9.62	13.9	21.7	13.4
Cer	12.7	0.00	3.37	8.77	6.96	12.8	10.8	16.4	3.76	8.37	8.56	29.8	.826	2.77	2.10
Div	15.2	2.80	0.00	8.32	5.32	15.1	11.8	21.5	8.85	6.42	8.71	31.3	3.24	5.44	5.31
ElF	16.5	7.85	7.55	0.00	3.34	20.9	29.0	11.8	9.62	4.13	6.05	31.4	10.9	12.6	6.56
Fvu	10.4	6.23	4.77	2.61	0.00	21.0	21.2	14.2	11.9	4.16	7.49	24.6	7.68	10.5	5.18
Is C	15.5	11.9	14.3	19.8	20.1	0.00	17.1	47.5	15.5	29.4	28.4	16.7	11.4	18.2	15.8
Kay	16.3	9.95	11.1	27.7	20.2	16.2	0.00	45.5	20.6	30.9	34.1	30.4	8.56	17.9	16.2
Lim	36.9	15.3	20.5	10.8	13.3	45.9	43.8	0.00	13.5	9.34	10.9	64.1	20.0	18.3	13.5
MoB	21.4	3.02	8.13	8.66	11.1	14.5	19.5	12.6	0.00	11.8	11.1	40.1	5.38	7.89	6.87
Par	20.8	7.50	5.74	3.27	3.45	28.1	29.5	8.43	10.8	0.00	1.28	42.1	11.1	8.50	5.75
RiC	23.4	7.63	7.92	5.09	6.65	27.1	32.6	9.90	10.1	.476	0.00	44.7	12.2	7.22	5.62
SMi	8.64	28.5	30.0	29.9	23.5	15.7	29.0	61.9	38.5	40.4	42.8	0.00	28.2	40.7	30.2
Tuk	12.9	.183	2.68	9.98	6.94	10.6	7.78	19.0	4.61	10.2	11.2	26.9	0.00	3.49	3.26
Ura	20.4	1.99	4.74	11.5	9.61	17.1	16.8	17.1	6.97	7.54	6.24	39.0	2.70	0.00	3.48
Val	12.3	1.32	4.60	5.59	4.39	14.8	15.2	12.5	5.96	4.84	4.66	28.7	2.47	2.59	0.00

Values above diagonal are F values, below are probabilities.

	Bus	Cer	Div	ElF	Fvu	Is C	Kay	Lim	MoB	Par	RiC	SMi	Tuk	Ura	Val
Bus	1.00	18.8	27.4	21.7	17.0	25.6	25.1	52.1	30.3	28.3	29.9	12.4	19.3	26.6	16.1
Cer	0.00	1.00	6.72	12.0	12.1	22.8	17.8	25.0	5.68	12.1	11.5	43.0	1.30	3.80	2.82
Div	0.00	0.00	1.00	14.2	12.2	35.9	25.5	42.1	17.0	11.7	14.5	56.8	6.56	9.26	8.83
ElF	0.00	0.00	0.00	1.00	5.04	32.3	41.9	16.0	12.9	5.30	7.29	40.3	15.2	15.5	7.90
Fvu	0.00	0.00	0.00	0.00	1.00	42.5	39.3	24.2	20.0	6.64	11.1	39.4	13.5	15.8	7.64
Is C	0.00	0.00	0.00	0.00	0.00	1.00	32.7	83.3	26.7	48.1	42.9	27.4	20.7	28.1	23.9
Kay	0.00	0.00	0.00	0.00	0.00	0.00	1.00	73.9	32.9	47.1	48.2	46.3	14.3	25.8	23.0
Lim	0.00	0.00	0.00	0.00	0.00	0.00	0.00	1.00	20.2	13.3	14.5	91.4	31.1	24.7	18.0
MoB	0.00	0.00	0.00	0.00	0.00	0.00	0.00	0.00	1.00	16.5	14.5	56.4	8.23	10.5	9.01
Par	0.00	0.00	0.00	0.00	0.00	0.00	0.00	0.00	0.00	1.00	1.62	56.7	16.2	10.9	7.25
RiC	0.00	0.00	0.00	0.00	0.00	0.00	0.00	0.00	0.00	.083	1.00	56.3	16.6	8.70	6.66
SMi	0.00	0.00	0.00	0.00	0.00	0.00	0.00	0.00	0.00	0.00	0.00	1.00	41.2	52.3	38.1
Tuk	0.00	0.22	0.00	0.00	0.00	0.00	0.00	0.00	0.00	0.00	0.00	0.00	1.00	4.84	4.43
Ura	0.00	0.00	0.00	0.00	0.00	0.00	0.00	0.00	0.00	0.00	0.00	0.00	0.00	1.00	4.19
Val	0.00	.001	0.00	0.00	0.00	0.00	0.00	0.00	0.00	0.00	0.00	0.00	0.00	0.00	1.00

Values above diagonal are upper confidence bounds, below lower bounds for 99% confidence.

	Bus	Cer	Div	ElF	Fvu	Is C	Kay	Lim	MoB	Par	RiC	SMi	Tuk	Ura	Val
Bus	0.00	22.0	23.6	29.2	17.6	25.2	27.1	60.2	36.1	35.6	40.7	16.0	22.2	35.6	22.4
Cer	6.67	0.00	5.32	14.4	10.9	19.2	16.7	25.3	6.34	13.6	14.1	46.1	1.45	4.90	3.78
Div	9.26	1.19	0.00	12.6	7.93	21.1	17.0	30.7	13.1	9.79	13.2	44.6	5.11	8.46	8.30
ElF	8.50	3.75	4.13	0.00	5.63	31.6	44.6	19.1	15.8	7.24	10.6	50.2	17.7	21.1	11.4
Fvu	5.58	3.18	2.60	.839	0.00	30.3	31.1	21.4	18.2	6.82	12.0	37.0	11.9	16.4	8.49
Is C	9.00	6.97	9.32	11.7	12.8	0.00	25.1	69.1	23.2	43.6	42.7	25.2	17.2	27.6	24.2
Kay	9.16	5.46	6.89	16.3	12.5	10.0	0.00	67.6	31.2	46.9	52.4	46.1	13.3	27.9	25.5
Lim	21.5	8.67	13.2	5.50	7.71	29.6	27.4	0.00	21.3	15.1	17.8	98.3	30.7	29.1	21.9
MoB	11.9	1.05	4.68	4.18	6.24	8.61	11.5	6.83	0.00	18.9	18.2	62.1	8.82	13.1	11.5
Par	11.3	3.64	3.01	1.01	1.37	17.4	17.8	4.18	5.57	0.00	2.38	66.0	17.7	14.3	9.92
RiC	12.5	3.59	4.34	1.91	3.22	16.4	19.4	4.91	4.98	-.28	0.00	71.3	19.7	12.5	9.91
SMi	4.06	16.7	19.6	16.8	14.2	9.26	17.5	37.6	22.8	23.5	24.4	0.00	43.5	64.8	48.6
Tuk	6.86	-.26	1.13	5.07	3.66	6.13	4.10	11.0	1.99	5.31	5.77	15.8	0.00	6.04	5.70
Ura	10.8	.416	2.29	5.61	5.10	10.0	9.46	9.35	3.18	3.42	2.54	22.3	.797	0.00	6.30
Val	5.97	.080	2.18	2.18	1.85	8.46	8.38	6.47	2.55	1.85	1.64	16.0	.653	.586	0.00

PROC IML routine for SAS. The program also corrects D^2 for bias, and gives the F statistic for the test of equality between centroids, and the corresponding probability of exceeding the observed F. The results for a run of this program for the *Zygodontomys* data are given in Table 11.

Asymptotic Tests

All of the tests of hypotheses in MANOVA (except for two groups and a few other special cases) and all, for principal components as well, even when multivariate normal assumptions are plausible, are asymptotic. That is, the tests' reported significance levels are only correct in the limit (as the sample sizes approach infinity). Thus, the significance levels are essentially correct only for very large samples.

Morrison (1976) gives asymptotic standard errors for eigenvalues and eigenvectors from which one can find confidence intervals and ellipsoids for large samples using standard normal tables. Maximum likelihood factor analysis produces goodness-of-fit tests that are asymptotically chi-squared tests for path models. They give approximately correct significance levels for large samples, again if the variables have a multivariate normal distribution. For smaller samples, the chi-squared statistic can serve as a goodness-of-fit criterion, whose probability level is uncertain, though other goodness-of-fit criteria may be more satisfactory for evaluation of the model.

Thus, when our data are samples from multivariate normal distributions there are a few special exact tests (most are not useful) such as T^2, and all others are approximate or asymptotic tests. Other approximate or asymptotic tests are not difficult to formulate; for example Reyment (1969) has developed a test for the equality of two eigenvectors from different data sets.

One of the main problems with multivariate analysis in morphometrics, especially when we have measured many variables, is that we seldom have enough data to adequately assess the form of the distribution that generated our sample. We know

that we have little power for detecting non-normality. Or even if multivariate assumptions are correct and our data is homogeneous with respect to age, sex and other sources of differences, it is difficult to verify the correctness of the distributional assumptions.

There are other problems with some of the common tests available. For example, Van Valen (1978) pointed out that Bartlett's test for equality of covariance matrices is as sensitive to non-normality as to heterogeneity, the thing it is suppose to be testing (the SAS Manual does point out this problem). Therefore, Van Valen recommends that this test not be used.

Re-Sampling Principal Components - the Jackknife and the Bootstrap

The problem of distribution assumptions, especially for small samples, has been addressed by a class of procedures called re-sampling schemes. The data themselves are repeatedly sampled to generate standard errors and estimate probabilities for confidence intervals, or to test hypotheses. Re-sampling schemes require intensive computation and only became a practical choice with the wide availability of fast, relatively cheap-to-use computers.

Two recently popular re-sampling techniques are the Jackknife and the Bootstrap. Gibson et al (1984) applied the Jackknife to principal component analyses of morphometric data on samples of common myna birds from 11 recently introduced populations in Hawaii, Fiji, Australia and New Zealand. Fourteen skeletal variables were measured on sample sizes varying from 17 to 50 birds of the same sex. Their results and a similar analysis of the *Zygodontomys* data are discussed below. Chatterjee (1984) presents the results of a Bootstrap applied to principal components in the social sciences.

It should be pointed out that all of these results, and the new ones given below, apply to "homogeneous" data sets, such as the fowl example, the *Zygodontomys* data, and the data studied by

Gibson et al. Samples of mixed taxa may behave quite differently.

Gibson et al (1984) estimated sample statistics using the Jackknife technique, where eigenvalues and eigenvectors in a principal component analysis are computed for all possible sub-samples generated by leaving one specimen out in each repeated computation. For a sample of 30 individuals one does a principal component analysis, i.e., computes the eigenvalues and eigenvectors, 30 times; once for each sub-sample of 29. Gibson et al. (1984) following Mosteller and Tukey (1977) show how to combine the statistics from the analysis of each sub-sample into a single estimate of the population parameters, each population eigenvalue and eigenvector element in this example; and how to find the standard error for all of these statistics. One computes the mean of the estimates from the sub-samples and applies a correction for bias (an equivalent but slightly modified formula from Gibson et al, 1984; and closer to the presentation in Efron, 1982) as follows:

Jackknife estimate = mean of sub-sample estimates + bias correction .

The bias correction for a total sample of n is:

Bias = n × (total sample estimate-mean of sub-sample estimates) .

The standard error is computed by calculating the standard deviation of the sub-sample estimates over the n sub-samples and then multiplying by (n-1) and dividing by the square root of n (in my reformulation).

There is a broader class of Jackknife techniques which consider all samples of k out of n. This requires yet more computation. These variations have not been used much. In my discussion k will always be n-1.

The Bootstrap procedure samples n out of n with replacement. For a sample of thirty individuals, each replicate sample selects 30 of the original observations with replacement. For $n = 30$ individuals there are 30^{30} patterns. One only need

generate a relatively small number of these possible patterns to reliably estimate a parameter and its standard error. The Bootstrap estimate and its standard error are just the mean and standard deviation of the relevant statistic over the replicated samples. One can get a good approximation to a 95% confidence interval using 1000 replicates (several techniques are given in Efron, 1982). For example, the 2.5 and 97.5 percentiles of the Bootstrap distribution provide one such set of estimates for the confidence limits. For 1000 replicates, the Bootstrap requires computation of the covariance matrix and its eigenvalues and eigenvectors 1000 times. Then the mean and standard deviation for all of these statistics are computed over the 1000 replicates. In order to estimate confidence limits from the percentiles, each of the $p(p+1)$ distributions of 1000 eigenvalues and eigenvector coefficients would have to be partially sorted to find the appropriate percentiles. This latter step was not done. All Bootstrap results reported here used 1000 replicates. Perhaps 100-200 replications would be sufficient to give an adequate estimate of a parameter and its standard error. Diaconis and Efron (1983) provide a clear description of the Bootstrap with applications and Efron (1982) has written a monograph on re-sampling which discusses both the Jackknife and the Bootstrap, as well as other re-sampling schemes.

I have run both the Jackknife and Bootstrap on all 276 observations of the fowl data, and then for various sized random samples to get an idea of the stability of the eigenvalues and eigenvectors as a function of sample size when sampling from a population; 276 is a vastly larger sample size than is usually available in most systematic studies. However, small random samples of 8, 15, 30 and so on can give some idea of what to expect for small samples of data as well-structured as the fowl data, as we observe more and more individuals.

I present the results for both the Jackknife and Bootstrap. Programs written in SAS PROC IML that generated these results are given in the software on one of the provided disks. This work was done on a large main-frame computer. It would be time-consuming on a PC, even on a fast one. The programs perform moderately well on the newest fast PC's. A compiled program written directly in a programming language such as Basic, Pascal, Fortran or C would be more efficient on a slower PC.

Asymptotic theory was also applied to the same data to see how well it performed compared to the Jackknife and Bootstrap. The asymptotic results are given first (Table 7). Calculations were done, using a PC SAS Version 6.03 PROC IML program, based on the formulae given in Morrison (1976). The asymptotic standard errors for the eigenvalues and eigenvectors are easily and rapidly obtained. They may give an adequate indication of the standard errors with far less computation time. The asymptotic standard error of an eigenvalue L is just $L (2/n)^{0.5}$ and the estimates of the eigenvalues are independently distributed (Morrison, 1976) if the population eigenvalues are different. An approximate standard error for the log of an eigenvalue is just $(2/(n-1))^{0.5}$ and is the same for all eigenvalues (Jolliffe, 1986). The formula for the asymptotic standard errors for the eigenvector elements is more complicated and involves all of the eigenvalues and eigenvectors. The eigenvectors are uncorrelated with the eigenvalues, but the elements of an eigenvector are correlated (formula (4) Section 8.7, Morrison, 1976). Eigenvector estimates for different principal components are also correlated with each other. However, only the standard errors for individual coefficients are given here in order to compare the results with the Jackknife and Bootstrap.

I follow Gibson et al. (1984) in studying the re-sampled statistic divided by its standard error, here designated T. T is used rather than the identically defined t they describe, so as not to confuse it with Student's t (see below). This provides a quick indication of the possible significance of our Jackknife estimate, for example, to test that an eigenvector element = 0. However,

for the Jackknife the distributions are highly skewed, more for the eigenvalues than the eigenvectors. As pointed out by Efron (1982), these T values will therefore not have a student's t-distribution as suggested by Mosteller and Tukey (1977). The distribution of the eigenvalue estimates using the Jackknife are a little less skewed when they are log transformed. If we construct t-like confidence intervals or use T in a student's t-test, the significance levels will be very unreliable. I use instead a very conservative probability statement derived from Chebyshev's inequality (Dixon and Massey, 1969) that says that for any probability distribution of x with mean μ and standard deviation σ:

$$P(|x-\mu| > k\sigma) < \frac{1}{k^2} .$$

Translated to our eigenvalue problem, if l_i is the ith eigenvalue for a sample estimating λi. SE l_i is the standard error of the ith eigenvalue and T is the number of standard errors for which one wants to make the probability statement, then:

$$P(|l_i - \lambda_i| > T\text{SE } l_i) < \frac{1}{T^2} .$$

Thus for $T=5$ and $\lambda_i=0$, the probability is less that .04 that l_i is more than 5 standard errors different from 0. We don't know the value of λ_i or the true standard error of l_i, but we can construct approximate confidence intervals for λ_i using the formula above substituting our estimate of the standard error for SE l_i. Then if the confidence interval contains 0 we accept the null hypothesis that $\lambda_i = 0$ (equivalent to using T to test the null hypothesis that $\lambda i=0$). Chebyshev's inequality provides a lower bound to a confidence coefficient and therefore an upper bound to the significance level for a corresponding test. It is very conservative, for example, even for a highly skewed distribution like a chi-squared distribution with 1 degree of freedom (Dixon and Massey, 1969). However, it would be incorrect to use Student's t, which may provide something closer to a lower bound for the probability. Neither approach is very satisfactory and my approach will have very little power for detecting true differences from 0.

The results were as follows:

For the fowl data (based on 276 individuals), Table 8 gives the eigenvectors and eigenvalues of the covariance matrix for the logged data. The correlations for the raw data are slightly different from Wright, but the pattern of coefficients is very similar, except for a coefficient for femur on the second component. Also the variances of the fowl measurements are not quite the same. However, the principal components are very similar, with minor discrepancies in coefficients.

The standard errors of the eigenvalues and eigenvector coefficients for the original chicken data (logged) based on asymptotic theory are given in Table 7. A ratio of the estimate to its standard error of $T = 5$ is probably not applicable here, and the sample is large enough that Student's t or normal z theory should apply. All coefficients for the first eigenvector have a T value greater than 12. For the second eigenvector all the T values are greater than 5 except for one at $T = 2.11$. It has a smaller coefficient than in Wright's analysis. For all the rest of the components all of the coefficients have $T > 6$ for the potentially interpretable contrasts and some are much larger. Small coefficients with absolute values less than .07 have T values less than 1.05.

When we look at the Jackknife and Bootstrap eigenvector element estimates, they differ only by a small amount in the third place for principal components one and two (Table 9). The third through sixth components differ in the second decimal place, but the patterns are the same and the interpretations would be the same. For principal component I the T values are mostly larger for the asymptotic results. All values greater than 5 for the asymptotic result are larger than 5 for the other methods except for the fourth component where the femur element is 4.62 on the Jackknife and Bootstrap and 6.97 for the asymptotic result. They are comparable for PC III and PC IV, but

occasionally a little larger. The results are again quite comparable.

It appears that the asymptotic standard errors give comparable estimates of the standard errors as compared to the Bootstrap and Jackknife, which require fewer assumptions, for the complete fowl data. They tend to be smaller for the first component and similar for the second; and then conservative or larger for the remaining components—but the results are very similar.

Single experiments were run for very small, medium and large sub-samples from the fowl data, in the range of those commonly used in systematic studies—up to 163 of the 276 in the sample, or approximately 60% of the complete data set. Only one method was used on each sample, but comparable sized samples were run for all three methods. An expanded study is planned. The asymptotic results are discussed first. For samples of $n = 12$, 15, 17, 28 and 33, principal component I was similar to the complete data, though with some different emphasis of variables. The $|T|$ values were > 5 for $n > 15$ except for the head measurements where they were > 3 for $n = 33$. For $n < 17$ some $|T|$ values were as low as 2. For components past the first, the patterns of coefficients were different from those for the complete data. $|T|$ values were only occasionally greater than 5, and not many were greater than 3. For $n = 59$ and 77, the pattern of coefficients is similar to the complete data, and $|T|$ values are greater than 3 or 4; with many greater than 5 for those greater than 5 for the complete data.

The sample for $n = 59$ gave results closer to the complete data than that for $n = 77$. Many duplicate random samples would have to be run for the same sample size to see clearer patterns, and to allow generalization of the results from this single set of experiments.

For $n = 108$, 143, and 159 the coefficients come closer to those for the complete data. The largest discrepancies are in the second component coefficients for limb measurements.

The Jackknife for $n = 11$, 12, 16, 21, 32 and 49 were similar to the asymptotic results for principal component I. However, the absolute value of T was typically in the range 2-5 for skull characters. T will be used as the absolute value of T, $|T|$, in the following discussion. The skull breadth coefficient T value was .30 for $n = 12$.

For $n = 49$, the T values were occasionally >5 and even if one used a cutoff of $T = 2$, the original pattern of the components past PC I was not recovered.

For $n = 75$ the pattern of coefficients is similar to the complete data, however those with T values >5 past the first are typically only >3 here, and some are smaller. For $n = 108$, 139, 156 and 163 there is some difference in results from the pattern based on the complete data, and values are judged to be significantly different from 0 if one bases this decision on $T = 4$ (with some minor discrepancies).

The Bootstrap for $n = 12$, 13, 16, and 24 showed similar results to the Jackknife, but with smaller T, much less than 5, on components 2 through 6. Head characters are sometimes not significant ($T < 5$) on principal component one.

For $n = 53$, T values are greater than 5 for all coefficients of eigenvector one. However, they only reach 4.5 and 7 on the Humerus and ulna variables for PC VI; similar to the pattern on VI for the complete data. The coefficients do not have quite the same pattern on the other components. For $n = 88$, principal component 2 has some significant coefficients; PC 3 and PC 6 are similar to the entire data; and 4 and 5 are not. For larger sample sizes PC I, V and VI show the same pattern of large T values; and the pattern of coefficients is similar to the complete data set.

Overall, it seems that a sample size in excess of 8 to 10 times the number of variables was required to recover the original interpretation of the principal components for the larger fowl data set. The Bootstrap appears to be the most conservative technique, the asymptotic method a little less so, and the Jackknife finds more significant coefficients for smaller sample sizes.

A graphical summary of the results of the sampling experiments is given in Figure 12. The number of T values greater than 5 out of 36 are plotted as upper case letters (A = Asymptotic; J = Jackknife and B = Bootstrap); while the number of $T < 2$ are plotted with the same lower case letters. The difference between these two entries for a specific sample size are then the number of T values between 2 and 5. There is a near linear increase of "significant" $T > 5$ for $n < 100$ as a function of sample size. Note that the asymptotic methods tend to give more large T values, the Bootstrap the least, and the Jackknife tends to be in between. For the "non-significant" T values <2, the results are consistent with the patterns above; however, the number fall off more rapidly and not linearly with sample size.

For the *Zygodontomys* data, asymptotic standard errors, Jackknife estimates and standard errors as well as Bootstrap estimates and standard errors were obtained for every sample. The results are summarized below and tend to agree with those of Gibson et al.(1984) especially for the elements of the eigenvectors. Some differences between these results and those of Gibson et al. will be pointed out. Programs written in SAS IML and on the accompanying disk can be used for analyses of other data sets.

Two points not dealt with by Gibson et al. should be mentioned. As eigenvectors are only defined up to a scale factor of $+1$ or -1, any given Jackknife or Bootstrap iteration may produce a vector which may have a pattern of signs that point in the opposite direction from that of the vector for the complete data. In that case, it is necessary to multiply the sub-sample vector by -1 to reverse the signs and to make it coincident. Therefore, the following rule was adopted: If the inner product of an eigenvector from the complete solution with the eigenvector from a Jackknife or Bootstrap sample was negative (the cosine of the angle between them would indicate they were in different quadrants), then the Jackknife or Bootstrap vector was multiplied by -1 to put it in the same quadrant.

Secondly, the average eigenvectors presented in Gibson et al. have values such that some of the coefficient values are clearly impossible as eigenvector elements. Some absolute values are larger than 3. The sum of squares of eigenvector elements conventionally sums to one, which means that all elements must be between plus one and minus one. We obtained numbers as large as those reported in Gibson, especially for the Jackknife. This effect was removed by standardizing the final bias-corrected Jackknifed eigenvector to have length one (by dividing the elements by the square root of the sum of squares of the elements). The standard errors were similarly corrected. This produced results comparable over the three techniques for the complete fowl data (see above), and this was used as a justification for the procedure in all subsequent analyses. The Bootstrap values stray less from the complete data values then do those for the Jackknife. They were similarly re-standardized. I am not sure why this happens and detailed examination of the effect of individual observations on the Jackknife results would be very time-consuming but worth while. This is akin to Krzanowski's search for outliers (discussed and cited in Reyment, this volume).

Asymptotic Results

The asymptotic standard errors for the eigenvalues and for the logs of the eigenvalues were computed for all of the *Zygodontomys* locality samples. The log results will be discussed here as they are comparable to the Jackknife and Bootstrap results. The standard error for the log of any eigenvalue is $(2/(n\text{-}1))^{0.5}$. All of the T values are large. Student's t or z would be inappropriate here as the samples

are too small, unlike the complete fowl data. The more conservative cutoff of $T=5$ is used for interpretation.

The eigenvector, coefficients of the first PC had T values > 10 for coefficients greater than .17 with very rare exception and most T values were > 5 for coefficients > .10. The exceptions occurred mainly for samples with $n < 35$. In the analysis of all of the 15 locality data sets, the two molar measurements (LM and BM1) and braincase related measurements (LIB and BB) had smaller coefficients for PC1 than the other characters, and therefore contributed less to PC1 as a "size" component. Based on their small absolute values and small T values, these are the variables that do not contribute significantly to "size." Exceptionally a T value is > 5 for these four variables. This occurs for relatively larger values of the coefficient (e.g., .08 for BM1 at El Frio). These coefficients were "not significant" ($T < 5$) for most samples.

When we look at the remaining principal components, values of $T > 5$ are very rare, seldom more than a few per data set from a given locality

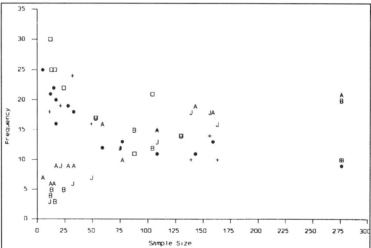

Figure 12. Comparison of sampling experiments on eigenvectors for Asymptotic, Jackknife and Bootstrap standard errors. The ordinate gives the number times out of 36 that T was > 5 (B for Asymp; Square for Boot; and Circle for Jack) and < 2 (Triangle for Asymp; * for Boot; and + for Jack) for each method for a range of sample sizes up to the complete data of 276 observations. T is equal to the observed value of a coefficient divided by its standard error.

and there is no obvious pattern. The larger T values tend to occur for single variable coefficients for specific components. One curious feature of the analysis is that Condylo-Incisive Length (CIL) dominates the last eigenvector for almost every locality sample. Its T value is frequently very large, as much as 52, larger than any other T value observed.

Results for the Jackknife on Zygodontomys Samples

The Jackknife estimates of the eigenvalues were never monotonic, decreasing from largest to smallest as a function of principal component number, though they decreased on the average with increasing principal component index. Frequently the third Jackknifed eigenvalue would exceed the second. Disparities of this type are common for the various locality samples. The ratio of the largest to smallest eigenvalue standard error for a locality was 4.95 for one sample, and for all others this ratio was less than 4. The approximate asymptotic standard error $(2/(n-1))^{0.5}$ is always contained within the range of the standard errors for the different Jackknifed estimates of eigenvalues for one locality. The results for the Jackknife eigenvector elements are consistent with the asymptotic results. For example, all T values >5 on PC1 for the asymptotic results are here also >5. The occasional values >5 for other principal components past one again form no obvious pattern, and tend to occur for single variables' coefficients for individual principal components. The larger T values do tend to occur for the same coefficients in both analyses, and they tend to be larger for the Jackknife results. The same large T value for CIL on the last eigenvector is also present here. I have no explanation for this result.

Bootstrap Results

The Bootstrapped eigenvalues always formed a monotonic decreasing series as a function of eigenvalue number as must occur for the complete data, and in contrast to the results for the Jackknife.

The logarithms of the eigenvalues tend to have smaller standard errors than those for the Jackknife. The average value is much closer to, and even less in some cases, than the approximate values given above under the asymptotic results, in contrast to the Jackknife results, which are larger.

The Jackknife and Bootstrap results are very similar for PC1's eigenvector. However, for PC2-PC12 the T values tend to be smaller for the Bootstrap. They very rarely exceed 5 except for the same curiously large coefficient for PC12 for CIL for most samples. The T values exceeding 3 for the Bootstrap frequently exceed 5 for the Jackknife.

Conclusions for the Zygodontomys Analysis and Comparison to Gibson et al. (1984)

The T values for all of the eigenvalues for all three analyses, asymptotic, Jackknife and Bootstrap, are large, almost always greater than 10. The T value can be thought of as a test statistic for the null hypothesis that an eigenvalue is 0. This null hypothesis would be rejected in all cases. This is just a verification that our data matrices are of full rank. This is not surprising as the number of cases is always roughly 2.5 to 5.7 times larger than the number of variables. Also the carefully selected measurements are not redundant and the variables are measuring different features of the skull. This result is in marked contrast to some of the results for the logs of the eigenvalues in Gibson et al. where the T values for the first eigenvalues exceeded 10 for only 2 out of their 11 samples. For the 2nd and 3rd principal component, T only exceeded 2 three times out of 22, and only exceeded one 10 times out of 22 for Jackknifed eigenvalues. In other words, eigenvalues past the first would generally be judged to be not significantly different from 0, in marked contrast to our results. I do not know why our results are so different, and do not believe that it is just because the data are different. We are probably computing something differently. We both, however, find similar discrepancies in the ordering of the eigenvalues. Gibson et al. only present results for the first three principal components and 4, out of their 11 third eigenvalues are larger than the second. I agree with them that this indicates an indeterminacy in the direction of components past the first. The Bootstrap results are different in that the eigenvalues always decrease monotonically as for the complete sample. However, adjacent eigenvalues are frequently very similar. An asymptotic chi-squared test for equality of consecutive eigenvalues is given in Morrison (1976:294). For the complete data set from each locality, eigenvalues 1 and 2 are always significantly different, eigenvalues 2 and 3 are not significantly different for any locality, and the hypothesis of equality of eigenvalues 2 to 6 is rejected for only 2 of the 15 localities.

The three methods agree very well for the eigenvector associated with the first principal component. However, for principal components 2 and beyond, the Jackknife finds the most significant coefficients and the Bootstrap the least. The asymptotic results fall somewhere in between. These results are somewhat different from the fowl sampling experiment, where the asymptotic results had more significant coefficients and the Jackknife fell in between. Thus, a person using one of these methods would possibly attempt to interpret more principal components using the Jackknife than the Asymptotic results or Bootstrap for our data.

Our results are similar to those given by Gibson where consistent patterns of coefficients (in terms of T values) are not obtained based on the Jackknife for principal components 2 and 3 for their comparable sample sizes and similar numbers of variables. Principal component 2 does show more large coefficients (as measured by T values) than does 3 for Gibson et al. However, in both studies we could not define a consistent pattern of associated variables over samples beyond the first principal component based on the coefficients assessed by the T values. This result is consistent with near equality of eigenvalues 2 and 3 for many localities.

None of these results for principal component analysis precludes the possibility of finding more than one significant biological pattern of association of variables, for example finding two or more correlated latent variables in a confirmatory factor analysis. The techniques are different in intent and therefore expected result.

Acknowledgments

Robert Voss, Department of Mammalogy, American Museum of Natural History, set a fine example of generous cooperation in providing the measurements for *Zygodontomys* before any results of his own analyses had been published. Patricia Escalante, Department of Ornithology, American Museum of Natural History, entered the data and did many of the preliminary analyses. The Department of Invertebrates at the AMNH has generously provided space and access to equipment during the earlier preparation of this manuscript. The City University of New York Academic Computer Center gave full access to its mainframe computer facilities, and the Academic Computer Center at Queens provided the PC version of SAS. Special thanks go to the University of Michigan and National Science Foundation for inviting me to the Michigan workshop and generously supporting me when I stubbornly extended my invited stay. The entire staff at Michigan and the participants made the workshop a worthwhile and memorable occasion. I wish to thank Richard Reyment, Paleontologic Institute, Uppsala University, Sweden for providing space and access to computer equipment during what was to be final preparation of the manuscript. The United States Geological Survey helped support the purchase of a sophisticated laptop computer that made continued work on the manuscript possible during extended travel. George Barrowclough, Robert Voss and an anonymous reviewer read a draft version of the manuscript and I thank them for their comments. Jim Rohlf gave useful suggestions for improving the manuscript. Annika Sanfilippo read a late version and her comments helped me to explain and hopefully clarify some of the multivariate jargon.

References

Atchley, W. R., C. T. Gaskins and D. Anderson. 1976. Statistical properties of ratios. I. Empirical results. Syst. Zool., 25(2):563-583. [See this and subsequent papers, and a number of other contributions subsequently dealing with ratios]

Bargmann, R. E. 1970. Interpretation and use of a generalized discriminant function. Pp. 35-60 in Essays in probability and statistics (Bose, R. C., I. M. Chakravarti, P. C. Mahalanobis, C. R. Rao and K. J. C. Smith, eds). The University of North Carolina Press, Chapel Hill.

Bentler, P. M. 1985. Theory and implementation of EQS: a structural equations program. BMDP Statistical Software, Inc., Los Angeles. (standalone program manual for structural relations—a little easier to use than LISREL, but does not include multiple samples).

Blondel, J., F. Vuilleumier, L. F. Marcus and E. Terouanne. 1984. Is there ecomorphological convergence among Mediterranean bird communities of Chile, California, and France. *in* Evolutionary biology, 18:141-213 (M. K. Hecht, B. Wallace and G. T. Prance, eds.). Plenum Publishing, New York.

Bookstein, F. L. 1982. Foundations of morphometrics. Ann. Rev. Ecol. Syst., 13:451-470.

Bookstein, F. L. 1989. "Size and shape": a comment on semantics. Syst. Zool., 38:173-180.

Bookstein, F. L. 1991. Morphometric tools for landmark data. Book manuscript, accepted for publication, Cambridge University Press.

Bookstein, F. L., B. Chernoff, R. Elder, J. Humphries, G. Smith, R. Strauss. 1985. Morphometrics in evolutionary Biology. The Academy of Natural Science of Philadelphia, Spec. Publ. No. 15, 277 pp.

Bookstein, F. L. and R. A. Reyment. 1989. Microevolution in Miocene Brizalina (Foraminifera) studied by canonical variate analysis and analysis of landmarks. Bull. Math. Biol., 51:657-679.

Campbell, N. A. and W. R. Atchley. 1981. The geometry of canonical variate analysis. Syst. Zool., 30:268-280.

Chatterjee, S. 1984. Variance estimation in factor analysis: an application of the bootstrap. Brit. J. Math. Stat. Psychol., 37:252-262.

Cock, A. G. 1966. Genetical aspects of metrical growth and form in animals. Q. Rev. Biol., 41:131-190. [very important paper on allometry and growth].

Dempster, A. P. 1969. Elements of continuous multivariate analysis. Addison-Wesley, Reading, Mass.

Diaconis, P. 1985. Theories of data analysis: from magical thinking through classical analysis. Pp. 1-36 in Understanding robust and exploratory data analysis (Hoaglin, D. C., F. Mosteller and J. W. Tukey, eds.). John Wiley & Sons, New York.

Diaconis, P. and B. Efron. 1983. Computer-intensive methods in statistics. Sc. Amer., 248: 96-108. (semi-popular article on the bootstrap).

Dillon, W. R. and M. Goldstein. 1984. Multivariate analysis. Methods and applications. John Wiley & Sons, New York.

Dixon, W. J. (Ed.) 1983. BMDP Statistical Software, 1983 revised printing. University of California Press, Berkeley.

Dixon, W. J. and Massey, F. J. 1969. Introduction to statistical analysis, third edition. McGraw Hill Book Company, New York.

Dunn, L. C. 1922. The effect of inbreeding on the bones of the fowl. Bulletin 152. Storrs Agricultural Experiment Station, pp.1-112.

Efron, B. 1982. The jackknife, the bootstrap and other resampling plans. SIAM Monograph 38. Soc. for Indust. and Applied Math., Philadelphia. (small monograph on the subject).

Feinberg, S. E. 1977. The analysis of cross-classified categorical data. The MIT Press, Cambridge, Mass.

Fisher, R. A. 1936. The use of multile measurements in taxonomic problems. Ann. Euqer., 7:179-184.

Flury, B. 1984. Common principal components in k groups. J. Amer. Statist. Assoc., 79:892-898.

Flury, B. and G. Constantine. 1985. The FG diagonalization algorithm. Algorithm AS 211, Appl. Stat., 34:177-183.

Flury, B. and H. Riedwyl. 1988. Multivariate statistics: a practical approach. Chapman and Hall, London.

Gabriel, K. R. 1971. The biplot graphical display of matrices with applicatin to principal componenet analysis. Biometrika, 58:453-467.

Gibson, A. R., A. J. Baker and A. Moeed. 1984. Morphometric variation in introduced populations of the common Myna (Acridotheres tristis): an application of the jackknife to principal component analysis. Syst. Zool., 33(4): 408-421.

Gnanadesikan, R. 1977. Methods for statistical data analysis of multivariate observations. John Wiley & Sons, New York.

Gower, J. C. 1966a. Some distance properties of latent root and vector methods used in multivariate analysis. Biometrika 53(3&4):325-338. [Classical paper on principal coordinates analysis].

Gower, J. C. 1966b. A Q-technique for the calculation of canonical variates. Biometrika, 53(3 & 4):588-590.

Greenacre, M. J. 1984. Theory and applications of correspondence analysis. Academic Press, London.

Hand, D. J. 1981. Discrimination and classification. John Wiley & Sons, Chichester. [best book on the subject]

Harris, R. J. 1975. A primer of multivariate statistics. Academic Press, New York.

Hawkins, D. M. 1974. The detection of errors in multivariate data using principal components. J. Amer. Statist. Assoc., 69:340-344.

Hawkins, D. M. 1980. Identification of outliers. Chapman and Hall, London.

Hoaglin, D. C., F. Mosteller and J. W. Tukey (eds.) 1985. Understanding robust and exploratory data analysis. John Wiley & Sons, New York.

Hocking, R. R. 1983. Developments in linear regression methodology: 1959-1983. Technometrics, 25:219-30.

Howarth, R. J. 1971. An empirical discriminant method applied to sedimentary rock classification from major element geochemistry. Math. Geol., 3:51-60.

Johnson, R. A. and D. W. Wichern. 1982. Applied multivariate statistical analysis. Prentice Hall, Englewood Cliffs, New Jersey.

Jolicoeur, P. 1963. The multivariate generalization of the allometry equation. Biometrics, 19:497-499.

Joliffe, I. T. 1986. Principal component analysis. Springer-Verlag, New York. [comprehensive up to date summary of PCA - but not much new].

Joreskog, K. G., J. E. Klovan and R. A. Reyment. 1976. Geological factor analysis. Elsevier, Amsterdam.

Joreskog, K. G. and D. Sorbom. 1979. Advances in factor analysis and structural equation models. Abt Books, Cambridge, Mass.

Joreskog, K. G. and D. Sorbom. 1985. LISREL VI. Analysis of linear structural relationships by maximum likelihood, instrumental variables, and least-squares methods. Scientific Software, Inc., Mooresville, Ind. [users manual for LISREL program for analyzing paths, factors, covariance and correlation structural relationships in one or more groups].

Kruskal, J. B. and M. Wish. 1978. Multidimensional scaling. No. 11 in Series: Quantitative applications in the social sciences. Sage Publications, Beverly Hills. [excellent paperback introduction to the subject; includes guide to then-available programs].

Krzanowski, W. J. 1979. Between-group comparison of principal components. J. Amer. Stat. Assoc., 74:703-707.

Krzanowski, W. J. 1988. Principles of multivariate analysis: a user's perspective. Oxford University Press, Oxford.

Little, R. J. A. and D. B. Rubin. 1987. Statistical analysis with missing data. John Wiley & Sons, New York.

Loehlin, J. C. 1987. Latent variable models: an introduction to factor, path and structural analysis. Lawrence Erlbaum Associates, Hillsdale, NJ.

Morrison, D. F. 1976. Multivariate statistical methods. Mc-Graw Hill, New York.

Mosimann, J. E. 1970. Size allometry: size and shape variables with characterizations of the lognormal and generalized gamma distributions. J. Amer. Stat. Assoc., 65:930-945.

Mosimann, J. E. and F. C. James. 1979. New statistical methods for allometry with application to Florida red-winged blackbirds. Evolution, 33:444-459.

Mosteller, F. and J. W. Tukey. 1977. Data analysis and regression. A second course in statistics. Addison-Wesley Pub. Co., Reading, Mass. [see especially pages 133-162 on the Jackknife; especially in discriminant analysis].

Mulaik, S. A. 1986. Factor analysis and psychometrika: major developments. Psychometrika, 51: 23-33.

Neff, N. A. and L. F. Marcus. 1980. A survey of multivariate methods for systematics. Privately published, New York.

Olson, E. C. and R. J. Miller. 1958. Morphological integration. Univ. Chicago Press, Chicago. [early ad hoc attempt at correlation structure analysis; lots of good data sets, e.g., see Zelditch].

Oxnard, C. 1973. Form and pattern in human evolution: some mathematical, physical and engineering approaches. The University of Chicago Press, Chicago.

Oxnard, C. 1975. Uniqueness and diversity in human evolution. University of Chicago Press, Chicago.

Petite-Maire, N. and J. F. Ponge. 1979. Primate cranium morphology through ontogenesis and phylogenesis, factorial analysis of global variation. J. Hum. Evol., 8:233-234.

Pielou, E. C. 1984. The interpretation of ecological data: a primer on classification and ordination. John Wiley & Sons, New York.

Rao, C. R. 1966. Covariance adjustment and related problems in multivariate analysis. in Multivariate analysis. (P. R. Krishnaiah, ed.) Academic Press, New York. [some aspects of power, sample size and number of variables in MANOVA].

Rempe, U. and E. E. Weber. 1972. An illustration of the principal ideas of MANOVA. Biometrics, 28:235-238.

Rencher, A. C. 1988. On the use of correlations to interpret canonical functions. Biometrika, 75 (2):363-365.

Reyment, R. A. 1969. A multivariate paleontological growth problem. Biometrics, 25:1-8.

Reyment, R. A. 1985. Multivariate morphometrics and analysis of shape. Math. Geol., 17(6):591-609.

Reyment, R. A., R. E. Blackith and N. A. Campbell. 1984. Multivariate morphometrics, 2nd. Edition. Academic Press, London.

Rindskopf, D. and T. Rose. 1988. Some theory and application of confirmatory second order factor analysis. Multivariate Behavior Research, 23: 51-67.

Rohlf, F. J. 1972. An empirical comparison of three ordination techniques in numerical taxonomy. Syst. Zool., 21:271-280.

Rohlf, F. J. and F. L. Bookstein. 1987. A comment on shearing as a method for size correction. Syst. Zool., 36:356-367.

Shea, B. T. 1985. Bivariate and multivariate growth allometry: statistical and biological considerations. J. Zool. Lond. (A) 206:367-390. [clear, interesting discussion; also includes size and shape discussion, and multiple group PCA discussions in growth context.]

Sjovold, T. 1975. Some notes on the distribution and certain modifications of Mahalanobis' generalized distance (D2). J. Hum. Evol., 4:549-558.

Sokal, R. R. and F. J. Rohlf. 1981. Biometry. The principles and practice of statistics in biological research. W. H. Freeman, San Francisco.

Thorpe, R. S. 1988. Multiple group principal component analysis and population differentiation. J. Zool. Lond., 216:37-40. [see references for applications.]

van Valen, L. 1978. The statistics of variation. Evol. Theor., 4:33-43.

Voss, R. S., L. F. Marcus and P. Escalante. in press. Morphological evolution in muroid rodents. I. Conservative patterns of craniometric covariance and their ontogenetic basis in the neotropical genus Zygodotomys. Syst. Zool.,

Werdelin, L. 1983. Morphological patterns in the skulls of cats. Biol. J. Linnean Soc., 19:375-391.

Werdelin, L. 1988. Correspondence analysis and the analysis of skull shape and structure. Ossa, 13:181-189.

Willig, M. R., R. D. Owen, and R. L. Colbert. 1986. Assessment of morphometric variation in natural populations: the inadequacy of the univariate approach. Syst. Zool., 35(2):195- 203. [naive consideration with some real data of relationship between statistical significance in ANOVA and MANOVA]

Wright, S. 1921. Correlation and causation. Jour. Agric. Res., 20:557-585.

Wright, S. 1954. The interpretation of multivariate systems. Pp. 11-33 in Statistics and mathematics in biology (O. Kempthorne, T. A. Bancroft, J. W. Gowen and J. L. Lush, eds.). The Iowa State College Press, Ames.

Wright, S. 1968. Evolution and genetics of populations. I. Genetic and biometric foundations. Univ. of Chicago Press, Chicago.

Zelditch, M. L. 1987. Evaluating models of developmental integration in the laboratory rat using confirmatory factor analysis. Syst. Zool., 36(4): 368-380.

Zelditch, M. L. 1988. Ontogenetic variation and patterns of phenotypic integration in the laboratory rat. Evolution, 42(1):28-41. [extensive application of LISREL to morphometric data].

Appendix: 1.

Data for *Zygodontomys* from Dividive. See Figure 4 for diagram of measurements.														
OBS	SEX	AGE	CIL	LD	LM	BM1	LIF	BR	BPB	BZP	LIB	BB	DI	LOF
1	F	2	23.00	6.20	4.05	1.30	5.05	4.05	2.25	2.55	11.50	4.25	1.20	8.50
2	F	2	24.20	6.45	4.30	1.25	5.25	3.90	2.50	2.75	11.45	4.50	1.25	9.00
3	M	2	24.90	6.95	3.90	1.15	5.70	4.15	2.50	2.50	11.95	4.20	1.30	9.65
4	F	2	24.65	6.75	4.10	1.20	5.65	4.25	2.60	2.35	11.80	4.20	1.30	9.05
5	M	3	24.75	7.00	4.00	1.20	5.30	4.15	2.50	2.90	11.20	4.50	1.30	9.05
6	F	2	25.50	7.15	4.10	1.20	5.60	4.45	2.75	2.85	11.65	4.45	1.40	9.85
7	F	2	25.30	7.35	4.05	1.30	5.85	4.35	2.55	2.70	11.95	4.25	1.40	9.40
8	F	2	24.75	6.85	4.10	1.20	5.55	4.90	2.65	2.60	12.30	4.55	1.35	9.35
9	M	2	25.05	6.85	4.35	1.35	5.50	4.65	2.65	2.50	11.85	4.80	1.40	9.40
10	M	2	25.15	6.85	4.10	1.25	5.65	4.40	2.45	2.65	11.70	4.45	1.45	9.40
11	F	2	25.20	6.90	4.20	1.25	5.80	4.30	2.50	2.60	12.80	4.60	1.45	9.90
12	F	2	25.70	7.45	3.90	1.30	6.05	4.35	2.60	2.55	11.95	4.30	1.40	10.05
13	F	3	26.15	7.30	4.30	1.35	5.60	4.85	2.85	2.90	11.70	4.65	1.45	9.55
14	M	2	26.05	7.20	4.50	1.40	5.75	4.70	2.50	2.90	12.15	4.90	1.25	9.65
15	F	2	26.00	7.15	4.40	1.30	6.15	4.35	2.60	2.80	11.95	4.85	1.60	9.60
16	F	3	25.75	6.95	4.60	1.35	5.60	4.70	2.75	2.85	11.45	4.65	1.45	9.40
17	F	3	25.90	7.35	4.25	1.25	6.30	4.50	2.65	3.05	12.35	4.50	1.50	9.90
18	M	3	26.25	7.25	4.15	1.25	6.05	4.50	2.65	3.20	12.35	4.65	1.45	9.70

19	F	3	26.60	7.15	4.55	1.25	5.75	4.75	3.15	3.25	12.05	4.45	1.60	9.70
20	F	3	26.95	7.65	4.00	1.25	5.70	4.95	3.15	2.85	12.55	4.60	1.60	10.05
21	F	3	26.75	7.55	4.25	1.25	5.55	4.75	2.75	2.70	12.10	4.65	1.50	9.75
22	F	3	27.15	7.50	4.20	1.20	6.00	5.50	2.90	3.10	12.00	4.80	1.55	10.20
23	M	2	27.10	7.65	4.25	1.25	5.60	4.65	2.95	2.75	12.40	4.75	1.50	10.10
24	M	2	26.80	7.40	4.30	1.30	5.95	4.50	2.70	3.10	11.60	4.65	1.55	9.75
25	M	3	27.25	7.30	4.50	1.40	5.80	5.10	2.75	2.90	11.85	4.65	1.75	9.95
26	F	3	27.70	7.65	4.35	1.35	6.25	5.25	2.85	3.10	12.60	4.70	1.60	10.50
27	M	3	27.70	8.00	4.50	1.35	6.15	5.10	2.85	3.15	11.75	4.35	1.55	10.25
28	M	2	27.15	7.85	4.20	1.25	6.20	4.65	2.80	2.85	12.00	4.35	1.55	9.85
29	M	2	25.90	7.20	4.40	1.30	5.85	4.60	2.60	2.75	12.30	4.75	1.40	9.90
30	M	2	26.90	7.50	4.30	1.30	6.10	4.75	2.90	3.30	12.25	4.55	1.50	10.30
31	F	4	27.95	8.20	4.10	1.25	6.55	4.95	3.25	3.10	12.50	4.90	1.65	10.25
32	F	2	27.70	7.55	4.20	1.25	5.75	4.80	2.80	2.90	12.55	4.35	1.50	10.40
33	M	3	28.00	7.85	4.35	1.25	6.75	4.75	3.10	3.25	12.20	4.85	1.50	10.25
34	M	3	28.40	7.90	4.10	1.30	6.05	5.10	3.10	3.05	12.80	5.10	1.65	10.40
35	F	4	27.00	7.70	4.25	1.30	6.25	5.00	3.25	3.25	12.60	4.60	1.65	10.40
36	F	2	27.80	8.05	4.35	1.25	6.45	4.65	3.00	3.00	12.80	4.55	1.45	10.30
37	F	3	27.50	8.15	4.15	1.30	6.30	5.25	3.20	3.40	12.50	4.95	1.45	10.25
38	F	3	28.20	8.05	4.35	1.25	6.30	4.65	2.90	3.55	12.25	4.45	1.60	10.75
39	M	2	28.60	8.25	4.30	1.20	6.45	4.90	2.75	3.00	12.60	5.00	1.65	10.85
40	M	3	27.50	7.90	4.25	1.35	6.30	5.35	3.05	3.05	12.55	4.90	1.65	9.70
41	M	4	28.20	8.20	4.10	1.25	6.30	4.95	3.35	3.20	12.75	4.75	1.75	10.50
42	F	3	28.35	7.80	4.50	1.30	6.05	5.45	3.25	3.40	12.15	4.90	1.75	10.65
43	F	3	28.05	8.05	4.15	1.20	6.55	4.70	3.05	3.10	12.70	4.80	1.55	10.45
44	M	4	28.20	8.20	3.90	1.20	6.25	5.00	3.05	3.40	12.35	4.45	1.65	10.35
45	F	3	28.25	8.20	4.15	1.25	6.30	5.25	3.55	3.10	12.45	5.05	1.65	10.30
46	F	3	28.00	7.90	4.10	1.25	6.20	4.85	2.90	3.40	12.30	4.80	1.45	10.75
47	M	3	28.10	7.90	4.20	1.20	6.20	5.15	3.00	2.90	12.40	4.80	1.65	10.75
48	F	3	28.30	8.00	4.30	1.30	6.30	4.80	3.10	3.25	12.55	4.80	1.70	10.70
49	F	4	28.45	7.75	4.25	1.30	5.95	5.35	3.20	3.50	12.30	4.75	1.70	10.70
50	M	4	28.55	8.35	3.95	1.20	6.70	4.95	3.10	3.05	12.40	4.75	1.65	10.40
51	M	3	28.50	8.30	4.40	1.30	5.75	5.20	2.85	3.25	12.40	4.95	1.65	1.070
52	M	3	29.10	7.95	4.25	1.20	6.40	4.90	2.85	3.25	12.00	4.75	1.65	10.90
53	M	2	28.15	8.20	4.25	1.35	6.60	5.30	3.05	3.00	12.70	5.10	1.55	11.45
54	M	3	29.20	8.55	4.10	1.35	6.75	5.60	3.25	3.10	12.35	4.70	1.65	10.70
55	F	4	28.70	8.30	4.10	1.30	6.90	5.45	3.20	3.20	12.70	4.65	1.70	10.75
56	M	3	29.10	8.40	4.20	1.25	6.15	4.75	3.00	2.80	12.45	4.90	1.50	10.55
57	M	3	28.60	8.35	4.15	1.30	6.85	5.30	3.20	3.25	12.10	5.25	1.55	10.55
58	M	3	28.80	8.35	4.35	1.30	6.80	5.05	3.15	3.20	12.80	5.00	1.65	10.70
59	M	4	29.45	8.45	4.30	1.35	6.45	5.15	3.15	3.35	12.45	4.85	1.75	10.80
60	F	3	28.60	8.20	4.30	1.25	6.45	5.05	3.05	3.30	12.75	5.00	1.50	10.10
61	M	4	29.25	8.55	4.50	1.30	6.25	5.15	3.15	3.20	11.80	4.95	1.70	10.75
62	M	3	29.85	8.35	4.65	1.35	6.85	5.40	3.05	3.20	12.50	5.00	1.80	11.25
63	F	4	29.20	8.75	4.15	1.30	6.90	5.60	3.30	3.70	12.50	4.90	1.55	10.55
64	M	2	29.65	8.65	4.35	1.25	6.85	5.80	3.15	3.45	12.10	5.35	1.70	10.95
65	M	4	30.10	8.80	4.15	1.30	6.45	4.95	3.20	3.35	12.65	4.65	1.80	11.05
66	F	3	29.85	8.60	4.50	1.25	6.40	5.55	3.25	3.45	12.75	5.30	1.75	10.95
67	F	4	29.95	8.55	4.15	1.25	7.15	4.95	3.15	3.40	12.50	4.75	1.70	10.85
68	M	3	30.70	8.90	4.40	1.25	6.30	5.75	3.45	3.45	12.45	5.15	1.85	11.45

Appendix: 2.

Programs used to generate some of the results in the paper and some additional routines are on a supplied disk in the directory LESWARE. All of the files in that directory are ASCII files.

List of programs:

PRIMER	SAS	SAS on a page
LESWARE	DOC	Documentation file
ZCONFD	FIN	CI's for Mahalanobis D2, etc.
ZDISTOUT	FIN	Example of output from above
ZBOOTIML	FIN	Bootstrap for Eigenvalues-vectors
ZDIVIN	FIN	Getting Zygo. Dividive data in
ZJACKIML	FIN	Jackknife for Eigenvalues-vectors
ZMST	FIN	Minimum spanning tree
ZASYMP	FIN	Asymptotic SE's for Eigenvalues-vectors.

Appendix: 3. Traditional Analysis of *Brizalina* Data

Bookstein (1991) has undertaken some analyses of a series of samples of fossil foraminifera, comparing standard distance techniques to some landmark analyses. I analyzed these data in the spirit of the present paper using the coordinates supplied in his book. There are 10 specimens from each of 5 stratigraphic levels from a bore hole in the Cameroons. These are benthic foraminifera and we would like to see how they differ or change at different level. The samples, 5 to 1, start at a depth of 1810 meters and occur approximately every 23 meters to 1716 meters.

Bookstein digitized 6 landmarks on scanning electron micrographs of the specimens (Figure 1). Two magnifications were used, and so the data with a magnification factor of 171 have to be multiplied by 1.71 to put them on the same scale. Two plots of landmark distance AB, AC, AF, CF, D_AC, and E_AF before and after correction are given in Figures 2 and 3. There is considerable heterogeneity before correction, and one specimen in

each of samples 3 and 4 appear to be outliers—perhaps wrongly assigned microspherics to the supposed megalospheric data set.

Of course the landmark-based shape comparisons do not require the magnification corrections, as they are scale free; but contamination by wrong morphs will not make a clean analysis.

Another problem with the data is that sample 1 is considerably more homogeneous than the other four samples. From the published statistics in a paper on the same material by

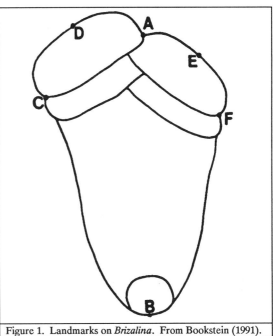

Figure 1. Landmarks on *Brizalina*. From Bookstein (1991).

Reyment and Bookstein (1989), it can be seen that samples 2-5 all have coefficients of variation similar to those obtained for the larger samples measured directly by Reyment. Bookstein had selected the material to digitize from the larger samples and apparently selected a very unlikely homogeneous sample of 10 from the larger sample available to Reyment.

Therefore, I have chosen to compare only samples 2-5 and have left out specimens 30 and 40 as potentially different polymorphs. Some of the

```
   5 +                          7 7     7      7 7 71   11
   4 +          Distance          1   71     711    11 7         1
   3 +           [AB]              7   7    7 7 1         7 7    1
   2 +                          7        7   7         7    71  1
   1 +                                        1 1 11 11
S
   ---+---+---+---+---+---+---+---+---+---+---+---+---+---+---+--
T  5 +              7  77 17  7    1    1
R  4 +          Distance    11   111 1 1    7
A  3 +           [AC]         7777      17
T  2 +            777 7 7           1 1
I  1 +                       111111
G
   ---+---+---+---+---+---+---+---+---+---+---+---+---+---+---+--
R  5 +         7  77  7    17
A  4 + Distance   71  1     71 1 111   7
P  3 +    [AF]    7   7777    7    117
H  2 +            77777 1 1
I  1 +              1 11111
C
   ---+---+---+---+---+---+---+---+---+---+---+---+---+---+---+-
   5 +                 77    7 777    7 11
L  4 +          Distance       7171 1  1    11 1         7
E  3 +           [CF]         7  77 7    11 7
V  2 +             777777        11
E  1 +                  1  1111  1
L
   ---+---+---+---+---+---+---+---+---+---+---+---+---+---+---+-
   5 +           777 77 1
   4 +           111 171       Distance
   3 +          77777711        [D_AC]
   2 +          7 777  11
   1 +             1111
   ---+---+---+---+---+---+---+---+---+---+---+---+---+---+---+-
   5 +          7717
   4 +         171711         Distance
   3 +         7777  1         [E_AF]
   2 +         771  1
   1 +         11111
   ---+---+---+---+---+---+---+---+---+---+---+---+---+---+---+-
       0   50 100 150 200 250 300 350 400 450 500 550 600 650 700
```

Figure 2. Plot of inter-landmark distances for *Brizalina* coordinates. Data not corrected for magnification. Units are arbitrary digitizing units. Specimens marked as 7 should be enlarged by 1.71.

methods described in Reyment's contribution to this volume would be worthwhile to try as well.

I used the 6 landmarks to generate 6 distance measures, namely AB, and CF as the nearest things to length and width, AC and AF as the ultimate and penultimate chamber diameters, and D_AC and E_AF as the heights of the triangles formed by DAC and EAF, respectively, as putative heights for these chambers.

A canonical variates analysis and the Mahalanobis D^2 computed for the four samples (2-5) indicated that there were no significant

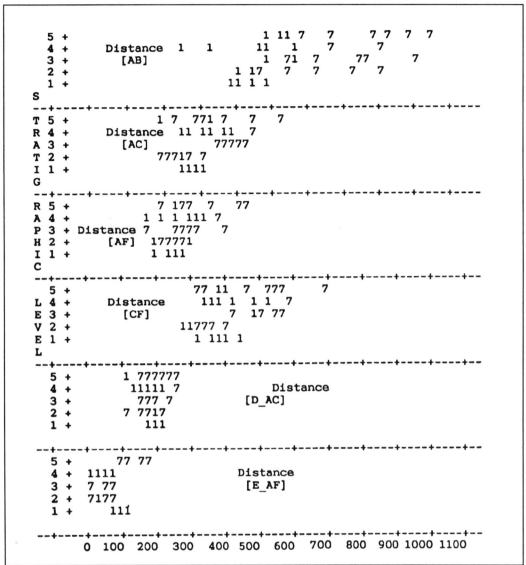

Figure 3. Plot of inter-landmark distances for *Brizalina* coordinates. Data corrected for magnification. Arbitrary digitizing scale increased due to magnification correction for specimens marked 7.

differences among the vectors of means. Therefore whatever differences are seen are those that would occur by random sampling from one biological population. It is not necessary to explain the differences as random changes since all of the results could occur by random sampling from one

population. In other words, the effects of sampling cannot be ruled out as the cause of the differences.

In two experiments, I mixed up the remaining 38 specimens and actually did a random sampling of four samples, 2 random samples of 10 specimens each and 2 random samples of 9 specimens each. The results were similar to those observed. The only indication in the original analysis of possible real differences among the samples was Roy's largest root criterion which was significant at the 10% level.

A robust fit of the specimens using Rohlf's GRF program, both to a consensus specimen for each sample separately and to the entire group of 50 specimens (no correction for magnification is necessary here), pointed to good conformity for the landmarks A,C and F. The most variable landmark is B, the position of the proloculus; followed by D and E, the two chamber height landmarks.

As the number of chambers is variable in these foraminifera, and one really does not know how many they have, there is biological support for the variability of proloculus position. We might suggest that the position, A, of the orifice relative to the sizes of the last two chambers is rather stable.

On the other hand, as pointed out by Bookstein (Reyment and Bookstein, 1989), the positions of D and E are the poorest landmarks, and they are somewhat arbitrarily determined as kinds of maxima. Therefore the variability may reflect this fact and also variability in attempting to locate a similar point over the specimens.

Note that I have hesitated to analyze the data stratigraphically. I see no reason to spend additional time on this set of data. I would have liked to have had more data, and feel the original larger data set of Reyment may have been more informative had all specimens been digitized.

In conclusion, this would not be a good data set to invest a lot of time in. Any analysis comparing methodology would have limited or no value.

Chapter 5

Reification of Classical Multivariate Statistical Analysis in Morphometry

Richard A. Reyment

Paleontologiska Institutionen, Uppsala Universitet
Box 558, S75122, Uppsala, Sweden

Abstract

The rapid spread of multivariate statistical methods in morphometrics, the study of morphological variability in living and fossil organisms, resulting from the ubiquitousness of computer programs for standard procedures, has created difficulties for the reification of vectors in principal component and canonical variate analyses. This is particularly so for situations in which the data diverge significantly from multivariate normality. Questions of stability of eigenvectors and the correct analysis of compositional data are taken up as well as the comparison of results of analyses obtained by different methods. An overview of multivariate procedures is given in which more important procedures are compared and contrasted. The methods are illustrated by means of data on microfossils.

Introduction

Reyment, Blackith and Campbell (1984) have given a detailed review of the history and conceptual background of *Multivariate Morphometrics*. I refer the reader to that volume for a general presentation of the subject, as well as a comprehensive bibliography.

The basic geometrical concepts underlying multivariate statistics are quite simple. One is concerned with finding a mathematical description, albeit idealized, of clusters of points in a space of two or more dimensions. For one population, we are interested in describing the geometrical properties of a sample forming a single cluster of points. In morphometric analyses on continuous variables it is usual to assume that these are multivariate normally distributed. The property of multivariate normality is a necessary requirement if hypotheses are to be tested, since the relevant statistical theory has been built around the requisite of normality. If the testing of hypotheses is not central to the investigation, then the nature of the distribution of the points is not decisive. In *Numerical Ecology*, for example, abundancy data (frequencies) are analysed by all the standard multivariate methods without untoward concern for theoretical niceties, since the principal aims of such studies are graphical displays.

Many problems arise in morphometrical work in which it is necessary to compare two or more clusters of points. This is a more complicated situation than for one sample, for we must consider not only the statistical properties of each cloud of points, but also relate the clusters to each other. This added element introduces further geometrical consequences.

The main theme of the present article is to discuss what can be done when the data do not conform with the theoretical requirements of multivariate-normally distributed variables in Cartesian space. Such data are far more common than is generally recognized. The most familiar deviation is when one or more outliers occur in a sample. This topic has attracted much attention of late, to which attests the growing number of articles and books on the subject. The *structure* of multivariate data can also be conclusive in an analysis. We shall be concerned with assessing the stability of eigenvectors in multivariate work, particularly in applications in which the elements of a vector are reified, that is, furnished with a morphometrical interpretation.

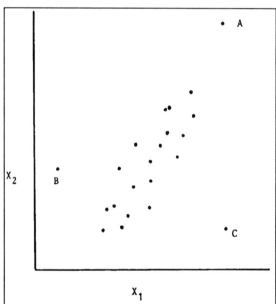

Figure 1. Three outlying points in relation to a set of bivariate normally distributed observations.

If our interest lies with **ordinating** our data, then the problems I have just outlined are usually not serious and for most purposes can be applied to "unsuitable" data without crippling effects. Ordination is basically graphically oriented with the end in view of representing the properties of a multivariate sample in just a few dimensions.

We shall also be taking a close look at the multivariate analysis of closed data sets, i.e., sets of measurements with a constant sum. *Compositional data* do not arise naturally in morphometrics but there are some special situations in which they do occur. As an example, consider the procedure known as "standardization" in which the measurements are reduced to a common scale of reference by dividing through by some constant.

The goal of this review is to present the various methods in general terms, to indicate the main assumptions, and to interpret (reify) results. We shall begin by taking a guided tour through the field of multivariate methodology.

The Panoply of Multivariate Methods

I give here a general overview of the entire field of multivariate statistical analysis in order to demonstrate the natural relationships occurring between the various methods. The subject can be approached by reference to Figure 1. This schematic diagram illustrates the connections between the most commonly employed statistical methods.

The Single Sample

The most popular methods have been devised for the analysis of a single sample of multivariate observations. The data will usually be in the form of a data matrix, here denoted as **X**, consisting of N rows (the number of observations) and p columns, the number of variables. The *mean vector* or *centroid* is the name applied to a vector composed of the means of each of the p variables. The analog of the variance of univariate statistics is called the *covariance matrix*, **S**; the principal diagonal of this matrix contains the variances of the p variables. The off-diagonal elements are the covariances for each possible pairing of the variables. Corresponding to this matrix there is the *correlation matrix*, **R**.

The most widely used method of multivariate analysis is that known as **principal component analysis**, based on the simple extraction of eigenvalues and eigenvectors of the correlation matrix or the covariance matrix. This is the most typical case of an R-mode analysis (R signifies the correlation matrix, the usual starting point for psychometrical studies). Principal component analysis is directed towards reducing multivariate data to a smaller number of composite dimensions, producing a graphical display of the transformed points, and attempting to interpret the elements of the eigenvectors.

Closely related mathematically to principal components is the procedure called **Factor Analysis**. Factor analysis saw the light of day in psychological connections to which the very name bears witness–the *factors of the mind*. In the form in which factor analysis is practiced in psychology, a good case can be made for the validity of the procedure. However, biologists, geologists, paleontologists, mining engineers, technometricians, etc., began to use the packaging of factor analysis in a quite different analytical situation from that of the psychometricians. Jöreskog et al (1976) devoted an entire book to the semantic confusion attaching to attempts to use factor analysis in descriptive natural science, but to little avail. The factor analysis of natural science is a variant of principal component analysis and was termed *principal component factor analysis* by Jöreskog et al. (1976).

Still within the framework of the analysis of a single sample, we have methods that can be usefully thought of as being inverse forms of principal component analysis and its surrogates. These are the Q-mode methods (so named by psychometricians because Q precedes R in the alphabet). Briefly, an R-mode method employs associations *between variables*, whereas a Q-mode method uses associations *between individuals*. Thus, when we speak of R-space, it is the space mapped out by the p variables to which reference is being made. When the allusion is to Q-space, it is the space formed by the N observations that is meant.

An early procedure put forward is that known as Q-mode factor analysis, in which the general methodology of inverted principal compo-

nent analysis, with divers factor-analytical appurtenances, is applied to a matrix of pseudo-correlation coefficients computed between the individual specimens of the sample. Principal coordinate analysis was developed by Gower (1966) in an attempt to stabilize Q-mode usage. It is, in essence, inverted principal component analysis and, although it can be applied to a standard matrix of correlations, it is most applicable in situations where some kind of similarity matrix linking the N specimens is involved. Gower (1971) proposed a quite general similarity matrix for use with principal coordinates which has the advantage of allowing the simultaneous treatment of data comprising observations on continuous, discontinuous, dichotomous, and qualitative characters.

We need to establish an important point of statistical procedure concerning the way in which the results of R- and Q-mode analyses can be used. The eigenvalues and eigenvectors of principal components are used in many aspects of multivariate work, including tests of significance and in developing statistical models. The eigenvalues and eigenvectors of Q-mode analyses do not have the same scope and it is not legitimately possible to attempt to reify the elements of the vectors. The aim of Q-mode analysis is graphical, with the end in view of providing an appraisal of the inter-relationships between the individual specimens of the sample and the possible existence of clustering in the data. An example will serve to illustrate this point. A biologically homogeneous sample (i.e., a sample comprising specimens from a single interbreeding species) can, on analysis, display groupings into several separated entities due to size and shape differences. Such a situation arises naturally in many species of marine ostracods. The separations indicated by the Q-mode analysis can supply a starting point for the recognition of sub-samples for further study.

Is there a simple mathematical relationship between Q-mode and R-mode? The answer is yes. The basic work on the subject is that of Eckart and Young (1936) who enunciated the theorem for the singular value decomposition of a matrix. This theorem states that *any rectangular matrix can be decomposed into three matrices*. In factor-analytical connotations, the singular value decomposition is usually known as the **basic structure** of a matrix. By means of a simple transformation, one can pass from a representation in Q-space to one in R-space, a facility that can be exploited for computational purposes.

There are two popular methods of multivariate data-analysis that make use of the singular value decomposition of a data matrix. The more widely known of these is the method called *correspondence analysis*. The basic features of correspondence analysis were recognized by M. O. Hirschfeld (later Hartley) in 1935. Since then the concept has been rediscovered many times and can be found as the discriminant analysis of contingency tables and, in ecological work, as reciprocal averaging. Later, the method was derived independently in France in an environment of its own so that the terminology of Benzecri (1973) is quite alien to Anglophones. This fact can give the impression that something quite new has evolved, but, as proved by Hill (1974), this is not so. Correspondence analysis is a useful graphical technique which allows simultaneous representation of Q- and R-mode relationships on the same plot. Greenacre (1984) shows how the method can be worked into standard multivariate connections, particularly canonical correlations, and ter Braak (1986) has produced a broadly conceived procedure for ecological work which he calls canonical correspondence analysis. The hallmark of correspondence is the method of scaling the axes used in the plots.

Gabriel (1968) produced a Q-R-mode development called the biplot in which the Q-R-mode duality forms the basis as in the foregoing case. The biplot is similar to correspondence analysis in the general approach used, but the graphical aims are different and there is an asymmetry in the way the rows and columns of the data-matrix are treated which is not found in correspondence analysis.

One of the ideas central to the graphical aspects of one-sample studies is that the distances between objects should not be seriously distorted by the reduction in dimensionality. An informative way of gauging distances between two-dimensional representations of a highly multivariate situation is to superimpose a **minimum spanning tree** on the plot. A good reference for trees is Gordon (1981).

Two Samples

The best known of all the multivariate methods is that known as the *linear discriminant function*, proposed by Fisher (1936) in answer to a taxonomical problem in botany. While Fisher in England was looking at the problem of **distinguishing** between two populations on the basis of multivariate samples, Mahalanobis, in India, was attempting to find a means of measuring multivariate statistical distance between two or more populations. Independently of British and Indian work, Hotelling in the United States was producing a test for ascertaining whether two multivariate samples were statistically different, an analogue of the univariate *t*-test. Surprising as it may seem today, the close connections between all of these methods, introduced about the same time, was not recognized and much printers' ink was expended and chauvinistic polemics vented before all the difficulties were ironed out. Briefly, the Mahalanobis generalized statistical distance measures the distance between two populations. Hotelling's T^2 tells us whether this distance is significant, and Fisher's linear discriminant function computed between the two populations provides a means of assigning a new specimen to either of the populations, with the understanding that it really belongs to one of them. The algebraical connections between the three procedures are straightforward.

The linear discriminant function and the generalized statistical distance, as originally proposed, take no account of structure in the data, apart from the requirement that the samples be drawn from multivariate-normally distributed populations. Thus, little attention was paid to geometrical considerations concerning the orientations of the sample ellipsoids and their inflations, robustness, etc. Anderson and Bahadur (1962) considered the generalized distance when the covariance matrices are unequal and proposed a multivariate version of the Behrens-Fisher univariate procedure. Non-linear observations require some special technique such as the *quadratic discriminant function* but the usefulness of this method for general work is not great as it does not yield a vector of reifiable discriminant coefficients.

Can the analysis of two samples by discriminant functions be related to the methods of the first section? There is one obvious connection, to wit, a geometrical interpretation in terms of the eigenvalues and eigenvectors of the individual samples. The information held in principal components can be applied to the interpretation of the geometry of clusters of points forming the data-ellipsoids. This is one good reason for letting principal component analyses form the basis of a large-scale study. To opt for a factor-analytical solution exclusively cuts you off from a wide range of developments that would otherwise be open to you.

Several Populations

The step from two to three or more samples is not as straightforward as one might hope. The reason for this is that the methods of calculation become more complicated, and it is no longer possible to use simple matrix manipulations for producing the analogs of discriminant function coefficients, for example. The method used for studying samples from k populations is conveniently referred to as *canonical variate analysis*, a set of interlocking techniques that encompass the multivariate analysis of variance, the graphical display of results (discriminant coordinates) and the properties of the eigenvalues and eigenvectors of the sum of within-groups matrices of sums of squares and cross-products, $\mathbf{W}$, and the corresponding between-groups sums of squares and cross-products matrix, $\mathbf{B}$. The sum of these two matrices gives $\mathbf{T}$, the total sum of squares and cross-products matrix. There are simple relationships between the eigenvalues of all combinations of any two of these matrices.

MANOVA, the multivariate analysis of variance, asks whether the centroids of the populations differ significantly; this is a simple generalization of one-way ANOVA. Multiple discriminant analysis looks to differentiating between samples and assigning new individuals to one of the constituent populations. The elements of the significant eigenvectors are the coefficients of the discriminant functions. For two samples, the linear discriminant function of the preceding section is obtained.

Relationships Between Sets

One may wish to ascertain whether a set of morphological measures is correlated with a set of ecological factors. Many attempts at doing this are made by means of principal components or one of its surrogates in which morphological and ecological measures are mixed in the same matrix. The technique of canonical correlation is more directly applicable, however (Cooley and Lohnes, 1971; Pimentel, 1979). Canonical correlation produces a set of correlations between various linear combinations of the variables of the two sets. As already noted, canonical correlations can be conveniently worked into the method of correspondence analysis (Greenacre, 1984; ter Braak, 1986).

When Things Go Askew

All that has been said in the foregoing sections applies strictly only when the data conform with the multivariate Gaussian model. Much morphometric material does just this and any deviations that occur are so slight as to be of no importance for the outcome of an analysis. Principal component analysis tends to be robust to minor deviations from the theoretical distribution. Canonical correlation is, on the other hand, sensitive to deviations; this is also true of canonical variates.

The question that now arises is, when do you know that your data are "dicey"? My advice is to start every investigation with a careful graphical appraisal of the data: histograms of each of the variables, scatter diagrams of all pairs. Explicit tests for multivariate normality are available, but these tend to be rather extravagant with respect to computing time and can only be realistically performed on a main-frame computer. A useful graphical technique is the Q-Qprobability plot, which can be applied to multivariate data via the Mahalanobis generalized distance. If the preliminary study indicates that your data conform reasonably well with the multivariate normal distribution, there is probably no need to become involved with any of the techniques discussed in the following pages. But please keep in mind that the fact that data can be shown to be univariate normal for each variable is no guarantee that they are together multivariate normal. For multivariate data, observations are often only found to be atypical when some value is considered in relation to the *other values of the same sample*.

Atypical Values

We begin by asking what is meant by an "atypical value." An atypical value is one that deviates markedly from its fellows but is not alien to the sample to which it belongs, in the sense that it is not a wrong observation. In other words, an atypical value can be biologically perfectly acceptable but inadmissible in the eyes of the statistician. The performance of a statistical procedure may be adversely affected by atypical values to the extent that a misleading result is yielded. As a consequence of this, analyses of different samples of the same material can suffer from poor repeatability.

The most immediate source of atypical values in a biologically homogeneous sample is **polymorphism**. Polymorphism in ornamental features is sometimes accompanied by polymorphism in the outline of the shell. This is particularly common among ostracods in which pleiotropic relationships occur frequently. For example, in the Paleocene species *Leguminocythereis lagaghiroboensis* Apostolescu, regular reticulation of the lateral surface of the carapace is accompanied by

the development of a strongly extended posterior in both males and females.

Campbell and Reyment (1980) analyzed polymorphism in a Late Cretaceous benthic foraminifer *Afrobolivina afra* Reyment. It was found that pseudomegalospheric individuals tend to distort bivariate scatter diagrams, principal component plots and canonical variate graphs.

Some statisticians advocate "trimming" of data, whereby markedly atypical values are deleted from the sample. This is not to be recommended in biological studies for the reason that important specimens may be removed from the analysis. However, the solutions available for "robustifying" a multivariate analysis can hardly be claimed to supply a perfect answer to the problem.

Robustifying procedures are usually constructed so that full unit weight is given to reasonable observations and reduced weight to atypical values. In multivariate work, this is done by utilizing the individual generalized statistical distances for each observation in relation to the centroid of the sample for assessing the weights and a robust average value and a robust estimate of the covariance matrix. Rhoads (1984, p. 248) pointed out that the main weaknesses in most multivariate analyses is their reliance on standard packages, which leads to a sterotyped treatment of morphometric problems. Seber (1984) has reviewed this question in the light of the reluctant acceptance of robust procedures by statistical practitioners.

Influential Observations

An influential value is one that does not show up as a multivariate outlier but which, nonetheless, would bring about a substantial change in the results of an analysis if it was omitted. This subject has been given close attention by Krzanowski (1987a, 1987b). This subject is taken up in the section on principal component analysis and cross-validation.

Reification and the Stability of Eigenvectors

In most analyses of morphometric distance-measures, the analyst wishes to provide the components of the eigenvectors with a biologically meaningful interpretation. This desire stems from a paper by Teissier (1938), who applied an early variant of principal component factor analysis (c.f. Jöreskog et al., 1976) to the study of sexual dimorphism in the crab *Maia squinada*. He interpreted the first principal component in terms of variation in size and shape, an idea that was resuscitated by Jolicoeur and Mosimann (1960). That the first principal component can often behave as an indicator of variation in size has been suggested by many other workers. Jolicoeur and Mosimann (1960) also claimed that the second principal component for distance-measures on an organism is a descriptor of differential changes in shape.

As far as my experience goes, the principal component interpretation of size and shape can be reasonable if the first two eigenvectors extract almost all of the variation in the covariance matric of logarithmically transformed variables. This implies that the covariance matrix is of rank two. If, however, the eigenvalues trail off slowly, the principal component method is not useful. Rao (1964) tried to provide the principal component approach with a theoretically satisfactory accoutrement but was only partially successful because of the difficulty in reconciling biological reality with statistical facts.

What happens if we do not have an ideally constituted data-set? This is a really difficult point and one that is not greatly loved by the main host of users of multivariate procedures. What is to be done with the myriad of data-analyses that are flawed in some manner and how can we judge whether one of these analyses is defective to such an extent that the reifications proposed are misleading? I regret to have to say that there is no patent answer to these vexing questions for the very reason I have already mentioned, namely, that standard packages take little or no account of

deviatory data (Rock, 1987). Fortunately, the effects of non-conformability with statistical theory on the ordination of multivariate data-points is much less severe. As Maurice Bartlett once said, any linear combination will ordinate multivariate data to some degree providing there is structure in the set of observations.

Campbell (1979, 1980a, 1980b) and Campbell and Reyment (1978) examined the question of stability in the coefficients of principal components and canonical variates. In canonical variate analysis, the question asked is, how stable are the elements of the canonical vectors in biological work if there is overweighting of redundant or near-redundant directions of within-group variation? One might ask what does it matter if the elements of the vectors are not stable. The answer to this is that if we sample repeatedly from the same source, we have a right to expect that a statistical method applied to these samples yields the same results within the limits of sampling variation. If something is unduly influencing the results, we can place little confidence in what we are doing. A good idea of the health of a sample can be obtained by a jacknifed principal component analysis, which will give indications as to the stablity of the eigenvectors. This has been used by Krzanowski (1987a, 1987b).

Canonical Variates and Ridge Regression. For the purposes of discussing stability, it is convenient to regard canonical variate analysis as a two-stage rotational procedure. The first rotation produces the principal components of the pooled samples via the within-groups covariance matrices. It is useful to transform the within-groups concentration-ellipsoid into the corresponding concentration-circle by scaling each eigenvector by the square root of the corresponding eigenvalue. The second rotation corresponds to a principal component analysis of the group means in the space of the orthogonal variables.

Consider now the variation along each orthogonalized value, i.e., principal component. Where there is but little variation between groups along a particular direction, and the *corresponding eigenvalue is small*, marked instability may result in some of the coefficients of the canonical variates.

The use of so-called *shrinkage* or ridge-regressional constants overcomes the instability in the coefficients. The shrinkage constants are added to the eigenvalues before they are used for standardizing the corresponding principal components. This maneuver has the effect of greatly expanding the variances of the orthogonalized variables.

Steps in the Calculations:

1. Begin with the usual within-groups matrix of sums of squares and cross-products matrix (SSQPR) **W** and the between-groups matrix of SSQPR, **B**. Matrices **W** and **B** are then standardized to the corresponding correlation matrices $\mathbf{R}_W$ and $\mathbf{R}_B$ by computing

$$\mathbf{R}_W = \mathbf{S}^{-1}\mathbf{W}\mathbf{S}^{-1} \tag{1}$$

and

$$\mathbf{R}_B = \mathbf{S}^{-1}\mathbf{B}\mathbf{S}^{-1}, \tag{2}$$

where the diagonal matrix **S** is the square root of the diagonal elements of **W**.

2. Compute the eigenvalues, e_i, and eigenvectors, u_i, of $\mathbf{R}_W$.

3. The eigenvectors are scaled by multiplying them with the square root of the corresponding eigenvalues. This results in within-groups sphericity.

4. Shrunken estimators are introduced by adding shrinking constants k_i to the eigenvalues e_i before the scaling exercise has been carried out. Thus,

$$\mathbf{K} = \text{diag}(k_1, ..., k_p)$$

and

$$\mathbf{U}^* = \mathbf{U}(\mathbf{E} + \mathbf{K})^{-1/2} . \tag{3}$$

5. Construct the between-groups matrix in the space of the within-groups principal components: note that the matrices **Q** and **U** are subscripted as in (3), to wit, $(k_1, ..., k_p)$.

$$\mathbf{Q} = \mathbf{U}^{*T}\mathbf{R}_B\mathbf{U}^* . \tag{4}$$

The ith diagonal element of **Q** in (4), q_i, is the between-groups SSQ for the ith principal component associated with $\mathbf{R}_B$. The diagonal elements, q_i, play an important role in the interpretation of the stability status of the canonical vectors.

6. Compute tr $\mathbf{Q}_{(0, ..., 0)}$ and then perform the eigen-extraction for **Q**. This yields the usual canonical roots, f, and canonical vectors for the principal components, $\mathbf{a}^U$. These canonical vectors are given by the relationship

$$\mathbf{c}^U = \mathbf{U}^*_{(0,...,0)}. \tag{5}$$

7. The shrunken estimators are found directly from the eigenvectors $\mathbf{a}^S$ of **Q** with

$$\mathbf{c}^S = \mathbf{U}^*\mathbf{a}^S \tag{6}$$

for the full range of k_i.

Equation (6) yields the desired canonical vectors.

It is often found that marked instability is associated with a small value of e_p together with a small value of the corresponding diagonal element of **Q**, q_p. If the eigenvector corresponding to e_p contains one or two dominant loadings, these variables tend to be the ones that are unstable under repeated sampling. A reasonable idea of possible instability in canonical vectors can therefore be obtained by inspecting the smallest eigenvector of the within-groups matrix of correlations.

Illustration

In order to illustrate the concepts involved in the foregoing account, I shall analyse data on a species of ostracods from the Cretaceous of the sub-surface of the Tarfaya Basin, Morocco. The measurements consist of length and height of the carapace and the distance of the adductor muscle field from the anterior margin and the dorsal margin on the species *Veenia rotunda*. There are four samples comprising sample-sizes 18, 8, 31, and 24 individuals. The basic statistics for these data are listed in Table 1, to wit, the between-groups SSQCP-matrix, the within-groups correlation matrix, the within-groups standard deviations and the sample means.

Let us first examine the eigenvalues and eigenvectors of the within-groups correlation matrix, listed in Table 2. As is usually the case in morphometric work, the first eigenvector has approximately equal elements. The second and third eigenvectors contain positive and negative elements, with some large loadings. The smallest eigenvector, which is connected to a very small eigenvalue, contains two quite large loadings, to wit, that for length of carapace and that for dorsal distance of the adductorial field; moreover, these loadings bear opposite signs. Bearing in mind a previous remark, we can begin to suspect that these elements will be unstable in the canonical variate analysis. This feeling is reinforced by the fact that q_4 is very small (Table 2).

The results of the standard canonical variate analysis are listed in Table 3 together with the same analysis, with the addition of a shrinkage constant to the contribution of the fourth principal component. Tr(Q) for the usual canonical variate analysis is 1.768 and for the adjusted analysis, 1.721, so the loss of discriminatory information is negligible. Consider now the standardized vector-elements. In the case of the first canonical vector, there is a drop from about 0.6 to 0.1 for length and a

Table 1. Basic statistics for four samples of *Veenia rotunda*.

Between-populations matrix (upper diagonal) and within-groups correlation matrix (lower diagonal).

	var1	var2	var3	var4
var1	1800.0	341.8	370.0	1403.0
var2	0.8354	393.9	174.5	656.2
var3	0.4755	0.4871	95.65	288.3
var4	0.9255	0.7694	0.4302	1095.0
Standard Deviations				
	7.090	2.017	1.608	6.293

Table 2. Principal component analysis of within-groups correlations and values of diag **Q**.

	eigenvectors			
	vector 1	vector 2	vector 3	vector 4
length	0.5510	-0.2293	-0.2154	0.7729
height	0.5233	-0.2323	0.8233	-0.1800
adductor-1	0.3714	0.9226	-0.1021	-0.0195
adductor-2	0.5335	-0.2845	-0.5143	-0.6081
eigenvalues	3.0100	0.6841	0.2407	0.0652
diag Q	0.8290	0.1484	0.7433	0.0474
	tr **Q** = 1.7681.			

N.B. adductor-1 denotes the distance from the anterior margin to the adductorial tubercle; adductor-2 denotes the distance from the dorsal margin to the adductorial tubercle.

corresponding increase for the fourth variable. The changes in the second and third elements are slight. Instability evidenced in the second canonical vector is not of any consequence.

Table 3. Canonical variate analyses for the first two canonical roots.

A. Usual analysis.

	canonical vectors (standardized)		
	vector 1	vector 2	
length	-0.5998	0.3420	
height	1.4400	0.4940	
adductor-1	0.1608	-1.1440	
adductor-2	-0.1344	0.0964	
canonical roots	1.6318	0.1332	0.0031

B. Analysis using shrinkage constant.

	canonical vectors (standardized)		
	vector 1	vector 2	
length	-0.1033	0.1457	
height	1.3420	0.5248	
adductor-1	0.1492	-1.1440	
adductor-2	-0.5334	0.0665	
canonical roots	1.5877	0.1327	0.0003

We can also briefly consider the example published by Campbell and Reyment (1978). Here, only the main points of the analysis are taken and for the full details, the reader is referred to the work cited.

Nine variables were measured on the test of the Maastrichtian foraminiferal species *Afrobolivina afra*. The smallest eigenvalue of the within-groups correlation matrix accounts for only 1.7% of tr $\mathbf{R_W}$. The corresponding eigenvector contains two large elements for characters 2 and 4. The corresponding between-groups SSQ for these principal components is $q_9 = 0.21$, which is relatively small. These three conditions are strongly indicative of the possibility of instability in the canonical vectors. This could be shown to be the case. Shrinkage of the contribution of the ninth principal component led to greatly improved stability in the canonical variate coefficients without marked loss of discriminatory information (Campbell and Reyment, 1978, p. 355).

In many canonical variate analyses of morphometric variables, an attempt is made to reify the coefficients, usually in terms of variation in shape and size. This idea carries over from what is often done in principal component analyses. Despite the fact that I have been guilty of attempting just this, I find it difficult, on mature consideration, to provide the reification of canonical vectors with an acceptable rationale unless all covariance matrices are equally inflated and identically oriented. This is almost never the case. Numerous examples of canonical variate reifications are reviewed in Reyment, Blackith and Campbell (1984).

Robust Estimation Procedures

Robust multivariate procedures are a simple modification of the standard methods of univariate analysis. The contribution of an observation to the statistic of interest is given full unit weight if it is a reasonable value; otherwise its contribution is downweighted.

A *robust estimator* can be defined by introducing a weight-function which depends on the discrepancy between an observation and some robust average value relative to a robust measure of scatter. This can be done by means of the **influence function** (cf. Hampel et al., 1986, p. 40) and bounding the influence of observations with unduly great discrepancies. The influence function allows a comparison of the results of an analysis with, and without, a suspected atypical observation. For multivariate data, the fundamental expression of discrepancy is the *Mahalanobis Generalized Statistical Distance*. If we let x represent the $p \times 1$ vector of sample-means, and $\mathbf{S}$ the sample covariance matrix, then the squared distance of the mth observation from the centroid of the observations is:

$$d_m^2 = (x_m - \overline{x})^T \mathbf{S}^{-1} (x_m - \overline{x});$$

$$m = 1, 2, ..., N . \tag{7}$$

Q-Q probability plots of the d_m^2 may be used to examine the assumption of a multivariate Gaussian distribution. The way in which the plot works is that atypical observations tend to deflate the covariances and inflate the variances, which will affect the generalized distance for that observation and hence distort the remainder of the plot (Seber, 1984, p. 165).

The robust estimation of multivariate location and scatter can be made efficiently by the method of M-estimation (Hampel et al., 1986, pp. 37, 100). The name derives from "generalized Maximum likelihood estimation." The appropriate measure of discrepancy is, again, the generalized statistical distance. Inasmuch as M-estimators have begun to appear more and more frequently in the statistical literature, I give below the main features of the methodology.

The usual method of estimation

We require the squared Mahalanobis statistical distances as in (7). A fixed proportion, z, (say 0.1) of the x_m with the greatest distances, are temporarily removed from the sample and the values of $\bar{x}$ and $\mathbf{S}$ recomputed (based thus on $N(1-z)$ individuals). One then iterates to a final estimate of $\mathbf{S}$: this procedure is known as multivariate trimming. It suffers from several drawbacks because it produces a biased estimate.

M-estimation

M-estimators offer a better approach. The mean can be estimated iteratively from

$$\bar{x} = \frac{\displaystyle\sum_{m=1}^{N} w_1 d_m x_m}{\displaystyle\sum_{m=1}^{N} w_1 d_m} \tag{8}$$

and the covariance matrix from

$$\mathbf{S} = \frac{1}{N} \sum_{m=1}^{N} w_2 d_m^2 (x_m - \bar{x})(x_m - \bar{x})^T \tag{9}$$

Here w_1 and w_2 are the appropriate weight-functions estimated by the method proposed by Maronna (1976).

The weak point of M-estimation is that there is a low level of "breakdown," that is, the proportion of outliers tolerated (Seber, 1984, p. 156). The method of estimation I use is that developed by Campbell (1979, 1980a, 1982), namely, the redescending psi-function. According to this procedure, one estimates the covariance matrix as in:

$$\mathbf{S} = \frac{\displaystyle\sum_{m=1}^{N} w_1^2 (x_m - \bar{x})(x_m - \bar{x})^T}{\displaystyle\sum_{m=1}^{N} w_1^2 (d_m) - 1} \tag{10}$$

Table 4. Robust analysis for samples 1 and 2 of *Veenia fawwarensis* from the Santonian of Israel. Usual correlations in upper diagonal, robust correlations in lower diagonal.

Sample 1

	var 1	var 2	var 3	var 4	var 5	var 6	var 7
var 1	1.000	0.617	0.270	0.892	0.816	0.836	0.616
var 2	0.583	1.000	0.347	0.727	0.896	0.692	0.868
var 3	0.209	0.338	1.000	0.157	0.320	0.417	0.222
var 4	0.875	0.636	0.059	1.000	0.883	0.812	0.737
var 5	0.850	0.746	0.255	0.869	1.000	0.818	0.828
var 6	0.806	0.500	0.372	0.735	0.726	1.000	0.624
var 7	0.547	0.569	0.095	0.629	0.568	0.352	1.000
means							
usual	1.030	0.557	0.221	0.451	0.428	0.481	0.439
robust	1.035	0.564	0.221	0.454	0.435	0.486	0.445
ratio of robust to usual variance							
	0.87	0.27	0.99	0.75	0.46	0.71	0.35

Sample 2

	var 1	var 2	var 3	var 4	var 5	var 6	var 7
var 1	1.000	0.753	0.539	0.871	0.938	0.885	0.680
var 2	0.760	1.000	0.760	0.952	0.863	0.604	0.703
var 3	0.526	0.710	1.000	0.776	0.521	0.278	0.299
var 4	0.934	0.846	0.487	1.000	0.912	0.683	0.653
var 5	0.955	0.800	0.529	0.945	1.000	0.851	0.757
var 6	0.917	0.732	0.649	0.857	0.870	1.000	0.691
var 7	0.587	0.668	0.551	0.578	0.568	0.547	1.000
means							
usual	0.996	0.549	0.228	0.444	0.413	0.468	0.466
robust	0.999	0.548	0.216	0.441	0.415	0.472	0.469
ratio of usual to robust variance							
	0.73	0.13	0.10	0.27	0.48	0.81	0.58

A comprehensive account of the field of robust statistics is given in Hampel et al. (1986).

Robust Canonical Variate Analysis

Details of how to program robust procedures of estimation in canonical variate analysis are given in Campbell and Reyment (1980, pp. 209-211). In the example illustrated in Figure 2a-c and Table 4, use is made of probability plots to unveil atypical values; here, the Wilson-Hilferty cube-root transformation of a gamma variable to Gaussian form is used. The ordered values of $d_m^{2/3}$ are plotted against the ordered quantiles of a Gaussian distribution.

It is necessary to be on the watch for the absolute influence of a differing observation, even if the divergent point may seem to have a large relative influence. Looks may deceive the eye and it is often found that such a point may have little effect on the statistics of interest. Figure 1 shows a tight, elliptical concentration of points with three outlying observations A, B, and C. The effect of point A is to inflate the variances of x_1 and x_2, but it does not markedly influence the correlation between these. Point B reduces the correlation and inflates the variance of x_1 but has little effect on the variance of x_2. Point C exerts little influence on the variances but it reduces the correlation.

Illustration

I shall illustrate robust canonical variate analysis by reference to a Santonian (Upper Cretaceous) species of ostracods from Israel, *Veenia fawwarensis*, recently studied in detail by Abe et al. (1988). The data comprise seven distance measures on the carapace of the ostracod (cf. Abe et al., Figure 1). These variables are (1) length of the carapace, (2) height of the carapace, (3) distance from the eye tubercle to the adductor tubercle, (4) distance from the adductor tubercle to the postero-ventral angle, (5) posterior carapace-height, (6) length of the hinge, (7) breadth of the carapace. There are three samples consisting of 32, 26 and 24 specimens, respectively. The analysis summarized in Table 4 and Figure 2a-c has the following points of interest: The first sample is reasonably multivariate normal, apart from one atypical value (Figure 2a). The second sample is also multivariate normal with the exception of two atypical values (Figure 2b). There are no atypical values in the third sample (Figure 2c).

How serious are the effects of atypical values on correlations and means? It will be seen in Table 4 that all means are influenced and the correlations more so. In the first sample, the ratio of the robust variance to the usual variance for variables 2, 5, and 7 differs in each case greatly from unity. We also

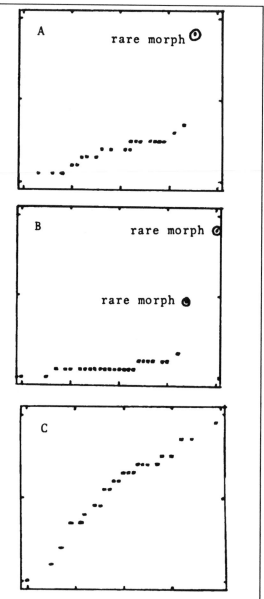

Figure 2. a. Q-Q probability plot for sample 1 of *Veenia fawwarensis* showing one atypical example. This example was identified as rare ecomorph of the species. Abscissa = normal order statistics; ordinate = ordered observations. $N=32$. b. Q-Q probability plot for sample 2 of the ostracod species *Veenia fawwarensis* displaying two atypical observations. These were identified as the rare ecomorph of the species found in sample1. Abscissa and ordinate as in figure 2a. c. The Q-Q probability plot for the third sample. The procedure did not indicate any atypical values to occur in the data. $N=24$.

observe that correlations with variable 7 tend to be influenced quite strongly. The ratios between robust and usual variances for sample 2 are strongly perturbed throughout. The estimates for robust and usual correlations, variances and means are identical for sample 3, which, as you will recall, contains no atypical values.

Table 5. Canonical variate analysis for *Veenia fawwarensis*.				
	First canonical vector		Second canonical vector	
	usual	robust	usual	robust
var 1	14.49	14.42	33.68	33.06
var 2	37.04	36.35	18.13	17.80
var 3	-20.42	-20.02	6.86	6.73
var 4	25.10	24.63	-35.50	-34.85
var 5	13.43	13.19	-22.08	-21.67
var 6	-9.42	-9.24	25.25	24.69
var 7	-45.19	-44.36	13.33	13.09
roots	0.7325	0.7325	0.4399	0.4399

Table 6. Canonical variate analysis for Dicathais, Campbell (1979).				
	First canonical vector		Second canonical vector	
	usual	robust	usual	robust
var 1	-0.50	-0.67	0.48	0.47
var 2	0.47	0.78	-0.99	-0.98
var 3	-0.35	-0.23	-0.31	-0.51
var 4	1.62	1.50	0.43	0.86
roots	2.13	2.35	1.68	1.87

It is instructive to see to what extent the atypical values influence the canonical vectors. The first two canonical vectors and canonical roots are listed in Table 5. It will be seen that the robust and usual estimates hardly differ from each other. This is not always so, however, as can be seen from an analysis for 14 samples of a species of the predaceous gastropod Dicathais from Australia and New Zealand analysed by Campbell (1979). Four morphologically diagnostic variables were measured on the shell of the species. The first two canonical roots and vectors are listed in Table 6. Both vectors are changed by robust estimation. In the first vector, variable 2 is fairly strongly affected. In the second vector, it is the fourth element that is most strongly altered.

It has been my experience that the canonical vectors are only slightly influenced by a small number of atypicalities, if the data are otherwise multivariate Gaussian. Canonical roots tend to be more markedly influenced, as for example in the case of Dicathais.

Robust Principal Components

Campbell (1980a) took up the subject of robust principal component analysis. Various technques are available such as the jacknife and the bootstrap (Efrom, 1979; Efrom and Gong, 1983). Robust estimation can be a useful complement to such procedures (Hampel et al., 1986).

A principal component analysis of the covariance matrix $\mathbf{S}$ (or associated correlation matrix $\mathbf{R}$) seeks a linear combination,

$$y_m = \mathbf{U}^T x_m,$$

of the original variables x_m such that the sample variance of y_m is a maximum. The solution is given by the eigen-reduction of S,

$$\mathbf{S} = \mathbf{U}\mathbf{E}\mathbf{U}^T.$$

Clearly, a direction u_i should not be determined by one or two atypical values. This argument is well known from the computation of correlation coefficients. In a situation in which all observations but one, an atypical value, form a tight ellipse, this rogue-observation may well determine the direction of the first eigenvector (cf. Figure 1). The first few principal components are sensitive to outliers that inflate variances and covariances and hence, correlations; the last few are sensitive to outliers that add "spurious dimensions" (Seber, 1984, p. 171). Although the subject does not seem to have been studied in the literature, it seems reasonable to assume that small atypicalities along the direction of the first principal axis will have little influence on the stability of the eigenvectors.

An obvious way of robustifying a principal component analysis is to replace $\mathbf{S}$ by its robust estimator $\mathbf{S}^c$, which is the solution by M-estimators. This procedure weights an observation according to its total distance d_m^c from the robust estimate of location, which can be decomposed into constituents along each eigenvector. Thus, if an atypical observation has a large component along one direction and small ones along the other directions, this situation can be accounted for by applying robust M-estimation of means and variances to each principal component. The result is that the direction cosines will be chosen to maximize the robust variance of the resulting linear combination.

Illustration

The example chosen here to illustrate robust principal component analysis is a sample ($N = 28$) of the Santonian ostracod species *Veenia fawwarensis* upon which seven distance-measures are available (Abe et al., 1988). The results of essential parts of the analysis are listed in Table 7. The calculations were made on the mean-centered covariance matrix of the variables. It was found that there are no differences at all in the elements of the first eigenvectors, but increasingly strong differences in subsequent eigenvectors. Thus, the effects of an atypical value (a rare morph) have been to leave the direction of the first principal axis of the ellipsoid of scatter untouched, but to influence the orientations of the other axes to varying

Table 7. Usual and robust principal components for ostracod data: selected results ($N = 28$).

	vector 2		vector 3		vector 4		vector 6	
	us.	rob.	us.	rob.	us.	rob.	us.	rob.
var 1	0.16	0.08	-0.26	0.14	-0.05	0.26	-0.10	0.35
var 2	-0.11	-0.17	0.63	-0.54	-0.14	-0.27	-0.27	0.29
var 3	-0.79	-0.55	0.11	0.23	0.35	-0.64	-0.03	-0.01
var 4	0.29	0.01	-0.06	-0.14	-0.22	0.39	-0.48	0.25
var 5	0.09	0.01	0.09	0.15	-0.20	0.13	0.83	-0.85
var 6	-0.03	0.37	-0.60	0.71	0.43	-0.16	-0.04	0.03
var 7	0.48	0.73	0.39	-0.30	0.76	-0.50	0.06	0.01

degrees.

Procrustean Methods

It is natural to enquire how well different methods of multivariate analysis applied to the same data duplicate each other. It is sometimes assumed that the scores produced by a canonical variate analysis should reasonably agree with those obtained by a principal component analysis of the within-groups matrix of SSQCP, although there is no intrinsic justification for this belief.

Because of indeterminancies in the results yielded by some scaling techniques, any attempt at comparing scores obtained by different methods must allow some kind of transformation of one of the sets of coordinates in order to seek as close a fit as possible to the other set of coordinates, which is held fixed- hence the nomenclatorial analogy to the activities of the mythical *Prokroustes*. A good account of Procrustean analysis is to be found in Gordon (1981, pp. 106-112). In the present connexion, we shall only be concerned with orthogonal Procrustean analysis: there are other varieties (Gower, 1987, p. 57).

Here we consider two *ordinations*, Y_1 and Y_2, which are two sets of points, $P_1, ..., P_N$ and $Q_1, ..., Q_N$ in r-dimensional space. The goal of orthogonal Procrustean analysis is to fit Y_2 to Y_1, using the translation and rotation appropriate to the rotation of rigid bodies such that

$$m_{12}^2 = \sum_{i=1}^{N} \Delta^2 (P_i Q_i) \qquad (11)$$

is minimized. The translation requires that the centroids of the two configurations be superimposed, which is realized by subtracting the column means from Y_1 and Y_2.

$$m^2_{12} = \text{Tr}(Y_1 Y_1^T + Y_2 Y_2^T - 2S) \qquad (12)$$

where S has zero elements except for along its diagonal, where we have that $s_1 >= s_2 >= .... >= 0$, and is the matrix of singular values of Y (cf. Jöreskog, Klovan and Reyment, 1976, chapter 2, section 11). The ordination will, however, usually be on different scales, so that it may be necessary to estimate some scaling factor to be applied to Y_2.

The scaling for fitting $\mathbf{Y}_1$ to $\mathbf{Y}_2$ is not the same as for doing the reverse fitting. Hence, the value of m^2_{12} in (11) is not the same as $m^2{}_{21}$. If, however, both $\mathbf{Y}_1$ and $\mathbf{Y}_2$ are normalized so as to have unit sum-of-squares, then

$$m^2_{21} = m^2_{21} = (1 - \mathrm{Tr}\, \mathbf{S})^2 \, ,$$

which is independent of the order of matching. This is sometimes referred to as Gower's m^2 statistic (Gordon, 1981, p. 110).

Procrustean methods have a useful part to play in multivariate morphometrics: firstly, for contrasting ordinations of the same data by different multivariate methods (cf. Gower and Digby, 1084, p. 22) and, secondly, for ascertaining how well distance-matrices for two, or more, different functional entities of an organism can produce an ordination. A useful theoretical paper is that of Sibson (1978) in which the goodness-of-fit between two configurations is treated. Procrustean analysis can sometimes provide an illustration of how answers differ for the same set of data and hence give a clue as to why this is so.

Illustration

A simple application of the technique to scores of principal component analyses obtained by robust estimation and by the usual method serves to illustrate the method. The data analysed are, as before, 28 specimens of the ostracod species *Veenia fawwarensis* from the Santonian of Israel, used in the example on robust principal component analysis. Seven variables were measured on the carapaces of the ostracods. These data are multivariate Gaussian, apart from the presence of an atypical value. The scores were computed in the usual manner.

Visual inspection of the scores (Table 8) might lead one to suspect that these coordinates differ markedly, although the patterns of signs are the same. The results listed in Table 8 (first 10

specimens only) show that the coordinates of set 1 on best-fit in two dimensions after Procrustean rotation differ only slightly from the coordinates of set 2 on best-fit on set 1 in two dimensions after Procrustean rotation. The close agreement in the two sets of coordinates is further documented by the small values of the residuals after fitting. This is

Table 8. Procrustean analysis of first 10 observations for sample analysed in Table 7: *Veenia fawwarensis* from the Santonian of Israel.

specimen	Usual scores coordinate pairs 1	2	Robust scores coordinate pairs 1	2	
1	0.07	0.20	0.41	1.10	
2	0.20	0.06	1.13	0.34	
3	-0.13	-0.14	-0.70	-0.78	
4	0.35	0.03	1.94	0.19	
5	0.23	-0.14	1.25	-0.23	
6	0.13	0.23	0.72	1.30	
7	0.25	0.06	1.38	0.35	
8	0.11	0.05	0.65	0.30	
9	0.40	-0.17	1.22	0.65	
10	0.37	0.26	2.08	1.43	
	configurations fixed		fitted		residuals
1	0.0258	0.0975	0.0310	0.0907	0.0085
2	0.0843	0.0354	0.0882	0.0275	0.0088
3	-0.0641	-0.0544	-0.0622	-0.0600	0.0056
4	0.1502	0.0241	0.1539	0.0147	0.0101
5	0.0960	-0.0558	0.0971	-0.0189	0.0370
6	0.0509	0.1139	0.0566	0.1068	0.0091
7	0.1046	0.0367	0.1084	0.0281	0.0094
8	0.0446	0.0331	0.0483	0.0257	0.0083
9	0.1278	-0.0666	0.0959	0.0527	0.1419
10	0.1615	0.1243	0.1673	0.1145	0.0114

a trivial demonstration, but it serves to make clear the utility of the method.

Compositional Data

Any vector x with non-negative elements $x_1, \ldots , x_p$, representing proportions of some whole is subject to the constraint

$$x_1 + \ldots + x_D = 1 \qquad (13)$$

In morphometric work, such a vector can arise if the observations have been constrained for the purpose of "standardizing" in the hope that this will make the observations size-equivalents. The classical reference is Weldon's appendix to Pearson (1897), dealing with the analysis of "spurious" shrimps.

Almost everybody will proceed to a routine multivariate analysis of compositional data, including many professional statisticians, sad to say. It is indeed lamentable that even quite recent statistical textbooks fail to present the perils attendant on the analysis of closed data-sets.

Another morphometric example concerns semi-qualitative observations on size. Eggs are graded into classes on their slipping through holes that provide the grounds for the grading. Bivalves are classified into size-ranges on a roughly administered sieving procedure (see Aitchison, 1986, pp. 18, 371).

Harking back to (13), we note that the compsosition indicated by this vector is completely specified by a d-part subvector $(x_1, ..., x_d)$ where d = D - 1. Thus,

$$x_D = 1 - x_1 - ... - x_D.$$

Hence, a D-part composition is essentially a d-dimensional vector. For the purposes of statistical work, Aitchison (1986, p. 27) defines the space appropriate to this d-dimensional vector as a d-dimensional simplex, embedded in D-dimensional real space. In terms of the standard notation of sets, this may be stated as

$$\mathcal{J}^d = ((x_1, ..., x_D): x_1 > 0, ..., x_D > 0;$$

$$x_1 + ... + x_D = 1) .$$

The natural sample-space of an $N \times D$ compositional data-matrix $\mathbf{X}$ is the d-dimensional simplex $\mathcal{J}^d$. Each row of $\mathbf{X}$ is a D-part composition which is represented by a point in $\mathcal{J}^d$. Hence, the compositions in $\mathbf{X}$ appear as N points in $\mathcal{J}^d$. Another geometric representation for $\mathbf{X}$ takes account of the fact that each column of $\mathbf{X}$ is an N-vector of positive numbers and so can be repre-

sented as a point in $\mathcal{R}_+^N$. Thus, the compositional data in $\mathbf{X}$ are represented by D points in $\mathcal{R}_+^N$. These two definitions are of consequence in applied multivariate work (R-mode versus Q-mode).

Aitchison (1986) provides a definition of covariance structure that is free from the deficiencies of the crude covariance structure for compositional data, to wit, the negative bias, diffficulties with respect to subcompositions and basis, as well as treatment of how to treat the problem of null correlation. His solution is to use the log-ratio covariance matrix. The sample log-ratio covariance matrix is defined as

$$\mathbf{S} = \text{cov}(\log \frac{x_i}{x_D}, \log \frac{x_j}{x_D}); \ i, j = 1, ..., d \qquad (14)$$

For the principal component analysis of compositional data, the centered log-ratio covariance matrix is to be recommended, since it allows a symmetric treatment of all D parts. This matrix is obtained by replacing the single component divisor x_D by the geometric mean

$$g(x) = (x_1, ..., x_D)^{1/D} ,$$

to give the $D \times D$ covariance matrix

$$\mathbf{G} = \text{cov}(\log \left[\frac{x_i}{g(x)} \right], \log \left[\frac{x_j}{g(x)} \right]); i, j, ..., D . \qquad (15)$$

This matrix is singular.

Log-contrast principal component analysis may be expressed in terms of the d positive eigenvalues of the centered log-ratio covariance matrix, aligned in descending order of magnitude, and with their corresponding eigenvectors (Aitchison, 1986, p. 190).

Illustration

An example of the analysis of compositional data is given in Table 9 in which the results obtained for correlations and principal components by the usual method and Aitchison's method are compared. The data used are artificial ($N=27$) and simulate standardized morphometric measures on an organism.

It will be seen that there are considerable differences in some of the correlation coefficients listed in Table 9. The only correlation coefficient to escape appreciable change is r_{24}. These differences carry over to the principal components but to a less marked degree. The angle between the first eigenvectors or the two matrices is about 10°. Hence, in the present example, the effects of closure on the principal component analysis are not

Table 9. Example of compositional data-analysis.

Normal correlation matrix (upper diagonal) and simplex correlations (lower diagonal): principal components of these matrices.

	var 1	var 2	var 3	var 4
var 1	1.	-0.3040	-0.2473	-0.3565
var 2	0.1376	1.	0.1210	-0.6101
var 3	0.2169	0.4553	1.	-0.4818
var 4	-0.6248	-0.7675	-0.7546	1.

(Significantly different correlations marked in bold print.)

PRINCIPAL COMPONENT ANALYSES

	Simplex			Usual		
variable	PC1	PC2	PC3	PC1	PC2	PC3
1	0.2423	-0.7974	0.2354	0.1345	-0.8358	0.1824
2	0.3472	0.5791	0.5423	0.4755	0.4915	0.5313
3	0.2740	0.1588	-0.8060	0.2287	0.1569	-0.8204
4	-0.8635	0.0595	0.0283	-0.8387	0.1874	0.1068

alarming and would not disturb the satisfactory reification of the eigenvectors.

Gower (1987) makes the point that for the purposes of ordination by principal components, the question of closure is immaterial, particularly if the data are not continuously distributed and multivariate Gaussian. For the ecological problems with which he was concerned, this is indeed so. However, the theoretical status of the argument given by Gower (1987) is flawed (Aitchison, *personal communication*, 1988).

Principal Component Analysis and Cross-Validation

We shall now pose the following questions:

1. Can one say anything more about *atypical values*, over what has already been mentioned?

2. What can be done about influential values? *Influential values* do not show up as multivariate outliers but nonetheless bring about a substantial change in the results of an analysis from which they are omitted.

3. Can one be specific about identifying *important variables* and, by the same token, *redundant variables*?

4. Can the dimensionality of a principal component analysis be determined?

The best way of approaching the foregoing questions is to introduce the morphometrically relevant aspects of the topic by reviewing the steps in the calculations (Krzanowski, 1987a, 1987b). The techniques involved are not new, being well known in psychometry and chemometrics (Kvalheim, 1987). The novel aspects brought in by Krzanowski concern the isolation of subtle atypicality, determining the dimensionality of a principal component analysis, and the multivariate recognition of influential values.

Steps in the Calculations:

1. Compute the eigenvalues and eigenvectors of the correlation or covariance matrix of the data matrix, **X**.

$$\mathbf{S} = \mathbf{VLV}^{\mathrm{T}}$$

and

$$\mathbf{V}^{\mathrm{T}}\mathbf{V} = \mathbf{I}_r,$$

or by the singular value decomposition

$$\mathbf{X} = \mathbf{UDV}^{\mathrm{T}},$$

where $d_i^2 = (N - 1)l_i$.

2. Compute the scores of the principal components

$$\mathbf{Z} = \mathbf{XV}.$$

3. For selecting the best number of principal components, one proceeds by means of the

Table 10: Principal components of the correlation matrix for 31 specimens of the Santonian ostracod species *Veenia fawwarensis* for seven distance-measures on the carapace (Abe et al.,1989).							
Eigen-vectors	1	2	3	4	5	6	7
var 1	0.435	0.078	-0.291	0.144	-0.291	0.779	0.091
var 2	0.375	-0.104	0.462	-0.674	0.345	0.237	-0.077
var 3	0.154	-0.894	0.180	0.233	-0.236	-0.040	-0.181
var 4	0.436	0.258	-0.151	0.004	-0.159	-0.295	-0.780
var 5	0.444	0.035	-0.052	-0.261	-0.474	-0.463	0.540
var 6	0.391	-0.195	-0.492	0.178	0.687	-0.180	0.179
var 7	0.323	0.279	0.634	0.608	0.138	-0.051	0.146
Eigen-values	4.4584	1.0455	0.7206	0.3986	0.1803	0.1080	0.0886

Size of the component-spaces being compared: residual sums of squares.

Variable deleted	PC1	PC1 + PC2	PC1 + PC2 + PC3
1	2.6072	2.6957	2.9646
2	2.6617	2.8959	6.0464
3	0.7044	**39.9630**	26.1433
4	2.8077	3.2786	2.7050
5	2.3767	2.3764	2.3894
6	2.8366	3.7992	5.6841
7	2.3889	4.3324	11.0116

criterion W_m, which is computed from the average squared discrepancy between actual and predicted values of the data-matrix by the method of cross-validation.

a) Subdivide **X** into a number of groups;

b) Delete each group in turn from the data and evaluate the parameters of the predictor from the remaining data and then predict the deleted values.

For practical reasons, the group deleted in the above procedure should be made as small as possible. In the illustration given below, I have made each group to be no more than a row of **X**. The method promulgated by Eastment and Krzanowski (1982) recognizes the need for allowing deletion of variables as well as of objects in order to manage the estimation of the data-matrix.

4. For determining the *influence* of each of the observations, one may proceed by computing critical angles between sub-spaces of a common data-space. The critical angle used here is interpretable as a measure of influence of each individual in the sample, with $t = \cos^{-1}d$, where d is the smallest element of the diagonal matrix **D** of step 1.

Large values of the angle are taken to indicate highly influential observations in the sample and hence observations deviating in some manner or other.

5. For scanning the data-matrix for variables supplying most information, one proceeds by the two-dimensional representation, whereby the variables are deleted one by one and the resulting residual SSQ examined. Small residual SSQ have slight effects on the principal component analysis; such a variable should be considered for removal from the analysis.

Table 10 lists the eigenvalues and eigenvectors of the correlation matrix for sample 1 of *Veenia fawwarensis*, with the obvious outlier removed. On the grounds of the Q-Q probability plot, we should hope that we have successfully identified all atypicalities in the data. In the case of the first vector, all variables are equally weighted, apart from the third element which, therefore, might conceivably be a candidate for deletion. The same table also contains the residual sums of squares when the 31 points on the first principal component are matched successively by Procrustean analysis, deleting each variable in turn. Thus column 1 of the latter representation signifies that the deletion of each variable in turn gives a relatively small residual sum of squares for variable 3 which indicates that this variable has least effect on the first principal component. The second column displays the situation in the first two planes, and the third column, that pertaining in the first three planes. All sums of squares are small for column 2 with the exception of the entry for variable 3. The column for the first three planes includes a fairly high entry for variable 7 in adddition to the high residual SSQ for variable 3.

Table 11. Choosing number of principal components to keep.

Fitted in the order 1 2 3 4 5 6 7.

Number of components	Value of PRESS-statistic	test-statistic
0	0.9677	0.0000
1	0.4695	**5.1288**
2	0.4276	0.4033
3	0.3404	**0.8644**
4	0.2965	0.3856
5	0.2798	0.1064
6	0.2706	0.0314

Table 12. Maximum angles for principal component planes: sample of *Veenia fawwarensis* from the Santonianof Shiloah, Israel.

Specimen deleted	1	2	3	-3	-2	-1
1	0.562	2.552	5.810	5.972	8.912	16.482
2	0.748	2.199	2.077	2.063	4.423	13.341
3	0.948	6.515	5.811	6.260	7.715	7.750
4	1.385	3.051	2.405	2.431	7.721	9.740
5	1.573	**17.182**	7.545	5.401	6.586	10.993
6	1.630	8.991	6.544	6.696	**15.636**	0.567
7	0.906	1.268	1.841	2.750	6.261	10.865
8	0.395	1.308	1.917	2.087	3.675	14.669
9	1.801	2.290	2.311	5.291	5.300	5.919
10	0.085	5.883	10.482	9.115	10.933	29.142
11	1.620	1.857	2.555	2.892	**15.703**	14.936
12	2.617	**17.821**	2.761	1.462	1.252	1.660
13	7.362	**18.777**	9.683	4.695	4.889	12.564
14	0.385	3.132	1.779	1.775	3.077	15.351

15-31 contain no markedly divergent values

What does this analysis tell us? It leads to the conclusion that it would not be admissible to remove variable 3 from the analysis on the grounds of its performance in the first principal component.

We now turn to the question of the "correct" number of principal components to maintain in the analysis. Krzanowski (1982b) advocates the use of the W_m-criterion, computed as indicated above. The aim is to estimate the number of statistically useful components, the rule being that values much less than unity are not likely to be associated with much information. In the present illustration, the test-statistic gave 5.1288 for one component, 0.4033 for two components, and 0.8644 for three components (Table 11). One could conclude here that

three principal components are probably significant. The reason why we find a non-significant value of the test-statistic sandwiched between two significant values is due to the fact that several eigenvalues do not differ greatly from each other.

The third topic we shall consider concerns the identification of atypical values by means of critical angles. Table 12 lists the largest critical angle between the plane defined by the first two principal components computed from the full sample and the planes defined by the first two principal components on deleting each sample member in turn. This information is listed in Table 12 under the heading "2". The column headed by a "3" gives the largest of three critical angles between the three-dimensional spaces defined by the first three principal components computed from the entire sample, deleting each observation in turn, and with subsequent replacement. The highest values in these columns indicate those individuals, the omission of which cause the greatest disturbances in the principal component analysis. These specimens are outliers of location or dispersion. The columns headed by negative numbers in Table 12 betoken the three smallest principal components. Thus, the column bearing the heading "-2" defines the plane of the two smallest principal components and analogously for the column headed by "-3". The values listed in the negative columns can usually be seen to encompass correlational outliers.

In the present illustration, we see that in the plane of the first two principal components, specimens 5, 12 and 13 (bold type in Table 12) deviate from the main corpus of the sample. These deviations seem to be connected with polymorphism in size, and they were not unveiled by the other method used earlier on, based on the generalized statistical distance as a measure of discrepancy. Directing our attention to the column headed "-2", we perceive that specimens 6 and 11 deviate from the rest of the material. The situation represented by point C in Figure 1 seems to occur here. The indications of the present example, which was not selected because of any special didactic properties, are that the method of cross-validation brings to the fore the need for a careful appraisal of the data-set before the full cannonade of multivariate analytical methods is brought to bear on a problem. No amount of intricate analysis can amount to anything much if the data are not suitable for the analyses performed upon them. It seems that the procedures recommended by Krzanowski (1987a, 1987b) hold promise for the future study of polymorphism in shape and size.

Concluding Comments

In the present note, I have been concerned with promoting a feeling for awareness of accuracy in analysing morphometric data. Obviously, one should begin with the measurements themselves. There is also the question of choosing the right multivariate technique for a particular analysis. Assuming that both of these questions have been attended to, there is a clear necessity of making a careful study of the empirical properties of a set of measurements, as almost all statistical methods in morphometrics stand or fall on this point. There can be little purpose in carrying out complicated computerized calculations if the outcome is doomed to absurdity from the word go.

My advice is to start, always, with a complete set of univariate analyses involving normality, atypicalities and standard statistics. Move then to graphical displays of the data on a pairwise basis. A scatter diagram can disclose much about the nature of a set of observations but cannot, of course, be the final court of appeal for multivariate observations. The next phase in the initial scanning of the data is to carry out a qualified search for atypical observations, as indicated in the present article. If you have biologically valid atypicalities, it is recommended that robust estimational procedures be used. It is important to avoid rushing into a standard analysis with scant regard for the nature of the data to be treated. In this connection, cross-validation has proven to be a valuable technique.

It is more difficult to be intuitive about the problem posed by instability in the elements of eigenvectors, especially if the aim of the investigation is to reify morphological variables. Notwithstanding the decision to embark forthright on a "robustified" analysis, it is good statistical practice to make an accompanying analysis by the corresponding standard procedure. It is only in this way that the existence of important differences can be brought to the fore.

References

Abe, K., R. A. Reyment, F. L. Bookstein, A. Honigstein, A. Almogi-Labin, A. Rosenfeld and O. Hermelin. 1988. Microevolution in two species of ostracods from the Santonian (Cretaceous) of Israel. Hist. Biol., 1: 303-322.

Aitchison, J. 1986. The statistical analysis of compositional data. Monographs on statistics and applied probability. Chapman and Hall, London, 416 pp.

Anderson, T. W. and R. R. Bahadur. 1962. Two sample comparisons of dispersion matrices for alternatives of immediate specificity. Ann. Math. Stat., 33:420-431.

Barnett, V. and Lewis, T. 1978. Outliers in statistical data. Wiley and Sons: Chichester, 365 pp.

Benzecri, J. P. 1973. L'analyse des correspondances. Dunod: Paris, 619 pp.

Bookstein, F. L., B. Chernoff, R. Elder, Humphries, J. Smith, and G. Strauss, R. 1985. Morphometrics in evolutionary biology. The Academy of Sciences, Philadelphia, Spec. Publ. No. 15, 277 pp.

ter Braak, C. A. J. 1986. Canonical correspondence analysis: a new eigenvector technique for multivariate direct gradient analysis. Ecology, 67:1167-1179.

Campbell, N. A. 1979. Canonical variate analysis: Some practical aspects. Ph.D. Thesis, Imperial College, London.

Campbell, N. A. 1980a. Robust procedures in multivariate analysis. I. Robust covariance estimation. Appl. Statist., 29:231-237.

Campbell, N. A. 1980b. Shrunken estimators in discriminant and canonical variate analysis. Appl. Statist., 29: 5-14.

Campbell, N. A. 1982. Robust procedures in multivariate analysis. II. Robust canonical variate analysis. Appl. Statist., 31:1-8.

Campbell, N. A. 1984. Some aspects of allocation and discrimination. Pp. 177-192 in Multivariate statistical methods in physical anthropology (van Vark, G. N. and W. W. Howells, eds.). Reidel, Dordrecht, 433 pp.

Campbell, N. A. and R. A. Reyment. 1978. Discriminant analysis of a Cretaceous foraminifer using shrunken estimators. Math. Geol., 10: 347-359.

Campbell, N. A. and R. A. Reyment. 1980. Robust multivariate procedures applied to the interpretation of atypical individuals of a Cretaceous foraminifer. Cret. Res., 1:207-221.

Cooley, W. W. and P. R. Lohnes. 1971. Multivariate data analysis. Wiley: New York, 364 pp.

Eastment, H. T. and W. J. Krzanowski. 1982. Cross-validatory choice of the number of components from a principal component analysis. Technometrics, 24:73-77.

Eckart, C. and G. Young. 1936. The approximation of one matrix by another of lower rank. Psychometrika, 1:211-218.

Efrom, B. 1979. Bootstrap methods: another look at the jacknife. Ann. Statist., 7: 1-26.

Efrom, B. and G. Gong. 1983. A leisurely look at the bootstrap, the jacknife and cross-validation. Amer. Statist., 37: 36-48.

Fisher, R. A. 1936. The use of multiple measurements in taxonomical problems. Ann. Eugenics London, 10:422-429.

Gabriel, K. R. 1968. The biplot graphical display of matrices with application to principal component analysis. Biometrika, 58:453-467.

Gnanadesikan, R. 1977. Methods for statistical data analysis of multivariate observations. Wiley and Sons, New York, 311 pp.

Gordon, A. D. 1981. Classification. Monographs on applied probability and statistics. Chapman and Hall, London, 193 pp.

Gower, J. C. 1987. Introduction to ordination techniques. Pp. 2-64 in Developments in Numerical ecology, (Legendre, P. and L. Legendre, eds.). NATO A 51 series, G 14, Springer Verlag, New York.

Gower, J. C. and P. G. and N. Digby. 1984. Some recent advances in multivariate analysis applied to anthropometry. Pp. 21-36 in Multivariate statistical methods in physical anthropology (van Vark, N. and W. W. Howells, eds.). Reidel: Dordrecht, 433 pp.

Hampel, F. R., E. N. Ronchetti, P. J. Rousseeuw, and W. A. Stahel. 1986. Robust statistics. Wiley and Sons, New York, 502 pp.

Jolicoeur, J. and J. E. Mosimann. 1960. Size and shape variation in the Painted Turtle. Growth, 24: 339-354.

Jöreskog, K. G., J. E. Klovan, and R. A. Reyment. 1976. Geological factor analysis. Elsevier, Amsterdam, 178 pp.

Krzanowski, W. J. 1987a. Cross-validation in principal component analysis. Biometrics, 43:575-584.

Krzanowski, W. J. 1987b. Selection of variables to preserve multivariate data structure using principal components. J. Roy. Stat. Soc. C, 36:22-33.

Kvalheim, O. 1987. Doctoral thesis in Physical Chemistry presented to the Faculty of Science, University of Bergen, Dec. 4th, 1987.

Maronna, R. A. 1976. Robust M-estimators of multivariate location and scatter. Ann. Statist., 4: 51-67.

Pearson, K. 1897. Mathematical contributions to the theory of evolution. Proc. Roy. Soc., 60: 489-498.

Pimentel, R. A. 1979. Morphometrics: the multivariate analysis of biological data. Kendall Hunt: Dubuque, Iowa, 278 pp.

Rao, C. R. 1964. The use and interpretation of principal components analysis in applied research. Sankhya, 26: 329-358.

Reyment, R. A., Blackith, R. E., Campbell, and N. A. 1984. Multivariate morphometrics. Second Edition, Academic Press, London, 232 pp.

Rhoads, J. G. 1984. Improving the sensibility, specificity, and appositeness of morphometric analyses. Pp. 257-259 in Multivariate methods in physical anthropology (van Vark, G. N. and W. W. Howells, eds.). Reidel: Dordrecht, 433 pp.

Rock, N. M. S. 1987. Robust: an interactive FORTRAN 77 package for exploratory data analysis using parametric robust and non parametric locations and scale estimates, data transformations, normality tests and outlier assessment. Comp. Geosc., 13: 463-494.

Seber, G. A. F. 1984. Multivariate observations. Wiley Series in Probability and Mathematical Statistics. Wiley and Sons, New York, 686 pp.

Sibson, R. 1978. Studies in robustness of multidimensional scaling: Procrustes statistics. J. Roy. Statist. Soc., B, 40: 234-238.

Teissier, G. 1938. Un essai d'analyse factorielle. Les variants sexuels de Maia squinaday. Biotypologie, 7: 73-96.

Part III
Section B

Methods for Outline Data

There are several approaches that can be used to deal with outline data (either complete closed outlines or open curves between two landmarks). They involve fitting some type of curve to the outline and then using the parameters of the curve for subsequent analysis.

- The most common approach is to fit some convenient, but arbitrary, function to the outline such as polynomials, trigonometric series (Fourier analysis of various types), splines, etc. and use resulting coefficients as variables for statistical analysis. Chapter 7 by Rohlf reviews some of the possibilities and points out that, while different methods can lead to different results, it is difficult to know how to choose among the alternative methods.

- Eigenshape analysis (Lohmann, 1983) represents what seems like a very different approach from that described above. In this case we do not fit a curve to the data—the function is constructed as a linear function of the observed data across one or more specimens. Lohmann (1983) calls it an "empirical function." Relatively few vectors (the parameters are just the projections of each object onto eigenvectors) are usually needed to obtain a good degree of fit—but each function is represented by as many numbers as one has data points. An interesting aspect of this method is that an ordination analysis is performed as part of the fitting process. Chapter 6 by Lohmann and Schweitzer reviews this method and its relationships to other methods for dealing

with outline data. Chapter 9 by Ray is an example of eigenshape analysis and also furnishes a critique of some of the operations performed in eigenshape analysis.

- Alternatively an a priori defined coordinate system can be used. For example, rather than using simple rectilinear coordinates to describe shell shape, Raup (1966) used parameters that reflected his understanding of the way in which shells grow. The advantage of the use of such coordinates is that it is easier to give direct geometrical and biological interpretation to the measurements. Chapter 18 by Ackerly in Part IV is in this tradition.

- The median axis methods described in Chapter 8 by Straney represent a very different approach. The outline is represented by line segments making up an internal line skeleton in the center of the figure together with the distance from each point on the skeleton to the nearest point on the outline. This information is sufficient to allow the original outline to be reconstructed. In morphometrics, however, this method is usually used simply because the structure of the line skeleton often provides suggestive information about homologies and can provide useful constructed landmarks.

References

Lohmann, G. P. 1983. Eigenshape analysis of microfossils: a general morphometric procedure for describing changes in shape. Math. Geol., 15:659-672.

Raup, D. M. 1966. Geometric analysis of shell coiling: general problems. J. Paleontol., 40:1178-1179.

Chapter 6

On Eigenshape Analysis

G. P. Lohmann and Peter N. Schweitzer

Woods Hole Oceanographic Institution
Woods Hole, MA 02543

Introduction

Paleontologists and biologists frequently encounter situations in which homologous regions of different specimens can be identified but homologous points cannot be specified. The margins of ostracode shells, for example, occupy the same general position and are produced by the same organ in many different taxa, but the shape of the margin varies considerably, forming a basis on which species and subspecies are classified (van Morkhoven, 1962). Statistical methods for treating these shapes provide a way to make systematic interpretations more objective and more easily understood by non-specialists. For the methods that have been commonly used, the most troubling aspect is how to incorporate homology into an otherwise purely geometrical description of form. In this paper we review one method that allows some information on homology to be preserved in a shape description and carried through the shape analysis.

The term "eigenshape analysis" was originally coined to call attention to the fact that collections of observed shape functions can be represented by a set of orthogonal shape functions derived from an analysis of the observations themselves (Lohmann, 1983). This decomposition of data functions into empirical orthogonal functions has been usefully applied in many different fields (Lorenz, 1959; Davis, 1976; Aubrey, 1979; Lohmann and Carlson, 1981; Aubrey and Emery, 1986; Mix et al., 1986).

Since it was introduced, eigenshape analysis has taken on a much broader meaning than was its original, limited intent. It has been misunderstood to include the choices, judgments, sampling strategies and technical details that were either required by the nature and aims of particular studies or dictated by equipment limitations. Here we wish to explain all aspects of this extended definition of eigenshape analysis, of which eigenfunctions themselves play a small part.

The methods that follow were chosen from among many possible ways to solve particular research problems, problems whose primary concern was the description of growth histories and developmental pathways from measurements of fossil populations. The validity and suitability of these methods should be evaluated in that context.

Description of an Individual Shape

We consider shape to be every aspect of an object's outline except its position, orientation, and scale. Size, however, is a fundamental quantity in our studies and an integral part of the way we have chosen to measure outlines, so we include scale in our description of their shapes.

Our studies deal with the incremental growth of three-dimensional objects, and we can expect them to have grown and changed their shape differently along different dimensions. With the technology currently available to us it is only practical to measure two-dimensional projected outlines of these objects. To approximate their full three-dimensional shape we use the expedient traditionally employed by taxonomists: we combine descriptions of their two-dimensional projections (Figure 1). These views are conventionally used by taxonomists to illustrate the three-dimensional morphology of these organisms. In a similar way, we approximate their three-dimensional shape by combining measurements of the outlines of these views.

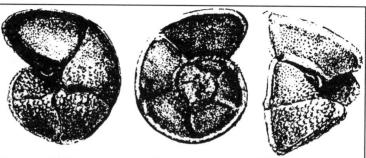

Figure 1. Spiral and edge views of the planktonic foraminifer *Globorotalia truncatulinoides*. These views are conventionally used by taxonomists to illustrate the three-dimensional morphology of these organisms. In a similar way, we approximate their three-dimensional shape by combining measurements of the outlines of these views.

An individual object is typically digitized at over 1500 points, including the two views needed for a quasi-three-dimensional representation. Since we expect to measure and study the shape of a great many objects, we need our shape descriptions to be as compact as possible. We achieve this in two ways, by reducing their detail and by adopting certain conventions.

The detail of the descriptions can be reduced considerably, to a few hundred points, without visibly changing the outlines. Our first convention is to interpolate every outline to a fixed number n of equally spaced points. This number is usually much smaller than the number of points generated by automatic digitizers (typically $n = 100$). While suitable interpolation methods can be found for even the most convoluted shapes (e.g., the methods presented by Evans et al. 1985), most biological outlines may be treated with simple linear methods.

The $\phi^*(\ell)$ outline transformation of Zahn and Roskies (1972) permits us to reduce shape descriptions still further and to distinguish various aspects of the outline's shape. Each closed, two-dimensional outline projected from an object of interest is described by calculating its net angular bend (ϕ^*) as a function of the distance along its perimeter (ℓ) (Figures 2 and 3). Note that there is an implicit 90° angular change in Figure 3 between the last point measured on one view and the first point measured on the other.

This transformation is as complete a description of the shape of an object's outline as are the rectangular coordinates of the points from which it was calculated. It also allows us to isolate three distinct attributes of an individual's outline, which we refer to as form, size, and angularity.

Form. In our usage, the form of an individual refers to all aspects of its outline that do not change as one varies the scale of its perimeter or the amplitude of the angular change ϕ^* of its Zahn and Roskies' shape function. Our measurement of the form of an individual's outline is its standardized $\phi^*(\ell)$ shape function; i.e., one that has been measured at n equally spaced points around an outline's perimeter and that has had the amplitude (or variance) of the values of ϕ^* rescaled to 1.0.

Size. We measure size in two different ways, as the length of the perimeter around the outline and as the area enclosed by the outline.

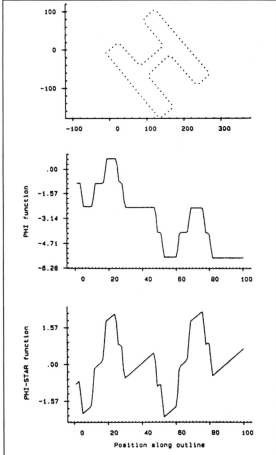

Figure 2. Angular change ϕ as a function of perimeter l around an outline of the letter H (after Zahn and Roskies, 1972). The net angular bend ϕ^* is the difference between the shape of the measured closed curve and a smooth circle.

The first measure of size (perimeter l) is an integral part of the Zahn and Roskies' shape function, $\phi^*(l)$. Because measurements are equally spaced around the outline, we can ignore the l part of $\phi^*(l)$, further reducing our shape description. By convention, we know n, the number of points on the outline. We only need to remember one other number to replace the distance S between adjacent, equally spaced measurements of ϕ^*.

$$S = l / n$$

Varying this parameter changes the size of an outline by changing the length of its perimeter.

We also measure size as enclosed area or volume. The area enclosed by a two-dimensional polygonal outline is simply and accurately calculated from its rectangular coordinates (Eves, 1972). The area A of a polygon defined by points (x_1y_1), (x_2y_2),..., (x_ny_n), is

$$A = \frac{1}{2}(x_1 y_2 + x_2 y_3 + ... + x_{n-1} y_n + x_n y_1$$
$$-y_1 x_2 - y_2 x_3 - ... - y_{n-1} x_n - y_n x_1)$$

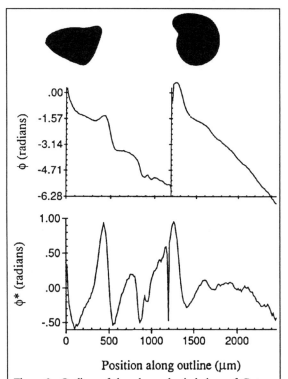

Figure 3. Outlines of the edge and spiral views of *G. truncatulinoides* transformed to $\phi^*(l)$, angular change f as a function of outline perimeter l. The three-dimensional shape of the shell is approximated by combining the shape functions for the edge and spiral views. There is an implicit 90° angular change between the last point measured on one view and the first point measured on the other.

In contrast, the accuracy with which the volume enclosed by a three-dimensional surface can be calculated depends on the complexity of the object's surface topography and on the detail with which we measure it. A simple approximation satisfies our needs. Volume V is approximated from the areas enclosed by our selected two-dimensional projections, as

$$V = (A_{edge} \times A_{spiral})^{3/4} \times G \ .$$

Usually the value for the coefficient G is set at 1.0, but if the object can be modelled as a simple geometric form, G can provide a closer approximation of its enclosed volume.

Angularity. Angularity refers to those aspects of an object's outline that change only by varying the amplitude of the angular part $\phi(*)$ of its shape function (Figure 4). Amplitude is calculated as the standard deviation of the observed values ϕ^*, before they are standardized. We refer to the resulting differences in their shape as differences in their "angularity."

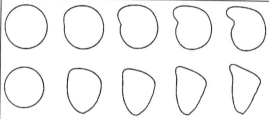

Figure 4. Outlines of the edge and spiral views of a single shell of G. truncatulinoides. All are constructed from the same ϕ^* shape function (i.e., they have the same form), but each is scaled by a different amplitude (after Lohmann, 1983, Figure 3). We refer to resulting differences in their shape as differences in their "angularity."

Objects become more angular as their surface features become more pronounced, increasing the amplitude of ϕ^*. They becomes less angular as surface features are smoothed, decreasing the amplitude of ϕ^*. As the amplitude of ϕ^* approaches zero, the angularity of a shape approaches zero and the object's outline becomes a smooth circle.

Since changing form can be expected to change the amplitude of ϕ^*, the angularity attribute has useful meaning only when comparing objects of the same form. Our usage of angularity refers to the degree to which the features of a particular form are accentuated, not to their spatial scale. A uniform change in the spatial scale of features

changes an object's size; disproportionate changes in their spatial scale affect its form.

In summary, we measure the shape of an outline at n equally spaced points around its perimeter, representing a three-dimensional object by combining the outlines of selected projections. Each outline is described by three attributes: (1) *Form* is the angular part of its $\phi^*(\ell)$ shape function, standardized to unit variance. (2) *Size* is measured two ways, as the distance between the uniformly spaced points along the outline's perimeter and as the area or volume enclosed by the outline. (3) *Angularity* is the amplitude of the observed ϕ^*.

As long as measurements are comparable among outlines (i.e., as long as measurements are made at points that can be reasonably compared), any shape function can be subjected to an eigen-analysis. Likewise, choosing to use eigenshape analysis does not dictate the shape function one must use.

Comparison of $\phi^*(\ell)$ with r(θ)

The radius function is one alternative representation of an object's outline. Points on the outline are represented by their polar coordinates from some center. One can subject the radius function to eigenshape analysis in exactly the same way as with the Zahn and Roskies' $\phi^*(\ell)$ function; the results of such an analysis are shown as pictograms by Scott (1980, 1981).

We know of no theoretical advantage for choosing r(θ), other than that one shape function may emphasize some aspects of shape differently than another. If these differences are understood, they may be desirable in some applications. We do not discuss these situations in detail.

There are two reasons why we declined to use the r(θ) shape function in our work:

Reason 1. The radius function requires that one choose a center from which to measure the lengths of r as a function of angle θ. This is either calculated or chosen as a landmark in the interior of the outline. The position that is determined becomes critical when the radius functions (or some transformation of them) are used to compare the shape differences among the outlines of several objects. Any variation in positioning the centers becomes an additional source of shape variation.

Suppose we calculated or chose a center whose position relative to the outline was highly variable in the sense that it lies on different biological structures on different specimens. Then the shape variations we describe statistically are largely attributable to the variable position of the center, though we have described them as changes in the outline. The resulting confusion becomes evident when we compare the outlines of specimens that (the analysis tells us) look different. In some cases, the outlines will visibly differ, while in others, the outlines look alike, but only the position of the center varies.

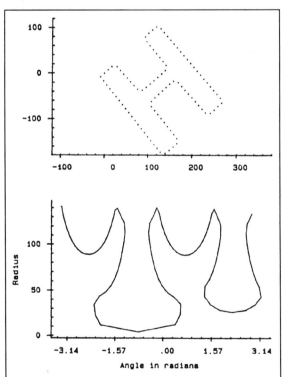

Figure 5. Description of the letter H by the $r(\theta)$ shape function. The joint mean of the x,y-coordinates was taken as the center from which to measure the radius r. Notice that r is frequently a multivalued function of θ.

If our center were indeed a homologous biological landmark, we could simply make a more complicated figure to portray the situation. But if the center we chose were some mathematical quantity (like the centroid or the point that minimizes the first Fourier harmonic), we could not make a valid biological interpretation at all. In effect, the analysis is of shape alone, though we might wish to impute biological meaning to the results.

Reason 2. For convoluted outlines, r is not a single-valued function of θ (Figure 5). This limits application of the radius function to simple outlines or to outlines that can be trimmed to a simpler shape.

In summary, there are two reasons for choosing the $\phi^*(l)$ shape function over $r(\theta)$: $\phi^*(l)$ does not require a center and it is not limited to simple outlines. But these arguments are not motivated by a limitation of eigenshape analysis; in practice, either shape function can be submitted to an eigenshape analysis.

Comparison of $\phi^*(l)$ with the Elliptical Fourier Transform

The elliptical Fourier transform has been shown to be an efficient means for measuring the shape of outlines (Rohlf and Archie, 1984). We see no theoretical advantage for choosing this representation over Zahn and Roskies' $\phi^*(l)$. However, each may offer practical advantages in certain situations.

For example, the trace of any outline measured with a video digitizer becomes noisier at the spatial scale of the pixels. This can be removed by simple smoothing or by constructing a filter tuned to the scale of the pixel noise. If the outline is collected as or transformed to a Fourier series, one might simply truncate the series at terms smaller than the scale of the pixels. This applies to all shape functions, but filtering might be more efficiently incorporated into an outline measurement procedure that begins with the outline's elliptical Fourier transform.

We effectively remove pixel noise and other small scale variations in a similar way, but in space rather than in the frequency domain. This happens when we interpolate outlines from the pixel scale in which they were collected to a lesser number of more widely spaced points, typically a reduction from 1000 to 100 points.

As pointed out by Rohlf (1986), eigenanalysis of a collection of shape functions or of the Fourier transforms of the same shape functions yields the same results. This holds true *only if none of the data are discarded*–nothing from the transformation of the shape functions to a Fourier series can be discarded (neither the phase associated with the amplitude of each harmonic term nor any of the terms themselves). However, because different shape functions can be expected to weight aspects of a shape's outline differently, eigenanalysis of different functional representations of the same shapes should not generally be expected to yield the same results.

Comparing Shapes

Our own research objectives do not require that we analyze or decompose the shape of an isolated, individual object into simpler elements as an end in itself. Rather, we need to determine the relationships among a collection of objects on the basis of similarities and differences in their shapes.

In our view, individuals are alike only if they are identical in form, angularity and size the correlation between their shape functions must be 1.0 and the amplitudes of ϕ^* and spacing between the n measured points around each outline must be the same. In evaluating the similarities and differences among individuals that are not identical, each of these attributes of shape can be considered separately.

Comparing Size and Angularity

Ways of comparing differences in the size and angularity attributes among a collection of objects need no elaboration other than to point out that such

comparisons are most useful when comparing objects of similar form (since changing form can, by itself, change our measures of size and angularity). Consider the following examples:

Example 1. Within a population of individuals of the same species (and therefore of similar form) there is often a strong relationship between overall size and developmental stage. Where this applies, the size attributes of shape can be used to arrange individuals along their developmental pathways and to measure the associated allometries that produce changes in form (Figure 6). The joint mean and covariance between size and edge/spiral ratios for each of 10 size fractions is indicated in Figure 6 by the 95% confidence ellipses. The shape of the shells from each size fraction is illustrated. This figure shows the allometric growth that accompanies the overall increase in size during development.

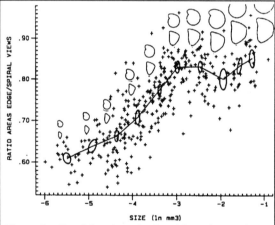

Figure 6. Size (plotted as the logarithm of approximate enclosed volume) versus.the ratio of the areas enclosed by outlines of the edge and spiral views of *G. truncatulinoides*. All shells are from one sample. The joint mean and covariance between size and edge/spiral ratios for 10 size fractions is indicated by the 95% confidence ellipses. The shape of the shells from each size fraction is illustrated. This figure shows the allometric growth that accompanies the overall increase in size during development.

Example 2. An ecophenotypic (environmentally induced) phenomenon we are studying, the secondary encrustation of the shell wall of foraminifers, changes the shapes of shells simply by

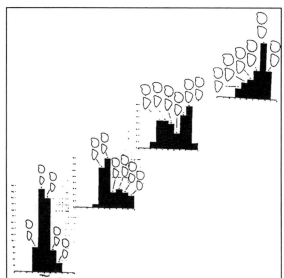

Figure 7.   Histograms of the average density (shell weight/volume) of shells of *G. truncatulinoides*. Shells are from different size fractions of the same sample. Addition of a calcitic crust over the shells increases their density and smoothes their outline. This figure shows the change from predominantly less dense, more angular shells in the smaller size fractions to predominantly more dense, smoother shells in the larger size fractions.

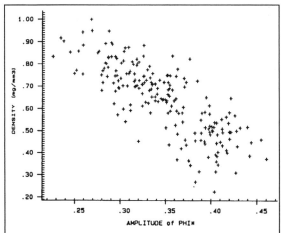

Figure 8.   Measure of shell "angularity" (amplitude of ϕ^*) vs. their average density. Addition of a calcitic crust both smoothes the shell's outline (reducing its angularity) and thickens the shell wall without appreciably increasing its volume (thereby increasing its density).

shells increases their density and smoothes their outline. The change is from predominantly less dense, more angular shells in the smaller size fractions to predominantly more dense, smoother shells in the larger size fractions. This smoothing is measured as a reduction in the amplitude of the ϕ^* shape function, the angularity attribute (Figure 8).

Although most useful among objects of similar form, comparisons of size or angularity do not require that one identify comparable features or regions among the objects. In contrast, meaningful comparisons of form depend critically on the correspondence of features, point for point.

Comparing Form

We use either the correlations or the covariances among shape functions as measures of their similarity in form. For these measures to make sense, shape functions must be comparable point for point, a condition that often does not occur in practice. In our studies on ostracodes and foraminifera, we have found that homologous regions along the outlines can be easily identified, but homologous points cannot. This distinction is important because it allows us to determine which situations will be best described by the shape function we have chosen. The ϕ^* shape function is best applied to the regions *between* homologous landmark points. Where there is only one consistently locatable landmark, the shape function simply describes the shape, but the information contained in that landmark is preserved.

Consider our studies of the shape changes that accompany growth in foraminifera and ostracodes. We can expect comparable points that are measured among individuals at the same stage of growth to closely approximate biological homology. On the other hand, since foraminifera grow by adding new chambers over old shell and ostracodes discard their old shells entirely and grow a new, larger one, we cannot expect all the measurements among different growth stages to be biologically homologous. With this qualification clearly in mind, comparable points among growth stages may

smoothing their outlines (Figure 7). This figure shows that the addition of a calcitic crust over the

specified and the shape differences measured between them may prove useful for exploratory analysis.

We treat the measurements of similarity in form among objects, the correlations among their shape functions, in different ways:

1. *Designated "Type Specimens."* If one has the insight to create or designate certain individuals from a collection as reference types T, one can measure the similarity of all the other objects Z to these types by calculating the correlation C between the reference shape functions and all the others.

$$C = T^T Z$$

One might then use these correlations as criteria for classification of the objects from the collection with one of the various types.

2. *Cluster Analysis.* Correlations among the objects in a collection can serve as the basis for a conventional cluster analysis (Figure 9).

3. *Eigenshape Analysis.* The matrix of correlations, **R**, among the objects, **Z**, can serve as the basis for an eigenanalysis. Note carefully how the shape functions are arranged in **Z**. Each column in **Z** contains a shape function that describes the form of an individual's outline with measurements of ϕ^* made at each of n points. Each row j in **Z** contains the values of ϕ^* that were measured at the jth point on each and every outline. Each column in **Z**, containing the values of ϕ^* that are associated with a single individual, is standardized to unit variance. In the terminology of Cooley and Lohnes (1971), columns of "standardized test scores" are our "standardized shape functions" and rows of "subjects" are our "homologous or comparable points among individuals." This procedure is computationally identical to the R-mode principal components analysis described by Cooley and Lohnes (1971).

R is a matrix of correlations among the standardized ϕ^* shape functions **Z**, so that

$$\mathbf{R} = \mathbf{Z}^T \mathbf{Z} \ .$$

R is decomposed to its eigenvectors, **V**, and eigenvalues, **S**, (with Cooley and Lohnes' recoding of Matula's subroutine HOW), so that

$$\mathbf{R} = \mathbf{V}\,\mathbf{S}\,\mathbf{V}^T \ .$$

The elements in the eigenvectors, **V**, weighted by $\mathbf{S}^{1/2}$ are the correlations (loadings) between the observed shape functions, **Z**, (test scores) and the eigenshape functions, **U** (principal factor scores), so that

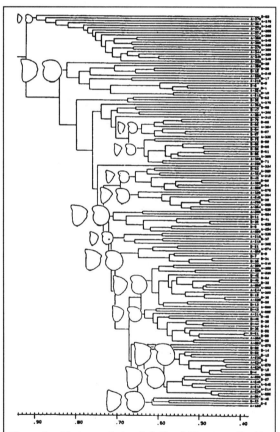

Figure 9. Cluster analysis of shells of *G. truncatulinoides* from a wide range of sizes (125-800μm sieve size) collected at one locality. Clustering was based on similarities in shape among the shell outlines. Shape similarities were measured as the correlations among the three-dimensional ϕ^* shape functions. The average shape of the shells from within some of the possible clusters is illustrated. This analysis ignores differences in size and "angularity."

$$U^T Z = S^{1/2} V^T \quad .$$

From this expression, the eigenshape functions, U (scores), are calculated from Z, S, and V, rearranged as

$$U = Z V S^{-1/2} \quad .$$

Because of the limits imposed by our available computer memory, we calculate our eigenshape functions from **R** following this procedure of Cooley and Lohnes, as just outlined. However, **U, V** and **S** can all be computed directly (and more precisely) as a singular value decomposition of the observed shape functions, **Z** (Golub and Reinsch, 1970),

$$Z = U S^{1/2} V^T .$$

This direct approach avoids the potential imprecision introduced by accumulating large sums of squares in the calculation of **R**. We present it here because it shows more transparently just what an eigenshape analysis is: Observed shape functions, **Z**, are decomposed into a set of empirical orthogonal shape functions (eigenshape functions), **U**, and associated set of weights $S^{1/2}V^T$. These weights are the correlations between the eigenshapes and the observed shapes,

$$U^T Z = S^{1/2} V^T \quad .$$

The eigenshape functions, **U**, account successively for the maximum possible proportion of variation in the collection of observed shape functions analyzed. No other set of shape functions can account for the variation more efficiently (Davis, 1976).

The column standardization is not strictly necessary. Alternatively, one can simply calculate the singular vectors of the covariance matrix

$$Y = (X - m) \quad ,$$

where m is the mean of the columns of **X**. The eigenshape functions are then calculated as

$$U_Y = Y V S^{1/2} \quad ,$$

from the singular value decomposition $Y = U_Y S^{1/2} V^T$. The advantage of this alternative is that the eigenshapes are optimized to account for both form and angularity (i.e., both the shape of ϕ^* and its amplitude). The disadvantage is that specimens with higher angularity are given more influence in determining the direction of the eigenfunctions. Note that the first eigenfunction of **Y** describes the variation among shapes rather than the mean shape. This latter approach was used in a study of the evolution of ontogeny in ostracodes (Schweitzer 1990; Schweitzer and Lohmann in press).

We use the eigenshape functions in three different ways:

1. *As an average shape.* In situations where measurements can be easily made and where the number of objects available for measurement is unlimited, it may be desirable to increase analytical precision by making repeated measurements on the same objects and to increase accuracy by measuring more objects. In either case, the measurements are usually summarized by their mean value. To summarize repeated measures of size, weight, or angularity we use the arithmetic mean. We can use the first eigenshape to summarize repeated measures of shape.

In accepting the mean as the best estimate of a population of values one assumes that the values averaged are roughly normally distributed about it. One expects the mean to describe the central tendency of a population, not an intermediate value between distinct subpopulations. The mean of repeated measurements of the same object is the best estimate of its value and the variation about that mean is noise associated with the measurement procedure. For a group of many objects, it may be difficult to show that the mean is the best summary (though it is usually clear when it is a poor one) and to distinguish noise from signal in the variation about the mean.

Aside from such considerations, one can use the mean in a different way, as a device to selectively emphasize variation among groups by ignoring variation within groups. Figures 10 and 13 are the first eigenshapes of shell outlines from different size classes. The starting reference point is shown for each plot. Since the first eigenshape accounts for most of the variance among a collection of similar objects, it describes features they all share, features that make them look the same, and it approximates the mean shape. The other eigenshapes account for the remaining variance contributed by features that make the objects look different. Our intention was to examine the shape

changes that accompany development. Not knowing at the outset how to distinguish developmental stages by shape, we have used size. For an individual, growth is a good measure of development and we can hope that this approximation extends to the population. The differences between the first eigenshapes of the size classes emphasize

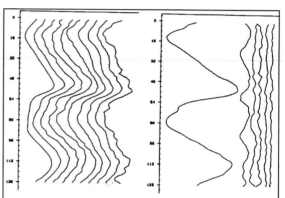

Figure 11. Zahn and Roskies' $\phi^*(l)$ shape functions (on the left) of the ostracode outlines shown in Figure 10 and their first 5 eigenshape functions (on the right). The values of ϕ^*, the net angular bend around each outline, are plotted on the horizontal axis. The vertical axis orders the 128 comparable points located around the perimeter of each outline. Each observed shape function is rescaled to unit variance. Each eigenshape function has been rescaled by its associated eigenvalue, the portion of the total shape variation it represents. The first eigenshape function approximates the mean of the shape functions, and its large amplitude reflects the overall similarity among the shapes. The remaining eigenfunctions describe differences among the shape functions along various orthogonal dimensions and are ordered by the decreasing amount of the shape differences (variance) they explain. Computed with programs OUTLINE and EIGENS.

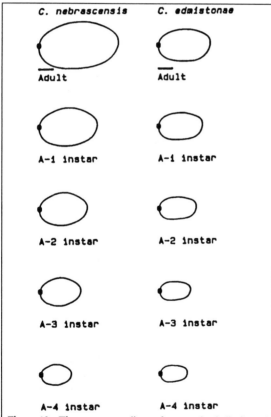

Figure 10. The average outlines of ostracode shells from 5 instars (growth stages) in each of two related species of *Cavellina* (from Schweitzer et al., 1986). The outlines are plotted to the same scale, the starting reference point is indicated, and each outline is drawn between 128 equally spaced points.

Figure 12. Shape oulines constructed from the 15 eigenshape functions plotted in Figure 11 (rebuilt and plotted by program CSHAPE). Only the first eigenshape resembles the ostracodes that were analyzed. Since it accounts for most of the "variance" among a collection of similar objects, it describes features they all share, features that make them look the same, and it approximates the mean shape. The other eigenshapes account for the remaining "variance" contributed by features that make the objects look different.

development. Figure 16 shows how well the first eigenshapes describe the shapes of *G. truncatulinoides* in each size class. The minimum at intermediate sizes reflects the increased variability associated with shell encrustation.

To study development, we chose to ignore the shape variation within size classes and focus on the changes in shape that accompany changes in size. Lohmann and Malmgren (1983) used a similar strategy to study geographic changes in shape of *G. truncatulinoides*. Shells were selected from a narrow size range to minimize developmental sources of shape variation and from a wide range of oceanographic environments to maximize biogeographic sources. The shape at each locality was summarized by a first eigenshape, and the differences between the first eigenshapes emphasize biogeographic changes in shape.

These uses of group means to focus on between-group variation does not deny that meaningful within-group variation exists, of course. It can be examined separately, as follows.

2. *As a basis for examining the relationships among objects in a reduced shape space.* As described, each threedimensional object has 200 measurements of form, 1 of size and 1 of angularity. Together, these attributes locate each object in a really big hyperspace. Because there is typically a great deal of similarity in form among the objects (they are usually the same or related species), there is a large degree of redundancy in the form measurements and most of the variation in form can be represented by a few eigenshape functions. No other basis functions preserve more of the original variation in as few dimensions.

As a reduced basis for examining the relationships among objects in shape space, the first few eigenshape functions preserve most of the observed shape variation while permitting the visual evaluation of the relationships among the objects. This is a particularly valuable feature when used in conjunction with dynamic computer graphics and projection pursuit tactics (Friedman and Stuetzle, 1982). These techniques enable one to "fly" through

shape space, searching for trends, subpopulations, end-members, and such (Figure 17). The projections shown in Figure 17 are three of many possibilities. Used in this way, as a method for constructing a reduced shape space, there is no

Figure 13. The average outlines (edge and spiral views) of the shells of *G. truncatulinoides* from 10 size-fractions of a single sample. The outlines are plotted to the same scale, the starting reference point is indicated, and the outline of each view is drawn between 100 equally spaced points.

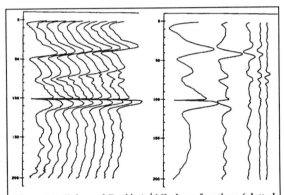

Figure 14. Zahn and Roskies' $\phi^*(t)$ shape functions (plotted on the left) of the combined edge and spiral views of *G. truncatulinoides* shown in Figure 13 and their first 5 eigenshape functions (plotted on the right). See Figure 11 for explanation.

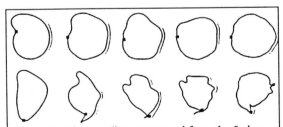

Figure 15. Shape outlines constructed from the 5 eigenshape functions plotted in Figure 14 (rebuilt and plotted by program CSHAPE). The implicit 90° turn between the combined outlines separates the edge and spiral views of the eigenshapes.

need to assume normality of the data distribution or to fear the orthogonality constraints placed on the eigenshape functions. The eigenfunctions simply provide a basis for viewing as much of the data as possible all at once.

3. *As components for modeling shape variation.* While the eigenshape functions may or may not parallel actual trends in shape variation or coincide with end-members or groups, those that do or that can be rotated near such positions (as in a VARIMAX rotation scheme) can be used to model continuous shape variation along such trends or between such groups. This is simply done by adding portions of the higher order eigenshapes to the first one (Figures 18, 19 and 20). The point is to gain some insight into the character of the shape differences in the collection analyzed. In Figure 19 the modeled variation reflects the tendency for the shell to open its umbilicus during development, from more compact juveniles to more open adults.

Cifelli (1965) determined that most of the morphologic variability he observed among living populations of *G. truncatulinoides* could be explained by the changes in shape of the shell that accompany their growth through ontogeny. By using eigenshape analysis in the three ways just described, we have characterized this developmental history from measurements of fossil populations (Figure 21).

Comparison with Fourier Harmonic Functions

The Fourier transform of an object's outline is a description of its shape (Figures 22 and 23). Shape analysis, meaning the procedure used to reduce and summarize observations, discards part of that description. Fourier shape analysis is optimized

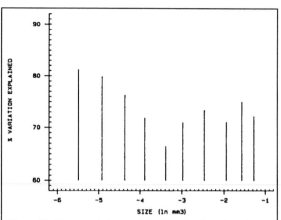

Figure 16. Shape variation explained by the first eigenshape of each of 10 size classes of *G. truncatulinoides* shells. The minimum at intermediate sizes reflects the increased variability associated with shell encrustation.

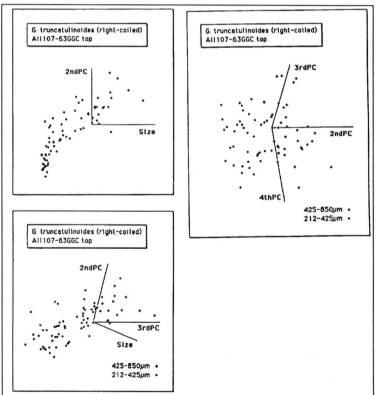

Figure 17. Shape variation among shells of *G. truncatulinoides* from the 212-850 μm sieve size in a single sample. The initial hyperspace that their shape differences define was reduced using eigenshape analysis, then rotated using dynamic graphics program MacSPIN. The projections shown here are three of many possibilities.

neither for explaining the variation among shapes nor for discriminating among groups of shapes. Note that we use efficiency and optimality interchangeably (see Lohmann, 1983; Aubrey and Emery, 1986b); the first n eigenfunctions portray the relationships among the shapes better than the first n of any other set of basis functions. By "portray the relationships better" we mean that the distances between shapes in the reduced space are (on average) closer to their distances in the original data space. For optimal explanation we must turn to eigenfunctions; for optimal discrimination we use discriminant functions.

For most studies, the choice of eigenfunctions over harmonic functions is simply one of efficiency. One can use the harmonic functions from Fourier transforms of individual shape functions as references by which to measure and to model their shape differences in the same ways that we use eigenshape functions.

The usual procedure in a Fourier shape analysis is to decide which harmonic terms are important for describing the observed shape variation (or discriminating among groups). For each harmonic selected, the harmonic amplitudes asso-

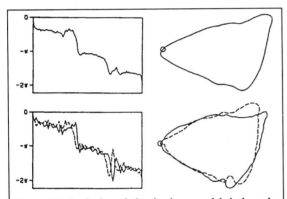

Figure 18. Synthetic variation in shape modeled along the dimension of the second eigenshape. The upper figures show the first eigenshape function (approximating the mean) and its reconstructed outline. The lower figure shows shape functions and the reconstructed outlines synthesized by adding and subtracting small amounts of the second eigenshape to the first.

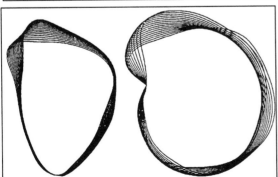

Figure 19. Synthetic variation in shape modeled as in Figure 18. These are edge and spiral views of the shell of *G. truncaltulinoides* and the modeled variation reflects the tendency for the shell to open its umbilicus during development, from more compact juveniles to more open adults.

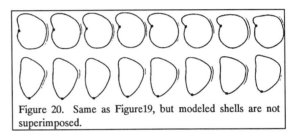

Figure 20. Same as Figure19, but modeled shells are not superimposed.

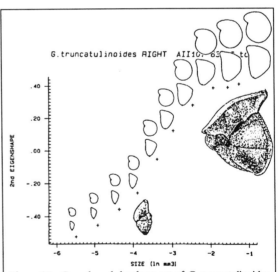

Figure 21. Growth and development of *G. truncatulinoides*, determined from their fossil shells. The shapes of the size classes are described by first eigenshapes. The second eigenshape function of these is a measure of their shape differences, the shape differences between size classes. This is modeled in Figures 18, 19 and 20. Cifelli's (1965) illustrations of juvenile and adult shells are shown.

ciated with the individual shapes are plotted against each other or some covariate. Zahn and Roskies (1972) used the Fourier amplitude coefficients from transformed $\phi^*(l)$ shape functions to characterize the shapes of written numbers and position them in a reduced space defined by harmonic functions. In a similar way, Schweitzer et al. (1986) used Fourier harmonic functions as a basis for viewing ontogeny in ostracodes, the shapes of their shells also having been described by Zahn and Roskies' $\phi^*(l)$.

These studies illustrate some of the advantages and disadvantages of harmonic shape functions over eigenshape functions: Eigenshape functions describe shape variation among a collection of individuals most efficiently. They are able to do this because they are defined by an analysis of the data they are meant to describe and are tailored to do just that. This was demonstrated by Schweitzer et al. (1986). The shape functions that most efficiently decompose the shapes of the ostracodes (Figure 12) are not harmonic functions (Figure 23) both the spacing and amplitude of the nodes on the outlines are irregular. In contrast, Zahn and Roskies (1972) desired a measure of shape that could be defined *before* it was to be applied, could be counted on to describe any shape, and was insensitive to the location of the points on the outline at which measurements would be made. The Fourier transform offers all this in its harmonic functions and power spectrum.

Most studies that have used harmonic functions to measure shape differences have employed the $r(\theta)$ radius function rather than Zahn and Roskies' $\phi^*(l)$ to describe the shapes themselves. This difference, introduced at the level of the data description, complicates direct comparison between the results of these studies and those that have used eigenshape analysis. One can, however, be confident in two expectations: (1) An eigenanalysis of either the shape descriptor functions themselves or their complete Fourier transforms will yield the same results (though most studies discard most of the Fourier transform and all but the first few eigenshape functions). (2) Fourier harmonic shape

functions cannot represent the observed shape functions more efficiently than eigenshape functions; usually they will be much less efficient.

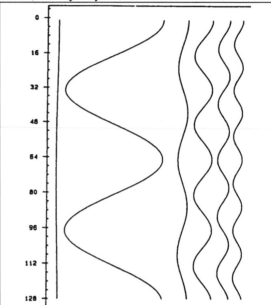

Figure 22. First five harmonic shape functions of the ostracode shell outlines shown in Figure 10, each scaled by the average variance it accounts for. Calculated from the Fourier transform of $\phi^*(l)$.

Figure 23. Shape outlines constructed from the first five harmonic shape functions.

While all studies that use Fourier shape analysis begin with a Fourier decomposition of observed shapes, they differ from each other and from eigenshape analysis primarily in the way they use those measurements.

Eigenshape Analysis of Frequency Distributions (Histograms)

Mathematically, this is essentially identical to the first step in the factor analysis of histogram plots of harmonic amplitude values shown, for example, by Healy-Williams et al. (1985). We object to this procedure only when it is applied to frequency distributions of similarity measures (such as harmonic amplitudes, correlations, and factor loadings).

If **H** is a matrix of the frequency values from a number of histograms, with each column in **H** representing a different histogram and each row in **H** representing a separate bin of comparable values among the histograms, then the matrix of covariances **R** among the histograms is

$$\mathbf{R} = \mathbf{H}^T \mathbf{H}$$

(**R** is a matrix of correlations if the histogram frequencies are standardized to unit variance).

As above, eigenfunctions and eigenvalues can be calculated either from **R** or from a singular value decomposition of **H**,

$$\mathbf{H} = \mathbf{U}\mathbf{S}^{1/2}\mathbf{V}^T.$$

U is a set of empirical orthogonal histograms (eigenfunctions or principal components of the histograms) and $\mathbf{S}^{1/2}\mathbf{V}^T$ are the correlations between them and the observed histograms. These can be rotated in a number of ways (e.g., the VARIMAX scheme used in the original CABFAC program by Klovan and Imbrie or its elaborations by EXTENDED CABFAC and EXTENDED QMODEL) to designate end-member histograms, and as long as the rotation scheme maintains the orthogonality of the components, they provide a suitable reduced basis for viewing the interrelationships among shapes of the histograms.

In principle, we have no quarrel with the orthogonal rotation of principal components (or eigenshapes, in the context of analysis of shape functions, of which a histogram is one). As discussed below, our objection to the work that has applied this procedure is in the construction of the histograms themselves.

Caveats

First, *do not attach statistical significance to the eigenshapes.* To compute the eigenfunctions of a set of shapes does not require probabilistic assumptions about the distribution of the shapes. Such assumptions arise when a researcher seeks statistical justification for ignoring the higher-order eigenfunctions (those associated with small eigenvalues) or for using the first eigenshape as the best estimate of the population mean. If all that we require is a way to plot the data that does not sacrifice much of its variance, statistical tests are inappropriate. Rather we can derive from the analysis quantitative measures that tell us how well our reduced space describes each shape. Practically speaking this is the ratio of a shape-function's Euclidean length in the reduced space to its length in the full space.

Second, *do not attach biological significance to the eigenshape functions themselves.* They provide a basis on which to plot the shapes and a compact description of the shape variation subtended by that basis, but nothing in the procedure guarantees that they embody biological phenomena. Like shapes constructed from harmonic functions, they are a set of orthogonal elements that can be added in various proportions to represent observed shapes. Unlike the shapes reconstructed from harmonic functions, they are derived empirically from an analysis of the observed shapes they are meant to represent.

Interpretations of the meaning of eigenfunctions can be made only by considering information that was not explicitly included in the shape analysis: by looking at the covariation between the eigenshapes and biological, geographical, or chemical information.

Third, *do not make histograms of similarity measures (loadings or correlations) between individuals and the eigenshape functions.* This objection applies equally to harmonic functions (the amplitudes of a single Fourier harmonic associated with a collection of shape functions) or any other reference shape functions. We know of no valid use for such histograms.

Consider the following example: Suppose we wish to examine the structure of a large collection of objects of many different sizes and shapes and search for subpopulations. First consider size. A size frequency distribution of the objects, a histogram of their sizes, provides an excellent visual description of the size structure of the collection. One might debate the optimum interval over which to sum the size frequencies, one might argue over what constitutes a distinct mode or subpopulation, or one might question whether the collection faithfully reflects the population from which it was drawn, but there should be no misunderstanding what a histogram of sizes is. It shows clearly and unambiguously the number of large and small objects in a collection.

Histograms of similarity measures, in contrast, are inherently misleading. Measures of shape, expressed as the degree of similarity of each shape with some reference shape, cannot be treated like size. Unlike measures of size, measures of shape do not have the same information content throughout their measured range. Values of shape similarity that approach identity (correlations approaching 1.0) tell us a lot about the shape of the measured object. Those approaching zero tell us very little. A histogram of similarities skewed toward 1.0 gives us a good idea about the shape of the objects in a collection the objects look like the reference shape they are compared to and they look like each other. A histogram of similarities skewed towards zero not only does not give us much information about the shape of the objects but, by careless analogy with size histograms, can misleadingly suggest that they comprise a subpopulation of similar-looking objects that do not look like the shape they are compared to. In fact, they may not look at all like each other. All we can say from such a histogram is that the objects in the collection it represents do not look like the reference shape to which they have been compared. As a subpopulation, their only common attribute may be their lack of similarity to some reference shape.

Most histograms constructed of shape similarity measures consist of values that fall somewhere between no similarity and identity. Typical studies using these constructions show patterns of variation among samples portrayed as a collection of histograms, some of which are skewed to the left, others skewed to the right. But these patterns formed by the distributions reflect differing contributions from measures which tell us a lot and from others which tell us almost nothing, and those that tell us nothing tend to associate objects which may have no shape similarity with each other at all. There is no valid procedure for picking more than one subpopulation from such a collection of histograms because the histograms can identify only one subpopulation.

A practical example is shown by Healy-Williams et al. (1985). How is one to interpret their figure 7? End-member 1 identifies samples that contain shapes that are somewhat triangular, since end-member 1 is a histogram of amplitude coefficients with high loadings on the third harmonic. In contrast, end-member 2 identifies samples whose only common characteristic is that they are *not* triangular. What interpretation can be made of a series of samples that progresses from having very little in common to having something in common? What interpretation can be made of end-member 3, which identifies samples whose shapes have less of harmonic 3 than end-member 1 but more than end-member 2? It is not an end-member at all, yet the factor analysis of the histograms has identified it as such. The pattern of variation in these histograms is misunderstood because the *lack* of shape similarity indicated by the amplitude histograms has been ignored.

This misleading use of histograms, while most often seen in studies that follow the methods of Ehrlich et al. (1980) for using Fourier harmonic functions to measure shape, applies to histograms of any similarity measure. Histograms of eigenshape amplitudes typically appear to be normally distributed about zero (for an example, see Malmgren et al., 1983, figure 7). In fact, this tendency only indicates that most of the individuals measured do not look like that eigenshape. The values for harmonic amplitudes where phase has been ignored are reported as positive numbers. When these values tend toward zero, the histogram is skewed to the left. Because phase is an integral part of the eigenshape function, the sign associated with its amplitudes has meaning and its values are reported as signed numbers. When these values tend toward zero, their histogram is skewed toward zero and appears normally distributed around zero. Like histograms of harmonic amplitudes, we claim that such constructions are misleading and uninter-

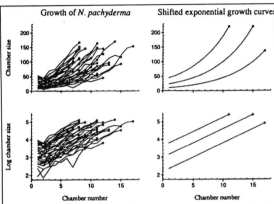

Figure 24. Growth curves for *N. pachyderma* and a simple model to explain them. Differences in the size of the initial chamber determine the point on a common growth curve at which an individual begins to grow, in effect, shifting the growth curve. The need to accommodate fewer chambers of similar size produces a wide range in external morphology. The shifted curves are all the same curve. From unpublished work with Michael Spindler (Alfred Wegener Institut, Bremerhaven).

pretable.

Finally, *do not assume just because the external morphology of an individual is determined by the processes by which it grew, that these processes can be recovered from a shape analysis.* Figure 24 shows growth curves for *Neogloboquadrina pachyderma*, a species of planktonic foraminifera with a variable external morphology. Differences in the size of the initial chamber determine the point on a common growth curve at which an individual begins to grow, in effect, shifting the growth curve. The need to accommodate fewer chambers of similar size produces a wide range in external morphology. The shifted curves are all the same curve. Any kind of shape analysis will attribute most of that variability to the number of visible chambers, i.e., the number of chambers in the outer whorl. This amounts to a redescription of the shapes that gives no insight into

their cause. The growth curves, constructed from internal measurements, suggest that variability in the external morphology of *N. pachyderma* can be explained by a simple mechanism: All individual *N. pachyderma* follow a common growth curve. But there is variation among individuals in the size of their initial chamber (their proloculus), and the size of this chamber determines where on the growth curve an individual begins to grow. While the growth curve is the same for all individuals, those that follow it from different points have different numbers of chambers when compared at the same size. The differences in chamber arrangement that are needed to accommodate this produce the differences we see in their external morphology. Shape analysis of the wide variation that this simple mechanism can produce will give no clue to its origin.

An Alternative to the Use of Covariance Matrices in Eigenanalyses of Shapes.

All of the published work to date that used eigenshape analysis has either followed Lohmann (1983) in fully standardizing the columns of the data matrix or has followed Schweitzer (1990) in using the covariance matrix. Here we present yet another method for coping with shape amplitudes in a way that emphasizes the differences between the covariance and the correlation matrices of shapes. Let $\mathbf{X}$ be a $n \times p$ matrix of ϕ^* shape functions whose columns have zero mean (because the mean of ϕ^* corresponds to the orientation of the shapes) but not unit variance. The transformation of the columns to unit variance is $\mathbf{Z} = \mathbf{X}\mathbf{A}^{-1}$ where $\mathbf{A}$ is the $p \times p$ diagonal matrix whose elements $a_1, ..., a_p$ are the amplitudes (standard deviations) of the columns of $\mathbf{X}$. Note that while $\mathbf{X}^T \mathbf{X}\frac{1}{n}$ is the matrix of covariances among shapes, $\mathbf{Z}^T \mathbf{Z}\frac{1}{n}$ is the matrix of correlations among shapes.

In the original formulation of eigenshape analysis, eigenvectors $\mathbf{U}$ are the left singular vectors of the matrix $\mathbf{Z}$, and are derived from the eigenanalysis of the matrix $\mathbf{Z}^T\mathbf{Z}$:

$$\mathbf{V}^T \mathbf{Z}^T \mathbf{Z} \mathbf{V} = \mathbf{S}$$
$$\mathbf{S}^{-1/2} \mathbf{V}^T \mathbf{Z}^T \mathbf{Z} \mathbf{V} \mathbf{S}^{-1/2} = \mathbf{I}$$
$$\mathbf{U} = \mathbf{Z} \mathbf{V} \mathbf{S}^{-1/2}$$
$$\mathbf{Z} = \mathbf{U} \mathbf{S}^{1/2} \mathbf{V}^T$$

since $\mathbf{V}^T \mathbf{V} = \mathbf{I}$ and $\mathbf{S}$ is diagonal.

One might ask why the eigenvectors are not computed from the singular value decomposition of $\mathbf{X}$ instead, as $\mathbf{X} = \mathbf{U} \mathbf{S}^{1/2} \mathbf{V}^T$. Then the reduced space would optimally describe the sum of squares of the shape functions $\mathbf{X}$ rather than that of the normalized shapes $\mathbf{Z}$.

By considering the geometry of the data space we can see why the shapes might be normalized prior to computing eigenfunctions. Each shape is described by 100 numbers (the angles of the ϕ^* function), and so each shape is a point in a 100-dimensional hyperspace. Some of the shape functions have more variance than others, because some of the shapes are more sharply angular than others. These are represented by points that are farther from the origin, since variance is directly related to the square of Euclidean length in vectors that have zero mean:

$$||x||^2 = ns_x^2 = \sum_{i=1}^{n} x_i^2 ,$$

if

$$\sum_{i=1}^{n} x_i = 0 .$$

Standardization to unit variance has the geometrical effect of projecting all of the points (each of which describes one shape) onto a sphere. Then all points are equidistant from the origin, and consequently each shape exerts the same influence on the directions of the eigenvectors. Shapes tend to "pull" the first eigenvector towards them in proportion to their amplitude, and by giving all of the shapes equal amplitude we release the first eigenvector from this bias.

But when we projected all of the shapes onto a sphere, we distorted the distances among them. Since the eigenvectors just rotate the data space, we can restore the original lengths of the vectors describing the shapes merely by multiplying the correlations between the shapes and the eigenshapes by the appropriate shape amplitude:

$$\mathbf{Z} = \mathbf{X} \mathbf{A}^{-1}$$
$$\mathbf{X} = \mathbf{Z} \mathbf{A}$$
$$= \mathbf{U} \mathbf{S}^{1/2} \mathbf{V}^T \mathbf{A} .$$
$$\mathbf{U}^T \mathbf{X} = \mathbf{S}^{1/2} \mathbf{V}^T \mathbf{A}$$
$$\mathbf{X}^T \mathbf{U} = \mathbf{A} \mathbf{V} \mathbf{S}^{1/2}$$

This is the matrix equivalent of converting a correlation to a covariance when one of the variables has unit variance:

$$\sigma_{xy} = \rho_{xy} \sigma_x \sigma_y \text{ where } \sigma_y = 1.$$

In summary, the eigenvectors that define the reduced space in which we view the data can be computed so that each shape has *equal weight* in the analysis. When we plot the data to see the relationships among shapes, we can plot the covariances between the shapes and these eigenvectors. By using covariance we restore to the points their original lengths. This allows us to distinguish among shapes that differ only in amplitude.

Acknowledgments

This work was supported by the National Science Foundation (Grants OCE82--14930, OCE84--10221, and OCE84--17040). This is Woods Hole Oceanographic Institution Contribution No. 7048.

References

Aubrey, D. G.   1979.   Seasonal patterns of onshore/offshore sediment movement. J. Geophys. Res., 84:6347-6354.

Aubrey, D. G. and K. O. Emery. 1986a. Relative sea levels of Japan from tide-gauge records. Geol. Soc. Amer. Bull., 97:194-205.

Aubrey, D. G. and K. O. Emery. 1986b. Relative sea levels of Japan from tide-gauge records: Reply. Geol. Soc. Amer. Bull., 97:1282.

Cifelli, R. 1965. Planktonic foraminifera from the western North Atlantic. Smithsonian Miscellaneous Collections, 148(4):36pp.

Cooley, W. S. and P. R. Lohnes. 1971. Multivariate data analysis. Wiley and Sons, New York, 364 pp.

Davis, R. E. 1976. Predictability of sea surface temperature and sea level pressure anomalies over the North Pacific Ocean. J. Phys. Oceanogr., 6(3):249-266.

Ehrlich, R., P. J. Brown, J. M. Yarus, and R. Przygocki. 1980. The origin of shape frequency distributions and the relationship between size and shape. J. Sedim. Petrol., 50(2):475-484.

Evans, D. G., P. N. Schweitzer, and M. S. Hanna. 1985. Parametric cubic splines and geologic shape descriptions. Math. Geol., 17(6):611-624.

Eves, H. 1972. Analytic geometry. in Standard mathematical tables (Selby, S. M., ed.). Chemical Rubber Co., Cleveland, 20th edition, 353 pp.

Friedman, J. H., and W. Stuetzle. 1982. Projection pursuit methods for data analysis. Pp 123-147 in Modern data analysis (Launer, R. L, and A. F. Siegel, eds.). Academic Press, New York.

Golub, G. H. and C. Reinsch. 1970. Singular value decomposition and least squares solutions. Num. Math., 14:403-420.

Healy-Williams, R. Ehrlich, and D. F. Williams. 1985. Morphometric and stable isotopic evidence for subpopulations of Globorotalia truncatulinoides. J. Foraminiferal Res., 15(4):242-253.

Lohmann, G. P. 1983. Eigenshape analysis of microfossils: A general morphometric procedure for describing changes in shape. Math. Geol., 15(6):659-672.

Lohmann, G. P. and J. J. Carlson. 1981. Oceanographic significance of Pacific Late Miocene calcareous nannoplankton. Mar. Micropaleo., 6:553-579.

Lohmann, G. P. and B. A. Malmgren. 1983. Equatorward migration of Globorotalia truncatulinoides ecophenotypes through the Late Pleistocene: Gradual evolution or ocean change? Paleobiology, 9(4):414-421.

Lorenz, E. N. 1959. Empirical orthogonal functions and statistical weather prediction. Report No. 1, Statistical Weather Forcasting Project, MIT.

Malmgren, B. A., W. A. Berggren, and G. P. Lohmann. 1983. Evidence for punctuated gradualism in the Late Neogene Globorotalia tumida lineage of planktonic foraminifera. Paleobiology, 9(4):377-389.

Mix, A. C., W. F. Ruddiman, and A. McIntyre. 1986. Late Quaternary paleoceanography of the Tropical Atlantic, 1: Spatial variability of annual mean sea-surface temperatures, 0-20,000 years B. P. Paleoceanography, 1(1):43-66.

Rohlf, F. J. 1986. The relationships among eigenshape analysis, Fourier analysis, and the analysis of coordinates. Math. Geol., 18:845-854.

Rohlf, F. J. and J. Archie. 1984. A comparison of Fourier methods for the description of wing shapes in mosquitoes (Diptera: Culicidae). Syst. Zool., 33:302-317.

Scott, G. H. 1980. The value of outline processing in the biometry and systematics of fossils. Paleontology, 23:757-768.

Scott, G. H. 1981. Upper Miocene biostratigraphy: Does Globorotalia conomiozea occur in the Messinian? Revista Espanola Micropaleo., 12:489-506.

Schweitzer, P. N. 1990. Inference of ecology from ontogeny of microfossils. Ph. D. Thesis, MIT/WH01, WH01-90-04.

Schweitzer, P. N., R. L. Kaesler, and G. P. Lohmann. 1986. Ontogeny and heterochrony in the ostracode Cavellina Coryell from the Lower Permian rocks in Kansas. Paleobiology, 12(3):290-301.

Schweitzer, P. N. and G. P. Lohmann. 1990. Life-history and the evolution of ontogeny in the ostracode genus, Cyprideis. Paleobiology, In press.

van Morkhoven, F. C.P. M. 1962. Post-Palaeozoic Ostracoda: Their morphology, taxonomy, and economic use. Elsevier, Amsterdam, 204 pp.

Zahn, C. T. and R. Z. Roskies. 1972. Fourier descriptors for plane closed curves. IEEE Trans. Comp., C-21:269-281.

Chapter 7

Fitting Curves to Outlines

F. James Rohlf

Department of Ecology and Evolution
State University of New York
Stony Brook, NY 11794-5245

Abstract

This paper reviews methods for describing the shape of a structure, either a complete outline contour or a segment between two landmarks, by fitting a function to the outline. The parameters of the fitted function are often used as variables in subsequent multivariate analyses in order to study patterns of variation and covariation in shape. These methods are of interest when there are few (if any) homologous landmarks on a structure or when the outline shape itself is of interest rather than its relationship to various landmarks. Computational formulas, plots, and simple numerical examples are given for most methods.

Introduction

When few or no landmarks are available, the shape of a structure is often captured by the coordinates of a sequence of points along its outline. The points may be equally spaced or else spaced more densely where there are more rapid changes in the curvature of the outline. But a large number of coordinates along an outline is not a very compact or efficient way to describe a shape since raw coordinates are expected to be highly correlated with each other. Thus they contain large amounts of redundant information. One solution is to transform the information in these coordinates into a more compact form. This paper surveys methods for fitting a function to the sequence of points and then using the parameters of the fitted function as descriptive variables for further multivariate analysis. A related approach is that of eigenshape analysis (Lohmann, 1983). It is not covered here because it is discussed elsewhere in this volume. Rohlf (1986) discusses some of its relationships to other methods.

Two major classes of outlines must be distinguished since they require different methods. First, there are outlines that correspond to *open contour curves* curves along an outline of a structure between two landmarks. An example would be the curve corresponding to a vein in an insect wing. The landmarks might be the origin of the vein and its intersection with the margin of the wing. Second, there are outlines that correspond to complete *closed contour curves*. An example would be the curve representing the entire outline of an insect wing. In most applications it is assumed that the closed contour begins with a homologous landmark. In some cases there is also an additional landmark that can be used as a reference point for the origin of a coordinate system or to rotate the structure into a standard orientation.

The account given below is organized according to whether one has an open or closed contour and with respect to aspects of the complexity of the curves. But these considerations are not sufficient to determine uniquely the most appropriate function to use in a given application. This review will be limited to only those functions that have been proposed for use in morphometrics. Thus it is a survey of what has been done rather than a compendium of all the functions that *could* be used in a morphometric study.

A simple example is furnished for most methods. The results are shown both graphically and as tables of numerical results. These should both provide a better understanding of the methods and serve as test data for the development of software to implement the methods.

Open Curves

Many functions are available which can be used to fit data points along a curve between two reference points. The choice among them depends, in part, upon the complexity of the outline. An important special case is when one coordinate, e.g., y, can be expressed as a single-valued function of the other coordinate, x. It may be necessary to rotate the objects into a standard orientation such as by placing the x-axis through the two reference points. Such data permit the use of especially simple methods. More complex curves require more complex methods.

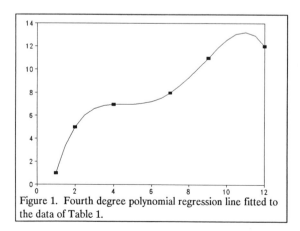

Figure 1. Fourth degree polynomial regression line fitted to the data of Table 1.

Simple Open Contours

The following methods assume that a curve can be oriented in a standard way such that the p y-coordinates along the outline of a structure correspond to a monotonically increasing set of x-coordinates. Figure 1 is a plot of the simple set of artificial data given in Table 1. These data will be used in a number of examples given below.

Table 1. Artificial data and solutions to 4th degree polynomial regression.

Observations		Solution	
X	Y	i	b_i
1	1	0	-6.73478
2	5	1	10.14688
4	7	2	-2.69606
7	8	3	0.29935
9	11	4	-0.01119
12	12		

Polynomials   The simple polynomial equation, $y = b_0 + b_1x + b_2x^2 + \cdots + b_{p-1}x^{p-1}$ can be fitted to a sequence of y-coordinates along a curve. If all of the p parameters, b_j are to be found, then the computations simply involve the solution of a set of simultaneous linear equations. In practice, many fewer parameters are usually sufficient to give a satisfactory fit. In such cases, the following least-squares procedure can be used to estimate the parameters:

1.   Form a $p \times 1$ matrix, **Z**, containing the y-coordinates of the p points along the curve.

2.   Construct a $p \times k$ matrix, **X**, with the second column containing the x-coordinates of the p-points along the curve. The first column is a vector of all 1s and the other columns, j, contain x_{j-1}^{i} As many columns, k,are included as parameters one wishes to fit.

3.   The vector of parameters, **B**, is computed as $\mathbf{B} = \mathbf{TZ}$, where $\mathbf{T} = (\mathbf{X^tX})^{-1}\mathbf{X^t}$. The details of the computations are described in many statistical textbooks (e.g., Sokal and Rohlf, 1981).

If the x_i are equally spaced, then simpler methods are available. For example, the columns of $\mathbf{X}$ can be filled with coefficients of orthogonal polynomials.

A simple example is shown in Table 1 where a fourth degree polynomial is fit to an artificial example with 6 data points (these same data points are used for the other examples furnished below). The R^2-value is 0.999997. While the fit is very good, the path taken by the curve between the observed data points seems unlikely (see Figure 1).

The parameters, $\mathbf{B}$, correspond to a linear transformation of the original y-coordinates. If the higher order terms are ignored (since they usually have very small coefficients), then the transformation corresponds to a nonrigid rotation followed by a projection onto a lower dimensional subspace. The effect of ignoring the higher order terms is to fit a smooth curve to the observed data. This is often desirable since at least some of the local irregularities of the curve correspond merely to digitizing error.

Cubic Splines A cubic spline is a series of cubic polynomial functions pieced together in such a way that the fitted function passes smoothly through all of the observed data points and is constrained so that their first and second derivatives are continuous throughout the range being fitted. Cubic splines can be fitted to data in which the x-coordinates of the points being fitted are monotonically increasing (the next section describes parametric cubic splines which do not have this limitation). Atkinson and Harley (1983), de Boor (1978), and Press et al. (1986) give helpful introductions to cubic splines. Atkinson and Harley (1983) provide a Pascal procedure that can be used to compute cubic splines. Press et al. (1986) furnish programs in both FORTRAN and Pascal.

Let y_i for $i = 1,...,p$ be a set of y-coordinates along an outline for a set of monotonically increasing x-coordinate values. Then for any interval, e.g., x_i to x_{i+1}, we can interpolate for y using the cubic polynomial

$$y = A y_i + (1 - A) y_{i+1} + C b_i + D b_{i+1}, \qquad (1)$$

where $A = (x_{i+1}-x)/\Delta x_i)$, $C = \frac{1}{6}(A^3 - A)(\Delta x_i)^2$,

$D = \frac{1}{6}[(1 - A)^3 - (1 - A)](\Delta x_i)^2$, and $\Delta x_i = x_{i+1} - x_i$.

The b_i (second derivatives of the interpolating polynomial function) are the parameters of the curve. By adding the equations $b_1 = 0$ and $b_p = 0$, it is possible to express the spline equations as $\mathbf{CB} = \mathbf{A}$ where

$$C = \begin{pmatrix} 1 & 0 & 0 & \cdots & 0 & 0 \\ \Delta x_1 & 2(\Delta x_1 + \Delta x_2) & \Delta x_2 & \cdots & 0 & 0 \\ 0 & \Delta x_2 & 2(\Delta x_2 + \Delta x_3) & \cdots & 0 & 0 \\ \vdots & \vdots & \vdots & \vdots & \vdots & \vdots \\ 0 & 0 & 0 & \cdots & 2(\Delta x_p + \Delta x_{p-1}) & \Delta x_p \\ 0 & 0 & 0 & \cdots & 0 & 1 \end{pmatrix} \qquad (2)$$

$$A = 6 \begin{pmatrix} 0 \\ \frac{\Delta y_2}{\Delta x_2} - \frac{\Delta y_1}{\Delta x_1} \\ \frac{\Delta y_3}{\Delta x_3} - \frac{\Delta y_2}{\Delta x_2} \\ \vdots \\ \frac{\Delta y_{p-1}}{\Delta x_{p-1}} - \frac{\Delta y_{p-2}}{\Delta x_{p-2}} \\ 0 \end{pmatrix}, \qquad (3)$$

$\Delta y_i = y_{i+1} - y_i$, and $\mathbf{B}$ is the column vector of parameters. This tridiagonal system of equations can be easily solved as $\mathbf{B} = \mathbf{C}^{-1}\mathbf{A}$.

Table 2. Fitting a cubic spline to the data of Table 1. **A** and **C** matrices used to estimate the coefficient vector **B**.

C						A	B
1	0	0	0	0	0	0	0
1	6	2	0	0	0	-18	-2.97519
0	2	10	3	0	0	-4	-0.07444
0	0	3	10	2	0	7	0.89926
0	0	0	2	10	3	-7	-0.87965
0	0	0	0	0	1	0	0

A simple example of a cubic spline is shown in Table 2 and plotted in Figure 2. The path of the function seems more reasonable than that shown in Figure 1.

Note that the use of cubic splines always results in as many parameters as one has data points. This is expected since the spline function is required to pass through all of the data points. Thus there is no reduction to a more compact set of parameters. However, one can resample the curve with fewer points and then check that the new interpolated function stays fairly close to the original points.

Complex Open Contours

When the outline curve is complex, the y-coordinates cannot be expressed as a single-valued function of the x-coordinates. In such cases other methods are needed. One approach is to express the x and the y-coordinates each as parametric functions of the cumulative chordal distance of each point along the contour. Parametric cubic splines are one example. Another approach is to fit a more complex function such as the Bezier curve (see below).

Parametric cubic splines   This method is as described above for cubic splines but with x and y-coordinates (or even z-coordinates) fitted separately with the cumulative chordal distance, t_i, of the ith point used as the abscissa. That is, in the equations of the previous section, replace Δx_i with Δt_i and then generalize the matrix, **Z**, to a $p \times n$ matrix with the columns corresponding to the x, y, and possibly z-coordinates of the p points along the outline. The solution matrix, **B**, will also have n columns.

Using the same artificial data as before, the vector, **T**, of chordal distances from the first point is shown in the first column of Table 3. The **C** matrix shown in the table is the same for both x and y-variables. The vector, **A**, is now a matrix with columns corresponding to the x and y-variables. The solution matrix, **B**, is shown in Table 4. The parametric cubic spline is plotted in Figure 3 for comparison with the previous figures.

Evans et al. (1985) describe parametric cubic

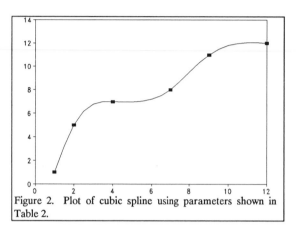

Figure 2. Plot of cubic spline using parameters shown in Table 2.

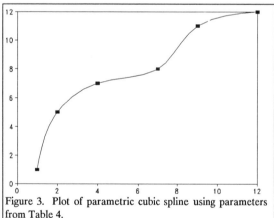

Figure 3. Plot of parametric cubic spline using parameters from Table 4.

Table 3. **T**, **C**, and **A** matrices for the computation of a parametric cubic spline for the data of Table 1.

T			C				A	
							X	Y
0	1	0	0	0	0	0	0	0
4.12311	4.12311	13.90307	2.82843	0	0	0	2.78743	-1.57821
6.95153	0	2.82843	11.98141	3.16228	0	0	1.44946	-2.34527
10.11381	0	0	3.16228	13.53566	3.60555	0	-2.36390	3.09494
13.71936	0	0	0	3.60555	13.53566	3.16228	2.36390	-3.09494
16.88164	0	0	0	0	0	1	0	0

splines and furnish examples of their application to microfossil outlines and to rock folds. Some useful general references include Spath (1974) and deBoor (1978). Gasson (1983) describes its application to computer graphics.

Table 4. Solution matrix, **B**, for a parametric cubic spline fitted to the matrices given in Table 3.

X	Y
0	0
0.16914	-0.05581
0.15410	-0.28365
-0.27681	0.38300
0.24838	-0.33067
0	0

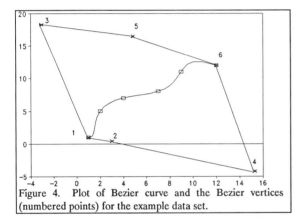

Figure 4.   Plot of Bezier curve and the Bezier vertices (numbered points) for the example data set.

Bezier curves   A Bezier polynomial (Bezier, 1970) of degree q is a function that can be made to pass through an arbitrary ordered sequence of $p = q+1$ points in a 2 or 3-dimensional space. The parameters of a Bezier curve are simply the locations of $q+1$ constructed points that correspond to the vertices of an open polygon that forms a convex hull around the observed data points. The first and last vertices are the same as the first and last data points and should correspond to morphological landmarks. The coordinates of the other vertices are parameters to be estimated. These vertices define a unique polynomial function that interpolates smoothly between the first and last vertices. As in the case of parametric cubic splines, the sequence of points can have a very complex trajectory since the polynomial is a function of an intrinsic parameter, t, which is the chordal distance along the curve (scaled to go from 0 to 1 as one goes from the first observed point to the last).

The vector of coordinates, **B**, of the Bezier vertices can be constructed as solutions of the equation $\mathbf{Z} = \mathbf{JB}$, where **J** is a $p \times (q+1)$ matrix of coefficients

$$J_{ij} = \begin{bmatrix} q \\ j\text{-}1 \end{bmatrix} t_{j\text{-}1}^{i}(1\text{-}t_i)^{q\text{-}j}+1, \tag{4}$$

$$\begin{bmatrix} q \\ j\text{-}1 \end{bmatrix} = \frac{q!}{(j\text{-}1)!(q\text{-}j+1)!}, \tag{5}$$

t_i is the normalized chordal distance from the first data point to the ith (the normalization is such that $t_1 = 0$ and $t_p = 1$) and **Z** is a matrix with rows corresponding to the coordinates of the p observed points and columns corresponding to the x, y (and possibly z) coordinate axes.

An advantage of the use of Bezier curves is that one can easily fit lower order polynomials, $p > q+1$, without having to discard data points. In this case matrix **J** will have more rows than columns and the first and last vertices will no longer correspond exactly to the first and last data points. A least-squares solution for this case can be obtained as

$$\hat{\mathbf{B}} = (\mathbf{J^tJ})^{-1}\mathbf{J^tZ}, \tag{6}$$

This minimizes

$$SS = tr((\mathbf{Z} - \hat{\mathbf{Z}})(\mathbf{Z} - \hat{\mathbf{Z}})^t), \tag{7}$$

where $\hat{\mathbf{Z}} = \mathbf{J\hat{B}}$ is a matrix of estimated coordinates of the $q+1$ vertices.

Table 5 Vector of normalized t values and **J** matrix for a 5th degree Bezier polynomial fitted to the example data set of Table 1.

t	J					
0.00000	1.00000	0.00000	0.00000	0.00000	0.00000	0.00000
0.24424	0.24656	0.39840	0.25750	0.08322	0.01345	0.00087
0.41178	0.07042	0.24649	0.34510	0.24159	0.08456	0.01184
0.59910	0.01036	0.07738	0.23126	0.34560	0.25823	0.07718
0.81268	0.00023	0.00500	0.04341	0.18833	0.40854	0.35448
1.00000	0.00000	0.00000	0.00000	0.00000	0.00000	1.00000

Table 5 furnishes the vector of normalized t_i values for the 6 data points from the artificial example and the **J** matrix for a 5th degree Bezier curve. Table 6 gives the constructed Bezier vertices. Figure 4 shows a plot of the Bezier curve and the Bezier vertices. The vertices are numbered in the order in which they are constructed. They are also connected by dotted lines to show how they form an envelope (something like a convex hull) around the observed data points.

Engel (1986) shows examples of fitting Bezier curves to shell valves of *Anodonta* and of human sagittal profiles of male and female skulls of various ages. Gasson (1983) describes

Table 6. Bezier vertices for the example data set.	
X	Y
1.00000	1.00000
2.99504	0.40721
-3.06162	18.29616
15.29614	-4.23698
4.85415	16.51657
12.00000	12.00000

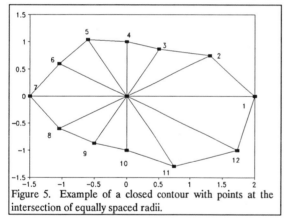

Figure 5. Example of a closed contour with points at the intersection of equally spaced radii.

some of the properties of Bezier curves and generalizations to Bezier surfaces.

Closed Contours

Data consisting of points along closed contours, complete outlines, have been used more commonly in morphometrics than open contours. If a homologous landmark along the outline is available, it is usually taken as the first point along the curve; it is treated as if homologous from specimen to specimen. If no landmark is available, an arbitrary point is taken as the starting point for the curve and some

adjustment must be made in the fitted coefficients to take this into account. As in the previous sections, it is useful to distinguish between the case of simple closed contours and the more general case of complex contours, whose outline can even appear to intersect in the view being analyzed. While many different techniques could be employed, some type of Fourier analysis is usually used.

Simple Closed Contours

The outline of an Ostracode was used in one of the first examples (Kaesler and Waters, 1972) of the use of Fourier analysis in morphometrics. In this and in many of the subsequent studies the outline was described in terms of the lengths of radii emanating from a central point (see Figure 5 for an example).

Fourier analysis of equally spaced radii For this method the data consist of the lengths of p equally spaced radii emanating from a central point within the object. This point is the origin of the polar coordinate system. In most early studies it corresponded to a morphological landmark. In more recent studies it is simply the centroid of the outline of the object. An example is shown in Figure 5. Fourier analysis consists of fitting the following function to the observed data

$$\theta = a_0 + \sum_{i=1}^{k} (a_i \cos(i\,\theta) + b_i \sin(i\,\theta)) \qquad (8)$$

where the angle, θ, varies from 0 to 2π; a_i and b_i are the Fourier coefficients for the ith harmonic; and k is the maximum number of harmonics computed ($k < p/2$). The least-squares estimates of the Fourier coefficients can be computed as

$$a_i = \sqrt{\frac{2}{p}} \sum_{j=1}^{p} \theta_j \cos(i\,\theta_j)$$

$$b_i = \sqrt{\frac{2}{p}} \sum_{j=1}^{p} \theta_j \sin(i\,\theta_j) \qquad (9)$$

for $i = 1$ to $p/2$. The zeroth harmonic coefficients are

$$a_0 = \sqrt{\frac{1}{p}} \sum_{j=1}^{p} \theta_j$$

$$b_0 = 0 .$$

(10)

This simple solution is possible since the independent variables in the equation for θ given above are orthogonal when the angles are equally spaced.

An example is furnished in Table 7. The angles, θ, from 0 to 2π are divided into 12 equal-angular steps of $\pi/6$. The lengths, ρ, of the radii, given in the second column, are relative to the location of a landmark near the center of the object. The Fourier coefficients are shown for harmonics 0 through the 6th.

Table 7: Lengths, ρ, of radii at an angle θ and Fourier coefficients for harmonics 0 through 6.

θ	ρ	i	a_i	b_i
0.00000	2.0	0	1.34167	0.00000
0.52360	1.5	1	0.65427	-0.20813
1.04720	1.0	2	0.85732	-0.42426
1.57080	1.0	3	0.08165	-0.20412
2.09440	1.2	4	0.08165	0.07071
2.61799	1.2	5	-0.12355	0.00400
3.14159	1.5	6	0.12247	0.00000
3.66519	1.2			
4.18879	1.0			
4.71239	1.0			
5.23599	1.5			
5.75959	2.0			
6.28319	2.0			

If a landmark is not used as the origin, the data must be transformed to polar coordinates relative to the centroid of the contour. This can be done by transforming the polar coordinates to a rectilinear coordinates (using the relationships $x = \rho \cos \theta$ and $y = \rho \sin \theta$), translating the origin to the centroid ($x' = x - x_c$, $y' = y - y_c$), and then converting back to the polar coordinate system ($\theta = \tan^{-1}(y'/x')$ and $\rho = \sqrt{x'^2 + y'^2}$). Parnell and Lestrel (1977) describe these and other manipulations that can be performed with such data. The coordinates

of the centroid of the contour should be found by a numerical integration, but they can be approximated by using only x and y. If the original angles were equally spaced, the transformed coordinates will no longer be equally spaced. One solution is to interpolate along the curve to obtain a new set of equally spaced values. A simpler solution is to use the methods of the next section which allow for unequally spaced values.

In order to compare the coefficients of one specimen with those of another, the angle of the starting radius must be the same for both. This means that the outlines must be oriented so that the intersection of the first radius with the outline must correspond to a homologous landmark. If this is not the case, then the rotational information [phase angle, $\phi = \tan^{-1}(b/a)$] in the coefficients must be ignored. This is easily done by combining the two coefficients for each harmonic into a single number, the harmonic amplitude, computed as $h_i = a_i^2 + b_i^2$.

These amplitudes are a measure of the amount of "energy" at each harmonic. Lohmann (1983) and Bookstein et al. (1985) point out that if a reference point can be defined, the information on phase angle should be retained. An alternative is to rotate each specimen so that it aligns with some standard. Lohmann (1983) rotated each specimen so as to maximize its correlation with a reference specimen. Ferson et al. (1985) aligned each specimen with its principal axes. Full and Ehrlich (1986) refer to these alignments as mathematical homology.

If the number of points happens to be a power of 2, it is possible to use the "fast Fourier transformation" algorithm (e.g., Cooley and Tukey, 1965) to perform the calculations. It should be noted that this is simply an efficient computational algorithm not another type of analysis. The same numerical values are obtained for the coefficients.

Fourier analysis of unequally spaced radii Fourier analysis can also be applied when the radii are not equally spaced. While most studies use equally spaced points, it is more efficient to collect points more densely in regions of higher curvature. Note that by using equally spaced radii the parts of the outline close to the center are sampled more densely relative to regions of the outline far from the center. The base and tips of a mosquito wing, for example, are sampled very poorly since the wings are very elongate. The computations are more complex than in the case of equally spaced points since the vectors corresponding to the observed values of $\cos(i\theta_j)$ and $\sin(i\theta_j)$ are not orthogonal. Least-squares estimates of the Fourier coefficients can be obtained using the methods of multiple regression analysis with the $\cos(i\theta_j)$ and $\sin(i\theta_j)$ used as independent variables. An example of their use is shown below for the analysis of the ϕ^* function.

Complex Closed Contours

The above methods cannot be used for more complex contours when one or more radii would intersect the contour at more than one point. The solution here is to express the contour as a parametric function of the cumulative chordal distance along the contour from some fixed starting point. Several such procedures have been proposed for use in morphometrics and are listed below.

Fourier analysis of tangent angles   Zahn and Roskies (1972) suggest expressing the contour in terms of the function:

$$\phi^*(t) = \phi(t) - t, \qquad (11)$$

where t is the cumulative chordal distance along the curve (scaled as if it were an angle ranging from 0 to 2π radians), $\phi(t)$ is the angular difference in orientation between a tangent at the starting point ($t=0$) and a tangent at a distance t along the outline. The last value (corresponding to the last tangent angle) is not computed since it is assumed to be identical to the first value. See Figure 6 for a plot of ϕ and ϕ^* based on the example data.

Care must be taken in the computation of the angles in order to obtain their proper sign. The function ATAN2 in FORTRAN (or its equivalent in other programming languages) that uses the signs of Δx and Δy rather than just their ratio, should be

Table 8. Zahn and Roskies (1972) ϕ^* values and their Fourier coefficients for the example data of Table 7.

t	ϕ	ϕ^*	i	a_i	b_i
0	2.32241	0	0	-0.34626	
0.71074	2.99739	-0.03575	1	0.17595	-0.10944
1.26974	2.87979	-0.71236	2	0.91564	0.26300
1.62813	3.07630	-0.87424	3	-0.15540	0.02055
2.04442	-2.35619	-0.43984	4	-0.02769	0.37984
2.47448	-2.22569	-0.73939			
2.99824	-0.91591	0.04663			
3.52200	-0.45831	-0.01954			
3.93829	-0.26180	-0.23932			
4.29668	-0.37940	-0.71530			
4.85568	0.29558	-0.59933			
5.56642	1.30900	-0.29665			
6.28319					

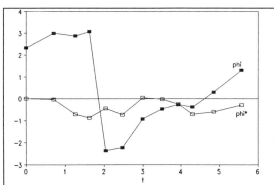

Figure 6. Plot of ϕ and ϕ^* as functions of cumulative chordal distance t.

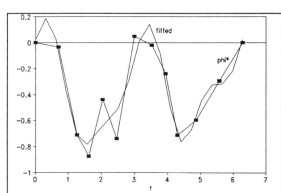

Figure 7. Plot of ϕ^* and an estimate of ϕ^* based on the first four harmonics.

used. This function is defined so that a circle will have $\phi^*(t)$ values equal to 0 for all values of t. In most applications the array of observed $\phi^*(t)$ values is adjusted, using interpolation, to correspond to equally spaced t-values. Fourier coefficients are then computed for the array of $\phi^*(t)$ values. Table 8 furnishes $\phi^*(t)$ values and their Fourier coefficients for the example data of Table 7 based on a least-squares analysis for the first four harmonics using the unequally spaced data. Figure 7 shows the fit based on the first four harmonics ($R^2 = 0.88$). This function has been used in many morphometric applications in recent years.

Elliptic Fourier analysis   This method (developed by Kuhl and Giardina, 1982) is based on the separate Fourier decompositions of the first differences of the x and y-coordinates (Δx_i and Δy_i) as parametric functions of the cumulative chordal distance, t, of the points around the outline. As above, the distance t is scaled to go from 0 to 2π radians. The Fourier coefficients for the kth harmonic of the outline's x-projection are (Kuhl and Giardina, 1982)

$$A_k = \frac{T}{2p^2\pi^2} \sum_{i=1}^{p} \frac{\Delta x_i}{\Delta t_i} \left[\cos\frac{2\pi k t_i}{T} - \cos\frac{2\pi k t_{i-1}}{T} \right]$$
$$B_k = \frac{T}{2p^2\pi^2} \sum_{i=1}^{p} \frac{\Delta x_i}{\Delta t_i} \left[\sin\frac{2\pi k t_i}{T} - \sin\frac{2\pi k t_{i-1}}{T} \right] \quad (12)$$

where p is the number of steps around the outline, $\Delta x_i = x_i - x_{i-1}$, Δt_i is the chordal distance of the step

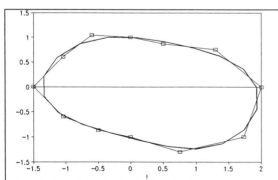

Figure 8. Plot of elliptic Fourier function based on the first 3 harmonics for the example data (squares connected with thin lines) of Table 7.

between points i-1 and i, t_i is the cumulative length of such steps up to step i, and T ($= t_p$) is the total length of the outline contour. Note that $\Delta x_1 = x_1 - x_p$. The constant term for x is

$$A_0 = x_0 +$$
$$\frac{1}{T} \sum_{i=1}^{p} \left\{ \left[\frac{\Delta x_i}{2\Delta t_i}(t_i^2 - t_{i-1}^2) + \left(\sum_{j=1}^{i-1} \Delta x_j - \frac{\Delta x_i}{\Delta t_i} t_{i-1} \right) \Delta t_i \right] \right\} \quad (13)$$

where the sums with limits of i-1 are defined to be zero when $i = 1$. This equation represents a numerical integration of $\frac{1}{T}\int_0^T x(t)\,dt$.

Given a set of elliptic Fourier coefficients, a curve can be drawn using the equations (for $t = 0$ to 2π) and n harmonics:

$$x(t) = A_0 + \sum_{k=1}^{n} A_k \cos kt_k + B_k \sin kt_{k-1}$$
$$y(t) = C_0 + \sum_{k=1}^{n} C_k \cos kt_k + D_k \sin kt_{k-1} \quad (14)$$

The coefficients for the y-projections, C_i, D_i, and C_0, are found in the same way (simply replace the x's with y's in the above equations). Thus four parameters are estimated for each harmonic. As a result one expects fewer harmonics to be necessary to describe a contour than are necessary using the methods described above.

Table 9 shows the elliptic Fourier coefficients for the example data of Table 7. Figure 8 shows a plot of the example data from Table 7 with the elliptic Fourier function, based on the first three harmonics, superimposed. Since the general shape of the outline of the example data is elliptical, the fit is very good even with very few harmonics. The harmonic amplitude, sum of the squared coefficients for each harmonic, is much larger for the first harmonic. Kuhl and Giardina (1982) show several examples of how the elliptic Fourier function can be fitted to very complex curves—including those that self-intersect. Rohlf and Archie (1984) found

Table 9. Elliptic Fourier coefficients for the example data of Table 7.			
i	A_i	B_i	C_i	D_i
0	0.31460	—	-0.09995	—
1	1.56468	-0.27246	0.00959	1.13297
2	-0.01796	0.00444	0.00135	0.03388
3	0.06718	-0.03061	0.08219	0.03999

this method to be effective and convenient in their studies. An example of its use to discriminate between two populations of mussels using a variety of multivariate analyses is given in Ferson et al. (1985).

Dual axis Fourier analysis This approach, suggested for use in describing shapes of 2-dimensional objects by Moellering and Rayner (1981, 1982), is closely related to elliptic Fourier analysis. In this method the x and y-coordinates are fitted directly by trigonometric series, rather than Δx and Δy, as a function of the cumulative chordal distance along the contour. This method does not seem to have been compared to elliptic Fourier analysis so its relative advantages or disadvantages are not known.

Parametric cubic splines Evans et al. (1985) show that one can fit parametric cubic splines to a closed contour. No special modification of the method for parametric open curves was made. The fact that the first point, x_1, y_1 was the same point as the last, x_p, y_p, was not taken into account.

Discussion

A variety of methods for describing the outline shapes in morphometrics, for both partial and complete outlines, are described above. As can be seen from the various tables, different numerical results are produced by the different methods. But there are simple relationships between the coefficients produced by some of the different methods. For example, polynomial regression, cubic splines, and Bezier curves based on the same set of coordinates will yield values for coefficients that differ by affine transformations (Rohlf, 1990). This means that statistical procedures that are invariant under affine transformations (such as generalized distances and discriminant functions) will yield equivalent results when based on data from these different methods. On the other hand, methods sensitive to affine transformations (such as principal components analysis or UPGMA cluster analysis of taxonomic distances) will yield different results for these methods. Methods based on parameters that differ by non-linear transformations, such as $\phi * (t)$ values versus ρ, are *expected* to yield different results.

These results indicate that the choice among the various method for describing an outline should not be determined by computational considerations since different methods can lead to different results. Unfortunately, it is not clear just how one can decide which method to use in a given application. This leads to an inherent arbitrariness in how one describes the shape of an organism. If one's purpose is to develop a method to discriminate between two or more forms that is not likely to be a problem. If, for example, one is able to develop a discriminant function that satisfactorily distinguishes two species then that is sufficient. If, however, one's purpose is to determine whether species A is more similar to species B or to species C, then arbitrary choices of different methods can lead to different conclusions.

Acknowledgments

This work was supported in part by a grant (BSR 8306004) from the National Science Foundation. This paper is contribution number 762 from the Graduate Studies in Ecology and Evolution, State University of New York at Stony Brook.

References

Atkinson, L. V. and P. J. Harley. 1983. An introduction to numerical methods with Pascal. Addison-Wesley, Reading, Mass., 300 pp.

Bezier, P. E. 1970. Emploi des machines a commande numérique. Masson:Paris.

Bookstein, F. L., B. C. Chernoff, R. L. Elder, J. M. Humphries, G. R. Smith, and R. E. Strauss. 1985. Morphometrics in evolutionary biology.

The Academy of Natural Sciences of Philadelphia, Philadelphia. Special Publ. No. 15, 277 pp.

de Boor, C. 1978. A practical guide to splines. Springer-Verlag, New York, 392 pp.

Cooley, J. W. and J. W. Tukey. 1965. An algorithm for the machine computation of complex Fourier series. Math. Comput., 19:297-301.

Engel, H. 1986. A least-squares method for estimation of Bezier curves and surfaces and its applicability to multivariate analysis. Math. Biosci., 79:155-170.

Evans, D. G., P. N. Schweitzer, and M. S. Hanna, 1985. Parametric cubic splines and geological shape descriptions. Math. Geol., 17:611-624.

Ferson, S., F. J. Rohlf, and R. K. Koehn. 1985. Measuring shape variation of two-dimensional outlines. Syst. Zool., 34:59-68.

Full, W. E. and R. Ehrlich. 1986. Fundamental problems associated with "eigenshape analysis" and similar "factor" analysis procedures. Math. Geol., 18:451-463.

Gasson, P. C. 1983. Geometry of spatial forms. Ellis Horwood Limited, Chichester, 601 pp.

Kaesler, R. L. and J. A. Waters. 1972. Fourier analysis of the ostracode margin. Geol. Soc. Amer. Bull., 83:1169-1178.

Kuhl, F. P. and C. R. Giardina, 1982. Elliptic Fourier features of a closed contour. Computer Graphics and Image Processing, 18:236-258.

Lohmann, G. P. 1983. Eigenshape analysis of microfossils: a general morphometric procedure for describing changes in shape. Math. Geol., 15:659-672.

Moellering, H. and J. N. Raynor. 1981. The harmonic analysis of spatial shapes using dual axis Fourier shape analysis (DAFSA). Geographical Anal., 13:64-77.

Moellering, H. and J. N. Raynor. 1982. The dual axis Fourier shape analysis of closed cartographic forms. The Cartographic J., 19:53-59.

Parnell, J. N. and P. E. Lestrel. 1977. A computer program for comparing irregular two-dimensional forms. Computer Programs in Biomedicine, 7:145-161.

Press, W. H., B. P. Flannery, S. A. Teukolsky, and W. T. Vetterling. 1986. Numerical recipes: the art of scientific computing, FORTRAN and Pascal version. Cambridge, New York, 818 pp.

Rohlf, F. J. 1986. The relationships among eigenshape analysis, Fourier analysis, and the analysis of coordinates. Math. Geol., 18:845-854.

Rohlf, F. J. 1990. The analysis of shape variation using ordinations of fitted functions. *In* Ordinations in the study of morphology, evolution and systematics of insects: applications and quantitative genetic rationales. (Sorensen, J. T., ed.) Elsevier, Amsterdam. In press.

Rohlf, F. J. and J. Archie, 1984. A comparison of Fourier methods for the description of wing shape in mosquitoes (Diptera: Culicidae). Syst. Zool., 33:302-317.

Sokal, R. R. and F. J. Rohlf. 1981. Biometry. Freeman, New York, 859 pp.

Spath, H. 1974. Spline algorithms for curves and surfaces. Utilitas Mathematica Publishing, Winnipeg, Canada, 198 pp.

Zahn, C. T. and R. Z. Roskies. 1972. Fourier descriptors for plane closed curves. IEEE Trans. Comput., C-21:269-281.

Chapter 8

Median Axis Methods in Morphometrics

Donald O. Straney

Department of Zoology, Michigan State University
East Lansing, Michigan 48824

Abstract

Median axes are geometric transformations of an outline that identify a branching set of points that constitute the middle of a form. At least five different algorithms exist for constructing median axes, but the line skeleton is of most immediate morphometric interest. Operational homologies between line skeleton branches and branch points can be established among reasonably similar forms, permitting branch points to be used as constructed landmarks. An example analysis is provided using line skeleton branch points to study complex outline shape variation in rodent bacula. Line skeleton measures are more informative than simple measures of length and width and provide nearly the same results as an elliptical Fourier analysis. Line skeletons are a promising approach to morphometric analysis of outline shape variation, even though there are some challenges involved in using them. Line skeletons can be the basis of a rich geometric analysis of outline shape, and can be used to identify phylogenetic characters, model outline shape analytically, and study correlations in shape between interacting structures.

Introduction

Some shapes of biometrical interest resist analysis with landmark-based methods. Most systematists can probably think of regions of their organisms where major differences exist between taxa that cannot be measured easily because appropriate landmarks are not located where obvious differences exist. In bivalve shells, for example, landmarks can be identified in the hinge region and on the internal face of the shell, but these do not help quantify differences in the shape of the shell margin. The bones of vertebrate skulls have obvious landmarks where three or more sutures meet at a point, but these landmarks are missing during ontogeny before the bones are in contact. Growth itself can complicate matters. Structures like bones, which grow by surface addition of matrix, can have surface landmarks at one age buried during growth. In these cases, surface landmarks at successive ontogenetic stages are not strictly homologous.

Three basic strategies exist for measuring forms that lack appropriately placed landmarks. A very appealing approach is to model mathematically the shape of an outline. The most widely used methods involve some form of Fourier modeling (Rohlf, this volume), but approaches based on other analytical functions have been proposed (e.g., Bookstein, 1978; Sampson, 1982). The utility of these methods is well illustrated elsewhere in this volume. I would note, though, that these methods are developed within a framework that need not include the concept of homology. Systematists will, in general, wish to cast their analysis of morphological difference and change in terms of comparable aspects of biological shapes. This can be difficult (but not impossible) within a modeling paradigm.

A second, very common approach is to use arbitrary "landmarks of convenience" by measuring longest, widest, shortest or thinnest dimensions between the extreme points on an outline. Alternatively, points might be chosen at specific fractions of outline arc length, and functions of the outline sampled at those points (e.g., tangent angle functions, Bookstein, 1978; eigenshape analysis, Lohmann and Schweitzer, this volume). Using arbitrary landmarks, however, will confound errors in locating specific points with interindividual variance, reducing the power of an analysis that compares groups. To be useful, convenience landmarks must be identified with a point-location error that is much smaller than the magnitude of group differences. This approach also is limited to comparisons of relatively similar shapes. Although the points might be chosen arbitrarily, morphometric analyses will still assume that distances between points or functions at points are in some sense homologous. If the shapes compared differ radically, the arbitrary landmarks can occur in different regions in different samples, and any analysis could misrepresent the biological basis of the differences between taxa.

The usefulness of convenience landmarks would be improved if a landmark construction scheme could be found that was precise enough. A third approach to analyzing outline shapes might be one that attempted to combine the precision of modeling with the comparability of operationally homologous landmarks. If we could find a mathematically precise way to identify constructed landmarks, not arbitrary ones of convenience, we could combine the best features of the other two approaches while possibly avoiding their limitations. To date, work along these lines has focused on median axis methods that geometrically transform an outline into a branching, tree-like graph that represents the points "in the middle" of an outline. Bookstein (1979) suggested that the branch points of a median axis might behave as if they were homologous points. These points can be constructed very precisely, to the precision of digi-

tizing the outline used to construct them. Although they lie within, rather than on, an outline, these points could be useful if they can be homologized between outlines. Here, I review methods of median axis construction and their use in several morphometric contexts. In particular, I explore Bookstein's suggestion that this construct can provide constructed landmarks useful for morphometrics of shapes that otherwise lack them. I also compare the information obtained from morphometric analysis of these constructed landmarks with the results of the other two approaches to analyzing the same outlines.

Calculating Median Axes

The literature on median axis methods refers to these under several different names: median axis, medial axis, symmetric axis, skeleton, and line skeleton. Unfortunately, these names are used interchangeably even though there are at least five distinct algorithms for constructing median axes which produce substantially different results. Although I will use a standardized taxonomy of median axis methods (and use the term "median axis" to refer to this class of methods generally), readers of the computational literature should be aware that usage is far from fixed.

Symmetric Axis

Blum (1967) introduced median axis methods to morphometrics in the general form of symmetric axes. For a continuous, smooth, closed curve, the symmetric axis is the continuous, branching curve that in some well-defined sense lies in the middle of the outline. The symmetric axis is the locus of points in the interior of the outline that are equidistant from points on the outline. Consider the outline as a wave front at some initial time, and imagine that this wave front moves inward toward the interior of the outline at a constant rate. The symmetric axis is the set of points where wave fronts from opposite sides first touch as they meet in the middle of the form. This is usually referred to as the "prairie fire" definition: if the outline curve marks the boundary of a patch of grass, and the wave front is a fire that starts simultaneously at each point along the boundary, the symmetric axis will be the points in the interior where the fire extinguishes itself. If the equation of an outline shape is known, an equation for its symmetric axis can be calculated precisely as an exercise in calculus. Since the equation of the outline curve is rarely known in morphometric applications, this definition does not lead to a useful morphometric algorithm.

An alternative definition for the symmetric axis, entirely compatible with the first, leads to the simplest algorithm for calculating any median axis, and illuminates some important properties of symmetric axes. Imagine inscribing circles that are tangent to an outline at two, three, or more points. The set of centers of these circles will be the symmetric axis (Figure 1) since the symmetric axis is equidistant from both sides of the outline. The centers of circles that are tangent to the outline at two points will represent points along the curves of the symmetric axis, while circles that are tangent at three or more points have centers where the symmetric axis branches. The symmetric axis will end at points that mark the centers of circles of arbitrarily small radius that touch the outline at two points. It is clear from the geometry of this definition that each point on the symmetric axis is associated with a width function, the radius of the circle centered at that point and tangent to the outline at more than one point. The width function is the distance from outline to symmetric axis, measured

normal to the outline. Each outline point is associated with a single axis branch or branch point.

This definition of the symmetric axis provides an easy algorithm for calculating the axis. A compass or set of circular templates can be used to identify points of the symmetric axis from an image of an outline. In practice, the outline should be enlarged sufficiently to minimize the errors of identifying tangent points of inscribed circles. Frequently, it will be sufficient to identify branch points of the symmetric axis. Care should be taken to identify all branch points; beginners frequently overlook some, such as the two in the rectangle in Figure 1.

Although this algorithm is extremely easy, it must be carried out by hand. Most work on median axis methods has sought computer-based algorithms that process digitized outlines. Computer analysis produces results no more precise than a careful use of a compass, but it does free the researcher from tedious manual labor. All computer algorithms find something like the symmetric axis by replacing the outline curve either by a polygon connecting digitized points or by a set of pixels sampled along the outline. In both cases, the approximation can differ radically from the symmetric axis of the original outline. Derived measures of the symmetric axis (Blum and Nagel, 1978; Bookstein, 1978) can generally be applied to these digitized approximations.

Medial Axes

The term "medial axis" is generally associated with Lee's (1982) algorithm for finding the symmetric axis of a polygon. His algorithm is the most efficient for this task and is limited in accuracy only by the arithmetic precision of the computer and the resolution of the display device (an X-Y plotter will provide much better results than most microcomputer raster displays). The medial axis of a polygon is a special case of a much more general construct, the Voronoi diagram, which is the basis of many algorithms for solving "nearest neighbor" problems (Preparata and Shamos, 1985).

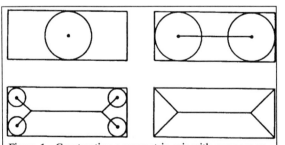

Figure 1. Constructing a symmetric axis with compass or circular templates. Centers of circles tangent to the outline at two points define branches of the axis. Circles tangent at three (or more) points define branch points of the axis.

Lee's medial axis is the symmetric axis of a polygonal boundary. If the polygonal outline is an approximation of a continuous outline, does its medial axis approximate the symmetric axis of the continuous outline? The medial axis does contain an estimate of the symmetric axis of the continuous outline approximated by the polygon, but it also contains elements not present in that symmetric axis (Figures 2a and 2b). Consider a convex vertex of the polygon (internal angle at the vertex less than 180°). We can certainly inscribe arbitrarily small circles within such a vertex that are tangent to the sides at only two points. The centers of these circles will define a branch of the medial axis that enters the vertex. Every convex vertex will have such a branch, and when an outline is approximated by a polygon of many sides, there is a veritable forest of such branches that do not exist in the symmetric axis of the associated continuous outline. It should be clear that the symmetric axis of a polygonal approximation of a continuous outline can be a poor approximation of the symmetric axis of the continuous outline.

Lee's algorithm is well suited for analyzing truly polygonal outlines. In morphometric studies where the polygon is an approximation of a non-polygonal shape, some pruning of medial axis branches is required to estimate the original symmetric axis. It would be worth some study to determine how medial axes could be restricted to provide a more useful basis for morphometrics. I am unaware of any formal treatment, but the following two methods are worth trying if one would like to use Lee's medial axes in a morphometric context.

1.   Montanari (1969) presents discrete analogs of the "velocity" of advance of the prairie-fire-like wavefront along the symmetric axis. This statistic could be used to discriminate among branches. Adjusting a threshold value for branch deletion could result in a close approximation to a compass-drawn estimate of the symmetric axis of the continuous outline.

2.   At the very least, branches that are defined only by contiguous edges of the polygonal estimate need to be pruned. This should produce a figure very similar to the line skeleton (see below). However, Lee's merging procedure will often construct additional connecting branches between the unwanted ones, and a threshold procedure may be needed to adjust the degree of contiguity appropriate to consider for pruning in specific instances.

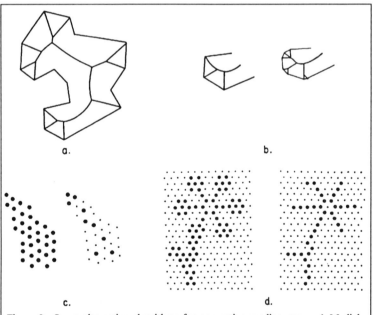

Figure 2. Some alternative algorithms for computing median axes. a) Medial axes computed for a polygon by Lee's algorithm. b) When a continuous outline is represented by a high order polygon, this algorithm adds branches of the medial axis that enter each convex vertex. As the limiting smooth curved outline is approached, the number of these branches increases. The symmetric axis of the smooth outline in this figure would contain only a single branch. c) Pixel approximation of a shape (right) and its digital skeleton (left). d) Pixel approximation (right) and homotypic thinning (right). (a after Lee; c and d from Serra).

Line Skeleton

Bookstein (1979) presents an algorithm that calculates from a polygonal approximation of a continuous outline an estimate of the original outline's symmetric axis. This construct, which Bookstein calls a line skeleton, is composed of line segments. As such, it is not the symmetric axis of the approximating polygon, which should have parabolic arcs in addition to line segments (Bookstein, 1979; Lee, 1982). But the line skeleton is not designed to be the symmetric axis of the polygon, as is the medial axis. Instead, it is constructed to estimate the symmetric axis of the original outline, an estimate that improves as the number of sides of the approximating polygon increases.

The width function for both line skeletons and medial axes is not a single-valued function like that of the symmetric axis. Each element of the line skeleton or the medial axis represents a segment along the bisector of the angle formed by extension of two opposite outline edges. Since the edges may occur at different distances from the vertex of this angle, each edge endpoint may be a different distance from the axis element. Bookstein (1979) discusses this ambiguity, and suggests the minimum distance from endpoint to skeleton (measured perpendicular to the edge) as an appropriate estimate of the width function to associate with each skeleton segment. Other measures of width can be appropriate in specific applications (Zablotny, 1988).

Line skeletons are currently the median axes of choice for morphometric analysis. The algorithm has been implemented by Bookstein and by Straney and Kriegel (software distributed with this volume). Further aspects of the usefulness of line skeletons is presented below. There are limitations on their utility however, which are also discussed below.

Digital Skeletons and Homotopic Thinning

Polygons are not the only approximators of continuous outlines. An unconnected set of points along the outline is also a good representation of outline shape, if points are sampled densely enough. When the curved outline is captured by a video digitizer, the outline and interior of a shape can be represented on a display screen by a set of pixels. A digital median axis can be computed for this pixel approximation simply by operating on the pixels themselves. Serra (1982) summarizes approaches to calculating digital median axes of such an outline set. In this digital domain, only one of the two definitions of the symmetric axis is appropriate. The digital axes are based on the tangent circle version of the symmetric axis; in digital situations, the prairie-fire definition produces a different, less well-behaved construct which will not be discussed here.

Two properties of the symmetric axis, its median position and its connectedness, cannot be achieved simultaneously in analyzing a pixel approximation. Serra (1982) presents two different algorithms that implement one or the other of these properties. The digital skeleton is the simplest to implement (see Appendix 2) and achieves the median property of the symmetric axis by uniting the results of a series of patterned erosions and dilations of the digital representation. The result is often a poorly connected set of points (Figure 2c). A second algorithm, homotopic thinning, produces a connected set of points that are not always in the middle of the outline (Figure 2d). Zhang and Suen (1984) present a more efficient thinning algorithm. Whether either approach will be useful in systematic morphometrics remains an unstudied question. I encourage morphometricians who analyze video images (Fink, Macleod, and Rohlf, this volume) to explore their utility.

Morphometrics of Line Skeletons

Because they provide close approximations of the symmetric axis of continuous outlines, line skeletons have been the algorithm of choice for applications of median axis methods in morphometrics. Bookstein (1979) analyzed an ontogenetic series of human mandibles and noted the stability and apparent homology of branch points of the line skeletons. Webber and Blum (1979) discovered invariant angles between branches of line skeletons in a similar growth series of human mandibles. Bookstein et al. (1985) measured angles between line skeleton branches of opercula in two species of centrarchid fishes and found that these measures clearly discriminated between species, although there was no apparent ontogenetic pattern of difference within species. There are a large number of other derived measures that can be taken on line skeletons (Blum and Nagel, 1978; Bookstein, 1978), but an important use of line skeletons in morphometrics will lie in the branch points they construct within landmark-free forms. Unlike biological landmarks, these points have no real existence. But as computed landmarks they may facilitate analysis of outline shape differences. To be useful, though, they must enable us to identify comparable points in different individuals, to establish a homology between points. This homology is not the same as for biologically real entities, but instead is an operational procedure that makes comparison possible.

To illustrate how line skeletons can be used to study shape differences, I present a subset of the results of a study of shape evolution of the baculum in spiny rats (genus *Proechimys*) (Straney and Patton, in prep.; Figure 3). Mammalian systematists have found bacular shape a useful source of taxonomic information, but the lack of any homologous landmarks on the bone has restricted morphometric study to simple length and width measures (e.g., Patton, 1987). Bacular shape is complex in *Proechimys*, and both intrapopulation variation and interspecific differences are often striking. This example provides a realistic picture of issues involved in using line skeletons for landmark-based morphometric analysis.

The sample analyzed here comprises bacula of 54 individuals representing 10 nominate species in 6 species groups within *Proechimys*. Thirty individuals represent an age series of *P. brevicauda* collected at one locality (Huampami, Rio Cenepa, Amazonas, Peru; age classes follow Patton and Rogers, 1983); samples of other species were indi-

viduals illustrated in Patton's monograph (Patton, 1987; they were all old adults). Scaled original camera lucida tracings made by J. L. Patton were digitized by sampling 150 to 350 outline points. Points were sampled inversely to the curvature of the outline: more points were taken in regions of the form where the outline was highly curved than in regions that were relatively straight. In some individuals, the proximal base of the baculum was irregular in outline where the corpora cavernosa attach. I sampled outline points in these individuals to smooth out such irregularities.

Identifying Landmarks

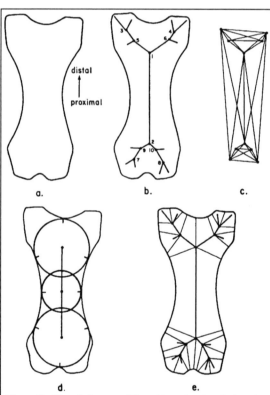

Figure 3. Line skeletons of *Proechimys brevicauda* bacula. a) Generalized outline. b) Generalized line skeleton (numbers index branch points discussed in text). c) Truss measurement scheme based on line skeleton branch points. d) and e) Steps in parsing the outline shape based upon outline points that define branch points of the line skeleton.

In general plan (Figure 3b), the line skeleton of *Proechimys* bacula contains a central segment that branches both distally and proximally where the baculum flares into expanded and somewhat flattened ends. Because these bacula are usually longer than they are wide, the central segment of the line skeleton is oriented in a proximodistal direction. Along the central segment, circles centered on the line skeleton can be drawn to touch both right and left portions of the outline at one point each. At the endpoints of this segment (points 1 and 2 in Figure 3b), such an inscribed circle will also touch a third point at the terminal end of the outline (Figure 3d). More terminal elements of the line skeleton will branch into the flared ends of the baculum. Points 1 and 2, therefore, are useful landmarks that mark the transition points between the central shaft of the baculum and its more elaborate ends. Indeed, the points on the outline where a circle centered at a branch point is tangent can parse the outline into regions associated with particular line skeleton branches (Figure 3e). This association is unambiguous and helps assign operational homologies to branches across specimens.

The ontogenetic series of bacula from *P. brevicauda* also helps clarify the homologies among line skeleton branches within the distal and proximal ends (Figure 4). There are two branch points present in all *P. brevicauda* distal end line skeletons (points 3 and 4 in Figure 3b). Terminal branches from these points penetrate into the corners defined by a flattening of the distal margin of the baculum. This distal edge is barely noticeable in the youngest individual in the sample, but is well developed by age class 5. Among age class 6 individuals, a second branch point appears on the distal end line skeleton (points 5 and 6 in Figure

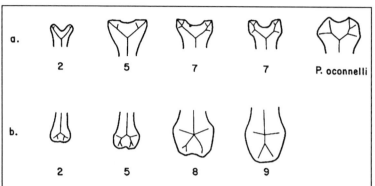

Figure 4. Representative age series demonstrating the development of the line skeleton within a) distal and b)proximal ends of *Proechimys brevicauda* bacula. Numbers refer to age classes.

3b). This point and its lateral branch appear when a distinct lateral flange is evident on the distal end. At its earliest appearance, the lateral flange appears to be positioned such that points 5 and 6 are coincident with points 3 and 4. With age, points 5 and 6 become distinct and are displaced proximally towards point 1 (there is some variation in the rate of displacement with age).

Some *P. brevicauda* (and most other species, Figure 5) lack a well-defined lateral flange, and therefore do not appear to have points 5 and 6 present on the line skeleton. Based on the ontogeny of this flange, I interpret these individuals to have undergone insufficient bone deposition to elaborate this flange and locate points 5 and 6 in these individuals coincident with points 3 and 4. Among the other 9 species (Figure 5), the distal flange is present only in one individual of *P. oconnelli* (Figure 4). In individuals without distal edges, I interpret points 3 and 4 to be coincident with the endpoint of the line skeleton in the distal end. The position of points 3 and 4 in the youngest *P. brevicauda* suggests strongly that, as the distal edge develops, points 3 and 4 move from a position very close to the skeleton end point to a position more proximal on the skeleton.

The proximal-end line skeleton has branch points that are analogous to those in the distal end (Figure 4). A proximal edge is present that defines branch points 7 and 8 (Figure 3b). It is more variable in its presence than is the distal edge, but appears to follow a similar ontogeny. A distinct shoulder appears on the otherwise smoothly bulbous base of the baculum by age class 5 in *P. brevicauda*. At its first appearance, it defines a branch point on the line skeleton that lies between point 2 and points 7 and 8 (Figure 4). With increasing age, the proximal shoulder moves distally, and branch points 9 and 10 follow. Passing through a stage when they are coincident with point 2, points 9 and 10 are located along the central branch of the line skeleton in most older individuals. As with the distal flange, some individuals lack a proximal shoulder. I have interpreted points 9 and 10 to be coincident with point 2 in these individuals, based on the ontogeny of these points in *P. brevicauda*. This is reasonable for all but the youngest *P. brevicauda*, but ontogenetic evidence is not sufficient to indicate where these points may be located in such a young individual.

Two individuals of *P. cuvieri* deserve special notice. The baculum of this species is only slightly longer than it is wide; in two individuals, the baculum is actually wider than it is long (Figure 5k and l). The line skeletons of these individuals has a central segment that is oriented from left to right, not proximodistally as in all other *Proechimys*. The endpoints of the transverse central segments are not in any sense homologous with points 1 and 2 in other individuals where they parse the baculum into left and right, not proximal and distal, ends. The difference in the

topology of the line skeleton in this species poses problems since the same landmarks cannot be found on all individuals. Because the topology of the line skeleton can change dramatically with such small changes in outline shape, this situation is likely to be encountered whenever the shapes studied are more square (or circular) than they are rectangular. In the present case, the individuals with the minority topology could be excluded from the analysis, or only those points present on all individuals might be analyzed. I have taken a third choice in the analysis to follow. Bacular length and width vary continuously in this species; individuals can be found that connect the topologies of Figure

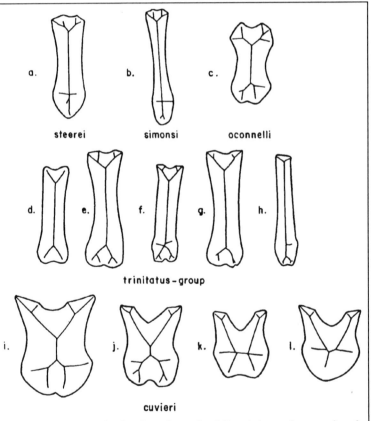

Figure 5. Representative bacula and associated line skeletons from species of *Proechimys*. *trinitatus*-group bacula represent the following nominate species: d) *P. chrysaeolus*; e) *P. guairae*; f) *P. magdalenae*; g) *P.trinitatus*; h) *P. hoplomyoides*. i-l) Four specimens of *P. cuvieri*.

5k and 5l in a continuous fashion. If we imagine transforming Figure 5i through Figure 5j into Figure 5k and 5l, keeping track of points 1 and 2 during the transformation, we would find that these points end up along the transverse central axis. I have constructed approximations of points 1 and 2 for these two individuals by making them coincident at a point halfway along the path from points 3 to 4. This procedure is in the same spirit of compromise between rigor and sample size as one utilized by Bookstein et al. (1985) in their line skeleton analysis of sunfish opercula.

A number of individuals in the sample of bacula had unique line skeleton branches produced by small irregularities on the outline, but only branches and branch points common to many individuals were considered potentially homologous. Because these common branches and branch points can be unambiguously linked to portions of the outline, the standard positional, relational and ontogenetic arguments for homology can apply to outline fragments and the branches and branch points they define. As in any morphometric situation, some branch points can be homologized only by invoking an ontogenetic hypothesis that must be confirmed separately. The branch points of line skeletons therefore seem capable of supporting landmark-based analyses of forms that otherwise lack appropriately placed landmarks.

Truss Analysis of Proechimys *Bacular Line Skeletons*

To illustrate the morphometric utility of line-skeleton-based landmarks, I have used the landmarks derived in Proechimys as the basis of a truss measurement scheme. In some ways, this is a very inefficient use of the line skeleton, since it ignores all but a very small part of the total skeleton. At the very least, measures of branch arc length and angles between branches could be included. The diversity of derived measures of median axes (Blum and Nagel, 1978) presents morphometricians with a rich source of information. The present analysis, however, does indicate the utility of line skeletons

when they are treated as simply as possible. More sophisticated analyses are likely to be even more informative.

Twenty-seven interlandmark distances were calculated (Figure 3c) and analyzed with an additional ten variables representing the width function of the line skeleton at each landmark. Principal components, extracted from the covariance matrix, were used to display patterns of variation among species (analysis of the correlation matrix yields very similar results).

Three components had eigenvalues greater than 1.0 and represent 65%, 23%, and 5% of the variation among individuals of all species. Figure 6 illustrates the variables that have high loadings (> 0.55) on each component. Component I reflects the distance between proximal and distal ends of the baculum, or the length of the central shaft of the bone. Component II represents variation in the relative position of landmarks within both the distal and proximal ends of the bacula. Individuals with high values of Component II have landmarks within both ends more widely separated than in individuals ranking low on this component. Four width function variables within the ends also contribute to this component. Component III reflects variation in the position of landmarks in the proximal end that is independent of the covariance between distal and proximal width indexed by component II.

A different set of principal components emerges from an analysis of the ontogenetic series of *P. brevicauda*. Within this species, four components (Figure 6) represent 53%, 11%, 9%, and 8% of the variation within the species. Component I is a general size component with high positive loadings for all but seven variables. It reflects the unsurprising fact that the baculum becomes larger in most dimensions with age. Component II indexes the non-ontogenetic variation in proximal flange position; there appears to be very little non-ontogenetic variation in distal flange position that is separable from the overall size effect. Component III is a measure of the width function at the proximal end of the line skeleton, and Component IV indexes variation in the length of the distal-most branch of the skeleton.

No single size factor appears in the interspecies analysis, even though that analysis contains the *brevicauda* specimens which by themselves display a strong effect of general size. The size axis within *brevicauda*, however, is evident in a plot of the first two interspecific principal components (Figure 7) as the major axis of the elliptical scatter of brevicauda. The interspecific analysis lacks an explicit size factor because the variation among species is greater than size-related variation within. This is a common result of multigroup principal component analyses and further analysis could follow the suggestions of Rohlf and Bookstein (1987).

Comparisons with Other Methods

How effective are line skeleton branch points at capturing information about outline shape varia-

tion? To justify the computational effort, we would hope they provide more information than do measurements between convenience landmarks. But line skeletons are approximations of the symmetric axis of an outline shape, itself approximated by a polygon. This extra level of approximation, and reliance on only a few constructed landmarks to capture all of the information on variation in outline shape, may make an analysis of line skeleton branch points less informative than a Fourier analysis of the entire outline. To evaluate the relative effectiveness of the line skelton for morphometric comparisons, both extremal measurements and Fourier analysis were conducted on the same specimens of *Proechimys* bacula.

Patton (1987) measured *Proechimys* bacula with calipers and recorded three extremal dimensions: maximal length, maximal distal width, and maximal proximal width. A plot of distal width versus bacular length (his Figure 12) provides a very similar picture of variation among species as does a plot of the first and second principal components of the among-species analysis in Figure 7. Component I varies with the length of bacula while Component II represents variation in the width of each end. Because interspecific differences in these two dimensions are large, it is no real surprise that caliper measures of analogous length and widths provide the same view of patterns of variation. The errors of caliper end-point placement are very small relative to the magnitude of shape differences, which are large enough to see by eye. Patton (pers. comm.) first noted this pattern of structured variation on visual inspection and chose his caliper measures to help convince readers of the differences between groups.

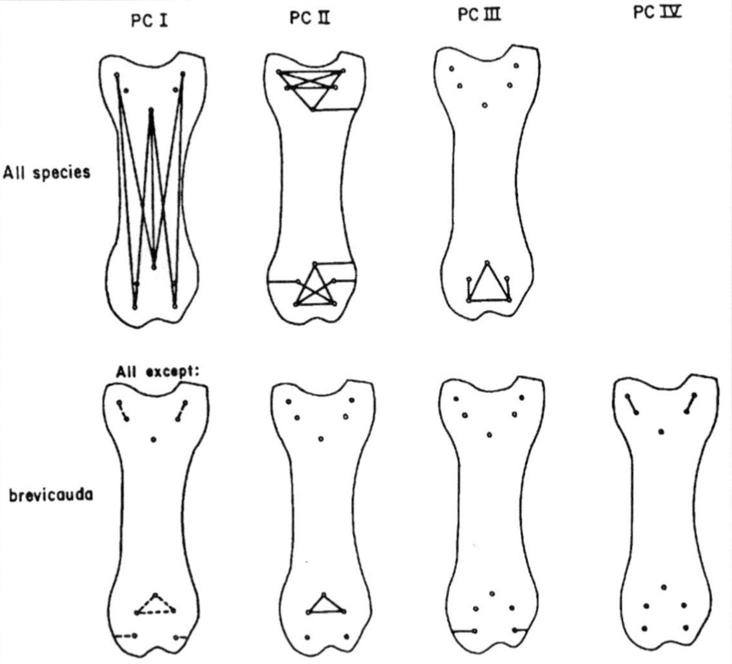

Figure 6. Principal component loadings for interspecific (top) and intraspecific (bottom) analyses of truss measurements of line skeletons of *Proechimys* bacula. Open dots represent landmarks; solid lines indicate dimensions with high loadings for a particular component (except in PC I, bottom, where all dimensions except the ones indicated in dotted lines have high loadings on this component).

While it is reassuring that analysis of line skeleton branch points confirms the subjective impressions of a good systematist, these points support an analysis that resolves more than an eye sees or calipers measure. Variation in bacular length and end width accounts at best for 88% of the variation in this sample of bones. (and only 61% if the more customary analysis of the correlation matrix is performed). Component III, with its complex covariation of proximal flange position and the relative roundness of the proximal end, would be difficult to measure using calipers. Studying the ontogeny of *P. brevicunda* with caliper measurements would be even more difficult. Any one

measurement would mimic the general size factor identified in Figure 6, but it is difficult to identify repeatable measurement schemes for measuring the other, independent patterns of variation in this sample. With Figure 6 in hand, it might be possible to find caliper measurements accurate enough to recover the same patterns of variation. But these would be unlikely measures to choose unless the patterns were known to be present. Certainly no one has thought to measure such dimensions during the history of study of bacular shape in this genus (Patton, 1978). The line skeleton analysis detects much subtler patterns of variation than do standard measurement schemes applied to this bone.

This comparison, however, may be unfair because Patton's three distances are competing with 37 based on the line skeleton. A more informative comparison is with the more analytical Fourier analysis. The files of digitized outlines used to construct line skeletons were used as input to Rohlf's elliptical Fourier program (Rohlf, this volume). Twenty-two terms were required to produce an acceptable fit of the Fourier reconstruction to the outline (maximum x or y deviation within $10\times$ digitizing error) in six individuals chosen to represent the range of shape variation in the sample. Reconstructions for all individuals used 22 terms and were made invariant to orientation and digitizing starting point, resulting in 86 estimated Fourier coefficients for each individual.

Reconstructions were not made invariant to size so that results would be comparable with the line skeleton analysis. Covariances among coefficients were calculated across specimens and analyzed by principal components analysis.

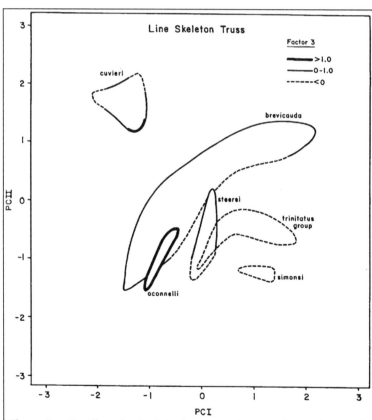

Figure 7. Two-dimensional plot of component scores for specimens of *Proechimys* bacula. Measurements were of truss elements connecting branch points of line skeletons and width functions at branch points. Components have been scaled to unit variance to facilitate comparison with Figure 8. Unscaled, a unit distance along Component I is three times as long as one along Component II.

The covariance matrix yielded three components accounting for 60.2%, 26.4% and 7.0% of the variation, respectively. Component I has a large loading only for coefficient A_0, the Fourier size index. *P. brevicauda* scores on this component have a correlation of 0.97 with scores on Component I of the ontogenetic line skeleton analysis, itself a size axis. Component II has high loadings for coefficient D_0 (the eccentricity parameter) as well as coefficients A_2 and (negatively) D_4. Examination of specimens of varying scores on Component II but with near-zero scores on Components I and III indicates long-thin bacula have high scores and short-broad bacula have low scores. Component III is highly correlated with four coefficients (A_1, D_1, A_4, and A_6). Specimens with near-zero values of the first two components vary along Component III in a pattern suggesting a contrast between proximal and distal width. Proximal ends are widest in specimens with high Component III scores, distal ends are widest in specimens with low scores, and both ends are nearly equal in specimens with near-zero scores. Figure 8 illustrates the distribution of specimens along these components.

These results cannot be compared directly with the results of the line skeleton truss. The two sets of components must first be transformed to be as alike as possible, since nothing in their extraction constrained them to be similar (Zelditch, DeBry, and Straney, 1989; Rohlf and Archie, 1984; Mulaik, 1972). Usually, one would transform the pattern matrix of variable loadings using some variant of Procrustean rotation to make one analysis match the other, or to make both match some independent criterion. I

have not followed this strategy because the variables in the two analyses are so fundamentally different. I see no way to directly transform one pattern matrix into the other. Further, I am uncertain how to specify a hypothetical target matrix for the Fourier analysis. Rohlf and Archie (1984) note reasons for viewing as a virtue the uninterpretability of the Fourier coefficients; in the present context, however, this is a liability. Interpreting the principal components of the Fourier data set relies on inspection of how they sort the specimens. Consequently, I have sought transformations of the distribution of specimens on the three components that make Figures 7 and 8 as similar as possible. Graphical transforms that bring

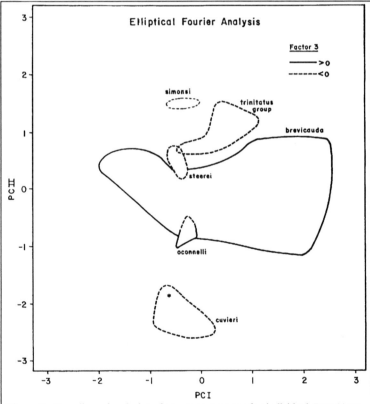

Figure 8. Two-dimensional plot of component scores for individual *Proechimys* bacula based upon elliptical Fourier coefficients. Components have been scaled to have unit variance to facilitate comparison with Figure 7. The unscaled unit distance in Component I is three times as large as for Component II. The asterisk labels one *P. cuvieri* with a score for PC III > 1.0.

Figure 7 into agreement with Figure 8 can then be applied to a vector plot of loadings of line skeleton variables on the original components to identity the transformed axes. These can then be compared with the interpretations of Fourier components to determine how differently the two methods describe shape differences within the sample.

Figures 7 and 8 have already been subjected to one transformation to make them comparable. Both figures have axes that are each standardized to have unit variance. In fact, the scale in each figure is different. Although the two analyses yield components that explain nearly the same proportion of the total variance (see above), the total variance in the line skeleton data is 22 times greater than in the Fourier coefficients. This difference is reflected in the distance between points in the two results. Mahalanobis D^2 distance between species centroids is 20.3 units greater in the line skeleton analysis. This scale difference, reflecting the difference in how the variables are measured, must be removed to make units comparable. When this difference is removed, the distances between centroids are very similar except those involving *P. oconnelli* (mean difference between line skeleton and Fourier distances = -0.074 without *oconnelli*, and distances involving this species differ by 0.980). This suggests that the main features of the distribution of points in Figures 7 and 8 are potentially more similar than is apparent.

The sign of Components I and II in the Fourier analysis (Figure 8) can be changed without altering the results, since components are unique up to a change in sign. This turns Figure 8 upside down and backwards. Superimposing Figure 7, and rotating it about the origin by 42° brings the species distributions into good alignment. *P. oconnelli* is placed differently in the two analyses. A small rotation of Component II and III of Figure 7 around Component I improves the visual fit of the distributions still further, but has little effect on *P. oconnelli*.

These transformations bring the taxa into remarkably good registration, but in doing so, the original components have themselves been transformed. The same rigid rotations of the component axes must be applied to a vector plot of variable loadings to interpret what aspects of variation among line skeleton measures the transformed component's index. Transformed Component I has moderate (0.4-0.6) positive loadings for most distances. Only distances between points 2, 9, and 10, and widths at points 3 through 8 (Figure 3b), have near-zero loadings. This component is easily interpreted as a size axis, as was Component I of the Fourier analysis. Component II is transformed into a contrast between measures of length with extreme negative loadings (-0.5 to -0.6) and measures of distal and proximal width with extreme positive loadings (0.4 to 0.6). This is the same variation in shape indexed by Fourier Component II. Component III has also become a contrast, but one between widths between distal landmarks (-0.3 to -0.4) and widths between proximal landmarks and width functions at distal end landmarks (0.3 to 0.5). This component is similar to Fourier Component III, except that it includes a tendancy for broad distal ends to be more sharply pointed which was not apparent in the Fourier analysis.

Principal component analysis of line skeleton and Fourier coefficient data sets therefore yield very similar results. Very nearly the same components can be found for both data sets, and the components distribute species in a similar way. The major difference between the original analyses is one of perspective: Figures 7 and 8 view essentially the same multidimensional clusters of data from different viewpoints. The reason for this difference lies in the distribution of the variance across variables. In the Fourier data set, the size coefficient A_0 is nearly twice as variable as the next most variable coefficient, and Component I is compelled to be aligned with its variation. Likewise, line skeleton measures of bacular length are slightly more variable than widths and therefore dictate placement of the first component along their axis of covariation, rather than along a size trend. This difference between analyses has little biological importance and reflects only the difference in statistical distributions of the variance of linear measures and the variance of coefficients in a trigonometric series.

The important difference between these analyses is their respective treatment of *P. oconnelli*. The line skeleton measures find bacula of this species to be less eccentric than does the Fourier analysis. I suspect this is because the line skeleton measures are more strongly influenced by the relatively flared ends of these bones than are the Fourier coefficients that load on the first three components. The line skeleton measures probably capture more shape variation in fewer orthogonal dimensions than do Fourier coefficients. This is certainly true of Component III, which contains information about distal end sharpness in the line skeleton analysis that seems absent in the Fourier analysis.

In conclusion, line skeleton measures do much better than expected compared with other approaches to landmark-free outline shape analysis. For this data set, they permit a much more subtle analysis than has been possible with convenience landmarks. Surprisingly, line skeleton and elliptical Fourier analysis of these bacula yield very similar results. The line skeleton landmarks index shape variation in 10 very limited regions of the outline. That this limited sample of the outline shape competes so well with an analysis of overall outline shape is unexpected. I see no basis in the results of these two analysis for preferring one over the other. I find the line skeleton analysis easier to interpret, since component loadings can be interpreted much more directly. Because they construct points that serve well as landmarks, line skeletons permit some types of analyses, such as shape coordinates, (Bookstein, this volume), not available with Fourier coefficients. Perhaps it is only the purpose of a study that can dictate the choice between these two techniques.

Limitations

I must leave the preceding optimism for line skeletons with some limitations on their usefulness. The most obvious problem they pose is in how they are calculated. Unlike other methods of shape analysis, line skeletons are calculated interactively, and often iteratively. Without some experience, it is difficult to digitize outlines and calculate a useful line skeleton on one unattended pass through a program. In the early stages, one will often go back to redigitize an outline to add points that will bring out a skeleton branch the program refuses to find, or to delete points that induce a confusing array of spurious branches. Straney and Kriegel's (1989) program includes an outline editor to make the learning process easier. Bookstein's (1979) algorithm is quite sensitive to starting position. For some shapes, some starting points may produce only parts of the line skeleton. The chaining protocol of this algorithm can have difficulty finding branches within some regions of an outline. In this case the outline must be redigitized with more or fewer points. The algorithm is also sensitive to minor digitizing errors, producing branches caused by dimples or pimples in the outline shape. Smoothing the outline or changing the threshold for diagnosing these irregularities (Straney and Kriegal, 1989) can prune these products of digitizing error. I find it extremely useful to sketch the symmetric axis of an outline with circular templates before beginning to digitize a set of new shapes. One must know what the skeleton *should* look like before using one in an analysis. Uncritical, casual use of a line skeleton program will not be informative. Care and patience are rewarded, not only with a useful line skeleton, but also with a solid understanding of the shape being studied.

If outlines are digitized too coursely, the locations of branches and branch points can be sensitive to the way outlines were digitized. The line skeleton algorithm assumes that the outline is captured as a polygon with many sides, and enough points must be taken to reduce the effect of digitizing error on branch and branch point location. For each set of new shapes, I generally digitize one individual at several different point densities and examine the (x,y) coordinates of branch points. It is then a simple matter to determine a threshold point density that will produce stable results. The end point of terminal branches, however, is always sensitive to the scale of digitization: terminal branches invade further into a region as point density increases. If the end of terminal branches is important for a morphometric analysis, that region of the outline must be digitized to the same relative point density in different individuals.

There are some shapes that measures of line skeletons will not analyze well. Sausage-like shapes, with smoothly covarying sides, will have a skeleton with only a single branch no matter how the outline snakes across a plane. Whatever is measured on this line skeleton could just as easily be measured on the original outline. Compact, nearly square or circular shapes pose a very different problem, and must be studied with caution. Very slight changes in these outlines can produce major changes in the topology of the skeleton. Analyzing line skeletons of populations of nearly circular shapes is very challenging, since the topological differences make identifying comparable branches or branch points very difficult (except in certain cases; see discussion of Figure 5i-l, above). To analyze these shapes well, one must be able to distinguish between topological differences induced by biologically meaningful changes in outline shape and those produced by unimportant, "random" perturbations. But the topologies of the line skeletons of other shapes can also vary, such as in the bacula of *Proechimys* (Figure 5). The ontogeny of the baculum was reasonably stable, however, and indicated that much of this variation was produced by a simple process of topological change. The ontogenetic model is what made the analysis of these line skeletons possible. In general, one will need some type of biological model that specifies how different topologies can be related or compared if the line skeletons are to be analyzed quantitatively. I am not aware of analytical

methods for studying topological variation directly. In extreme cases, where topological differences cannot be made comparable, the shapes may be analyzed more easily by techniques like Fourier analysis.

Finally, median axes currently can be calculated only in two dimensions. While Bookstein's (1979) algorithm might be extended to a third dimension, the extension might be easier to make with pixel-based image analysis algorithms. Median axes of three-dimensional forms projected to two dimensions will inevitably lose some information. Results should be viewed carefully for potential artifacts of projection or perspective.

Other Uses of Line Skeletons

Recognizing Phylogenetic Characters

Phylogenetic analysis of biological shapes without suitable landmarks can be as challenging as a metric analysis. Identifying consistent patterns of character transformation, and communicating them effectively, is a cognitive task that shares many features with a morphometric analysis. Just as line skeletons can be used to find operational homologies between regions of a complex outline for morphometric study, they can be a useful aid in the more visual task of character analysis. Oxnard (1972) was the first to use median axis methods as an aid to phylogenetic study.

Mammalian systematists have used features of bacula shape to diagnose taxa, but generally there has been little attempt to connect intertaxon differences in transformation series. For example, the several recent treatments of bacula shape within echimyid rodents focus on descriptive differences, but not in an explicit phylogenetic context (Didier, 1962; Patton and Emmons, 1985; Patton, 1987). Line skeletons of these bacula supported a detailed morphometric analysis of patterns of variation, and they also facilitate the identification of transformation series among taxa.

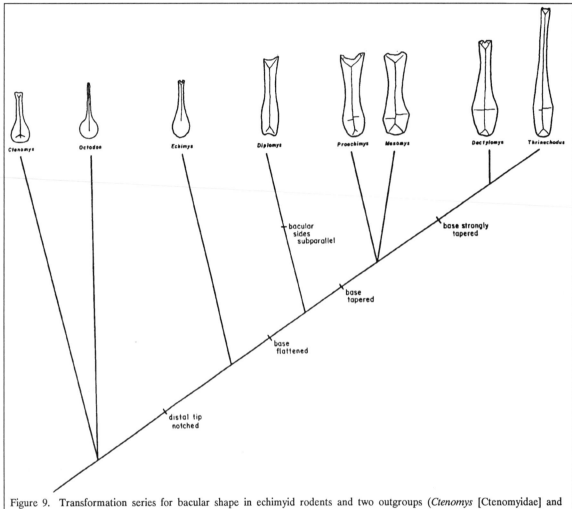

Figure 9. Transformation series for bacular shape in echimyid rodents and two outgroups (*Ctenomys* [Ctenomyidae] and *Octodon* [Octodontidae]). Outlines are drawn with line skeletons.

Figure 9 illustrates the shape of the baculum in six genera of the Echimyidae, plus two outgroups, the Octodontidae and Ctenomyidae. I have examined bacula from several species of *Ctenomys*, *Echimys*, and *Proechimys* but illustrate here what I interpret to be the primitive bacular shape for each genus (Straney and Patton, in prep., analyze phylogenetic patterns within *Proechimys*; I assume that generic limits are accurate). Examining the bacula, it was easy to note that echimyids generally possess a shallow notch on the distal tip of the bacula. But I had questions about the shape of the proximal end

that were difficult to resolve. The proximal end of the baculum in *Echimys* appeared intermediate between the bulbous, flask-shaped base in *Octodon* and *Ctenomys*, and the tapered base of the remaining echimyids. Within the latter group, there appeared to be similarities between *Thrinacodus*, *Dactylomys*, and *Mesomys* in the shape of the proximal flange, but I was uncertain how the bacula of *Proechimys* and *Diplomys* could be related to these three.

After calculating the line skeletons of the bacular outlines, I could resolve these uncertainties easily. The intermediacy of *Echimys* is more apparent than real. The proximal end of the baculum in this genus shares a noticeable feature with *Octodon* and *Ctenomys*: the base of the baculum is so evenly curved that it is virtually circular (secondarily flattened in *Ctenomys*). The line skeleton in these forms ends abruptly at an end point that is the center of a circle that describes nearly the entire proximal base of the baculum. The smoother transition from base to shaft in *Echimys* makes the baculum appear similar to other echimyids, but the strikingly simple shape of the base, shared with the outgroups, defines the primitive bacular shape for the family.

The remaining echimyids share a feature I had not noted in examining the outlines themselves. All display line skeletons that branch at the proximal end, due to a flattening of the base. *Diplomys* is unique among these flat-based taxa in having subparallel bacular sides; the remaining four genera have bacula that bow outward in a proximal flange. I interpret the inward tapering of the proximal base as a synapomorphy uniting these genera. *Dactylomys* and *Thrinacodus* carry this transformation one step further. In these genera, the proximal flange is located more distally, with a longer proximal taper, than in *Proechimys* and *Mesomys*. I had originally thought that *Mesomys* shared a proximal shape with *Dactylomys* and *Thrinacodus*, but the line skeleton indicates the position of the proximal flange in this genus is most like *Proechimys*. One feature is shared by *Mesomys*, *Dactylomys* and *Thrinacodus*: in all three, the proximal flange is more sharply angled than in *Proechimys*. This is apparent in the outline, and the deeper penetration of the flange by the line skeleton in these genera confirms that the flange is more abruptly angled than in the more smoothly sloping *Proechimys*. However, this trait is highly variable within genera, and even between sides within individuals. This variability is not particularly clear on the outlines, but the line skeleton makes it obvious that this similarity is of little systematic utility.

Figure 9 summarizes the steps in the transformation series deduced from the line skeletons of these bacula. It confirms the reality of the subfamily Dactylomyinae (including *Dactylomys* and *Thrinacodus*), established on cranial evidence (Woods, 1984), and suggests that bacular morphology may be useful in resolving relationships within the Echimyinae (comprising the remaining four echimyid genera). Figure 9 is not a phylogeny of the echimyids; that would require more than a single transformation series. But it does provide information more useful in constructing a phylogeny than the vague uncertainties that result from examination of the outlines alone. I find line skeletons useful in providing an additional basis of comparison for constructing transformation hypotheses. They help clarify patterns of variation in complex shapes that might otherwise go unnoticed or be misinterpreted. Line skeletons will probably have little utility for someone with extensive experience in character analysis of a particular structure. But for beginners, or for someone initiating a phylogenetic analysis of a previously unstudied structure, line skeletons can serve as a very useful tool in identifying characters and understanding their evolutionary connections.

Line Skeletons as Coordinate Systems

Median axes provide a convenient and intrinsic coordinate system for describing outline shape. With symmetric axes, each point along the axis is associated with a unique width that identifies specific points on the outline on either side of the axis. If this width is plotted as a function of distance along the symmetric axis, the resulting two-dimensional plot will represent the closed outline shape as a single curve. Because the width function is symmetric about the symmetric axis, this plot of width versus axis arc length effectively folds the outline along the symmetric axis and then straightens the axis to create an abscissa. With line skeletons and medial axes, the width function is not uniquely defined and some care is required in defining width functions to permit the axes to serve as coordinate systems that will meaningfully and consistently represent outline shape.

Zablotny (1988) used line skeletons to provide an intrinsic coordinate system in his study of the ontogeny of outline shape of the fifth ceratobranchial bone in bluegill and pumpkinseed sunfishes. As juveniles, these species have similar zooplankton diets, but adult pumpkinseeds switch to feeding on snails when body length exceeds approximately 45 mm, while adult bluegills continue to feed on zooplankton and aquatic insects (Werner and Hall, 1976; Mittelbach, 1984). The tooth-bearing pharyngeal bones used by these centrarchids to process food also diverge between these species with age: in pumpkinseeds these bones, and particularly the fifth ceratobranchial, become large and stout to be effective in crushing the shells of ingested snails (Figure 10). To quantify the ontogenetic changes in ceratobranchial shape in these species, Zablotny calculated line skeletons of ceratobranchial outlines (Figure 10). These line skeletons are particularly simple, unbranched curves in all but the oldest pumpkinseeds (where a minor branch appears that penetrates the medial lobe of the ceratobranchial). The width function used was not the standard Bookstein function, but rather the width from axis to outline measured perpendicular to the axis. This permitted an unambiguous characterization of the shape of the medial portion of the outline as a function of line skeleton arc length. It also allowed him to ignore the minor side branch that was present in relatively few individuals.

A fifth degree polynomial function provided an excellent fit to the points of the width versus arc length plots of each individual. Zablotny expressed

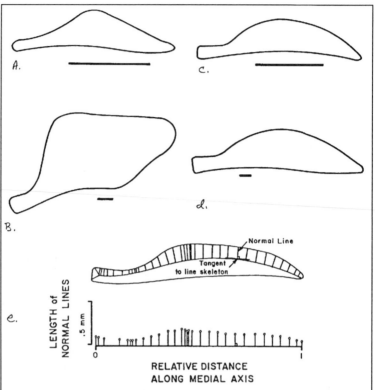

Figure 10. Outlines of sunfish fifth ceratobranchial bones. a) Juvenile pumpkinseed. b) Adult pumpkinseed. c) Juvenile bluegill. d) Adult bluegill. e) Width function used to produce intrinsic coordinate system, and an example of such a coordination of outline shape. (After Zablotny).

the polynomial regression in geometric form (Mortenson, 1985) to ensure the computational independence of its terms and to permit much easier identification of the regions of the curve affected by changes in particular coefficients. By regressing each coefficient in each species against a body size predictor (standard length), he demonstrated that the ontogenetic trajectories in bluegill and pumpkinseeds were linear functions of size. The ontogeny of both posterior width and the change in curve slope at the anterior end did not differ between species. The ontogenetic basis for ceratobranchial divergence in these two species lay in differences between species in anterior width, outline slope at anterior and posterior ends, and the change of outline slope at the posterior end. Zablotny concluded that ceratobranchial shape ontogeny began diverging between species at approximately 35mm. standard length, before the functional diet shift occurred. He was also able to use the regression equations to model the ontogeny of ceratobranchial shape in these species (Figure 11).

Using the line skeleton as the basis for an intrinsic coordinate system provides a straightforward quantitative analysis of shape ontogeny in this case. Ceratobranchial outlines could have been analyzed directly using Fourier analysis, but the results were particularly easy to interpret in terms of the geometric polynomial regression. This use of median axis methods may have broad applications in modeling outline shape change in analytical terms. There is one potential advantage to this approach: if the topology of the median axis is similar in all individuals, the outline may be parsed into regions (e.g., Figure 3e) to be analyzed separately. In some applications, this may be a desirable feature not available with Fourier analysis.

Analyzing Parallel Curvature Between Two Structures

In a study of mechanical isolation in grasshoppers of the genus *Barytettix*, Bennack (1988) was interested in quantifying the match between the surface

shape of male genitalia and the shape of the female bursa during copulation. Cohn and Cantrall (1974) had suggested genital shape in these grasshoppers produced mechanical isolation through a lock-and-key mechanism. Under this hypothesis, there should be regions in both male and female genitalia where surface shape is similar enough to permit close matching that is required for successful insemination. The surface of the matching portions of the genitalia should display an identical pattern of curvature as a function of arc length at an appropriate scale.

Rather than calculate curvature and arc length, and establish algorithms for determining an appropriate scaling for comparison, Bennack made use of the properties of median axes to quantify the degree of fit between genitalia. He serially sectioned genitalia *in copulo* and calculated line skeletons of the region between male and female structures at specific section levels (Figure 12). He ignored portions of the line skeleton produced by the outline of only male or only female structures. His measure of the degree of matching between male and female shape was the variance of the Bookstein width function over the line skeleton. He reasoned that in areas where male and female shapes matched, the outline curves should be parallel and the width function should be nearly constant. Regions of poor fit between genitalia, on the other hand, should be characterized by width functions that varied greatly along the line skeleton. The variance of the width function was useful in determining the influence of fit between genitals on insemination success. Bennack was able to conclude that coevolved shape similarity between male and female genitalia was unimportant in determining reproductive success in intertaxon crosses; body size differences between mates was so much more important in influencing the reproductive outcome of matings that there is little reason to invoke a

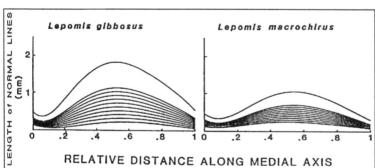

Figure 11. Use of an intrinsic coordinate system based on a line skeleton to model the ontogeny of ceratobranchial bone shape in two species of sunfishes. Outline shape was modeled with a fifth-order geometric polynomial; coefficients of this model (estimated by polynomial regression) from each individual were regressed against standard length and the resulting equations used to construct model outline shapes at specific standard lengths. (From Zablotny)

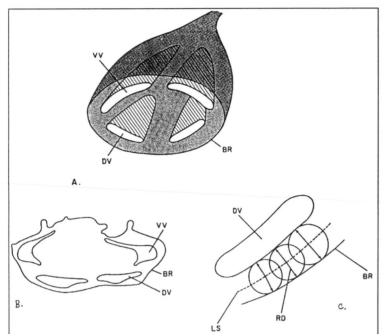

Figure 12. Analysis of genital shape in copulating grasshoppers of the genus *Barytettix*. a) Diagrammatic view of genitals *in copulo*. b) Tracing of serial section through the genitalia of copulating individuals. c) Details of calculation of the line skeleton between male and female genetalia *in copulo*. Abbreviations: BR, bursa of female; DV, dorsal valve of male genitalia; LS, line skeleton between male and female genitalia; RD, radii of circle defining line skeleton, the width function at a particular line skeleton point; VV, ventral valve of male genitalia (male genitalia are rotated 180° upon insertion).

lock-and-key mechanisms in these grasshoppers.

In this case, the simple geometry of the line skeleton provided a straightforward method for answering an otherwise computation-intensive problem. Bennack's approach may prove useful in studies of other interacting-structure shape problems, such as joint surface shape comparisons. Bookstein (1978) has outlined a more sophisticated approach to width function analysis that deserves serious consideration. The large number of derived measures of median axes (Blum and Nagel, 1978) may hold other, equally simple solutions of complicated shape comparisons.

Conclusion

Median axes are well-defined, geometrical objects that can be calculated for any closed outline shape.

They are useful in morphometric studies of shapes that otherwise lack appropriate landmarks. The branch points of median axes, in particular, can be homologized operationally across forms by the same logic that is applied to more traditional, non-computed landmarks. When used as a basis for a truss measurement scheme, the branch points of a line skeleton are more informative than an analysis based upon simple landmarks of convenience and yield results nearly the same as elliptical Fourier analysis. They also provide a more interpretable analysis of patterns of variation, and provide insights that might be difficult to quantify in other ways.

The geometric richness of median axes offers many opportunities for novel approaches to analysis of complex shapes. The uses discussed here (parsing outline shape for phylogenetic analysis and several applications of width functions) barely begin to explore the types of geometrical comparisons that could be made between taxa using median axes. Because the geometry of median axes is relatively straightforward, there should be additional situations where they facilitate analyses.

Acknowledgments

Fred Bookstein was very helpful in the early stages of my interest in line skeletons. I particularly appreciate the access he granted to his line skeleton program. I relied heavily on Robert Kriegel's expertise while preparing my own implementation of this algorithm, and we were aided by suggestions from James Zablotny and Dan Bennack. Jeannette Boylan assisted in digitizing outlines and Yining Luo was extremely helpful with many of the analyses included here. Both he and Miriam Zelditch helped me overcome some of the challenges of analyzing line skeletons. I am particularly grateful to James Patton for the *Proechimys* baculum data and his insights on this genus. This work was supported by a grant from the National Science Foundation (BSR-8400544).

References

Bennack, D. E. 1988. The lock and key hypothesis of mechanical reproductive isolation: Variation in genitalic fit and its influence on spermatiphore transfere in the grasshopper *Barytettix humphreysii*. Ph.D. dissertation, Michigan State University, East Lansing, Michigan. 148 pp.

Blum, H. 1967. A transformation for extracting new descriptors of shape. Pp. 362-380 *in* Models for the perception of speech and visual form (W. Whaten-Dunn, ed.). MIT Press, Cambridge, MA, 470 pp.

Blum, H. and R. N. Nagel. 1978. Shape description using weighted symmetric axis features. Patt. Recogn., 10:167-180.

Bookstein, F. L. 1978. The measurement of biological shape and shape change. Lecture Notes in Biomathematics, Vol. 24. Springer-Verlag, New York, 191 pp.

Bookstein, F. L. 1979. The line skeleton. Comp. Graph. Image Proc., 11:123-137.

Bookstein, F. L., B. Chernoff, R. L. Elder, J. M. Humphries, Jr., G. R. Smith, and R. E. Strauss. 1985. Morphometrics in evolutionary biology. The Academy of Natural Sciences of Philadelphia, Spec. Publ. No. 15, 277 pp.

Cohn, T. and I. Cantrall. 1974. Variation and speciation in the grasshoppers of the Conalcaeini (Orthoptera: Acrididae: Melanoplinae): The lowland forms of western Mexico, the genus *Barytettix*. Memoirs, San Diego Society of Natural History, Number 6, 131 pp.

Didier, R. 1962. Note sur l'os penien de quelques rongeurs de l'Amerique du Sud. Mammalia, 26:408-430.

Lee, D. T. 1982. Medial axis transformation of a planar shape. IEEE Trans. Pattern Anal. Mach. Intell., vol. PAMI-4:363-369.

Mittelbach, G. G. 1984. Predation and resources partitioning in two sunfishes (Centrarchidae). Ecology, 65:499-513.

Montanari, U. 1969. Continuous skeletons from digitized images. J. Assn. Comput. Mach., 16:534-549.

Mortenson, M. E. 1985. Geometric modeling. John Wiley & Sons, New York, 763 pp.

Mulaik, S. A. 1972. The foundations of factor analysis. McGraw-Hill, New York, 452 pp.

Oxnard, C. E. 1972. Some problems in the comparative assessment of skeletal form. Pp. 1-23 *in* Symposium on human evolution (M. H. Day, ed.). Taylor and Francis, London.

Patton, J. L. 1987. Species groups of spiny rats, genus *Proechimys* (Rodentia:Echimyidae). Fieldiana (Zoology), 39:305-346.

Patton, J. L. and L. H. Emmons. 1985. A review of the genus *Isothrix* (Rodentia:Echimyidae). Amer. Museum Novitates, 2817:1-14.

Patton, J. L. and M. A. Rogers. 1983. Systematic implications of non-geographic variation in the spiny rat genus *Proechimys* (Echimyidae). Zeitshrift fur Saugertierkunde, 48:363-370.

Preparata, F. P. and M. I. Shamos. 1985. Computational geometry: An introduction. Springer-Verlag, New York, 390 pp.

Rohlf. F. J. and J. W. Archie. 1984. A Comparison of Fourier methods for the description of wing shape in mosquitos (Diptera: Culicidae). Syst. Zool., 33: 302-317.

Rohlf, F. J. and F. L. Bookstein. 1987. A comment on shearing as a method for "size correction". Syst. Zool., 36: 356-367.

Sampson, P. D. 1982. Fitting conic sections to "very scattered" data: An iterative refinement of the Bookstein algorithm. Comp. Graph. Im. Proc., 18:97-108.

Serra, J. 1982. Image analysis and mathematical morphology. Academic Press, London, 610 pp.

Webber, R. L. and H. Blum. 1979. Angular invariants in developing human mandibles. Science, 206(4419):689-691.

Werner, E. E. and D. J. Hall. 1976. Niche shifts in sunfishes: Experimental evidence and significance. Science, 191:404-406.

Woods, C. A. 1984. Hystricognath rodents. Pp. 389-446 *in* Orders and families of recent mammals of the World (S. Anderson and J.K. Jones, Jr., eds.). John Wiley & Sons, New York, 686 pp.

Zablotny, J. E. 1988. A non conventional morphometric technique for measuring ontogenetic shape changes of the fifth ceratobranchial in two species of centrarchid fishes. M.S. thesis, Michigan State University, East Lansing, Michigan, 82 pp.

Zelditch, M. L., R. W. DeBry, and D. O. Straney. 1989. Triangulation measurement schemes in the analysis of size and shape. J. Mamm., 70:571-579.

Zhang, T. Y. and C. Y. Suen. 1984. A fast parallel algorithm for thinning digital patterns. Comm. ACM, 27:236-239.

Appendix: An Algorithm for Calculating Digital Skeletons on Microcomputers

The availability of reasonably priced image processing systems for microcomputers makes accessible to most systematists the technology for computing digital skeletons. Serra's (1982) algorithm for calculating digital skeletons is here translated from his symbols into pseudocode and has been changed slightly to make use of an iterative procedure. To implement this algorithm, an image processor must be able to dilate and erode a high contrast (black-and-white) image on three "pages" of memory (A, B, and C; page C may be a hard-disk drive if speed is of no concern). Images captured in continuous grey-scale form should be processed so the entire shape of interest, and its entire interior, have grey-scale pixel values at one end of the continuum, and the rest of the screen has pixel values at the opposite end. Whether black or white values are assigned to image or background will depend on how a particular image processor implements erosions and dilations. Use of this algorithm with unfiltered grey-scale images may be entertaining only.

In the algorithm, $f(A)$ means that the image transform f should be applied to the image in page A; $B := f(A)$ means the result of $f(A)$ should is assigned to page B.

```
begin
    {Calculate the first element of the digital
    line skeleton}
        A := original image;
        B := erode(A);
        B := dilate(B);
        C := | A - B |;
    {Calculate the remaining elements}
        repeat
            A := erode(B);
            B := erode(A);
            B := dilate(B);
            C := an OR combination of C
                 and | A - B |;
        until the image in B is composed of
        pixels all of background grey-scale
        value;
end.
{Page C will contain the digital skeleton}
```

Digitization of an image may cause aliasing of outline edges. In some applications of this algorithm, I found it necessary to edit out points of the digital skeleton that were produced by such artifacts. The digital skeleton is a set of unconnected pixels that might be useful as landmarks, or that can be connected to provide an analog of the line skeleton.

Chapter 9

Application of Eigenshape Analysis to Second Order Leaf Shape Ontogeny in *Syngonium podophyllum* (Araceae)

Thomas S. Ray

School of Life & Health Sciences
University of Delaware, Newark, Delaware 19716

"One size fits all" - Fruit of the Loom

Abstract

Eigenshape analysis is applied to characterization of changes in form that occur during second-order development in *Syngonium podophyllum*. Second-order development refers to changes in the mature forms of successively produced serial homologous organs, in this example, the forms of successive leaves on the shoot. Leaf outline and landmark coordinates were acquired on the BioSonics OPRS system. Outline coordinates were converted to ϕ^* functions by the ES-TRANS program. The ϕ^* functions were processed by the EIGENS program which computes an R-mode principal components analysis of a matrix of shape functions, and which outputs ϕ^* functions for the first five eigenshapes and a listing of the loadings on the first five eigenvectors for each of the 86 leaves of the sample. Algorithms are provided for reconstructing the outlines from the ϕ^* functions and the eigenshapes. The influence of the weighting of each of the five eigenshapes on the geometry of the outline is explored. A size-shape space for characterizing second-order developmental trajectories is chosen, based on the length of the leaf perimeter and the first two eigenvectors. The eigenshapes are used to synthesize simulated second-order developmental trajectories of plants with allometric and with metamorphic shoot development. The normalization of Lohmann and Schweitzer is questioned. A modification of the eigenshape technique is proposed which will allow the integration of outline and landmark approaches, and which will allow eigenshape analysis to be applied to non-closed as well as closed curves.

Introduction

The Biological Problem

Owing to the continued activity of meristems, plant development can be considered to occur at several hierarchical levels. At the lowest level, which I call first-order development, primordial tissues develop into mature organs (internodes, leaves, flowers). This is the level of plant development that is most comparable to development in the majority of animals, which leads from the zygote to the mature organism, and is deterministic, in contrast to the higher levels of development. Meristems repeatedly produce primordia, with the result that a succession of organs is strung together in a linear series producing a shoot. The process which produces the shoot, that I call second-order development or shoot development, is relatively indeterminate and open, consisting of a serial repetition of the first-order process. Various aspects of shoot development in the Araceae have been discussed by Ray (1981, 1983a, b, 1986, 1987a, b, c, 1988, Accepted a, Accepted b) and Ray and Renner (Accepted a). The present paper explores the use of image acquisition technology and eigenshape analysis (Lohmann, 1983; Lohmann and Schweitzer, Chapter 6) to analyze second-order development of leaves.

Materials & Methods

Specimens

Leaf samples were collected from *Syngonium podophyllum* Schott var. *peliocladum* (Schott) Croat (voucher: Barry Hammel 12787, Missouri Botanical Garden), in May 1988 at Finca El Bejuco in the low-land rain forests of the Sarapiqu'i region of Costa Rica. Leaves were collected in four groups totaling 86 leaves. One group consisted of 29 leaves gathered from various indi-vidual plants, and were selected to represent the full range of variation in leaf form for the species. Each of the other three groups was collected along shoots of individual plants. The three plants used to collect the leaves in these three groups were all growing on the trunk of the same tree, and may have been three shoots of the same clone. The three shoots contained 16, 17, and 24 leaves. Before collecting the leaves from each shoot, the leaves were numbered along the shoot, in order from bottom to top. Where leaves were missing from a node, the node was counted in the numbering, so that gaps appear in the number se-ries where leaves were missing. An identifying letter (z, b, c, or d) was placed on each leaf to indicate which shoot it was collected from.

After the leaves were marked, they were cut off the plant at the point of insertion of the petiole. The stems and the youngest leaves were left on the tree. All leaves were placed in a plant press and dried.

Data Acquisition - Hardware

The dried leaves were taken to the NSF-sponsored Morphometrics Workshop in Ann Arbor, where the outlines were digitized using the BioSonics OPRS system of Norman MacLeod. The leaves were placed with the under-side facing the camera, and the outline was traced counter-clockwise, with the

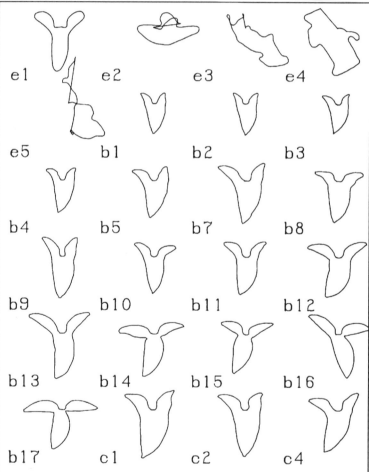

Figure 1. Plots of the ϕ^* functions of the first five eigenshapes and the 86 leaf out-lines (e1 to c4). These figures represent outlines interpolated from the original coordinate data by the ES-TRANS program.

point of insertion of the petiole as the starting point. The OPRS system then sampled and recorded the x,y coordinates of 128 "equally" spaced points around the outline, saving this data in a file with the letter and number label for each leaf. In addition, the coordinates of four landmarks on the leaf outline were recorded: the point of insertion of the petiole, the leaf tip, the tip of the right lobe, and the tip of the left lobe. The output file from the OPRS system lists coordinates using an arbitrary scale, but it also puts out a scaling factor for each sample so that the arbitrary units can be adjusted to measurements in the units defined at the time the data were acquired. The data files from the OPRS system are included on the software distribution disk.

Data Analysis - Software

Outline coordinate data were transformed into ϕ^* functions (Zahn and Roskies, 1972) using the ES-TRANS program of Norman MacLeod. ES-TRANS was developed by modifying the OUTLINE program of Schweitzer and Lohmann to fit it specifically for the output from OPRS. ES-TRANS takes output of outline coordinates in the OPRS format, makes the proper adjustment for scaling, and outputs the ϕ^* functions. In the implementations of OUTLINE and ES-TRANS that have been provided, the outline is divided into 100 roughly equally spaced steps by interpolating between the x,y coordinates in the raw data file. The ϕ^* functions put out by these programs consist of 100 measures of normalized angular deviations from a circle (Zahn and Roskies, 1972; Lohmann, 1983; Lohmann and Schweitzer, 1988). Also see below for more specific details on ϕ^* functions.

The file of ϕ^* functions for all 86 leaves was then processed by the EIGENS program of Schweitzer and Lohmann which performs an R-mode principal components analysis using the HOW routine (Cooley and Lohnes, 1971). EIGENS also puts out a listing of the loadings (projections) of each of the individual specimens on the first five eigenshapes (eigenvectors) and generates ϕ^* functions for the first five eigenshapes.

The loadings of the individual leaves on the five eigenshapes are output by the EIGENS program into two files. According to the DEMO.CMD file of Schweitzer and Lohmann, the

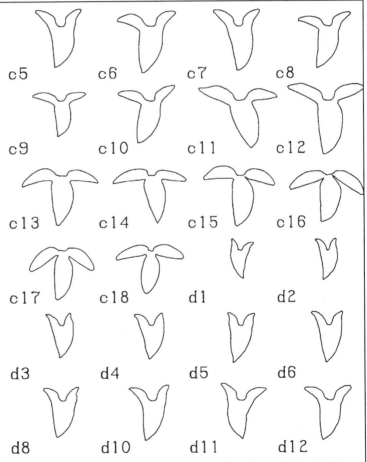

Figure 2. Plots of the ϕ^* functions of the first five eigenshapes and the 86 leaf outlines (c5 to d12). These figures represent outlines interpolated from the original coordinate data by the ES-TRANS program.

SVD.XU file contains the "covariances between shapes and eigenshapes," and the SVD.VS file contains the "correlations between shapes and eigenshapes." The loadings in the SVD.VS file are for functions with normalized amplitudes, the SVD.XU file provides loadings on functions which have been de-normalized. The loadings from the SVD.XU can be used to reconstruct outlines from the eigenshapes, and they have been used as the loadings throughout the analysis in this paper.

In presenting the results of the analysis, the leaf outlines were reconstructed by reading the ϕ^* functions and converting them to x, y coordinates to be displayed on the screen by Generic CADD or sent to a Hewlett Packard pen plotter. I include, on disk, the code for a routine **phi()**, in the C language, which draws a single outline according to the specified position, direction, orientation, step length, and amplitude. Lohmann and Schweitzer distributed FORTRAN source code for their PLTSHAPE program at the workshop. The RE-BILD routine in that program performs essentially the same function as my **phi()** routine. In addition, the five eigenshape functions were used to synthesize leaf outlines given various selected loadings on the eigenvectors (eigenshapes). The **eigen()** routine, included on disk, performs this synthesis of a single form according to the specified position, direction, orientation, step length and five loadings.

Results

Analysis of the 86 leaf outline samples by the EIGENS program showed that the first five eigenshapes accounted for 71.5%, 11.5%, 3.6%, 2.3% and 2.1% of the variance successively. The sample label, lobe lengths, leaf area, perimeter, five loadings, and the amplitude for each

leaf are reported in the file TABLE_1, on disk. The ϕ^* functions for the first five eigenshapes were output by EIGENS to the SVD.U file, which is in the same format as the output files from the ES-TRANS program. The first five eigenshapes and the 86 leaf shapes (drawn from their ϕ^* functions) are plotted in Figures 1-4. The first eigenshape represents an average shape; the remaining eigenshapes represent deviations from the average shape, and so would not be expected to resemble the leaf form.

Geometric Interpretation of Eigenshapes

While it seems imprudent to attempt to place a biological interpretation on the individual eigenshapes, it does seem appropriate to interpret the eigenshapes geometrically. It should be possible to understand the influence, on the geometry of the

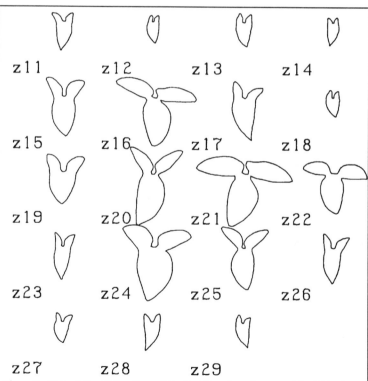

Figure 3. Plots of the ϕ^* functions of the first five eigenshapes and the 86 leaf outlines (z11 to z29). These figures represent outlines interpolated from the original coordinate data by the ES-TRANS program.

outline, of changing the weighting of each eigen-shape. To this end I have selected two leaves (d1, d26) which represent extremes of shape variation. For these leaves (Figures 5-6), I have plotted the outline as reconstructed from the ϕ^* function and as synthesized from the five eigenshapes weighted by the five loadings for that leaf. Below these shapes I have plotted the ϕ^* function with varying amplitudes, using five equally spaced values through the observed range of amplitudes in the sample of 86 leaves. Below these five shapes I have plotted outlines synthesized from the eigenshapes by varying each of the loadings individually, again using five equally spaced values through the observed range of loadings. While varying each loading, I keep the remaining four set to the values for that leaf as listed at the top of the figure.

Varying the value of the amplitude alters the degree to which the shape deviates from a circle, a characteristic which Lohmann and Schweitzer (Chapter 6) call angular-ity. If the amplitude were set to zero, the resulting shape would be a circle. Amplitudes are greater for the more lobed leaves, and increas-ing the value of the amplitude of any leaf increases the degree to which the resulting outline is lobed. How-ever, increasing the amplitude can also cause the outline to collapse, because some portions of the outline deviate inward from a circle. This inward bending is enhanced as the amplitude is increased.

Varying the weighting of the first eigenshape is equivalent to varying the amplitude of the average outline, and has an effect very simi-lar to varying the amplitudes of individual outlines, causing the leaf shape to vary from cordate to hastate

to three-lobed. This is an approximate description of the primary change in form taking place during shoot development in this species. Taking the weighting above the range observed in the data causes a collapse of the figure similar to what is observed in the amplitude series.

Varying the weighting of the second eigen-shape primarily affects the orientation of the lobes of the leaves. With low values, the lobes point forward, and as the value increases the lobes bend toward the rear. This is an interesting observation, as the problem of articulation is a general one in the characterization of shape. That the articulation

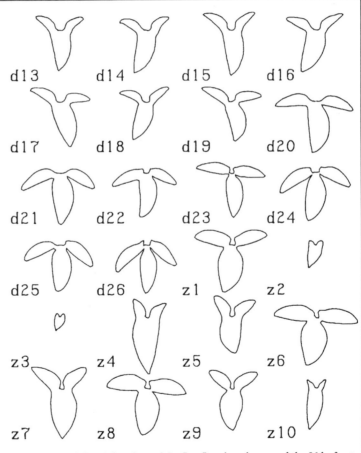

Figure 4. Plots of the ϕ^* functions of the first five eigenshapes and the 86 leaf out-lines (d13 to z10). These figures represent outlines interpolated from the original coordinate data by the ES-TRANS program.

of a structure could be isolated to a single variable in the characterization of the shape seems like a useful feature.

It is much more difficult to interpret the influence of the third eigenshape. In most leaves, large negative values result in inappropriate rounding or extra bulges in the lobes. The leaves with the most negative values of the third loading fall into two groups, those with low first loadings (z2, z3, z10, z12, z14, z28), and those with high first loadings and bulges at the base of the lobes (d25, z6, z8, z16, z21). What these two groups have in common is that there is a bulge or leaf tip at a position on the outline closer to the petiole than in most leaves. The third eigenshape apparently acts to compensate for the lack of homology in the position of the lobe tips, by creating an extra bulge near the petiole when needed.

The fourth eigenshape seems to have an influence very similar to that of the third. In fact, the tracings of the two eigenshapes (e3, e4) are very similar. High values of the fourth loading are found in leaves with bulges near the base of the lobes (d24, d25, c14, c17). Leaves with these bulges tend to have either a large negative value for the third loading or a high value for the fourth. An interesting example is the series d24, d25, d26, which all have bulges on the lobes, but which alternate between having high fourth and low third loadings. The only leaf to combine both low third and high fourth loadings is z10 which is one of the most

abnormal leaves in the sample. Low values of the fourth loading tend to be associated with fat central lobes (z5, z6, z8, z9, z15, z16, z19, z20, z21, z22, z24, z25). The fattening influence of low fourth loadings can be seen in Figures 5-6 where this value is

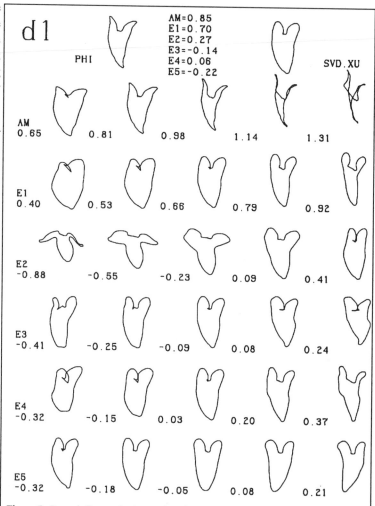

Figure 5. In each figure, the interpolated outline of the leaf as reconstructed from the ϕ^* function is represented in the upper left. In the upper right is the leaf outline synthesized from the first five eigenshapes as weighted by the loadings on the first five eigenvectors found in the svd.xu file. The values of the amplitude and the first five loadings are listed in the top center. Below this, the outline is reconstructed from the ϕ^* function with five values of the amplitude, evenly spaced over the observed range of amplitudes in the sample of 86 leaves. Below this the same is done for each of the five loadings. As each loading is varied individually, the remaining four loadings remain fixed at the values listed at the top.

varied. The third eigenshape also seems to have a similar influence, but to a lesser extent and in the opposite direction (fat central lobes are associated with high third loadings).

The influence of the fifth eigenshape seems fairly clear, it affects the handedness of the leaf. With low fifth loadings the central lobe swings to the right, and with high values the lobe swings to the left.

Developmental Trajectories

One of the principal objectives of this study is to use eigenshape analysis to portray second-order ontogenetic trajectories of leaf shape. In order to search for possible size-shape spaces in which to view these trajectories, I plotted each of the loadings against one another, and against the amplitude and three measures of size: the length of the central lobe (computed from the coordinates of the petiole insertion and leaf tip landmarks), the leaf area, and the length of the perimeter. I examined all two-dimensional projections of the size-shape space onto the axes of the variables just listed.

These observations suggest that the most appropriate size-shape space for examining second-order development would be that made up of the leaf perimeter and the first two eigenvectors. Examples of second-order trajectories in two projections of this space can be seen in Figure 7. These synthetic examples illustrate how leaf shape ontogeny can be characterized, and how one can recognize the difference between the second-order trajectories of allometry and metamorphosis.

Simulation of Second-Order Ontogeny

My second principal objective for using eigenshape analysis is to be able to synthesize the complex and subtle transformations in leaf form that occur in second-order development for use in simulations, and as a graphic device to illustrate different classes of shoot development. This has been done in Figure 7, where I illustrate the difference between allometry and metamorphosis of leaf form in shoot development. Eigenshape analysis

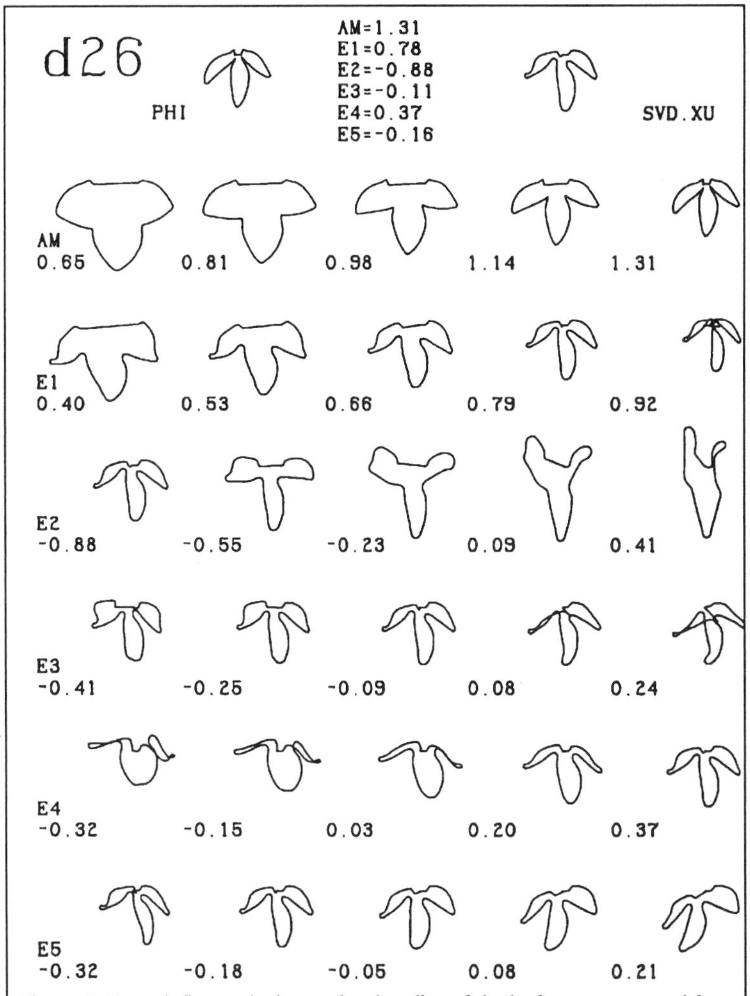

Figure 6. In each figure, the interpolated outline of the leaf as reconstructed from the ϕ^* function is represented in the upper left. Details as in Figure 5.

appears to be a handy tool for making the point graphically.

identical to the R-mode principal components analysis described by Cooley and Lohnes (1971)."

Discussion

Normalization

In order to reconstruct the outline coordinates from Lohmann and Schweitzer's ϕ^* function, each value in the function must be multiplied by the amplitude, because they were all divided by the amplitude when the functions were normalized. Lohmann and Schweitzer (Chapter 6) state "Amplitude is calculated as the standard deviation of the observed values of ϕ^*, before they are standardized." The source code for the ES-TRANS program indicates that amplitude, **PS**, is computed by the following formula:

$$\mathbf{PM} = \Sigma \phi(i) / n \qquad (1)$$

$$\mathbf{PS} = (\Sigma(\phi(i) - \mathbf{PM})^2 / n)^{1/2} \qquad (2)$$

where $\phi(i)$ are the individual values of the ϕ^* function before normalization, and n is the number of values in the function (100). The value PS is recorded in the output of ES-TRANS and is referred to as the amplitude.

Lohmann (1983) states "Note that the amplitude of $\phi^*(l)$ is subsequently standardized to unit variance as a consequence of computing correlations between shapes." This is apparently a reference to the fact that, in performing a principal components analysis from a correlation matrix, the values are first normalized. Lohmann and Schweitzer (Chapter 6) state that their procedure is "computationally

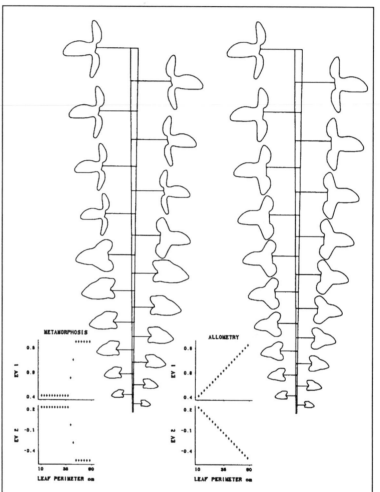

Figure 7. Representations of isometry, allometry and metamorphosis in second-order leaf development. In the drawing of a plant on the left, leaves increase in size without changing shape, isometry, for eleven segments. Then metamorphosis causes an abrupt change in shape, involving only two leaves. After the abrupt change, isometric development resumes for seven more segments, during which leaves again increase in size without changing shape, but fixed at the new post-metamorphosis shape. In the drawing of the plant on the right, successive leaves develop allometrically. Leaf form changes gradually as leaves increase in size. The lowest leaf is based on the loadings of leaf z2 and the highest leaf is based on the loadings of leaf d20. The intervening leaves are based on an evenly stepped linear interpolation between the loading values for the two leaves. Below and to the left of each plant, the second-order ontogenetic trajectory through the size-shape space is plotted for two projections: leaf perimeter and the loading on the first eigenvector (above), and leaf perimeter and the loading on the second eigenvector (below). Leaf perimeter measures size and the loadings measure shape.

However, the traditional normalization for an R-type analysis is made across specimens for each measure, not across measures for each specimen (Cooley and Lohnes, 1971; Pimentel, 1979). Their methods amount to an R-type principal components analysis performed with a Q-type normalization.

Normalizing the amplitude causes all N points to be equidistant from the origin of the p axes. All points are projected onto the surface of a hyper-sphere whose center is the origin of the p axes of angle measures; however, the N points will not be distributed over the entire surface of the hyper-sphere. The entire surface of the hyper-sphere represents all possible shapes with normalized amplitudes. Any real sample of specimens will represent only a small sub-set of all possible shapes, so the points in p space will be restricted to a patch on the surface of the hyper-sphere.

While the first eigenvector will pass through the centroid of the patch on the hyper-sphere (unless the patch is significantly larger than the radius of the hyper-sphere), subsequent eigenvectors will not pass through the swarm of points at all. The eigenanalysis of the data swarm takes place by rotation of a set of Cartesian axes which are located at the origin. The sample centroid will be inside the hyper-sphere, close to its surface, and therefore displaced from the origin by a distance of slightly less than the radius of the hyper-sphere. All eigenvectors after the first will be parallel to tangents to the principal components of variation of the data swarm, which curves with the surface of the hyper-sphere.

In the technique of Lohmann and Schweitzer, the first eigenvector locates the sample centroid, not the first principal component of variation. The first principal component of variation is represented by the second eigenvector. As points deviate from the sample centroid, the swarm curves toward the second and higher eigenvectors. The plot of EV1 against EV2 in Figure 8 shows that as we move out from the origin along the second eigenvector, the data swarm bends around the surface of the hyper-sphere. If there is enough variation, the data swarm will eventually become perpendicular to the second eigenvector, resulting in a loss of information. The second eigenvector loses ability to represent the variation in the periphery of the data swarm. Thus the first principal component of variation is not well represented by its associated eigenvector.

Any projection on the first eigenvector and another eigenvector shows the bending of the data swarm caused by projection on the hyper-sphere. This bending is most apparent in the projection on the first two eigenvectors (Figure 8). This same bending was apparent in the MacSpin demonstration at the workshop. This demonstration showed a data swarm projected on the first three eigenvectors, being rotated. The swarm showed a strong bending as a result of being wrapped around the hyper-sphere. It is an unfortunate property of the spherical projection that the greater the variation along any principal component, the less efficiently will the associated linear eigenvector account for the variation, because the greater will be the bending. If angles had been normalized across samples, the sample centroid would have been located by the normalization step. This would leave the first eigenvector free to represent the first principal component of variation from the sample centroid. The latter approach would make more efficient use of the eigenvectors. In addition, normalization of angles across samples would not project the data onto the surface of a hyper-sphere; therefore the linear eigenvectors could more efficiently account for the variation. Lohmann and Schweitzer (Chapter 6) state, in reference to their eigenshape functions, "No other set of shape functions can account for the variation more efficiently." This is not true because of their inefficient application of Cartesian eigenanalysis to a spherical distribution, and because the first eigenvector locates the sample centroid rather than the first principal component of variation.

Given the lack of resolution toward the periphery of the swarm caused by bending around the hypersphere, and the inefficient use of eigenvectors, I would like to see compelling reasons for the normalization of amplitude. I see two motivations for this procedure:

1) To be able to describe an individual's outline with three attributes: form, size, and angularity. This seems a reasonable way of breaking down the attributes of outlines described with ϕ^* functions. However, I see no biological use for the isolation of angularity. Lohmann and Schweitzer (Chapter 6) give the example of the secondary encrustation of the shell wall of foraminifera, which changes the shapes of shells by smoothing their outlines. I doubt that a close look at the data would show a change in angularity with absolutely no change in form. At any rate, angularity of outlines can always be calculated, adjusted, or compared, without normalization of amplitude.

Decomposition of outlines into the three components, form, size, and angularity is arbitrary. The angularity attribute is peculiar to the ϕ^* representation of outlines, and arises as a result of the "removal of a circle step." If outlines were represented by an $r(\theta)$ function or a tangent angle function (Zahn and Roskies, 1972; Bookstein, 1978; and see below, "Proposal of a New Technique") there would be no angularity component to be isolated. It can be seen in Figures 5-6 of this paper, and in Figure 4 of Lohmann and Schweitzer (Chapter 6), that

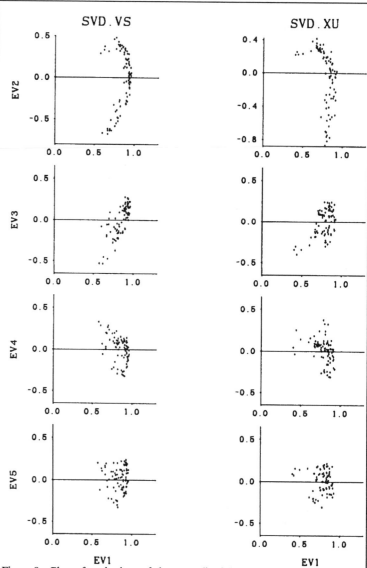

Figure 8. Plots of projections of the normalized (SVD.VS) and un-normalized (SVD.XU) data on the first eigenvector and the next four eigenvectors. The normalization by amplitude causes all N points to be projected onto the surface of a hyper-sphere, making all points the same distance from the origin. The two axes of each plot have been set to the same scale in order to preserve the circular curvature of the surface of the sphere. In the projection of the surface of the hyper-sphere onto two dimensions, some of the points appear closer to the origin than the radius of the sphere because they are bent around the surface of the sphere on some of the collapsed axes. However, even in this projection, no point can appear farther from the origin than the radius of the sphere. Therefore the margin of the data swarm farthest from the origin traces the circular outline of the sphere. This is most apparent in the plot of the first two eigenvectors, where the variance is the greatest.

variation in amplitude results in significant variation in the outline. I think it is unwise to exclude such significant variation in shape from the principal components analysis.

2) To give each individual shape an equal weight in determining the orientation of the first eigenvector. This is not accomplished by their normalization, because eigenvectors are oriented by minimizing the sum of the squared residual distances of the points from the eigenvectors (with the additional constraint that eigenvectors are orthogonal). Therefore points further from each eigenvector in the data swarm continue to carry more weight in determining the direction of eigenvectors (including the first). They have only removed the influence of amplitude, they have not given all points equal weight.

The sample centroid could be located by a simple average of angles across samples, rather than an eigenvector. In a simple average, individual points have an equal weighting, unlike in the process that determines the orientation of eigenvectors. Normalization of angles across specimens would provide a relatively unbiased sample centroid, would not project the data onto the surface of a hyper-sphere, would make more efficient use of eigenvectors, and would not exclude a significant component of shape, angularity, from the eigenanalysis.

Outlines and Landmarks, Proposal for a New Technique

Lohmann (1983) and Lohmann and Schweitzer (Chapter 6) developed and applied eigenshape analysis to microfossils lacking well-defined landmarks on their outlines. However, the leaves of my study have four well-defined landmarks on their outlines (the point of insertion of the petiole and the tips of the three lobes). The principal weakness of the eigenshape method, as I have implemented it, is that it makes the assumption that the nth segment of each outline is homologous to the nth segment on each of the other outlines. This assumption is clearly violated by the specimens of

my sample. In the smallest leaves, the tips of the poorly developed lobes occur proportionately closer to the petiole along the outline than in the larger leaves where the lobes are better developed; therefore the leaf tip landmarks do not line up in the analysis. The third and fourth eigenshapes compensate for this movement in the position of the lobe tips by producing bulges in the outlines at various positions as needed; however, this is neither an elegant nor a biologically meaningful solution to the problem.

It is feasible to modify eigenshape analysis in such a way as to treat landmarks on the outline in a more realistic manner. The obvious solution is to mark the landmarks around the outline, and to divide the portion of the outline occurring between any pair of adjacent landmarks into the same number of segments on each specimen. This would insure that real landmarks on the outline line up in each specimen.

The principal problem that would be created by this approach is that the step size would vary between different pairs of landmarks. However, in this example, only four different lengths are involved. In general, for closed curves, the number of different lengths involved would be equal to the number of landmarks. The lengths could be entered into the matrix and subjected to principal component analysis along with the angle data. Another problem that would arise from this modification of the technique derives from what Lohmann (1983) describes as "removing a circle" from the $\phi(l)$ function in the construction of the $\phi^*(l)$ function. Because the values of the ϕ^* function are the angle by which the nth step of the outline differs from the nth step of a circle, there is also an assumed homology between the sample outline and that of a circle. As long as the outline is a closed curve and both the outline and the circle are divided into equal numbers of equal length segments, this is not a problem. However, when the lengths of the segments on the sample outline vary, it becomes less clear how to determine the homology of a segment of the outline with a segment of a circle.

The most obvious solution is to proportion the steps around the circle in the same way they are proportioned around the outline; however, this would vary from specimen to specimen. It would probably be necessary to partition the perimeter of the circle in the same proportions as an average outline, in which case it is not clear that the average steps of the circle would be homologous to any of the steps of the outlines.

A better solution would be to dispense with "removing a circle." When normalization is applied in the traditional manner in principal components analysis, it involves the calculation of the mean value of each of the independent measures, in this case, the mean across the 86 leaves of each of the 100 values of the ϕ^* function. This results in a sample centroid function which is an average outline. Successive eigenshapes would describe deviations from the sample centroid basis shape rather than from the circle. Starting with a basis shape derived from the data rather than an arbitrary geometric form strikes me as more in the spirit of eigenshape analysis.

By eliminating the "circle removal" step we not only remove a source of uncertainty when we move to an analysis with unequal length steps, but we remove the limitation that our "outline" must be a closed curve. With this limitation removed, it would be possible to apply eigenshape analysis to the outlines of appendages, where the outline is not closed, or in general to any contour bounded by homologous landmarks. What I am proposing is that the ϕ^* function be replaced by the ϕ function (Zahn and Roskies, 1972), which simply records the angular direction in the plane at each step (with the angle of the first step, or the line through two landmarks, set to zero). In addition, step lengths would be recorded and appended to the function. Bookstein (1978) has proposed a similar approach in which N landmarks on the outline are sampled by N paired measurements of arc-length and tangent angle, forming a $2N$ vector. This technique could be applied to any outline or contour of any shape, closed or not, and so would have more general

applicability. Some experimentation and further thought should make this approach workable. This modified eigenshape analysis will provide a technique which nicely combines outline and landmark approaches.

In this study, eigenshape analysis has shown itself to be very useful, and potentially more so if landmarks could be treated more realistically. Most importantly, this experiment with eigenshape analysis in *Syngonium podophyllum* has shown that it is feasible to characterize complex and subtle second-order developmental trajectories of leaf form. However, it remains to be seen how the technique will work when applied to other leaf forms. For example, how will the technique cope with second-order leaf series involving differing numbers of lobes? Also, the technique may be inappropriate for some leaf forms, such as the leaves of some Monstera species which include varying numbers of holes as important features of shape. It will probably be necessary to use a diversity of analytical techniques in order to study second-order development of leaf shape in a wide variety of species.

Acknowledgments

I thank Norman MacLeod and Peter Schweitzer for generously providing source code for their eigenshape analysis programs. I thank Fred Bookstein, F. James Rohlf and Peter Schweitzer for their critical reviews of several versions of this manuscript, and for their extensive e-mail communications, without which this paper would not have been possible. I thank the Systematic Biology Program of the National Science Foundation for its support for attending the Morphometrics Workshop. Contribution No. 126 from the Ecology Program, School of Life & Health Sciences, University of Delaware.

References

Bookstein, F. L. 1978. The measurement of biological shape and shape change. Springer Verlag, Berlin, 191 pp.

Cooley, W. S., and P. R. Lohnes. 1971. Multivariate data analysis. Wiley & Sons, New York, 364 pp.

Lohmann, G.P. 1983. Eigenshape analysis of microfossils: a general morphometric procedure for describing changes in shape. Math. Geol., 15(6): 659-672.

Pimentel, R. A. 1979. Morphometrics, the multivariate analysis of biological data. Kendall/Hunt Publishing Co., Dubuque, Iowa, 276 pp.

Ray, T. S. 1981. Growth and heterophylly in an herbaceous tropical vine, *Syngonium* (Araceae). Ph.D. dissertation, Harvard University.

Ray, T. S. 1983a. *Monstera tenuis*. Pp. 278-280 in Costa Rican natural history (D. Janzen ed.). University of Chicago Press, Chicago, 816 pp.

Ray, T. S. 1983b. *Syngonium triphyllum*. Pp. 333-35 *in* Costa Rican natural history (D. Janzen ed.). University of Chicago Press, Chicago, 816 pp.

Ray, T. S. 1986. Growth correlations within the segment in the Araceae. Amer. J. Bot., 73(7): 993-1001.

Ray, T. S. 1987a. Cyclic heterophylly in *Syngonium* (Araceae). Amer. J. Bot., 74(1): 16-26.

Ray, T. S. 1987b. Leaf types in the Araceae. Amer. J. Bot., 74(9): 1359-1372.

Ray, T. S. 1987c. Diversity of shoot organization in the Araceae. Amer. J. Bot., 74(9): 1373-1387.

Ray, T. S. 1988. Survey of shoot organization in the Araceae. Amer. J. Bot., 75(1): 56-84.

Ray, T. S. Accepted a (in revision). Foraging behavior in climbing Araceae. J. Ecol.

Ray, T. S. Accepted b (in revision). Metamorphosis in the Araceae. Amer. J. Bot.

Ray, T. S., and S. Renner. 1990. Comparative studies on the morphology of the Araceae. A. Engler, 1877. Translation with an introduction, updated nomenclature, and a glossary. Englera 12. 140 pp.

Zahn, C. T., and R. Z. Roskies. 1972. Fourier descriptors for plane closed curves. IEEE Trans. Comp., C21(3): 269-281.

Part III
Section C

Introduction to Methods for Landmark Data

Fred L. Bookstein

Center for Human Growth and Development
University of Michigan, Ann Arbor, Michigan 48109

Abstract

Morphometrics is the biometric study of effects upon form. Its statistics proceeds by special adaptations of path analysis, the typical mode of quantitative multifactorial modeling, applied to carefully selected geometrical quantities. When these quantities represent the locations of landmark points, it becomes possible to match computed diagrams of effects against potential explanations of their mechanisms. That is, landmarks are the points to which typical epigenetic explanations refer. They bridge the geometric record of biological forms separately, the modeling of relations between pairs of forms as deformations, and the statistics of whole samples of forms in one single algebraic technology.

The varieties of landmarks accord with the sorts of explanations that apply to their locations. The most useful class are individually recognizable points on boundaries between regions of distinct histology. Such points can be imagined to respond to growth processes or other adaptations at any distance. Landmarks that lie in planes of symmetry may be analyzed using only two Cartesian coordinates. Somewhat less powerful are local geometric features of extended tissue boundaries, such as points of sharpest curvature ("corners") on curves, or surfaces that might serve as points of application of net biomechanical vectors even when traces of the relevant processes are not visible. A third,

weakest category of landmark points are characterized by their relationship to structures at a distance, such as endpoints of the greatest diameter of a form. Any statistically detectable "effect" upon points like these must be construed as applying to the entire length of the defining extremal property.

These strategies of morphometrics are unaligned with the goals of phylogenetic reconstruction. The proper role of morphometrics in biology does not include the provenance of putative measures of "similarity" or "distance" for purposes of phylogenetic inference, and landmark data should never be turned to that purpose.

Landmarks

Morphometrics is different from other biometrical methods. Its distinctiveness owes to the triple role played by landmarks. Their geometry supplies a language for describing individual forms, pairs of forms, and effects upon entire samples of forms. Morphometrics begins by exploiting landmarks under the first heading in gathering data; proceeds directly to the third heading to compute versions of all the classic path models; then reverts to the second when findings are interpreted, wherever possible, as if they specified processes relating single pairs of forms.

It is useful to begin the discussion by examining one classic sort of statistical analysis that landmark-based analysis does *not* resemble: it is not in any way an indirect measurement of true parameters, whether morphological "distance" or anything else. The metaphor of estimating "true values" arises in the classical method of least-squares, which already existed in recognizable form by 1810 (cf. Stigler, 1986). The method is designed for data in the form of "observations," such as the position of a planet at a particular time in the eyepiece of a telescope at a particular spot on the earth. Such observations are of no particular scientific meaning one by one. But in sufficient quantity, they may be combined in an "estimate" of the true orbital parameters of the celestial object(s) involved. First, the true Newtonian equations of motion of the planet are linearized in the vicinity of the true parameters, and the errors in all observed measurements combined into a single error for each equation. Second, these equations are repeatedly summed after multiplication by the (known) coefficient of each (unknown) parameter in turn, yielding as many "normal equations" as there are unknowns. Third, the normal equations are solved to supply estimates of the true orbital parameters that have, under certain probability models for measurement errors, their own desirable properties. In summary, observations (positions in a telescope) are combined into parameters (invariants of motion-potentials, momenta, etc.) which are conceptually different and physically constant.

In the modification of this model for biometric use, at the hands of Francis Galton, Karl Pearson, and Sewall Wright, the conceptual separation of data (measures of organisms) from parameters (path coefficients) was maintained. But the tie between "parameters" and exact laws was replaced by a vague notion of causality. Recall that in the classic contexts for which least-squares methods were developed—principally celestial mechanics and geodesy—there is no mention of causality. Newton's Laws don't "cause" the planets to take the orbits they take, but instead describe those orbits for all time, once "initial conditions" are set.

By comparison, in Galton's equally classic analysis of the inheritance of height, there is no law regulating height of the offspring under any circumstances. Instead, we may believe that offspring height is "determined" by a collection of numerous "causes," of which parental height is one. The purpose of the biometric regression is to compute not a physical constant, like the momentum of Jupiter, but instead a path coefficient the value of which is subject to further explanations, such as by evolutionary argument or by an auxiliary computation based in gene frequencies. The regressions are based on a presumption of causality, but end up interpreted functionally instead, in terms of mechanisms rather than coefficients.

As applied to morphometrics, the incompatibility between these two invocations of least-squares methods went unnoted until Huxley's seminal work of 1932. At the crux of the problem is the manner in which explanations of these coefficients were eventually to refer back to the data they were supposedly accounting for. Astronomers, after all, know exactly what object they mean by "Jupiter." They agree regarding the precise manner in which observations are to be compared from night to night, telescope to telescope, and regarding the roster of parameters relevant to celestial mechanics. Biologists must substitute contingencies for both of those intellectual luxuries; and while regression substitutes coefficients for constants, it is not at all obvious what substitutes for "pointing the telescope." Yet if we cannot say where we are pointing our measuring instrument, we cannot relate regression coefficients to meaningful explanations.

In Galton's analysis of height, for instance, by what principle can we assume that net height, from the floor to the crown, is the subject of regulation by any explicable biological process? Huxley reminds us that net height is actually the integral of all its *differentials*, its little elements, along the diversity of organs that make up the path taken by the yardstick. In principle, a different "parameter," a different covariance, should be computed for each infinitesimal segment of that path. To assert that a covariate of an epigenetic process affects "height" we must inspect the way in which it affects all the components of height; otherwise the explanation is wrong even when the regression is computed correctly.

Hence emerged the classic *growth-gradient* models that were applied to axial forms from invertebrates through man (cf. these views in Boyd, 1980, or Bookstein, 1978). The analogy on which this work is based relates the reading of a ruler placed upon an organism to the reading of a reticle placed upon the eye piece of a telescope. The elemental size of a bit of tissue, identified by its pair of endpoints, is presumably analogous to the visual separation of a pair of celestial objects, like Jupiter and Saturn. At root, then, the analogy stands or falls on the appropriateness of the axial element—the segment between two "landmarks"—for bearing the biological explanation (in this case, *allometry*, the effect of size on shape) beyond the regression coefficient.

The fundamental difficulty of morphometrics is already patent in this relatively early analysis. Even when we actually mark particular bits of tissue over ontogeny, as by following natural variations of pigmentation, the resulting regression coefficients do not necessarily support any interpretation in terms of explanations of form (cf. Bookstein, 1978). In root tips, mammalian bone, and several other common examples, explanations "move over" tissues: a process applying at one place at one time applies at "another" place at a later moment in ontogeny. In fact, the relevance of biological explanations to line-elements, or any other coordinate system applied to a form, is logically circular. No laws control our decisions about matching locations for the ends of the ruler separately prior to carrying out our regressions. Rather, it is the regression analysis itself, perhaps an exploration of allometry in one guise or another, that is to justify (however retroactively) the comparisons of segments among these tentatively labelled endpoints. To the extent that the "lengths" covary with

other measurable entities—weight, other lengths, fitness, habitat, histology, date—we assert these endpoints to be "landmarks." If the distances didn't manifest interesting covariances, they would be useless as landmarks, no matter how clearly or objectively they had been observed.

Then landmarks delimit our explanations of effects upon form; their role in investigations is not just to *be* there but to encourage hints about the processes that *put* them there. Except in one dimension, the axial case Huxley treated, there is no easy way to automatically generate morphometric "variables" (Bookstein, 1978). In two or three dimensions, the construction of a space in which to sift through covariances so as to uncover possible biological meaningfulness without any geometric bias requires algebraic tricks, for instance, the scheme of "shape coordinates" to be introduced below. Yet most of the shape variables explicitly required for fair coverage of any set of landmarks will prove uninteresting. Even in one dimension, the set of possible objects of a subsequent biological explanation includes all of the distances between pairs of landmarks; all of these are equally meaningless prior to a sophisticated study of their covariances among themselves and with exogenous factors, and all but a few will remain meaningless composites even after the study is completed.

Thus in all cases there are more morphometric variables than one can use in realistic descriptions of particular effects upon form. The logic by which these alternatives are sifted has nothing to do with the usual reasoning patterns of systematics, and morphometrics should not be thought of as a branch of systematics, or even as a tool of phylogenetic reconstruction, except insofar as it produces residuals from coherent explanations to which the incoherent explanation of taxonomy ("they are different taxa") applies as a sort of default. Many quantifications of organismic form may be both accurate and taxonomically useful and still be indistinguishable, for our purposes, from meristics, colors, or variables of any other disorganized class. I do not include such quantifications among the morphometric methods. Although they appear to suit the etymology, they are not conformal with the context of biological explanation: one cannot interpret their covariances. For instance, a list of elliptic Fourier coefficients serves the biologist essentially as a list of ingredients serves the supermarket shopper. One can duplicate the outline (the product), but it remains wholly unclear why just those ingredients are there. In the absence of an epigenetic theory, we can have no expectations about the "causes" or "effects" of these ingredients; we only know that products with paprika tend to have pictures of Hungary on the box, etc. Analogously, certain sorts of "similarities" between evolutionary recipes might express some combination of common causes of the lists in question. The methods of taxonomy attempt to separate these causes into those which express common ancestry, case by case, versus those which do not. Whether or not that separation is even a well-posed problem (I believe it is not), *as evidence for establishing common ancestry* morphometric methods are not necessarily either better or worse than other methods. That is not their purpose at all.

The information we morphometricians are after is not similarity due to common ancestry but the other part, the covariances with explanatory factors[1]. The word "factor" is used here in Sewall Wright's sense: a factor is a quantity measurable (directly or indirectly) upon single organisms and serving as a predictor in a path model. There, at last, we can revert to a theoretical grounding stated first and most passionately by D'Arcy Thompson in

---

[1] This search is often enough successful that, in general, it is unwise to presume a computed morphometric "distance" to at all approximate a measurement of taxonomic distance; there are too many other good explanations of similarity (see Bookstein, 1991). I believe it is unreasonable, for instance, to compute RFTRA distances and pass them through any sort of dendrogrammatic analysis, as is unfortunately exemplified in the chapters by Chapman, Riedel, and Reilly which follow.

the well-known passage from his *On Growth and Form* (1917:275-276) that I shall spare the reader the chore of rereading here. Thompson asserts that our descriptions of the processes that regulate form tend usefully to proceed in terms of the sizes and shapes of parts relative to others, and so would conduce to descriptions of relations among those forms in terms of geometric deformations. Conversely, deformations, as extended systems of distributed change, are a likely substrate in which to search for evidence of explicable biological processes.

It is one implication of this position that Huxley's basic model of "growth-gradients," which came 15 years later, would get its geometry wrong: the issue is one of size, shape, and relative position, not merely of a scalar "length" along a conventional axis.

Many investigators after Thompson attempted to discover methods for objectively seeking biologically meaningful descriptions within the mathematical space of deformations. As it happens, the algebra of landmark points provides the solution to that problem. That is, the statistics of the landmark locations are also the statistics of all models of deformation driven by the landmarks (cf. Bookstein, 1987)— but that is not the issue here. Rather, the question at hand is the manner in which the model of deformation can usefully be applied to multiple sets of data, not just single pairs of forms. On this matter contemporary morphometricians are approaching a consensus. The theory underlying landmark-based morphometrics currently runs somewhat as follows: We know that epigenetic processes regulate form as it unfolds and as it is arrived at in adulthood. We understand, too, certain influences on these processes. Some of the covariates of form are global, like net food intake; others are extremely local, as in the hypermorphosis resulting from muscular exertion or in the sequelae of particular mutations. Over this variety of causes and covariates of form, and in functional explanations throughout the huge diversity of taxa of which we have knowledge, it is found that the elements to which explanations point can often be identified with specific organs or tissues, and, further that they may usually be identified with specific *boundary points* of those organs or tissues. By a "boundary point" is meant a location at which tissue types are juxtaposed with other tissues or with the surrounding medium. Then landmarks are the boundary points, between organ and organ or between organism and environment, at which our epigenetic explanations adhere.

This ideal type for landmarks will be searched for throughout the remaining chapters of this Proceedings just as they are devoutly sought throughout (the valid) part of contemporary morphometric literature. **Landmarks are the points at which one's explanations of biological processes are grounded.** They are sign posts which the organism conveniently erects to ease our task of being functional or evolutionary biologists while remaining biometricians. In that capacity they ease the phylogenetic reconstruction task by specifying those deformations that are not of concern; that is their sole function in phylogenetic reconstruction.

I have claimed that landmarks are located to ease the task of biological explanation, and I seem to be justifying this claim by noting that many of the explanations of form we accept today as epigenetically valid seem to invoke deformations of the locations of landmarks. All this would seem to involve only the psychology of the biological profession. There is an additional advantage of landmark data, however, which in my view supersedes the preceding difficulty, by permitting us, at last, to transcend the context of classic biometry (Galton, Pearson, Wright) with its arbitrary separation of path coefficient from datum. In an appropriate multivariate context, to be reviewed presently, landmarks permit the biologist to circumvent this distinction. (Though D'Arcy Thompson suspected it, it has been formally recognized only in the development of morphometrics in the last decade or so.) The space of "configurations" in which landmark locations are recorded is, at the same time, the space of possible depictions of effects upon form and the space of statistics of deformations relating those forms. Thus *landmark-based morphometrics is the embodiment within biometrics of the functional form of biological explanation.* That is, the same statistical formalism can be used to delineate *explanations* as was used to gather the individual instances subject to explanation. The landmarks link three separate scientific thrusts: (a) the geometry of data, (b) the mathematics of deformation, and (c) the explanations of biology. For instance (Bookstein et al., 1985), any instance of size allometry (c), computed in the appropriate multivariate way, can be immediately interpreted as a deformation (b), or can be equally effectively used to "grow" any single form into others (a).

The importance of this formal property cannot be overstated. It applies to all the statistical manipulations of landmarks from Procrustes fits (which represent the no-deformation) through my relative warps. By contrast, the outline methods criticized by Bookstein et al. (1982, 1985) and by Rohlf (this volume) are incapable of leading to biological explanations. Landmark-based explanations can be localized, whereas those based on Fourier coefficients cannot be. For example, the location of the warp in Figure 5 of the chapter entitled Higher-Order Features of Shape Change for Landmark Data may be taken as the center of the quadrilateral of landmarks. The same objection applies to any other technique of integral measures, from eigenshapes to body weight: the language of *their* statistics is not a biometrical language, whereas the language of landmark statistics, the language of variables selected by virtue of their covariances with deformation, *is.* In morphometrics, the same landmarks that we locate on the image are "put there," or, rather, put nearby (there is always residual), by the processes we use to explain their patterns. The name of a species does not account for differences among configurations of landmarks on a page, but the biomechanics of a jaw joint can. We draw the effect of that biomechanical explanation using vectors, or describe its effect on shapes using tensors; likewise, we draw the pattern of morphometric covariances with size using vectors, and interpret it as a tensor field of allometric growth. But we cannot draw the change of a Fourier coefficient, or of an eigenshape, or of a net weight, using vectors: there is no place to put the arrowheads. For that matter, one can draw change of a single interlandmark distance, or of an angle or ratio among landmarks, in a great many metrically equivalent ways, as one or another landmark or subset of landmarks is "moving". The methods of conventional multivariate morphometrics (cf. the chapters by Marcus and Reyment, this volume), like the other integral methods, are generally not supportive of subsequent biological explanations. (See Bookstein and Reyment, 1989; for an exception, consider the "truss" method, Bookstein et al. 1985.)

In so radically empirical a context, the meaning of *homology* becomes far different from the usual meaning, and different, too, from the meaning attributed to it in the essays of the last Part of this Proceedings. "Homology" is detectable only by its opposite, for which let me coin an abominable neologism, "heterology." If it were a word, "heterology" would mean an ineluctable interference, an intellectual jamming of the tie between morphometric pattern and epigenetic description: they would be "different words," as the etymology might imply. We do not draw any conclusions about the relation of shapes between the fish jaw and the human jaw, for instance, because they are heterologous. In other words, "homology" is a residual category (which is why it is so very difficult to talk about it). The term applies to similarities of form and covariation with exogenous variables for which arguments in favor of heterology are not considered definitive. In the absence of a continuous fossil record, then, "homology" is just another name for effective morphometrics. In my 1978 book I argued that, for statistical purposes, the classic notion of homology reduced to the applicability of the deformation model; I maintain the same position now.

From this point of view, the class of phenomena considered "homologous," and thus subject to morphometrically-based evolutionary explanation (using landmarks or otherwise), will enlarge or contract over the years as phylogenies come and go in the journals. Morphometrics cannot settle these arguments, nor has it anything to say about "true phylogeny" at all.

The three principal types of points that are frequently usable as landmarks correspond to three ways in which are grounded the epigenetic explanations that motivate the measurements in the first place.

1. Discrete juxtapositions of tissue types. This category includes points in space at which three structures meet, such as the bridge of the nose in humans; branching points of tree structures, whether in two dimensions or three; centers or centroids of "sufficiently small" inclusions, preferably convex, such as the vertebrate eye or the nuclei of the brain; and intersections of extended curves with planes of symmetry. Such points can be modeled as displaced in any (geometrical) direction by relative growth immediately adjacent or at a distance (the difference is not statistically detectable—see Bookstein, 1987).

Landmarks of this first category can enter into many familiar sorts of biometrically valid functional explanations. Among the alternative accounts for deformation, for instance, is the conservation or optimization of biomechanical strength or stiffness under systematic changes of load; another is the biomathematical efficiency of sensory systems; yet another, the bioenergetics of propulsive systems. In particular, the statistics of landmarks are the same as the statistics of descriptive finite element schemes, which rely on landmarks to quantify strains in a manner comparable from form to form. Another common suggestion is the conservatism of enclosing structures under changes of their contents (the "functional matrix" hypothesis of Melvin Moss). When a lobe of a brain expands in ontogeny, for instance, there is induced a deformation of the surrounding bone. Although the brain behaves approximately like a fluid, expanding directionlessly, the enclosing bone must adjust to several considerations other than mere hydrostatic pressure, and will generally respond by a shape change that is not isotropic (cf. the example for rat calvarial growth in Bookstein, 1991). In other examples (cf. Bookstein, 1985), localized changes near the margin of an organ are propagated only a short way across it, leading to equally localized changes among landmarks exactly upon the margin. All such explanations are most persuasive within single ontogenies or functional cycles; their extension to comparisons across organisms involves inferences about fitness gradients, inferences that rely on faith to a greater or lesser extent.

2. Maxima of curvature or other local morphogenetic processes. These include tips of extrusions and valleys of invaginations. Landmarks of this second sort often serve as points of application of real biomechanical forces, pushes and pulls. Included are tips of predatory structures—claws and teeth, for instance—and tips of bony processes where muscle attachments may be centered. Landmarks of this sort may also signify a response to a bulge or other radial phenomenon at some distance from the geometrical boundary under study. As reviewed in Bookstein et al. (1985), one cannot, in principle, discriminate displacements of such landmarks lateral to the boundary direction from combinations of normal displacements outward to one side of the landmark, inward to the other; the more complex explanations of the latter category are less credible in general, but may be more valid in particular cases. Landmarks of the first category may enter into explanations of this second sort as well: for instance, tips of incisors.

3. Extremal points. These are points the definitions of which refer to information at diverse, finitely separated locations. This category, commonest in multivariate morphometrics, incorporates endpoints of diameters, centroids, intersections of interlandmark segments, points farthest from such segments, constructions involving perpendiculars or evenly spaced radial intercepts, and the like. Points taken as "farthest" from other points, or as

"endpoints" of a diameter of the form (i.e., as farthest from a point which is farthest from them), are rarely very useful as landmarks. Although the statistical methods following attach vectors to them just as if they had two or three real coordinates, their displacement is meaningful principally in a single direction representing the length ("size") of the defining segment; the other direction is badly confounded by unmeasured aspects of local shape. For an example that supports an appropriate explanation nevertheless, see Bookstein and Reyment, 1989. Similar difficulties apply in the case of points computed as intersections of a contour with perpendiculars to a chord: series of such points represent functions, not landmark configurations, and their statistics must be reinterpreted appropriately.

Landmarks upon the symmetric axis (see Straney, this volume) are hybrids of the second and third categories above. Endpoints (centers of curvature of boundary segments having locally extreme values of curvature) are at a distance from the boundary equal to that minimum radius of curvature along a vector aligned with the boundary normal. If the morphogenetic process regulating the form in this region is boundary-driven, as it may be for relatively small, heavily sculpted forms like teeth, the displaced point is less useful than the boundary point which it represents; if the process is driven from an interior center, like a bulge, the better point for explanations is the medial point. In contrast, triple points of the medial axis represent organizations of three or more boundary arcs at considerable separation. When these stand for biomechanical integration, as in the human mandible (Bookstein et al., 1985), they are likely to be more informative than the boundary points they summarize; otherwise, they are not likely to be useful.

Some other landmarks of the third category are associated with the convenience of extracting data from plane representations of solid form. The plane involved may be a homologously oriented plane of a section of the form, typically a plane of symmetry, or instead the apparent "plane" of a photograph or drawing. In space, certain structures are curves, and other structures, such as pairs of surfaces, may abut along curves. The points where curves begin and end in space are fine landmarks; but in plane descriptions, additional apparent landmarks arise where these curves appear to touch the apparent boundary of the organism. In reality such a "landmark" is merely a point at which the tangent to the surface passes through the point of view of the pictorial representation. This is obviously not a local characterization of the "landmark" unless the view is of a plane of symmetry. Even less landmark-like is the bulge-point of a surface in such a view; it corresponds to no recognizable location on the actual surface material.

When data are derived from radiographs instead of surfaces, the value of a two-dimensional representation is somewhat enhanced. Landmarks digitized in two dimensions from projection images typically have at least two-thirds of the information available in three dimensions. Of course, this compromise is not necessary. Landmarks can be digitized directly in three dimensions or reconstructed there from multiple projections (cf. Grayson et al., 1988). Any explanation of a two-dimension analysis of a three-dimensional configuration must acknowledge, of course, the absence of information normal to the plane of the representation.

Examples of all these types of landmarks occur throughout these Proceedings. The following paragraphs serve as a sketchy review of some of the usual possibilities.

Marcus. (Chapter 4)   The data of this didactic chapter are not submitted to landmark analyses; nevertheless, it is instructive to survey the measurements Marcus reports. For the fowl data, it is likely that lengths of limb bones involve landmarks of Category 2, maxima of curvature relevant to biomechanical explanations, while the skull length and width are presumably of Category 3 only. In the *Zygodontomys* form, the posterior skull width does not appear to me to involve landmarks at all; the distance LIB is based on landmarks of

Category 3; distance LOF is again not based on landmarks at all (as its endpoints are on different structures, chosen to have nearest neighbors that are farthest apart between structures); distances LD, BR, and BPB could be modified to be landmark-based by referring to "centers" of the structures they span; incisal length LIF links landmarks (root tip and apex), while molar length and width are of Category 3, extremal points; DI seems not to be landmark-based (note that there is nothing recognizable about the left endpoint of the vector in the figure); and lengths BZP and CIL appear not to be landmark-based in the lateral view. On the whole, Marcus's example supports the assertion here that traditional morphometrics is not particularly conducive to explanations of the patterns of covariance it uncovers.

Reyment. (Chapter 5)   No argument is apparent in this chapter that morphometric data might be thought to have anything to do with biology at all. The ostracode measurements, for instance, are labelled as "var 1" through "var 7," guaranteeing that biological interpretation is impossible.

Lohmann and Schweitzer. (Chapter 6)   For some landmarks appropriate to *Globorotalia*, see the chapter by Tabachnick and Bookstein, this volume. Bookstein (1986, 1991) shows that there appear to be usable landmarks in the eigenshapes of these shells; the landmark configuration correlates more strongly with latitude and with size than do the eigenshape scores used by Lohmann in his original publication. These landmarks are of the second category, "corners" of the form. For an assortment of landmarks upon an ostracode, *Veenia*, see Abe et al. (1988).

Straney. (Chapter 8)   The landmarks of this chapter are the end-points and triple-points of the medial axis of an outline viewed normal to its plane of symmetry. This particular pattern of points is not particularly well-suited for trussing, and the origin of the landmarks in various features of the symmetric axis does not seem to have been taken into account in the truss scheme. In particular, it would have been interesting to substitute the "corners" of the outline for the centers of curvature that the end-points represent. Compare Tabachnick and Bookstein (1990).

Chapman. (Chapter 12)   In the starfish example, the innermost landmark points may be of Category 1, histologically recognizable points, or of Category 2, inward-pointing "corners". The outer ring of points, "tips" furthest from a central ring, are of Category 3 only, unless it can be argued that their azimuthal coordinate (the direction in which they are displaced from the center) is not a taphonomic artifact. It is not at all clear, in any case, how the pentagonal symmetry is to be managed. The landmarks for *Diodon* and *Mola* were originally selected by me in 1977 from Thompson's figures, not from specimens. They were intended to be of Category 1 (juxtapositions of different tissue types along the midsagittal plane) except for the landmark on the body outline nearest the eye, which is of Category 3, having one coordinate only. The selection was not based on any knowledge of the processes actually delimiting ranges of fin-rays in fishes, and might have been improved had I known what I was doing. The figure of mine reproduced by Chapman invisibly incorporates some additional landmarks of Category 3, called "helping points" in Bookstein et al. (1985), for the purpose of specifying curvatures of boundary arcs between its Category 1 landmarks. The computations underlying the deformations depicted not only treat these as explicitly of Category 3 but also suppress them in the diagrams. The labelled points of the saurian skulls in Chapman's final examples all appear to be of Category 1, juxtapositions of discrete structures; but some cautions are generally in order here. The structures themselves need to be "the same" from instance to instance; the visible sutures cannot be a haphazard selection from a larger set of possibilities. This is known to be a problem for some fish skull bones, for certain bones of the primate skull, and elsewhere. The question of their homology is, as argued in the text, essentially a statistical problem. I must reiterate that no theory justifies the summary of residuals from the Procrustes fit in a net distance

measure, and that it is particularly unwise to submit such a measure to any sort of taxonomic analysis.

Tabachnick and Bookstein. (Chapter 13) There are two schemes of landmarks in this chapter, corresponding to two views of the same organisms at approximately 90°. No attempt is made here to correlate landmarks between the views. In the spiral view, the landmarks are the hybrid of categories 1 and 3 reviewed earlier. They lie on well-defined space curves (juxtaposition of successive chambers) where those curves intersect the edge of the regression of the form in the chosen view. In the apertural view, landmarks on the aperture itself appear to be of Category 1, juxtapositions of structures. The spiral point *s* is likewise of Category 1, as the earliest recognizable point of the organism. Landmarks labelled *w* are, again, hybrids of categories 1 and 3; they are loci at which well defined curves have tangents perpendicular to the chosen plane of view. In both views, that plane is defined by structures at a distance from the landmarks in question, contributing to their unreliability. None of these problems interferes with the application at which these data were aimed, a consideration of the role of holotypes in delimiting the modes of morphospace.

MacLeod and Kitchell. (Chapter 14)   There are two different types of landmarks in the data of this chapter; they are indicated using different plotting symbols in Figure 1. Points 5, 8, and 11 are juxtapositions of different structures. I consider these to almost attain Category 1, good landmarks, as the plane in which they are viewed at the periphery is characterized by reference to the structures on which these points lie. Other landmarks, as noted by the authors, are in Category 3, extremal points (widest, highest). The analysis might have proceeded more efficiently with respect to the centers of these chambers than by exploiting the greater number of these peripheral points.

Lindberg. (Chapter 15) There are only three independent landmarks in this study. Two are the ends of the long axis of the organism, presumably falling into our Category 1 (as they seem quite well-defined — head and tail — in the lateral view). The third is either of the points on the lateral margin chosen at an axial distance corresponding to the aperture. While individually these are of Category 3 only, they have two effective Cartesian coordinates: one of aperture position, the other of width. (In this dorsal view, the aperture has only one effective coordinate, being restricted to the midline.) Effects upon them can be interpreted in all directions even though they are not proper landmarks (compare the example of the cranial base in Apert syndrome, Bookstein, 1991). The resulting analysis of shape has only two geometric degrees of freedom and can be summarized in printed scatterplots without loss of information.

Reilly. (Chapter 16) Of the points Reilly gathered on these skulls, numbers 2, 3, and 6 appear to be of Category 1 (juxtapositions of structures). Numbers 4, 5 and 7 appear to be of Category 2 (sharp corners), and number 1 appears to be Category 3 (an extremal point), the point of the orbital margin closest to the midline. Again, I must disapprove of the computation of the RFTRA distance or any postprocessing of this value by UPGMA or any other algorithm from numerical taxonomy.

Ackerly. (Chapter 18) The points referred to here as "landmarks," while very interesting, do not fit into the approach described in this chapter. They are direct observations of the growth process: local extrema of a certain radial rate of accretion. The statistics of such points are not those of the landmark-based methods: for instance, distances between them are likely to be heterologous features of form, they may increase or decrease their number over ontogeny, and so on. Nevertheless, because points like these can be observed easily in accretionary processes, they may support cogent biological explanations, such as models for the direct regulation of growth rate around a closed boundary. These models will involve an algebra different from that of landmark locations.

Sanfilippo and Riedel. (Chapter 19) Landmark A should be taken not as an extremal point but instead as the intersection of the organism's axis of rotation with a certain region of surface. To the extent that the axis is well-defined, so is this landmark, which tends toward Category 1. Nevertheless, the center of this terminal sphere might have been a better choice. Landmark B is the center of a somewhat less localized structure, the distal aperture, and so falls in the second category above. Of the remaining landmarks, C, E, and G potentially fall in Category 1, anatomical landmarks, by the device of intersecting the form with a plane of symmetry. (In radiolarians, the symmetry is radial, and so any plane including the axis will do.) Landmarks D and F are in Category 3, points of greatest width, presumably not recognizable from a study of the patches of surface in which they lie. To the extent that the chambers of this form are restricted to sectors of spheres, the model of shape change as deformation cannot apply, so the analysis of these points would be equivalent to a simpler conventional multivariate morphometric analysis of the centers of the spheres, their radii, and the axial (or lateral) locations of their circles of intersection. The UPGMA trees based on RFTRA distances between forms are meaningless, as argued in the text.

The remainder of the discussion of landmarks in these Proceedings is divided into two parts. First is an exposition of the Procrustes method for standardization of size and orientation between homologous landmark configurations. The computations of this method result in one vector per landmark per specimen of "residuals" from a model declaring there to have been no shape change from the other form (if there be two) or about a mean form (if there be more than two). Such residuals may be meaningful if and only if there is sound reason for believing the model of no shape change, as in the study of deviations from bilateral symmetry (cf. Smith, Crespi, and Bookstein, submitted). These scatters are also useful in preliminary studies of the unreliability of diverse landmarks and in scanning for blunders in the digitizing. None too soon, although not until after some applications, we quit the arena of Procrustes analyses for the specifically biometrical context of extracting features of shape difference and variability. My introductory discussion of these methods argues the uselessness of the Procrustes step, explains the uniform component of shape difference or shape variability (and introduces a new, easy approximation to it in the form of a factor score), and presents a recently developed analogue to principal-components analysis for the context of landmark data: an algorithm for the largest-scale features of the deviation of the data from uniformity (itself representing "infinite scale" in this context). The systematist may think of this approach as a praxis for generating large-scale "characters" in a semi-automatic way not too crucially dependent on the actual choice of landmarks underlying the computations. These new methods are exemplified in two applications papers. I have restricted my editorial license to this single paragraph.

References

Abe, K., R. Reyment, F. L. Bookstein, A. Honigstein, and O. Hermalin. 1988. Microevolutionary changes in *Veenia fawwarensis* (Ostracoda, Crustacea) from the Cretaceous (Santonian) of Israel. Hist. Biol., 1:303-322.

Bookstein, F. L. 1978. The measurement of biological shape and shape change. Lecture Notes in Biomathematics, vol. 24, Springer-Verlag, New York.

Bookstein, F. L. 1985. A geometric foundation for the study of left ventricular motion: some tensor considerations. Pp. 65-83 in Digital Cardiac Imaging (A. J. Buda and E. J. Delp, eds.). Martinus Nijhoff, The Hague.

Bookstein, F. L. 1986. Size and shape spaces for landmark data in two dimensions. Stat. Sci., 1:181-242.

Bookstein, F. L. 1987. Describing a craniofacial anomaly: Finite elements and the biometrics of landmark location. Am. J. Phys. Anthropol., 74:495-509.

Bookstein, F. L. 1991. Morphometric tools for landmark data. Book manuscript, accepted for publication, Cambridge University Press.

Bookstein, F. L., B. Chernoff, R. Elder, J. Humphries, G. Smith, and R. Strauss. 1982. A comment on the uses of Fourier analysis in systematics. Syst. Zool., 31:85-92.

Bookstein, F. L., B. Chernoff, R. Elder, J. Humphries, G. Smith, and R. Strauss. 1985. Morphometrics in evolutionary biology. The geometry of size and shape change, with examples from fishes. The Academy of Natural Science of Philadelphia, Spec. Publ. No. 15, 277 pp.

Bookstein, F. L., and R. Reyment. 1989. Microevolution in *Brizalina* studied by canonical variate analysis and analysis of landmarks. Bull. Math. Biol., 51:657-679.

Boyd, E. M. 1980. Origins of the study of human growth. University of Oregon Health Sciences Center.

Grayson, B., C. Cutting, F. L. Bookstein, H. Kim, and J. McCarthy. 1988. The three dimensional cephalogram: Theory, technique, and clinical application. Am. J. Orthod., 94:327-337.

Huxley, J. 1932. Problems of relative growth. Methuen and Co., London.

Smith, D., B. Crespi, and F. L. Bookstein. Asymmetry and morphological abnormality in the honey bee, *Apis mellifera*: Effects of ploidy and hybridization. Evolution, submitted for publication.

Stigler, S. H. 1986. The history of statistics: the measurement of uncertainty to 1900. Harvard University Press, Cambridge, MA.

Tabachnick, R., and F. L. Bookstein. 1990. The structure of individual variation in Miocene Globorotalia, DSDP Site 593. Evolution, 44:416-434.

Thompson, D'A. W. 1917. On growth and form. Abridged and edited by J. T. Bonner. University Press, Cambridge.

Chapter 10

Rotational fit (Procrustes) Methods

F. James Rohlf

Department of Ecology and Evolution
State University of New York
Stony Brook, NY 11794-5245

Abstract

This is a survey of methods which superimpose configurations of landmarks from different specimens so that differences in landmarks can be discovered. The strengths and weaknesses of this approach (versus fitting deformational models) are discussed.

Introduction

This chapter is concerned with the problem of comparison of configurations of landmarks in two or more specimens. One approach is the use of *deformation models* (in the spirit of the transformation grids of Thompson, 1917). This approach is characterized by the depiction of the overall form of one organism as a continuous deformation of another, reference, organism. These can be implemented, for example, through the use of thin-plate splines (Bookstein, 1989). This approach is described by Bookstein in Chapter 11.

An alternative approach is that of *rotational fit (superposition) methods* in which the homologous landmarks of one organism are superimposed on those of another so as to optimize some measure of goodness of fit. This corresponds to fitting a very simple model (taking into account only translation, rigid rotation, scale, and possibly uniform shape change) for the differences in landmark configurations. Small scale (local) changes in the relative positions of the landmarks in different organisms are detected by studying the residuals from the fit. The approach is analogous to that used in regression when one fits a straight line through some points and then studies the residuals for evidence of curvilinearity. Large residuals or a nonrandom pattern of the residuals can suggest more appropriate—but more complex—models. Goodall (1990) and Goodall and Bose (1987) give extensive discussions of the statistical consequences of these models. Olshan et al. (1982) describes the application of these methods to outline data.

The primary purpose of this chapter is to review and compare various types of superimposition methods. But the discussion section (below) responds to some of the criticisms detailed by Bookstein (1991).

Methods

This section describes different approaches to determining the "optimal" superimposition of two organisms, the incorporation of affine transformations in the fitting process, and the generalizations to the fitting of more than two organisms. The account given below provides only a general summary since Rohlf and Slice (1990) describe many of the methods in detail.

Least-squares

Given two organisms represented by sets of x,y-coordinates of landmarks, Sneath (1967) investigated the problem of finding the optimal translation, rigid rotation, and scale change for the coordinates of one organism in order for it to be superimposed on another. He proposed the use of a least-squares criterion to measure the resulting lack of fit. Gower (1971) further developed Sneath's (1967) method and expressed the operations in terms of matrix operations as described below. Least-squares methods have also developed independently in the field of factor analysis. Ahmavaara (1957), Mosier (1939), and Hurley and Cattell (1962) are early references (but these are for the more general case of affine transformations, see below).

Let X_1 and X_2 be $p \times k$ matrices giving the k-dimensional coordinates of the p landmark points on organisms 1 and 2, respectively. The goal is to transform the matrix X_2 such that the sum of squared differences between the corresponding elements of it and X_1 is as small as possible. The method consists of the following steps:

1. Center both sets of coordinates at the origin by expressing their x,y-coordinates as the matrix of deviations, X'_1 and X'_2, from their respective means along each axis. This is the translation step. It can conveniently be done by premultiplying X by $(I - P)$, where I is a $p \times p$ identity matrix and P is a $p \times p$ matrix with every element equal to $1/p$.

2. Scale both organisms to the same size. It is convenient to follow Gower's (1971) suggestion of dividing each organism's x,y-coordinates by the square root of the sum of the squared distances of each landmark to the centroid. Thus $X'_i = (I - P) X_i / s_i$, where

$s_i^2 = \mathrm{tr}((I - P) X_i X_i^t (I - P))$. Bookstein (1991) calls s_i "centroid size."

3. Rotate, and possibly reflect, the second organism so that the new locations of its landmarks, $X_2^* = X'_2 H$, have minimal sum of squared deviations from the locations of the corresponding landmarks on the reference organism. The rotation matrix that accomplishes this, $H = V S U^t$, can be computed from the singular-value decomposition (Eckart and Young, 1936; Press et al., 1986) of the matrix $X_1'^t X_2 = U \Sigma V^t$. The matrix S is diagonal with elements ± 1. The signs are the same as those of the corresponding elements of Σ. For $k = 2$ dimensions, the matrix H is of the form

$$\begin{bmatrix} \cos \theta & -\sin \theta \\ \sin \theta & \cos \theta \end{bmatrix}.$$

While most published applications in morphometrics only consider coordinates in a 2-dimensional space, the above procedure can also be used for multivariate data in three or more dimensions.

Resistant Fit

Siegel and Benson (1982) made the important observation that the use of a least-squares criterion for the optimal superimposition of two specimens usually results in a general lack of fit at most points—even if the specimens being compared are identical except at a few landmarks. This makes it difficult to accept the residuals at face value and give them what may seem to be an obvious interpretation. Siegel and Benson proposed an alternative "resistant-fit" method, that is based on a nonparametric analog of least squares regression. Their method is better able to reveal differences between two organisms when the major differences are mostly in the relative positions of a few landmarks.

The computational steps are analogous to those of the least-squares approach but are somewhat more difficult to describe. Each step is designed to do the same as one component of the least-squares procedure. They are done, however, in a way that is more resistant to the effects of displacements in a minority of the data points. There is no obvious goodness-of-fit criterion that is being maximized or minimized. Since the method is discussed by Chapman in Chapter 12, it is described here just for completeness and for consistency.

The goal is to find a matrix $X_2^* = 1_{p \times 1}(\tilde{\alpha}, \tilde{\beta})$ + $\tilde{\tau} X_2 \tilde{H}$, such that corresponding entries in X_1 and X_2^* are very similar for most of the landmark points. The matrix $1_{p \times 1}$ is a $p \times 1$ matrix of all 1s. The parameters in the above equation are estimated one at a time as follows:

1. A least-squares fit as described above is performed. This will take care of the possible need for reflection and ensure that the subsequent steps begin with the two specimens approximately aligned.

2. The scale factor, $\tilde{\tau}$, is computed by first finding for each point, i, the median of the ratio of the distance from i to each of the other landmarks in the two specimens. Then $\tilde{\tau}$ is taken as the median of these medians (a *repeated median*).

3. The rotation angle, $\tilde{\theta}$, that the rotation matrix $\tilde{H}$ is based upon, is computed in a similar way. Let θ_{ij} be the angle necessary to rotate a line connecting points i and j in specimen 2 in such a way that it becomes parallel to the corresponding line in specimen 1. The median of such angles is found for each point i and then $\tilde{\theta}$ is taken as the median of these medians.

4. The vector of translation parameters, $(\tilde{\alpha}, \tilde{\beta})$, is estimated using ordinary medians of the differences in the x and y-coordinates of

specimens 1 and 2 *after* the above scaling and rotation operations have been applied.

The above procedure, which seems somewhat ad hoc, works quite well (several examples are given in Chapter 12 by Chapman). See Siegel (1982a, 1982b) for more detailed information about the algorithm and examples.

The algorithm given above is just for 2-dimensional data but an analogous algorithm can be developed for 3-dimensional data.

Affine Transformations (Oblique Procrustes)

Goodall and Green (1986) have suggested allowing the use of affine transformations when fitting one set of landmarks to another. Affine transformations can be interpreted as scale changes along orthogonal axes (for example, a magnification along one axis and a contraction along another). Bookstein (Chapter 11) refers to this as uniform shape change. The methods described in the previous section permit only changes in overall scale.

The solution to the fitting problem for the least-squares criterion was developed by Hurley and Cattell (1962) and by Ten Berge and Nevels (1977). Cattell and Khanna (1977) and Gower (1984) reviewed the general problem of transforming one set of coordinates into another when orthogonality constraints are not imposed. The least-squares method consists of the following steps:

1. The coordinates of the two specimens are first translated to place their centroids at the origin.

2. The matrix of coordinates of the landmarks of the second organism, after rotation and dilatation to fit the first organism, is given by $X_2^* = X_2' H^*$, where the affine transformation matrix is

$$H^* = (X_2'^t X_2')^{-1} X_2'^t X_1' . \qquad (2)$$

This is the equation for the least-squares estimates of the partial regression coefficients in a multivariate multiple regression analysis. There is no explicit scaling step since this is taken care of by the multiplication by the inverse matrix.

The transformation matrix, $\mathbf{H}^*$ can also be expressed in terms of a singular value decomposition,

$$\mathbf{H}^* = \mathbf{U} \Sigma \mathbf{V}^t. \tag{3}$$

Multiplication of $\mathbf{X}_1'$ by $\mathbf{H}^*$ corresponds to the rotation of the landmarks through the angle θ (determined by the matrix $\mathbf{U}$) to align the deformation axes with the coordinate axes, followed by the multiplication of the resulting landmark coordinates by the scaling factors p and q (diagonal elements of Σ) in the x and y directions, respectively, and the rotation of the configuration back through the angle ψ (determined by the matrix $\mathbf{V}$) to realign the fitted organism with the reference. The net change in the orientation of the fitted organism as $\theta - \psi$. However, θ is a function of the placement of the fitted organism on the coordinate system and ψ will be a function of the placement of the reference organism. Unless, both organisms were initially placed in some meaningful orientation neither these angles nor their difference will be biologically interesting by themselves. For a fixed reference, differences in ψ between different fitted configurations will identify differences in the direction of principal strains.

The resistant fit method can also be generalized to include an oblique rotation in order to best fit one configuration of landmarks to another. Rohlf and Slice (1990) (based on a suggestion by Goodall, 1983) estimated the elements of the transformation matrix $\mathbf{H}$ as the repeated medians of the elements of transformation matrices computed from all possible triplets of landmarks (or from a sample of triplets when there are a large number of landmarks).

The direction and magnitude of a dilatation can be shown graphically as a *strain cross*—a pair of orthogonal axes with lengths proportional to the two eigenvalues computed above and orientated so that the longer axis is parallel to the direction, ψ, of maximum stretching. If the eigenvalues are identical then the change is isotropic and the angles are not uniquely defined. See Bookstein (1982, 1984) for a more extended description. Goodall and Green (1986) give convenient equations for the relationships between the canonical parameters, the strain cross parameters, and the singular value decomposition described above.

Superposition Based on Thin-plate Splines

Since Bookstein's (1989) thin-plate spline method includes parameters for translation, oblique rotation, and scale change, it can be used as an affine rotational fit procedure by simply ignoring the fact that the higher order terms have also been fit (as suggested by Bookstein in Chapter 11).

The transformed coordinates for specimen 2 can be computed as follows. Let

$$\mathbf{Q} = \mathbf{V} \, \mathbf{K}^{-1} \, \mathbf{P} \, (\, \mathbf{P}^t \, \mathbf{K}^{-1} \, \mathbf{P})^{-1}, \tag{3}$$

where $\mathbf{Q}$ is a 2×3 matrix, $\mathbf{V}$ is a $p \times 3$ matrix with the first column containing 1s and the other two columns containing the x and y-coordinates of the landmarks in specimen 1, $\mathbf{P}$ is the corresponding matrix for specimen 2, and $\mathbf{K}$ is a $p \times p$ matrix of r^2 log r^2 values based on the distances, r, between all pairs of landmarks in specimen 2. See Bookstein (1989, 1991) or Chapter 11 for more information about this matrix. The matrix $\mathbf{Q}$ can be partitioned by columns as $(\mathbf{T} \mid \mathbf{H})$, where $\mathbf{T}$ has 1 column and corresponds to the translation effect and $\mathbf{H}$ is a 2×2 affine rotation matrix. The transformed coordinates for specimen 2 are then

$$\mathbf{X}_2^* = \mathbf{T} + \mathbf{H} \, \mathbf{X}_2^t \tag{4}$$

This approach fits the positions of the landmarks in specimen 2 to the positions in specimen 1 with the effects of the higher order terms (the principle warps of Chapter 11) held constant. In the usual Procrustes methods the effects of any higher order terms are ignored.

Other Superposition Methods

Several other superposition methods have been proposed. Bookstein and Sampson (1990) and Bookstein (1991) describe a generalized least-squares method that takes into account different variabilities at different landmarks and the tendency for neighboring landmarks to covary as a consequence of changes between them and the baseline.

Mardia and Dryden (1989) have developed a maximum-likelihood approach to the estimation of the uniform component. Both methods allow testing for the significance of the uniform component and of the residuals from the model.

More Than Two Specimens

There are several approaches that can be taken to deal with the problem of superimposing $n > 2$ specimens in a study. These include:

1   Match all specimens against a single reference specimen.

2   Match specimen 1 against 2, 2 against 3, etc. This assumes that there is a logical sequence (such as in developmental studies).

3   Match all $n(n-1)/2$ pairs of specimens against each other and then analyze a matrix containing some measure of difference (see below) between each pair of specimens.

4   Compute an "average" (or consensus) configuration and then match all n specimens against this average configuration.

The choice among these alternatives will depend upon the structure of the data. The first three require no special methods—just appropriate software. Two methods have been developed to construct average configurations of landmarks: least-squares and resistant fit. The least-squares solution was developed by several workers (Gower,

1975; Kristof and Windersky, 1971). A generalization for the resistant-fit method was reported by Rohlf and Slice (1990). Their method is patterned after the computational approach of Gower (1975). The steps are as follows:

1.   Center and scale the coordinates for each of the n specimens. Call the results X'_i.

2.   Set matrix Y to the coordinates of the first specimen and then do a least-squares fit of all other specimens, X'_i, to this initial reference specimen.

3.   Compute a new Y as an average (median in the case of resistant-fit) of the rotated specimens from step 2.

4.   Rotate the current X'_i to fit the present Y matrix using least-squares or resistant fit methods.

5.   Form a new Y from the current X'_i.

6.   If the change in the results is larger than some tolerance then go back to step 4, otherwise stop.

Note that the stopping criterion in the resistant fit method is the lack of change in the consensus configuration, not a measure of the degree of fit of the objects to the consensus. This is because no criterion of fit is being explicitly minimized. The least-squares procedure uses a change in the residual sums of squares.

For IBM PC compatible microcomputers a computer program, GRF, that performs the computations described above is available from author and Dennis Slice.

Procrustes Distances

In addition to the superposition itself, it is useful to have a quantitative measure of the degree of fit between two or more specimens. These are of interest both for testing whether the observed differences are larger than what one would expect just from digitizing error and for looking for patterns in the relative degrees of differences among three or more specimens.

Coefficients

While there are various indices that could be used to measure the degree of fit of two superimposed specimens, most studies have simply used the sum of squared differences in the coordinates of the specimens 1 and 2 (after specimen 2 has been optimally superimposed on 1), A problem with this simple approach is that the distance from specimen 1 to 2 need not be the same as that from 2 to 1. A solution is to first scale both specimens so that their centroid sizes are equal (Gower, 1971). This is an obvious choice since it is the criterion being minimized in the least-squares methods. This quantity can be expressed in several forms, for example:

$$m_{12} = \sqrt{\frac{1}{2} \text{tr}((\mathbf{X'}_1 - \mathbf{X}_2^*)(\mathbf{X'}_1 - \mathbf{X}_2^{*t}))}. \tag{4}$$

Sneath (1967) divided by p (to make it an average) rather than 2. Gower (1971) showed that $m_{12} = \sqrt{1 - \text{tr} \Sigma}$, where Σ is the matrix of eigenvalues from the singular-value decomposition described above. For $n > 2$ specimens the overall fit can be measured by summing m_{12}^2 over all comparisons with the consensus configuration.

Kendall (1984) suggested the use of the coefficient ρ where

$$\cos^2 \rho = \frac{X_{1i}X_{2i}{}^c \ X_{1i}{}^c X_{2i}}{X_{1i}X_{1i}{}^c \ X_{2i}X_{2i}{}^c}, \tag{5}$$

X_{1i} and X_{2i} are the x,y coordinates, treated as a complex number $(x+iy)$, for the ith landmark in the first and second specimens (each of which have been centered on the origin), and ^{c} stands for the complex conjugation operation. The quantity $1 - \cos \rho$ is idential to the Procrustes distance coefficient (Kendall, 1984).

Goodall (1990) proposed the multivariate Procrustes statistic:

$$S = \mathbf{E}^t (\mathbf{X}_1 - \mathbf{X}_2^*) \mathbf{T} \mathbf{H}, \tag{6}$$

where $\mathbf{X}_1$ is the matrix of coordinates for the reference specimen, $\mathbf{X}_2^*$) is the translated, rotated, and/or scaled estimate based on the second specimen, $\mathbf{T}$ is the orthonormal matrix needed to rotate the reference specimen to the second, and the $\mathbf{E}$ and $\mathbf{H}$ matrices are from a singular-value decomposition of $\mathbf{X}_2$. Basically, S is simply a $p \times k$ matrix of differences between the reference specimen and the fitted second specimen but with its rows and columns aligned with the row and column eigenvectors of the second specimen.

Shape space

Kendall (1984) uses ρ, see above, as a measure of shape distance between pairs of configurations of landmarks in a *shape space*. For the case of $p = 3$ points in $k = 2$ dimensions, Kendall (1984, 1985) shows that the shape space corresponds to the surface of sphere with radius 0.5. He also shows that, when such triplets of bivariate landmarks are sampled randomly and independently from the same normal distribution, the density of points in the shape space is uniform. For $p > 3$ the models are much more complex. Kendall (1981) and Small (1987, 1988) give very useful general surveys of shape spaces and statistics for comparing shapes, especially for the case of $p = 3$ points.

Due to the great complexity of shape spaces for $p > 3$, Goodall (1990) suggests that practical statistical models and techniques for shape comparison must be made in Euclidean space (but in a way that when projected into shape space are faithful to the statistical analyses one would like to perform in shape space).

Based on their S statistic (see above), Goodall and Bose (1987) and Goodall 1990) have developed a series of statistics based on T^2-tests, ANOVA, and multiple regression for testing for shape change in both cross-sectional and longitudinal studies, and with various assumptions about the error structure of the residuals. Bookstein and Sampson (1990) present slightly different tests for the same purposes.

Analyses of Matrices of Procrustes Distances

Given an index that measures the degree of agreement between two specimens, it is natural to consider using it to compare all possible pairs of specimens (e.g., Sneath, 1967). If there are more than just a few specimens, the distances can be placed in a matrix and subjected to cluster or ordination analyses. Gower (1975) shows that, with proper scaling, the distances define a metric and can be analyzed by metric or non-metric scaling methods. Since that time this approach has often been used (three examples are included in the present volume: Chapman, Chapter 12; Reilly, Chapter 16; and Sanfilippo and Riedel, Chapter 19).

Bookstein (in the introduction to part III C of this volume) strongly criticizes such uses of Procrustes distances. One problem is the fact that the residuals from an orthogonal Procrustes fit are a function of differences due both to uniform shape change and to local deformations. In his model (see Chapter 11) these anisotropy and bending energy are not commensurate and cannot reasonably be combined into a single coefficient. Using Procrustes distances after an affine fit would eliminate this problem. But a consequence would be that rather dissimilar looking organisms differing only by a uniform shape change would be considered identical

Another problem is the fact that the Procrustes distance is probably best thought of as a measure of residual variance. Thus it is most appropriate when the differences among organisms is similar to digitizing error (independent homogeneous random error across landmarks). But in many studies large systematic differences are found. It is also unreasonable to expect that the residuals at nearby landmarks are independent.

Spatial (in terms of the position on the organism) autocorrelations need to be taken into account.

Perhaps the largest problem is simply that combining all shape differences loses so much information. From a single distance coefficient one cannot distinguish between the case of modest differences at most landmarks versus a large distance at only a single landmark. One also loses information about the landmarks at which pairs of organisms differ. Usually such information is of more interest to a morphometrician than the overall degree of difference between a pair of organisms.

The problem is similar to that in numerical taxonomy where one combines different kinds of differences across a diversity of characters in order to construct a tree placing similar (minimum distance) organisms together or a tree such that distances between organisms along its branches are as short as possible. To compensate for this loss, the resulting trees are often imbedded in ordination spaces or "decorated" with information about implied changes in character states along the branches of the tree.

Displaying Results

Since inferences about the shape differences between the forms being compared are usually based on an examination of residuals, graphical techniques for displaying residuals are very important. Chapman (Chapter 12) shows examples of differences between two forms using vectors to represent the residuals.

When many specimens are superimposed, the scatter at each landmark can be summarized by performing principal components analyses at each landmark and then plotting equal frequency ellipses and the principal component axes. These usually show the magnitude and orientation of the residual scatter very well. Rohlf and Slice (1990) furnish several examples. Figures 8 and 9 of Lindberg (Chapter 15) show these for each of several groups of specimens that were superimposed.

Discussion

Bookstein (1991) describes a number of problems with the use of superposition methods to detect and describe shape change.

1. *The existence of uniform shape change may be difficult to detect by an examination of residuals.* This should not be a problem since it is easy to include affine transformations in the fitting process. But the presence of uniform shape differences can usually be detected quite easily from the systematic pattern of the residuals. For example, the landmarks at opposite sides of the configuration having vectors pointing towards each other and the landmarks at 90° from them in the configuration having vectors pointing away from each other. Figures 11 and 12 of Chapman (Chapter 12) are good examples as is Figure 9 (and perhaps 8) of Lindberg (Chapter 15). Bookstein's shape coordinates will show an analogous pattern of displacement in the presence of uniform shape change (displacements running parallel above and antiparallel below an object and with a magnitude proportional to their distance from the baseline. Of course one must be looking for these effects to see them.

2. *The displacement of a single landmark causes the centroid of the configuration to be displaced and thus creates apparent displacements at all landmarks.* While true for least-squares fitting methods, the problem is less severe than one might think. Bookstein (1991) shows that if one landmark is displaced by a vector v then after superimposition the residual at the landmark that was displaced will be $v(p-1)/p$ and $-v/p$ at the other landmarks. As the number, p, of landmarks increases, the relative lengths of the residuals at the displaced and undisplaced landmarks increases linearly so that it will be clear as to which landmark was actually displaced. Figure 6 of Chapman (Chapter 12) is a good

example for 10 landmarks. Resistant-fit methods are, of course, designed to avoid such problems.

3. *Procrustes superposition is an intentional misspecification of a model for shape differences.* True, but the fitting (and sometimes rejecting) of simple null models before considering more complex ones is one of the two main strategies in applied statistical studies. Unless one knows enough to put forward specific a priori hypotheses to be tested, the only alternative to these simple models is to deal from the beginning with a more complex model incorporation all possible patterns of variation (see Chapter 11). To choose Procrustes methods is reasonable if one means to state that one doesn't expect findings from those more complex analyses, and if one checks that expectation against the Procrustes residuals; it is not reasonable if it merely asserts that one does not understand the more complex methods.

4. *Superposition of centroids of configurations (necessary to achieve a least-squares fit) is biologically unreasonable.* It is true that the means of the configurations must align perfectly as do the median x,y-coordinates after rotation in the resistant fit method. But in order to superimpose, some part of the configuration must align.

5. *Procrustes superpositions are ill-suited for diagnosing most effects on biological shape as they are observed in practice.* If complex deformations are the most common type of shape change, then this is the most important criticism. When differences consist of mixtures of displacements of isolated landmarks and local deformations it becomes much more difficult to make inferences about the patterns of shape changes from a visual examination of the residuals. Figures 15 and 16 of Chapman (Chapter 12) are good examples of more complex differences in

shape where a deformational model might yield a clearer picture. But the patterns shown in his other figures are much simpler and are easy to describe using Procrustes analysis (especially if affine transformations are allowed).

Superposition methods seem most useful in two situations. The first case is when the major differences are limited to just a few landmarks after any uniform shape differences have been removed. In this case the pattern of residuals is very easy to recognize and interpret (especially when "resistant" methods are used). The second case is when the pattern of deviations across landmarks seems random (similar to what one would expect as a result of digitizing error). The level and orientation of the variability at each landmark need not be homogeneous, however. The study of asymmetry between right and left bee wings by Smith et al. (1990) is a good example. In such cases the squared Procrustes distance is better thought of as a residual variance that can be partitioned in various ways.

Acknowledgments

The helpful comments by Dennis Slice on a earlier version of the paper are gratefully acknowledged.

This work was supported, in part, by the National Science Foundation under grant BSR 8306004. This is contribution number 744 from the Graduate Studies in Ecology and Evolution, State University of New York at Stony Brook.

References

Ahmavaara, Y. 1957. On the unified factor theory of mind. Ann. Acad. Sci. Fenn., Ser. B, 106, Helsinki, 176 pp.

Bookstein, F. L. 1982. Foundations of morphometrics. Ann. Rev. Ecol. Syst., 13:451-470.

Bookstein, F. L. 1984. Tensor biometrics for changes in cranial shape. Ann. Human Biol., 11:413-437.

Bookstein, F. L. 1989. Principal warps: thin-plate splines and the decomposition of deformations. I.E.E.E Trans. Pattern Anal. Mach. Intelligence, 11:567-585.

Bookstein, F. L. 1991. Morphometric tools for landmark data. Cambridge: New York, *in press*.

Bookstein, F. L. and P. D. Sampson. 1990. Statistical models for geometric components of shape change. Communications in Statistics: theory and methods, *in press*.

Cattell, R. B. and D. K. Khanna. 1977. Principles and procedures for unique rotation in factor analysis. Chapter 9 in Enslein, K. and A. Ralston (eds.) Mathematical methods for digital computers, vol. III. Wiley-Interscience:New York.

Eckart, C. and G. Young. 1936. The approximation of one matrix by another of lower rank. Psychometrika, 1:211-318.

Goodall, C. 1990. Procrustes methods in the statistical analysis of shape. J. Royal Statistical Soc., Ser. B, *in press*.

Goodall, C. R. and A. Bose. 1987. Procrustes techniques for the analysis of shape and shape change. Pp. 86-92 in R. Heiberger (ed.) Computer science and statistics: proceedings of the 19th symposium on the interface. Alexandria, Virginia: Amer. Stat. Assn.

Goodall, C. R. and P. B. Green. 1986. Quantitative analysis of surface growth. Bot. Gaz., 147:1-15.

Gower, J. C. 1971. Statistical methods of comparing different multivariate analyses of the same data. Pp. 138-149 in F. R. Hodson, D. G. Kendall, and P. Tautu (eds.) Mathematics in the Archaeological and Historical Sciences. Edinburgh Univ. Press: Edinburgh

Gower, J. C. 1975. Generalized procrustes analysis. Psychometrika, 40:33-51.

Gower, J. 1984. Multivariate analysis: ordination, multidimensional scaling and allied topics. Pp. 727-781 in E. Lloyd (ed.). Handbook of applicable mathematics, Volume VI: Statistics. Wiley:New York.

Hurley, J. R. and R. B. Cattell. 1962. The Procrustes program: producing direct rotation to test a hypothesised factor structure. Computers in Behavioral Sci., 7:258-262.

Kendall, D. G. 1981. The statistics of shape. Pp. 75-80 in V. Barnett (ed.) Interpreting multivariate data. Wiley: New York, 374 pp.

Kendall, D. G. 1984. Shape manifolds, Procrustean metrics, and complex projective spaces. Bull. London Math. Soc., 16:81-121.

Kendall, D. G. 1985. Exact distributions for shapes of random triangles in convex sets. Adv. Appl. Prob., 17:308-329.

Kristof, W. and B. Wingersky. 1971. Generalization of the orthogonal Procrustes rotation procedure to more than two matrices. Proc. 79th Ann. Convention, APA, pp. 89-90.

Mardia, K. V. and I. L. Dryden. 1989. The statistical analysis of shape data. Biometrika, 76:271-281.

Mosier, C. I. 1939. Determining a simple structure when loading for certain tests are known. Psychometrika, 4:149-162.

Olshan, A. F., A. F. Siegel, and D. R. Swindler. 1982. Robust and least-squares orthogonal mapping: methods for the study of cephalofacial form and growth. Amer. Jour. Physical Anthropology, 59:131-137.

Press, W. H., B. P. Flannery, S. A. Teukolsky, and W. T. Vetterling. 1986. Numerical recipes. Cambridge Univ. Press: New York, 818 pp.

Rohlf, F. J. and D. Slice. 1990. Extensions of the Procrustes method for the optimal superimposition of landmarks. Systematic Zool., 39:40-59.

Siegel, A. F. 1982a. Robust regression using repeated medians. Biometrika, 69:242-244.

Siegel, A. F. 1982b. Geometric data analysis: an interactive graphics program for shape comparisons. Pp, 103-122 in Launer, R. L. and Siegel, A. F. Modern Data Analysis. Academic Press:New York.

Siegel, A. F. and R. H. Benson. 1982. A robust comparison of biological shapes. Biometrics, 38:341-350.

Small, C. G. 1987. A survey of shape statistics. Pp. 1-9 in Proc. of the section on statistical graphics, 1987 Annual Meeting of the Amer. Statistical Assn.

Small, C. G. 1988. Techniques of shape analysis on sets of points. Internat. Statistical Rev., 56:243-257.

Smith, D., B. Crespi, and F. L. Bookstein. 1990. Asymmetry and morphological abnormality in the honey bee, *Apis mellifera*: effects of ploidy and hybridization. Evolution, *submitted*.

Sneath, P. H. A. 1967. Trend-surface analysis of transformation grids. J. of Zoology, 151:65-122.

Ten Berge, J. M. F. and K. Nevels. 1977. A general solution to Mosier's oblique Procrustes problem. Psychometrika, 42:593-600.

Thompson, D. W. 1917. On growth and form. Cambridge:London, 793 pp.

Chapter 11

Higher-Order Features of Shape Change for Landmark Data

Fred L. Bookstein

Center for Human Growth and Development
University of Michigan
Ann Arbor, Michigan 48109

Abstract

This chapter reviews a decomposition of shape change or shape variation into geometrical and statistical components which, together, often support useful interpretations. The techniques apply to data in the form of homologous landmark locations. All the shape features discussed here are independent of the decision as to whether to standardize position, orientation, and scale in any way and, if so, in what way that standardization is carried out. In particular, the features that describe the shape relations of a pair of landmark configurations are the same as those that describe the residuals from the fit of either to the other by an isotropic Procrustes transformation, as discussed elsewhere in this volume. The nonuniform features are the same whether or not a uniform part is estimated and corrected first. For drawing shape changes, I will use my *two-point registration*. Shapes of triangles are expressed as shape coordinate pairs of one landmark in a coordinate system defined by fixing two others along a baseline at Cartesian locations (0,0) and (1,0). More general landmark reconfigurations are treated as vectors of displacement of all but two of the landmarks after reduction to sets of shape coordinate pairs one by one. Multivariate analyses of such representations, properly
interpreted, are very nearly independent of baseline.

The present chapter is, intended mainly to sketch the assumptions, computations, and dogmata underlying the two examples to follow in this Section. It is no substitute for a detailed study of the full statistical feature space for landmark-based shape. These matters were the subject of several lectures at the Michigan workshop and also of Chapter 7 in the preliminary draft of my Morphometric Tools for Landmark Data (Bookstein, 1991). Portions of the same material have appeared in Bookstein (1985, 1986, 1987, 1989a, 1989b) as well.

A Summary of the Basic Ideas

- Any change of shape for a configuration of landmarks has a *uniform part* and a *non-uniform part*, and any observed sample of landmark configurations incorporates variation of both the uniform part and the non-uniform part about a mean configuration.

- The uniform and the non-uniform parts of any change or scatter represent complementary subspaces of the full vector space of shape variation.

- The uniform part may be imagined as the change or variation of a "typical" triangle, rigorously interpolated or extrapolated so as to apply to every landmark triangle in the same way.

- A purely uniform transformation leaves parallel lines parallel. In the two-point registration, all landmarks are displaced by multiples of a single vector; each multiplier is proportional to the landmark's distance from the baseline.

- There are various ways of *estimating* the uniform part of a shape change that is not exactly uniform. The most convenient is as a *factor score*, an average of all landmark shifts weighted by their distance from the baseline.

- To any sample of shapes corresponds a two-dimensional distribution of this uniform factor score. It may have up to two *uniform statistical components*, which are eigenvectors of uniform shape variation with respect to the anisotropy metric (log ratio of principal strains).

- To any transformation of landmarks there is a *bending energy*, which may be thought of as the net energy required to bend an infinite, infinitely thin metal plate over one set of landmarks so that its height over each landmark is equal first to the *x*-, then the *y*-coordinate of the corresponding landmark in another set. Uniform transformations involve tilting and re-rolling this plate, not bending it, and so require zero bending energy.

- Any single non-uniform transformation may be expressed as a finite sum of *principal warps*, eigenfunctions of the bending energy corresponding to Procrustes-orthogonal displacements of the "metal plate" at the landmarks. These warps emerge in descending order of an eigenvalue, bending energy per unit summed squared Procrustes displacement, which can be identified with the inverse *geometrical scale* or *information localizability*.

- Because the uniform part of a mixed transformation can be defined in many different reasonable ways, the non-uniform part is like-wise not unique. However, all of its variants have the same bending energy.

- A sample of shape changes, or their residuals after subtraction of an estimated uniform part, may be usefully decomposed into a series of *relative warps*, which are eigenvectors of the variance-covariance matrix of landmark coordinates with respect to bending energy. These are analogous to ordinary principal components, in that they emerge in order of their power to account for transformations of landmark locations distributed as widely as possible over the form. They are calibrated by eigenvalues which represent variance per unit bending energy.

- The uniform transformations can be thought of as the zeroth of these relative warps, with eigenvalue infinite and zero bending energy per unit variance.

- The metrics for the uniform and non-uniform parts of shape change or variation are wholly incommensurate: for the uniform part, an anisotropy; for the non-uniform part, an energy. Describing the "magnitude" of a change of landmark configuration requires at least three "distances": anisotropy, bending energy, and also a size difference score. There is no good way to combine these into one single metric; the hope of a meaningful unitary matrix of distances between shapes is vain.

- Instead, the combination of two uniform components with some number of relative warps provides a useful feature space in which to search for evidence of diverse morphogenetic processes at multiple geometric scales.

The Simplest Example: Transformations of a Square

I shall assume that the reader agrees with my much-published view, beginning with Bookstein (1984), about how best to carry out the multivariate statistical analysis of a triangle of landmarks. The analysis of shape variability for a triangle reduces to a scatter of single pairs of shape coordinates. Distance in this shape coordinate plane is proportional to *log-anisotropy*, log of the ratio of diameters of the ellipse into which any circle is taken by the simple shear (cf. Figure 3) consistent with the landmark locations. All the conventional sorts of biometric hypotheses dealing with covariates of shape or shape change, such as size or group, may be rigorously tested in this space, and any effects found as displacements or trends in the shape coordinate plane may be interpreted immediately as specific scalar shape variables: ratios of homologously measured distances aligned with the principal strains of the effect construed as a deformation.

The question naturally arises as to whether the general change of shape for more than three landmarks can be described in equally simple language. The problem is clear in the context of even the very simplest configuration of four landmarks, the square in Figure 1. Consider two transformations of the square, one to a parallelogram and one to a kite-shaped object. Let us inspect the effects of these changes upon the two triangles into which we can divide the square. Figure 1a shows this analysis for the transformation to a parallelogram. The analyses of shape change for the two triangles agree regarding the principal directions and principal strains of the shape change. But in the change of square to kite (Figure 1b), the analyses of the shape change for the two triangles are somewhat in disagreement. The direction of greater strain for each is the direction of lesser

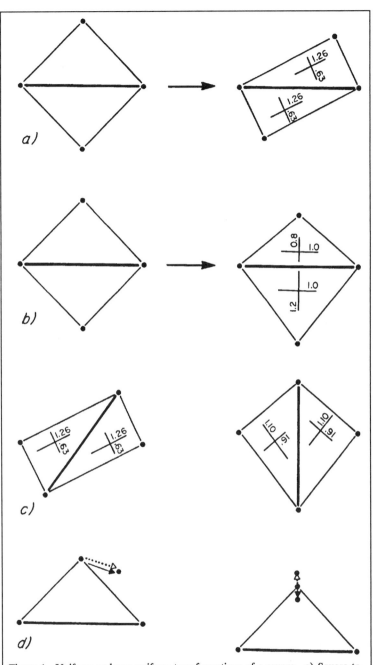

Figure 1. Uniform and non-uniform transformations of a square. a) Square to parallelogram: the strain tensors agree between the triangles. b) Square to kite: the strain tensors reverse greater and lesser principal strains. c) The same for a different starting triangulation. d) When the two strains are taken to refer to the identical starting triangle, the implied landmark displacements are equal for the change to the parallelogram, but opposite for the change to the kite.

strain for the other. When we switch to the other triangulation (Figure 1c), the agreement or disagreement continues.

It seems that the transformation of square to parallelogram can be described by a single triangle in some sense in which that of square to kite

cannot. We can attempt to quantify this, and, in fact, we arrive at the actual uniform and non-uniform spaces of shape change–for squares only!– if in each of Figures 1a and b we rotate one of the triangles by 180° around the baseline so that the starting positions of the third landmark are now superimposed in the same location. Then in the transformation of a square to a parallelogram (Figure 1d) the resulting displacements of the "same" point are *identical*, while in the transformation of square to kite, they are *opposite*. This agreement or disagreement of features, which is the same regardless of original triangulation (cf. Figure 1e), suggests a decomposition of any observed change of shape of a square of landmarks into two parts (Figure 1f). One part, representing the difference of the displacements of the two movable landmarks (which becomes their *average* after one is "flipped"), is the square-to-parallelogram part of the transformation, the same no matter how one triangulates the form. The other part, incorporating the average of the moving landmarks *before* flipping (e.g., their difference after flipping), seems to represent the pure contradiction between the alternate triangulations, likewise in a way that here seems independent of the triangulation.

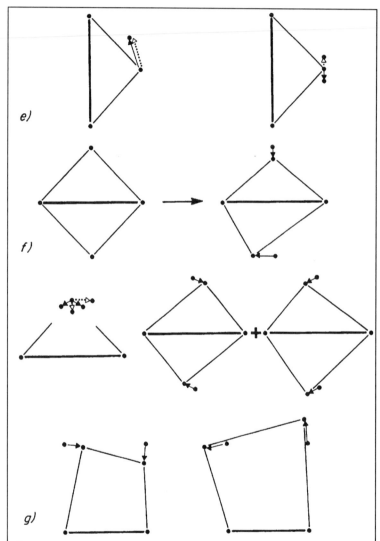

Figure 1 (continued). Uniform and non-uniform transformations of a square. e) The situation is the same for a different starting triangulation. f) For a square (but only for a square!), the uniform part of any shape change may be identified with the difference of the displacements of either internal triangulation, and the non-uniform part with their sum. g) The square to hite transformation to baselines along edges of the square.

A suggestive visualization of this distinction treats differences in the landmark locations between the square and the outcome form as if displacements were perpendicular to the picture rather than within the plane of the paper. Then the uniform transformation (Figure 2a) appears to involve a distorted square which is *tipped* with respect to the original picture–it looks like a projected image of the original form–whereas the non-uniform transformation (Figure 2b) appears to bend the square. In neither case can the transformation be "localized" to any single landmark or subset of landmarks. Both appear to be distributed evenly over the whole set of four. We will return to this rather potent metaphor shortly.

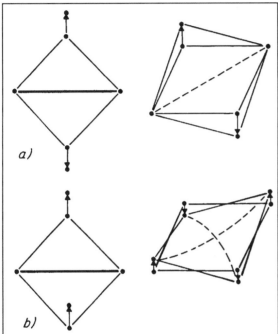

a)

b)

Figure 2. Metaphor for landmark displacements perpendicular to the plane of the starting square. a) Uniform transformation, square is tipped and "foreshortened." b) Non-uniform transformation, square is irrevocably bent.

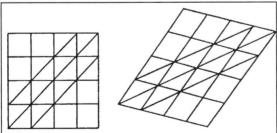

Figure 3. Uniform transformations leave parallel lines parallel.

The apparent agreement of these analyses between the triangulations corresponds to invariance of one's biometric analysis under such changes (Bookstein, 1987). If the square and the kite represented mean forms in two populations, conventional statistical tests of the difference between the groups would yield exactly the same significance levels whether based on optimal distance-ratios from the first triangulation or on those from the second. Different variables would be involved, of course, in discriminating the kite from those involved in discriminating the parallelogram.

The General Picture of a Uniform Transformation

We need to generalize the preceding discussion so as to refer to any number of landmarks located anywhere, not just four changing from the form of a square. A very useful model for this generalization is that shown in Figure 3: the class of transformations for which the shape change tensors computed from all triangles of landmarks are the same in their lengths and orientation upon "tissue". It can be shown that such transformations are simply the *uniform* or *affine transformation* that keep parallel lines parallel and preserve ratios of lengths measured in the same direction. These transformations take circles to ellipses whose axes are the principal strains of the transformation. For a review, see Bookstein et al., 1985.

In an arbitrary superposition, such as one resulting from a best-fitting isotropic Procrustes transformation, it is not at all clear when a transformation is uniform in this sense (Figure 4a). Things are much clearer when superposition is by means of shape coordinates to any baseline pair of landmarks (Figure 4b). In the shape coordinate plane, a uniform transformation displaces all landmarks in the same direction, by multiples of a single vector. Landmarks at the same height are displaced by the same vector regardless of their location along the baseline. The distance by which each landmark is displaced is proportional to the distance that landmark began above the baseline.

(Landmarks below the baseline are displaced by multiples of the *opposite* of that vector, corresponding to their negative distance "above.") All this corresponds perfectly to what we already noted about the square in Figure 1.

When a transformation is in fact uniform, there is no disagreement about what uniform transformation it is. When a transformation differs from the uniform, either by mere digitizing noise or by additional biologically real features, it is no longer obvious what we should consider to be the uniform "part" of the shape change. Several of us in morphometrics are working on this problem from different points of view.

The best solution, in my opinion, is that which I have proposed in the course of my papers on the thin-plate spline (e.g., Bookstein, 1989a). Any reconfiguration of landmarks in two dimensions can be uniquely expressed as the sum of a uniform transformation together with vector multiples of the function $r_i^2 \log r_i^2$, where r_i is the ordinary distance to the ith landmark of one form. The origin of these strange functions will be revealed presently. Because this decomposition is exact, it requires no decision about the direction in shape space along which to measure the "error of fit" to the uniform transformation which is to be minimized. The transformation is linear in the coordinates of any set of landmarks—indeed, it is expressed by the last three rows of the matrix L^{-1} in the next section. The resulting uniform part of each transformation may be expressed as a shape scatter in the usual way, by reference to its

effect on a standard triangle, and biometrics proceeds from there.

In this approach, the non-uniform part of the interpolation function is confounded with the uniform part. *If it were known* a priori that the non-

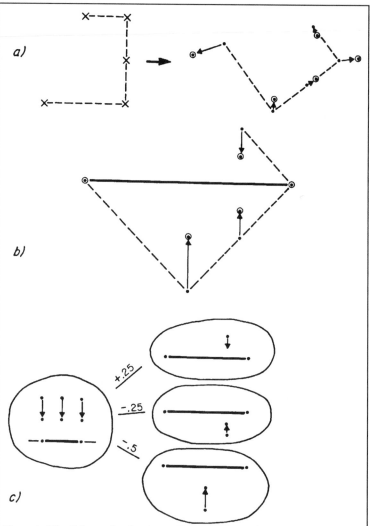

Figure 4. Visualizing and estimating uniform transformations as factors. a) It is very difficult to detect evidence of uniform transformations in the residuals left by fitted models of scale change. b) In the shape coordinate plane, a uniform transformation displaces all landmarks in a single direction by amounts proportional to (signed) distance from the baseline. c) One may thereby estimate the uniform part of any observed shape change using the usual formula for estimating factor scores (see text).

uniform part is meaningless, pure noise, then one might wish to estimate a uniform part which is the "closest" to the actual transformation for some reasonable measure of nearness, averaging away the apparent non-uniformity as best one can instead of compensating for it. Rohlf (Chapter 10) and Goodall (1989) both present least-squares methods for this computation under slightly different assumptions about the structure of error. Bookstein and Sampson (1987) present another, lifting the assumption that error is independent at the several landmarks, and Mardia (1989) suggests a maximum-likelihood technique.

The papers in this section use a much simpler method than any of these others. The scheme of Figure 4b, in which each vector of displacement is proportional to one single vector according to a known multiplier, is exactly analogous to the usual scheme of estimation of factor scores (Figure 4c). In the usual factor model, if a factor score is postulated to predict each of a family of indicators with known regression slopes and independent regression errors, then the best estimate of the factor score itself, given only its sequelae, is proportional to the average of the observed outcomes, each weighted by the inverse of its own error of prediction. (No factor-analyst would let me fail to remind the reader that this is true only if those errors are, in fact, uncorrelated–only if there is no *secondary* factor structure.)

We can apply this model to our landmark data by treating the uniform part as a vector-valued factor score, the *uniform factor*, which predicts each observed displacement of shape coordinates via a regression coefficient that, for each landmark, equals its distance from the baseline. The "error of prediction" of each displacement by this common score is technically unobservable, but can be guessed as approximately the same for each landmark (perhaps on the assumption that they are expressing the same digitizing noise); then the relative precision of each landmark-specific "prediction" is directly proportional to its distance from the baseline. The factor score is the sum of all these

inverse predictions, divided by a suitable constant. (The exactly analogous formula for factors not referring to landmarks can be found on page 89 of Bookstein et al., 1985.)

There results the following formula for the estimated uniform transformation underlying any observed change of shape coordinates (x_i,y_i) -> $(x_i,y_i) + (\Delta x_i,\Delta y_i)$:

$$(\bar{u}_x,\bar{u}_y) = \Sigma_i \bar{y_i}\,(\Delta x_i,\Delta y_i) \,/\, \Sigma_i \bar{y_i}^2 \ \ .$$

Here the sums are taken over all the landmarks; each weight is the mean vertical shape coordinate (relative distance from baseline) for its landmark. If the context is that of the description of shape variation rather than shape change, then the Δ's here should be the deviations of the observed shape coordinates from their sample means. If the transformation is indeed uniform– that is, if each $(\Delta x_i,\Delta y_i)$ equals $y_i\alpha$ for some common vector α–then these formulas (without the bars over the y's) recover the vector α exactly whatever the distribution of ordinates y_i. If the transformation is not exactly uniform, then this estimate will disagree slightly with estimates to other baselines and with estimates provided by Procrustes algorithms, my own projection algorithm, or the exact spline fit. In my view the simplicity of the formula above more than compensates for its not being an embodiment of any least-squares optimum.

The uniform component arrived at by this or any other estimation rule may be considered as if it were indeed the observed effect of the shape change or variation in question on one big, fuzzy triangle. For a single shape change, it may be re-expressed in terms of its principal strains by the construction in Bookstein et al., 1985. For a sample of forms, there results a sample of these estimated uniform "factors," which may be scattered for the cases of a sample, scanned for outliers, regressed into exogenous variables to find shape trends, or referred to a conventional component analysis of their own–which turns out to be with respect to anisotropy, a sensible choice (see below). There result up to two *uniform statistical components* of this uniform factor. Each may be inspected, just as any other principal component may be inspected, to see if it suggests some underlying biological process.

Pictures of Bending Energy

One might imagine "the" non-uniform part of a transformation to be the residual reconfiguration of landmarks left after one has undone the effects of a uniform part fitted to the data by my factor approximation, Rohlf's least-squares algorithm, or any other. It is not a trivial task to unearth a descriptor of such a residual that is independent of the algorithm used for producing the uniform part whose residual it is. Instead, one needs a method which extracts non-uniform parts of observed shape changes or variations *directly*, without requiring the (arbitrary) projection onto a fitted uniform part as an intermediate step. Such a procedure is available, I noted above, in the course of the decomposition of any observed change of landmark configuration as a *thin-plate spline*.

The general theory of thin-plate splines is somewhat technical (cf. Bookstein, 1989a, 1991), and it is inappropriate to review it here in any detail. Briefly, let $P_1 = (x_1 y_1)$, $P_2 = (x_2 y_2)$, ...,$P_n = (x_n y_n)$ be n points in the ordinary Euclidean plane according to any convenient Cartesian coordinate system. Write $r_{ij} = |P_i - P_j|$ for the distance between points i and j, and $U(r)$ for the function $r^2 \log r^2$. Define matrices

$$K = \begin{bmatrix} 0 & U(r_{12}) & \dots & U(r_{1n}) \\ U(r_{21}) & 0 & \dots & U(r_{2n}) \\ \dots & \dots & 0 & \dots \\ U(r_{n1}) & U(r_{n1}) & \dots & 0 \end{bmatrix}, n \times n;$$

$$P = \begin{bmatrix} 1 & x_1 & y_1 \\ 1 & x_2 & y_2 \\ .. & .. & .. \\ 1 & x_n & y_n \end{bmatrix}, 3 \times n;$$

and

$$L = \begin{bmatrix} K & | & P \\ P^T & | & 0 \end{bmatrix}, (n+3) \times (n+3),$$

where ^{T} is the matrix transpose operator and 0 is a 3×3 matrix of 0's.

Let $V^T = (v_1, ..., v_n)$ be any n-vector, and write $Y^T = (V^T \mid 0\ 0\ 0)$. Define the vector $W^T = (w_1, ..., w_n)$ and the coefficients a_1, a_x, a_y by the equation

$$Y^T L^{-1} = (W^T \mid a_1, a_x, a_y) .$$

Use the elements of $Y^T L^{-1}$ to define a function $f(x, y)$ everywhere in the plane:

$$f(x,y) = a_1 + a_x x + a_y y + \sum_{i=1}^{n} w_i U(|P_i - (x,y)|) .$$

Then the following three propositions hold:
1. $f(x_i, y_i) = v_i$, for all i.
2. The function f minimizes the nonnegative quantity

$$I_f = \oint \oint_{\Re^2} \left[\left[\frac{\partial^2 f}{\partial x^2} \right] + 2 \left[\frac{\partial^2 f}{\partial x\, \partial y} \right]^2 + \left[\frac{\partial^2 f}{\partial y^2} \right]^2 \right]$$

over the class of such interpolants. This is a constant multiple of the physical bending energy of an infinite, uniform, thin metal plate originally flat and level and now constrained to pass through all the points (x_i, y_i, v_i). The function f in fact gives the actual form of that plate, as it takes a position which minimizes precisely this energy.

3. The value of I_f is proportional to

$$W^T K W = V^T (L_n^{-1} K L_n^{-1}) V ,$$

where L_n^{-1} is the upper left $n \times n$ subblock of L^{-1}. This form is zero only when all the components of W are zero: in this case, the computed interpolant is $f(x,y) = a_1 + a_x x + a_y y$, a linear function.

In the present application we take V to be the $2 \times n$ matrix

$$\mathbf{V} = \begin{bmatrix} x'_1 & y'_1 \\ x'_2 & y'_2 \\ \dots & \dots \\ x'_n & y'_n \end{bmatrix}$$

where each (x'_i, y'_i) is a point "homologous to" (x_i, y_i) in another copy of $\Re^2$. The resulting function f now maps each point to its homologue (x'_i, y'_i) and is least bent (according to the measure $\mathbf{I}_f$, integral quadratic variation over all $\Re^2$, computed separately for real and imaginary parts of f and summed) over all such functions. In effect, our metric is the bending energy of a four-dimensional thin plate: two dimensions of plate, displaced in two "other" perpendicular directions.

It is instructive to view the form of the plate for the transformation of square into kite. Figure 5a shows the interpolation, Figure 5b the plate. It is the energy of this bending that is proportional to the quadratic form in point (3) above. In this figure one can finally see how it is that the non-uniform transformation is indeed localized: while the tilt of Figure 2a extrapolates out to infinity unchanged, the spline of Figure 5b goes asymptotically flat a short distance from the landmarks involved.

This particular transformation involves changes in one shape coordinate only (the ordinate). In general, bending energy derives from both shape coordinates. Figure 6a shows five landmarks, Figure 6b the two-dimensional spline interpolant f just introduced. Figures 6c and 6d, separately, show the x- and y-components of this interpolant after the uniform part is graphically suppressed. The energies of these two are 0.0205 and 0.0225 (in arbitrary units). This example is discussed at much greater length in Bookstein, 1989a.

The bending energy may be imagined as a metric (a distance measure) on shape space (cf. Bookstein, 1991, Appendix 2). Landmark configurations that differ by a uniform transformation are at distance zero from one another in this metric: bending energy zeroes out all transformations that

exactly fit any combination of those simple models. We already know how to describe those changes of configurations, however: by a combination of changes of position, orientation, and scale, together with a uniform "distance" measured as the logarithm of the ratio of the principal strains. Just as change of position is incommensurate with change of orientation (centimeters and degrees don't mix), and just as both are incommensurate with log anisotropy, so all of these natural metrics for uniform changes are incommensurate with the bending energy introduced here.

In most multivariate statistical applications, the appropriate picture of a distance measure is a generalized ellipsoid. "Statistical distance" in all directions is variously proportional to the Euclidean distances of the "natural" descriptor space. (Such metrics arise, for instance, when one refers to the difference between population mean forms in units of within-sample covariance, the underpinnings of Hotelling's T^2.) By contrast, the bending energy is a *deficient metric*. Its picture is a *cylinder* in landmark configuration space, not an ellipsoid. The *generators* of the cylinder–the "straight lines" on it–are in fact the sets of all transformations derived from a given non-uniform warping by application of any additional uniform transformation. All such additional transforms are at the same "distance" (bending energy) from the starting form.

The axes of this cylinder are the *principal warps* of the landmark configuration (Bookstein, 1990). These are eigenvectors of the bending energy with respect to summed squared landmark displacement in their original coordinate system. Each principal warp specifies the displacement of each landmark by a particular distance (positive or negative, summing to zero) in an unspecified direction that is the same for all landmarks (Bookstein, 1989c). The cylinder pairs these axes into circles of equivalent bending in any direction of the plane. The first principal warp represents the pattern of displacements having largest bending energy per unit root-mean-square landmark displacement; usually, it is the relative displacement of the two landmarks closest together with only small contributions from the others, which are effectively "at infinity." At the other extreme, the last principal warp is the largest-scale nonlinearity that can be considered to leave landmarks at infinity fixed: it usually looks just like the square-to-kite transformation in Figure 5a.

Relative Warps as Eigenfunctions with Respect to Bending Energy

In ordinary principal components analysis, the principal components are the directions in feature space which have the successively greatest variances as ratios of their "lengths" in a geometry for which direction cosines in all directions are weighted equally. This is almost always unreasonable in practice, as everybody fails to consider whether lengths in different directions of feature space really ought to be considered as calibrated by Euclidean distance regardless of direction. Nevertheless, in the analysis of the uniform factor I introduced above, the uniform components are taken as principal components in just that way: directions in the space of the uniform factor which have the greatest and least sample variance per unit length. This is justified here because in that space we know the meaning of length: it is exactly proportional to anisotropy, our preferred measure of the extent of a shape change, and as such really is independent of direction.

In the method of relative warps, the eigenanalysis of the observed variance-covariance matrix of shape coordinates is taken with respect to the bending-energy matrix

$L_n^{-1}KL_n^{-1}$ described above. We extract components of purely non-uniform shape variation, the relative warps, as directions in shape space of successively greatest variances in relation to bending energy. The computation of these directions by conventional matrix operations is a bit delicate (see

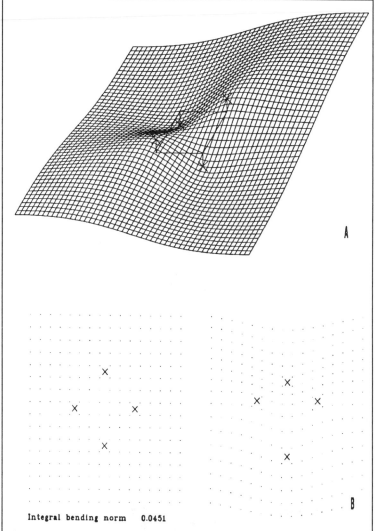

Integral bending norm   0.0451

Figure 5. The thin-plate spline for the square-to-kite transformation. a) The interpolant, treated as a vertical displacement over corners of a square: up at the ends of one diagonal, down at the ends of the other. Note that the bending is localized in the region of the four landmarks, though it is evenly distributed over the four. b) The equivalent three-dimensional "plate."

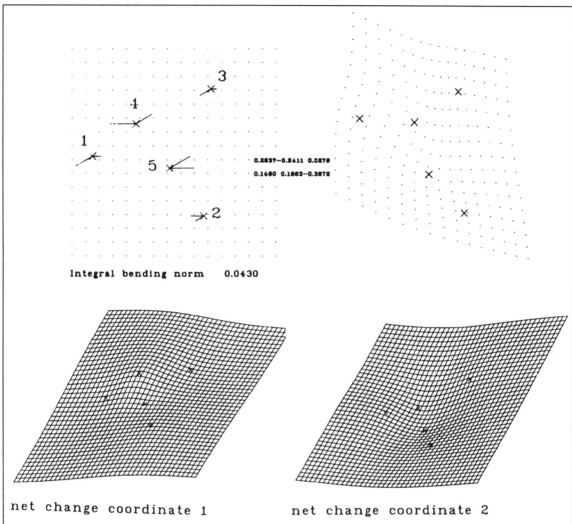

Integral bending norm   0.0430

net change coordinate 1   net change coordinate 2

Figure 6. A more general example of the thin-plate spline. a) Two sets of landmarks. b) The thin-plate interpolant. c) The equivalent three-dimensional plate for the *x*-component, with the uniform part visually suppressed. The bending is quite localized, representing the relative displacement of the central two landmarks from one another with respect to the corners of the triangle as "infinity." d) The same for the *y*-component. The bending here is larger-scale, describing the joint displacement of the two central landmarks from the outer corners.

Bookstein, 1991)–and will not be reviewed here. A program which computes them has been included on the set of software accompanying this publication.

In the relative warps, it was my intention to provide the strongest possible analogue to the notion of principal component for the very highly structured data of landmark locations. Intuitively, a principal component is attempting to find the dimensions of variability that the whole list of variables hold "most in common." The ordinary first principal component conflates covariances among all pairs of the original measures into that axis of the covariance ellipsoid which has the greatest length. In the translation into landmark-based morphometrics, "length" remains sample variance of coordinates, but we must be careful about what is meant by "most in common." In the metaphor of bending energy, reducing the scale of a set of landmarks by half multiplies the bending energy of a given set of displacements fourfold. Recall that the dimensions of shape space may be ranked in terms of intrinsic bending energy, the "size" of the square-to-kite transformation that best suits them. For a single set of landmark displacements, large regions have the lowest bending energy, and small ones the greatest.

The analysis by relative warps, which I have proposed, weights sampling variance inversely to this apparent "scale" of geometric nonlinearity before searching for the series of successive extrema which are the eigenvectors, the relative warps. A two-up-two-down transformation (Figure 5a) on the farthest corners of the form, for instance, will need only one-fourth the sampling variance of the same features restricted to a quadrant to emerge as the first relative warp. In this way, the relative warps pull out geometrically orthogonal dimensions of nonlinear shape variability in order of variance scaled inversely by feature size. This ensures at least one aspect in which the relative warps do not generalize the conventional components. There, analysis of the covariance matrix is independent of the sample means. But the matrix for bending energy is a function of the mean landmark configuration—the same landmark variance-covariance matrix leads to different relative warps as the variance is taken about different mean forms.

The uniform transformations have bending energy zero, and thus have "infinite" shape variance per unit bending energy regardless of direction. In that sense, they may be considered as the "zeroth" eigenvectors of this system. (They cannot be computed in that fashion, however, as we could not identify which direction held the largest variance of the uniform factor without reference to anisotropy, which is a different metric entirely.)

Extended examples of the complete analysis of a system of landmarks by its decomposition into the uniform statistical components, the relative warps, and their correlations are the concern of the paper by Tabachnick and myself later in this volume, and will not be repeated here. It is useful, however, to attempt a certain clarification of nomenclature:

The higher-order features of shape to which the title of this chapter refers apply to samples of shapes and, equally, to residuals from their analysis by any combination of partial fits to changes of size, orientation, and position. Any single shape change, and likewise any sample of deviations of shapes from a sample mean, may be usefully decomposed into two parts: *uniform* and *non-uniform*. The uniform part for each change or deviation of shape is representable as a single vector of length 2, the *uniform factor*. There are several ways of estimating it which differ only in statistical details. A sample of these factors may be usefully expanded in terms of its (first and second) *uniform statistical components*, each of which is an eigenvector of the observed sampling variability of the uniform factor with respect to its *anisotropy*. Complementary to the subspace of uniform transformations, those described exhaustively by their uniform parts in this sense, is the subspace of *non-uniform transformations*, those having no uniform part at all, according to whatever estimation routine one prefers. Sample variability of this non-uniform part is usefully considered in terms of the (first, second, etc.) *relative warps* of the sample, which are the first few eigenvectors of the observed sample covariance matrix with respect to the deficient metric that is *bending energy*. The uniform factor itself may be thought of as the first pair of eigenvectors of this computation, having "infinite" relative eigenvalue, owing to the fact that uniform transformations have no bending energy. These two metrics, bending and anisotropy, are *incommensurate*; it requires values for both of these distances to describe the "magnitudes" of any shape transformation. (And therefore it requires at least three distances to describe the general reconfiguration of landmarks: not only these two shape metrics, but a third, for size change, as well.)

Concluding Comment

Nothing here is meant to imply that other features of shape space might not be interesting in particular applications, only that one needs a reason to look at them. I have published several examples of the reporting of shape changes by specific features of change associated with reasonable biological processes. Of the conventional models for shape change, the uniform model sometimes fits real data (Bookstein, 1987), and the model of growth-gradients is very compatible with the highest-order relative warps as described here. In comparison, a shape change that is truly limited to one single landmark moving upon a background of all the others in an unchanging configuration, such as is postulated by the robust Procrustes methods, is smeared out into a series of relative warps by this method–so much for nonlinearity of the smallest quadrilateral around it, so much for the second-smallest, and so on–and thereby becomes unrecognizable. In one study of the rat cranium (Bookstein, 1989b) the first two relative warps are explicitly identified with the two dimensions of an obvious candidate for explanation of form change, the rigid motion of the vault of the skull with respect to the cranial base. Abe et al. (1988) find that a cubic growth-gradient accounts cleanly for some changes in a lineage of ostracodes. In another cranial data set (Grayson et al., 1985), the appropriate explanation of an observed difference between typical syndromal and normal forms of the human cranial base hinges upon one single landmark that is being pushed in two directions by two separate sequelae of the abnormality. In Tabachnick's chapter below, the first uniform component and first relative warp correlate 0.96, and together embody Raup's parameter θ for a spiral form in one view; the same uniform component in another view is interpreted quite differently. In all these papers, shape coordinates were computed first, and then configurations of their differences were inspected to see what simple features were suggested.

The method of shape coordinates permits both the inspection of single residuals and the construction of large-scale gradient patterns. I recommend the formal consideration of all of these components and their correlations as a necessary step in the morphometric analysis of any set of landmark data, regardless of whether the intended end-point of the investigation is an understanding of process. Any fitting of landmark data to restricted models, with or without a uniform part, by Procrustes methods or any other, is useless without a close inspection of the geometric and statistical covariances of the residuals it induces. When that latter analysis is done according to the decomposition I am recommending here, the appropriate fits are optional, a-posteriori approaches to summaries of effects already noted. They are properly taken as confirmatory, not exploratory, techniques.

Acknowledgment

The preparation of this chapter and the development of new methods it puts forward were supported in part by United States National Institutes of Health grant GM-37251 to the University of Michigan (Fred L. Bookstein, Principal Investigator).

References

Abe, K., R. Reyment, F. L. Bookstein, A. Honigstein, and O. Hermalin. 1988. Microevolutionary changes in *Veenia fawwarensis* (Ostracoda, Crustacea) from the Cretaceous (Santonian) of Israel. Hist. Biol., 1:303-322.

Bookstein, F. L. 1984. Tensor biometrics for changes in cranial shape. Ann. Hum. Biol., 11:413-437.

Bookstein, F. L. 1985. Transformations of quadrilaterals, tensor fields, and morphogenesis. Pp 221-265 *in* Mathematical essays on growth and the emergence of form (Antonelli, P. L., ed.). University of Alberta Press.

Bookstein, F. L. 1986. Size and shape spaces for landmark data in two dimensions. (With discussion and rejoinder). Stat. Sci., 1:181-242.

Bookstein, F. L. 1987. Describing a craniofacial anomaly: finite elements and the biometrics of landmark location. Am. J. Phys. Anthropol., 74:495-509.

Bookstein, F. L. 1989a. Principal warps: thin-plate splines and the decomposition of deformations. I.E.E.E. Trans. Pattern Anal. Mach. Intell., 11:567-585.

Bookstein, F. L. 1989b. Comment on D. G. Kendall, "A survey of the statistical theory of shape." Stat. Sci., 4:99-105.

Bookstein, F. L. 1990. Four metrics for image variation. in Proceedings of the XI international conference on information processing in medical imaging (Ortendahl, D. and J. Llacer, eds.). Alan R. Liss, Inc, New York.

Bookstein, F. L. 1991. Morphometric tools for landmark data. Accepted for publication, Cambridge University Press, Cambridge.

Bookstein, F. L., B. Chernoff, R. Elder, J. Humphries, G. Smith, and R. Strauss. 1985. Morphometrics in evolutionary biology. The geometry of size and shape change, with examples from fishes. The Academy Natural Science of Philadelphia, Spec. Publ. No. 15, 277 pp.

Bookstein, F. L., and P. Sampson. 1987. Statistical models for geometric components of shape change. Proceedings of the section on statistical graphics, 1987 Annual Meeting of the American Statistical Association, pp. 18-27.

Boyd, E. 1980. Origins of the study of human growth. University of Oregon Health Sciences Center.

Goodall, C. R. 1989. WLS estimators and tests for shape differences in landmark data. J. Roy. Statist. Soc., submitted for publication.

Grayson, B., N. Weintraub, F. L. Bookstein, and J. McCarthy. 1985. A comparative cephalometric study of the cranial base in craniofacial syndromes. Clef. Pal. J., 22:75-87.

Mardia, K. V., and I. Dryden. 1989. The statistical analysis of shape data. Biometrika, 76: 271-281.

Chapter 12

Conventional Procrustes Approaches

Ralph E. Chapman

Scientific Computing, A.D.P., Rm. EG-15
The National Museum of Natural History
The Smithsonian Institution
Washington, D.C. 20560

Sleepless Nights in the Procrustean Bed
- Harlan Ellison, 1984

Abstract

Conventional Procrustes methods allow the analysis of morphology through the superimposition of one morphology onto another using the positions of landmark points. These methods are most useful for the comparison of pairs or small groups of specimens when it is of interest to describe one morphology in terms of deformation from another. When population analyses are of interest, generalized methods tend to be more appropriate. Least-squares methods are most appropriate when change is more general and not localized, whereas robust methods are unexcelled in demonstrating localized change when it is present. Examples are given that demonstrate the relative strengths of the two approaches and comparison made with the results of a biorthogonal analysis. The algorithms used in calculating two forms of conventional Procrustes analyses, Least-Squares Theta-Rho-Analysis (LSTRA) and Resistant-Fit Theta-Rho-Analysis (RFTRA) are discussed in detail.

Introduction

Landmark methods of shape analysis include the most powerful morphometric procedures for determining biologically relevant patterns of morphological variability and change. These techniques make use of the relative positions of landmarks or homologous points (h-points) for their calculations. Included are many conventional multivariate analyses (e.g., Neff and Marcus, 1980), the powerful suite of methods included under the general heading of tensor or finite element techniques discussed in Tobler (1977, 1978), Bookstein (1977, 1978, 1986; Bookstein, et al. 1985, and references therein), Goodall and Green (1986) and Moss et al. (1987, references therein), and Procrustes techniques that work by superimposing two or more specimens onto one another.

As Goodall and Bose (1987) suggest, Procrustean methods fall into three major categories. The first includes very simple methods that map two or more morphologies by making a simple baseline in all specimens of equal size and with identical coordinates (see Benson, 1982a; Bookstein, 1978, 1986, for discussions). The second group involves an optimization of the fit by least-squares or related approaches (see Gower, 1970, 1975; Sneath, 1967; Siegel and Benson, 1982, for discussions). Finally, Goodall and Bose (1987) recognize a class of methods that make use of robust methods for making the superimposition while attempting to highlight localized differences in form. The most commonly used method of this last approach utilizes repeated medians and includes the method Resistant-Fit Theta-Rho-Analysis developed by Siegel and Benson (1982).

Separate from this classification, Procrustes analyses can also be divided into two philosophical categories, conventional Procrustes methods (i.e., ordinary Procrustes analyses by Goodall and Bose, 1987) and generalized approaches. The former superimpose one or a series of specimens onto a single or base specimen. The latter provide superimpositions based on a consensus specimen and were introduced by Gower (1975); they have been discussed subsequently by Goodall and Bose (1987), Rohlf and Slice (1990) and Chapter 10 of this volume).

Herein, I will concentrate on conventional Procrustes methods, which can include simple models (e.g., Bjork analysis, see discussion in Benson, et al., 1982) but I will concentrate on a robust method, Resistant-Fit Theta-Rho-Analysis (RFTRA), and a corresponding least-squares method (LSTRA). RFTRA and LSTRA refer to specific algorithms used by the Smithsonian morphometric community and are used in their abbreviated forms for convenience and consistency with other works (e. g., Benson, 1967, 1976, 1982a, b; Benson et al., 1982; Siegel and Benson, 1982; Chapman, in press; Chapman and Brett-Surman, in press). Statements made about LSTRA and analyses made running LSTRA also should apply to conventional least-squares approaches developed by other researchers (e. g., Huffman, et al., 1978; Sneath, 1967). RFTRA refers to a single robust procedure that utilizes repeated medians, which was developed by Siegel and Benson (1982) and used by Siegel (1982) and Benson et al. (1982).

As with many forms of shape analysis, conventional Procrustes methods were inspired by the works of D'Arcy Thompson (e.g., 1942). Least-squares approaches were first developed within this context by Sneath (1967) in an effort to quantify Thompson's transformation grids, applying the method to the study of primate skulls. The general approach was developed further through the works of Gower (1971, 1975), who formalized the least-square algorithm in matrix terms and discussed generalized solutions. The least-squares approach was used later by Benson (1976) studying ostracode morphology, and was discussed in general terms and with hypothetical figures by Huffman et al. (1978). Resistant-Fit Theta-Rho-Analysis was developed by Siegel and Benson (1982), Siegel (1982), Benson et al. (1982) and Benson (1982a, b; 1983), who used Sneath's (1967) primate examples, further hypothetical figures, examples from the Ostracoda, and even human caricatures made by Leonardo da Vinci. Finally, applications to dinosaurs are available (Chapman, in press; Chapman and Brett-Surman, in press).

Questions and Capabilities

Conventional Procrustean methods provide the researcher with the superimposed fit of one constellation of h-points or landmarks onto another. The landmarks are equivalent or corresponding points found on all specimens being studied. In their most useful form, landmarks are homologous points or h-points. In other forms, they can be analogous points determined either functionally or geometrically. As with all forms of shape analysis, these methods are designed to document interesting patterns of morphological change and variability; specifically, asking how two forms differ, based on selected points.

The question is answered by superimposing the figures using either the LSTRA or RFTRA algorithm and noting differences in landmark positions, or outlines and accessory figures carried along with the fit. The differences are quantified as vectors of change for each landmark from the point on the base specimen (the one fit to) to the corresponding point on the superimposed specimen. Vector directions and magnitudes can be examined and further quantified to give insight into the nature of the superimposition. The average of the squared magnitudes of the vectors provides a distance coefficient that can be used for comparison within single studies and for use with auxiliary techniques (e.g., cluster analyses and ordinations) to elucidate patterns further.

Conventional Procrustes methods provide a number of advantages. These include:

- adaptability to a wide range of studies, ranging from simple comparisons of pictures or specimens with illustrations, to studies of large suites of specimens within a phylogenetic framework, to detailed studies of population variability and evolution. Because of the availability of distance coefficients indicating the level or goodness of fit of the two constellations of landmarks, mapping methods are among the best techniques available for the examination of morphological trends within a temporal-spatial distribution (see, e.g., Benson, 1976, 1982b).

- intuitive understanding. Most morphologists can interpret the graphics immediately and obtain answers to the questions of interest. The mathematical manipulations also are relatively simple, helping to prevent errors in interpretation.

- efficient processing, allowing researchers to use RFTRA and LSTRA as exploratory tools. Within this context, they are especially useful in developing characters for use in standard phylogenetic studies because they allow researchers to focus on particular landmarks.

- allowing the researcher to examine shape differences after removing overall size following the algorithm relevant to the Procrustean method being applied.

- for RFTRA and other robust methods, providing an unexcelled insight into patterns of localized deformation. The use of robust procedures allows RFTRA to fit following a majority of the points and to concentrate the change in one or a few h-points, if that is where the differences lie. The other method, LSTRA, will tend to partition the differences among all the h-points. Selection between RFTRA and LSTRA depends on how biologically reason-

able localized change is within the application being studied. Using both methods often provides an excellent indication of the relative importance of localized change.

- allowing a combination of landmark analysis with the analysis of change in outlines (see below). During superimposition, only landmarks are used to make the fit, but outlines as well as other accessory figures can be superimposed using the same transformation coefficients, and the differences noted between specimens.

- quantifying populations. Multiple specimens of a single population can be superimposed onto a single or base specimen and an average individual for that population generated for comparison with others. This approach is best made using generalized techniques, but the conventional methods provide especially interesting results when the specimens being studied have an ordered nature (e.g., ontogenetic series). In such cases, it is worthwhile to apply both conventional and generalized methods.

- analyzing the differences in small or selected regions, especially with RFTRA. This has been stressed before in character development, but it also applies to specimens with missing sections (especially important in vertebrate paleontology), and to studies concentrating on functional complexes. Experience suggests that a relatively small number of additional general landmarks provides an overall fit that is moderately to greatly congruent with the one obtainable using larger numbers. This approach should be especially effective if combined with other methods (e.g., biorthogonal analysis of Bookstein, 1978) also useful in this context.

Limitations

No single method can solve all problems of interest; certain characteristics make conventional Procrustes methods unsuitable as the sole technique in some contexts. For example, the distance coefficients are phenetic and correspond directly to other coefficients such as euclidean distance. Consequently they are not as useful for the construction of phylogenetic trees as are cladistic methods. They are useful, however, for analyzing morphological variability within a predefined phylogenetic framework either by superimposing results on pre-existing trees or helping to evaluate competing trees. As mentioned earlier, these methods are useful for helping develop and evaluate potential characters for phylogenetic analyses.

In the study of allometry, Bookstein et al. (1985) point out that RFTRA (and LSTRA) cannot be used directly. Conventional Procrustean methods were not developed to analyze allometry directly. But, within the framework of an allometric study, they can provide insight into the morphological effects of the allometry documented using more conventional approaches.

Conventional Procrustean methods are less useful for population studies. Generalized methods are the better technique in that they do not require a base specimen to be selected. As mentioned above, conventional studies within this framework can provide interesting results in selected cases, but the generalized approach should be superior in most cases.

Robust procedures work best when most of the differences between the superimposed specimens are concentrated in less than half the landmarks. When the differences are extensive and widely spread, LSTRA may provide the optimal solution. Examination of the differences between the two analyses, however, frequently will provide insight into the quality or the distribution of the landmarks chosen, or will suggest important areas for further analysis. For those comparisons where localized change is unreasonable (e.g., deformation in a closed system, or balloon model) the robust model is inappropriate.

One of the strengths of Procrustes methods — that they provide a direct vector record of the changes that occur from one form to another — can also be the source of weakness. The vector patterns often can be complex and misleading. The general assumption of such an analysis is of isometric change. This can produce apparently large-scale and complex patterns from simple expansion and contraction. Further, the assumption of localized change made when applying robust methods such as RFTRA can cause a misleading result if the actual deformation is not localized.

Doing an Analysis

Applying RFTRA or LSTRA takes relatively little time and the steps necessary to do an analysis are summarized below.

1. As in any analysis, a first and major step is the development of the exact questions to be asked, and specifically for these analyses, exactly what type of output from RFTRA or LSTRA can contribute significantly to answering these questions. All subsequent work relies heavily on this step.

2. Next, one must select the specimens and illustrations/photographs to be used. This commonly is time-consuming and may require modifications in step 1. To run analyses, specimens are needed that exhibit adequate numbers of landmarks, especially in critical areas.

3. Selection of h-points or landmarks is the next step and one which has a strong effect on the analysis. The philosophy behind h-point selection varies with the questions being asked (Chapman, in press). If the researcher is interested only in documenting how a reconstructed illustration differs from a photograph of the original specimen on which it is based, then landmark selection can be quite flexible. However, for analyses made within a phylogenetic framework, the landmarks should be chosen to represent, as much as possible, truly homologous points. This has been stressed by Bookstein et al. (1985) who discuss this problem in detail and present interesting options. This is one area that deserves considerable attention in all detailed applied studies and on which more theoretical work needs to be done.

4. After landmarks are chosen, they must be identified for each specimen. Additional figures (e.g., outlines) must be selected for each, if of interest.

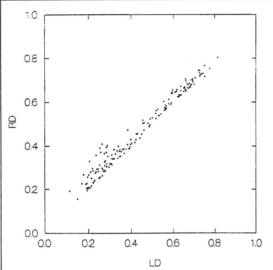

Figure 1. Bivariate scatter plot of 172 distance coefficients calculated for 19 ornithopod dinosaurs. Data from Chapman and Brett-Surnam (in press). Comparison is for LSTRA, LD, and RFTRA, RD, distance coefficients.

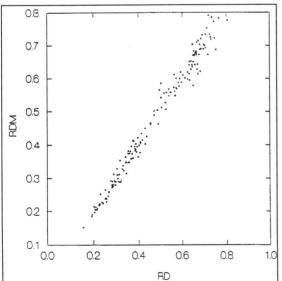

Figure 2. Bivariate scatter plot for data as in Figure 1 for RFTRA, RD, and median RFTRA, RDM, distances.

5. Next, one must digitize the illustrations, photographs or specimens. Generally, the first two are used because of current restrictions to two-dimensional space. However, with three-dimensional digitizers becoming more readily available, the use of actual specimens, even for two-dimensional studies, is now reasonable. Further, three-dimensional studies are the next type to be developed. This step typically goes quickly, and many specimens can be digitized per day. The additional figures, such as the outlines, take most of the digitizing time, which explains why many exploratory analyses reduce the number of accessory figures, or eliminate them altogether.

6. One must select the connections to be made between landmarks to provide "skeleton" or polygonal figures. There is no single correct series of connections. The researcher must balance each figure between adding connections to provide additional information and avoiding too many connections which will tend to obscure patterns. A typical philosophy is to make connections following established functional groups and highlighting areas of

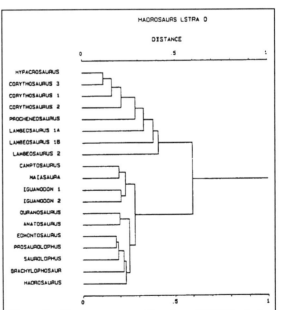

Figure 3. Dendrograms resulting from UPGMA cluster analysis using LSTRA distance, RD, using data described in Figure 1.

critical interest. A series of polygonal diagrams can be made for each series of specimens and these can be used experimentally in the initial output stages.

7. The analyses can now be made. Programs for LSTRA can be adapted from Huffman et al. (1978), and Siegel (1982) presents options for both LSTRA and RFTRA. A series of programs is available

from the author for most DOS microcomputers. Programs typically are run to perform single comparisons, produce graphics, output vector magnitudes and directions and plot these values, produce distance matrices for series of specimens, or do outline studies. Other programs allow an average figure to be calculated for a series of related specimens.

8. A wide variety of output formats is available. Standard graphics typically provide superimposed polygonal figures, plotted vectors of change, superimposed outlines and accessory figures, or combinations of these (see Figures 1,3). The user can also plot histograms of vector magnitudes, or histograms or rose diagrams of vector magnitudes and directions. Distance coefficients can be used in a wide variety of contexts, and matrices can be outputted to files for use with various software packages.

9. The final stage is interpretation of the results in light of the original questions being asked and, frequently, reformulating questions and starting over at stage 1.

For individual comparisons, entire studies can frequently be completed in less than an hour or two. Large scale studies will take much longer, although most of this time is spent in selecting h-points and specimens and digitizing. Analysis time (stage 7) tends to be short, even with a relatively slow CPU.

Mechanics of the Analysis

Both forms of mapping, LSTRA and RFTRA, function by calculating four coefficients. These are used to transform the two-dimensional coordinates of points that are part of the superimposed figure to their optimal fit onto the base specimen. In doing this, assuming the points on the base specimen have coordinates (x,y) and those on the superimposed specimen have coordinates (u,v), the latter are transformed to new coordinates (u',v') by the following,

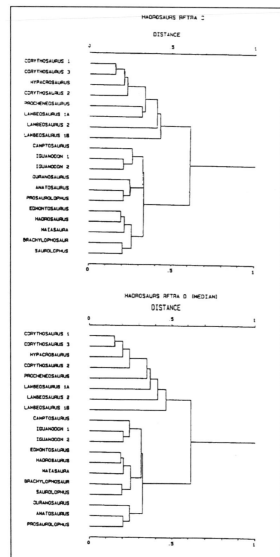

Figure 4. Dendrograms resulting from UPGMA cluster analysis using RFTRA distance (top), and RDM (bottom), based upon data described in Figure 1.

$u' = T_1 + T_3 u - T_4 v$

$v' = T_2 + T_4 u + T_3 v$ (1)

where T_1-T_4 are the four transformation coefficients. Of these, T_1 provides horizontal translation

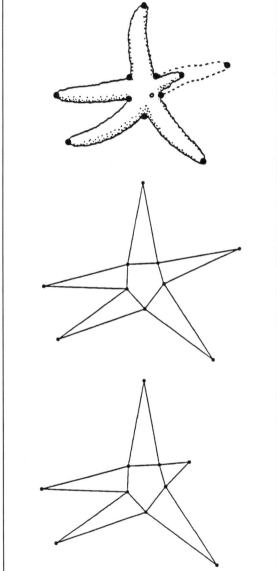

Figure 5. Illustration of the "Pinocchio effect" using a hypothetical starfish in two modes, normal and with part of one leg amputated. Top: original figure; middle and lower: landmarks on the two forms.

(along the x or u axis), T_2 vertical translation (along the y or v axis), and T_3 and T_4 provide a combination of scaling and rotation changes. T_3 equates to $S \cos \theta$ and T_4 to $S \sin \theta$. S is the scale factor that expands or contracts the size of the superimposed figure to fit the base specimen. The value θ, the rotation angle, is the degree of rotation necessary to fit the superimposed specimen optimally to the base. Equation (1) reduces to the equation,

$$\begin{bmatrix} u' \\ v' \end{bmatrix} = \begin{bmatrix} T_1 \\ T_2 \end{bmatrix} + S \begin{bmatrix} \cos \theta & -\sin \theta \\ \sin \theta & \cos \theta \end{bmatrix} \begin{bmatrix} u \\ v \end{bmatrix}$$ (2)

which has been used in equivalent form by Sneath (1967), Huffman et al. (1978), Siegel & Benson (1982), and Siegel (1982).

The differences between LSTRA and RFTRA result from the way the four transformation coefficients are estimated. LSTRA approaches utilize standard least-squares algorithms to minimize the sum of the squared differences between corresponding landmarks or h-points. Typically this is done in a single step, including the estimation of values for T_1 to T_4 (for source code see Huffman et al., 1978; Siegel, 1982).

In RFTRA the coefficients are estimated separately. The first calculation is to find the scale factor, S, which is estimated by using algorithms based on equations found in Siegel and Benson (1982). For each pair of corresponding h-points, a single scale value is calculated by

$$\text{scale value} = \frac{\text{distance in base specimen}}{\text{distance in superimposed specimen}}$$ (3)

More specifically, for h-points i and j, a single scale value, s_{ij}, is calculated by

$$s_{ij} = \sqrt{\frac{(x_j - x_i)^2 + (y_j - y_i)^2}{(u_j - u_i)^2 + (v_j - v_i)^2}}$$ (4)

For a comparison with n h-points, the global scale factor, S, is estimated using all possible $n(n-1)/2$ comparisons by taking repeated medians using,

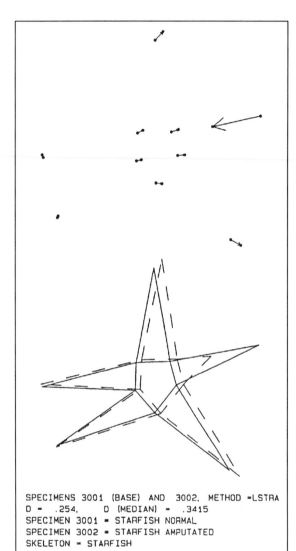

Figure 6. Results of LSTRA of starfish shown in Figure 1. Top: landmarks and vectors. Bottom: with polygonal diagrams superimposed. Relevant data are in the legend.

The determination of the rotation angle, θ, is done in much the same way, facilitated at first by an initial least-squares fit (see Siegel and Benson, 1982). Individual rotation values are calculated for each pair of h-points. This is done by calculating the angle needed to rotate a vector passing through the two h-points on the superimposed specimen to the corresponding vector on the base specimen. These rotation values θ_{ij} are then used to calculate the global rotation factor, θ, using repeated medians as in Equation (5) by the equation

$$\theta_i = \underset{i \neq j}{\overset{\text{MED}}{}} (\text{MED } \theta_{ij}) \tag{6}$$

As mentioned above, the scale factor, S, and this rotation angle, θ, are then used to calculate T_3 and T_4 as $S \cos \theta$ and $S \sin \theta$, respectively.

The final two coefficients, T_1 and T_2, are used to translate the superimposed specimen in two-dimensional space. They are estimated using only simple medians,

$$T_{1i} = \text{MED } (x_i - T_3 u_i + T_4 v_i) \tag{7}$$

and

$$T_{2i} = \text{MED } (y_i - T_4 u_i + T_3 v_i) \tag{8}$$

These equations are direct alterations of Equation (1). Simply, they determine the amount of translation necessary for each h-point of the superimposed specimen to be moved on top of its corresponding h-point on the base. The overall translation is then determined as the median of these values, one for each dimension. A detailed discussion of the repeated median approach and robust methods can be found in Siegel and Benson (1982).

The distance value between the two specimens is calculated after superimposition and in the same way for both LSTRA and RFTRA. The value is the average of the squared distances between corresponding h-points divided by a general size estimator. The equation used is

$$S_i = \underset{j \neq i}{\overset{\text{MED}}{}} (\text{MED } s_{ij}) \tag{5}$$

where MED is the median. Here, the median scale value for each h-point is determined. The grand median of all these medians is then used to provide the scale factor, S.

Table 1. Correlation matrix among standard LSTRA distance, RFTRA distance, and median RFTRA distance coefficients.

	LSTRA	RFTRA	mediam RFTRA
LSTRA	1.000		
RFTRA	0.987	1.000	
median RFTRA	0.977	0.881	1.000

$$D = \frac{\sqrt{SW/n}}{S_9} \qquad (9)$$

where D is the distance coefficient, SW the sum of the squared distances between corresponding h-points, n the number of h-points, and S_9 a size measure. The latter keeps the coefficient from varying in value due solely to different base specimen sizes. In past applications this has been calculated as the mean distance between each h-point, on both specimens, and the center of form.

An alternative now also calculated for RFTRA is the median distance rather than the mean. Except for small differences due to rounding errors in computation, RFTRA distance coefficients always will be equal to or larger than corresponding LSTRA coefficients. The distance coefficients are, as expected, very highly correlated. An example data set using 19 ornithopod dinosaur crania (171 distances in all) taken from Chapman and Brett-Surman (in press) provides an indication of how the standard LSTRA distance coefficients (LD), the standard RFTRA distance coefficients (RD), and the median RFTRA distance coefficients (RDM) intercorrelate within a large data set. The correlation matrix is shown in Table 1. Bivariate scatter plots of LD versus RD and RD versus RDM are given in Figures 1 and 2, respectively. The result of using these different coefficients with a UPGMA cluster analysis is demonstrated in Figure 3 for LD, and Figure 4 for RD (top) and RDM (bottom). For large studies with multi-specimen comparisons, LSTRA is the recommended procedure. However, as is apparent from Figures 1-4, the results vary little with variation in the coefficient used. In general, LD is

reported now for LSTRA analyses, and RD and RDM for RFTRA analyses. A more detailed study of the behavior of these coefficient is currently underway as well as the development of new size estimators. It is important to note that the D coefficient provided by the program given by Siegel (1982) does not factor out the effects of differences in size of the base specimen.

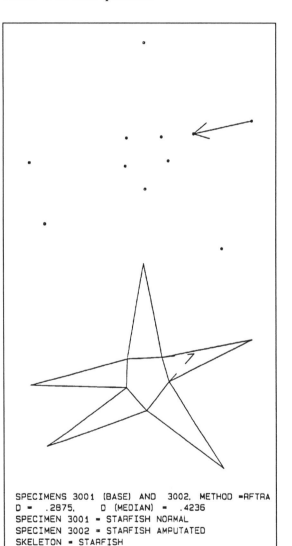

SPECIMENS 3001 (BASE) AND 3002. METHOD =RFTRA
D = .2875.   D (MEDIAN) = .4236
SPECIMEN 3001 = STARFISH NORMAL
SPECIMEN 3002 = STARFISH AMPUTATED
SKELETON = STARFISH

Figure 7. Results of RFTRA of starfish shown in Figure 1. Top: landmarks and vectors. Bottom: with polygonal diagrams superimposed. Relevant data are in the legend.

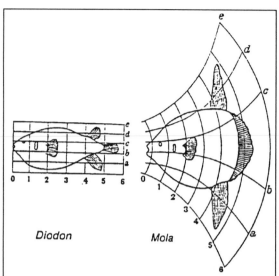

Figure 8. D'Arcy Thompson's grid-transformation analysis of *Diodon-Mola* transition taken from Bookstein et al. (1985).

Examples

The differences between LSTRA and RFTRA can be illustrated by a hypothetical example, much as has been done by Siegel and Benson (1982) and Benson et al. (1982). As an extreme example of what I will term the "Pinocchio effect," I will use a case of extreme localized deformation. The model of deformation used here is a starfish that has had a single leg amputated. Figure 5 illustrates the original figure with normal and amputated leg indicated (top), and polygonal figures for the normal (middle) and amputated (bottom) state. The results obtained from a LSTRA analysis are given in Figure 6, which provides a vector-landmark diagram (top) and polygonal figure comparison (bottom). The equivalent diagrams for a RFTRA analysis are presented in Figure 7. The results show that the least-squares algorithm distributed the lack

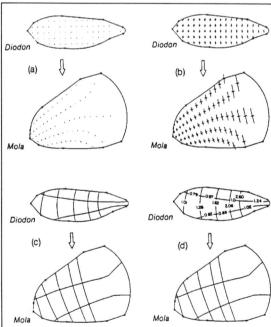

Figure 9. Biorthogonal analyses of *Diodon-Mola* transition taken from Bookstein et al. (1985).

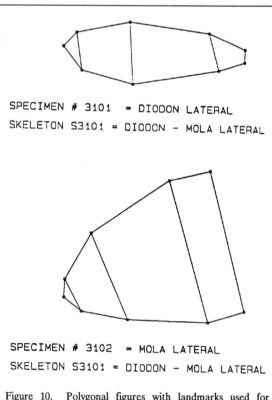

SPECIMEN # 3101 = DIODON LATERAL
SKELETON S3101 = DIODON - MOLA LATERAL

SPECIMEN # 3102 = MOLA LATERAL
SKELETON S3101 = DIODON - MOLA LATERAL

Figure 10. Polygonal figures with landmarks used for LSTRA and RFTRA analysis of *Diodon-Mola* transition.

of fit among the landmarks, whereas the RFTRA analysis represented the fit correctly.

A second example uses the classic *Diodon - Mola* series discussed in detail by Bookstein (1978) and originating with the form-transformation grid presented by D'Arcy Thompson (1942; Figure 8). The results of the biorthogonal analysis performed by Bookstein (1978) are given in Figure 9, demonstrating the strength of that method for illustrating shape gradients. These data have been analyzed using both LSTRA and RFTRA for comparison. Figure 10 presents polygonal diagrams for the two forms as used in the analyses. The LSTRA and RFTRA results are presented as in Figures 6 and 7, in Figures 11 and 12, respectively. This case of extreme deformation provides an opposing example to the previous one. The vector arrow graphics provide a useful complement to the

finite element output. As expected, the LSTRA output is most directly compatible with the finite element results because of the large overall change which is not localized. The RFTRA fit suggests that the dorsal sections expanded more than the ventral sections, but the LSTRA results suggest a more even deformation. Intuition suggests that results from finite element and LSTRA fit the actual data better. However, as I suggested earlier, the difference between RFTRA and LSTRA suggest that further analyses be made using landmarks not limited to the outline.

A final example using dinosaurs is given in Figures 13-17, taken from Chapman (in press). The

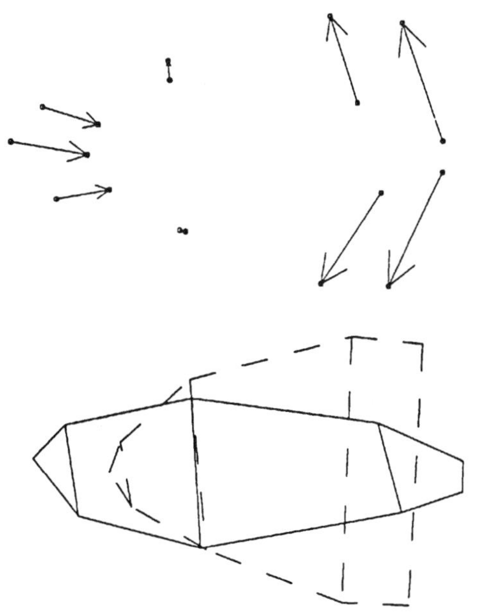

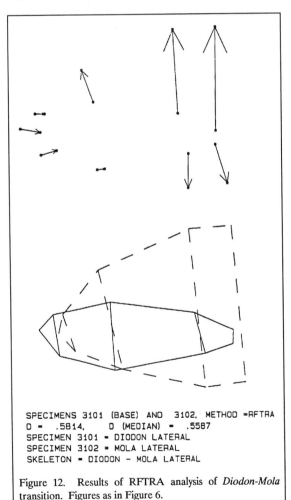

SPECIMENS 3101 (BASE) AND 3102, METHOD =LSTRA
D = .5185,   D (MEDIAN) = .5311
SPECIMEN 3101 = DIODON LATERAL
SPECIMEN 3102 = MOLA LATERAL
SKELETON = DIODON - MOLA LATERAL

Figure 11. Results of LSTRA analysis of *Diodon-Mola* transition. Figures as in Figure 6.

SPECIMENS 3101 (BASE) AND 3102, METHOD =RFTRA
D = .5814,   D (MEDIAN) = .5587
SPECIMEN 3101 = DIODON LATERAL
SPECIMEN 3102 = MOLA LATERAL
SKELETON = DIODON - MOLA LATERAL

Figure 12. Results of RFTRA analysis of *Diodon-Mola* transition. Figures as in Figure 6.

question here is how two carnosaurs, *Allosaurus* and *Tyrannosaurus*, differ in cranial morphology. This would be done ideally within a functional or systematic framework, using various views (if relevant). As an example we will use a simple lateral view of each. The illustrations used are given in the top and middle of Figure 13, and the landmarks chosen are illustrated on *Allosaurus* in the bottom of that Figure. For this example, landmarks were taken at the intersection of sutures, fenestrae, and teeth. Landmark selection in vertebrate crania is

far from clear-cut and is an area needing far more philosophical discussion and detailed developmental analysis because there are potential problems associated with most usable points. In this sense, the ostracode pore patterns used by Benson (1976, 1983), being extensions of the central nervous system, are superior for direct phylogenetic applications. However, these landmarks are certainly useful for these examples and for functional interpretation.

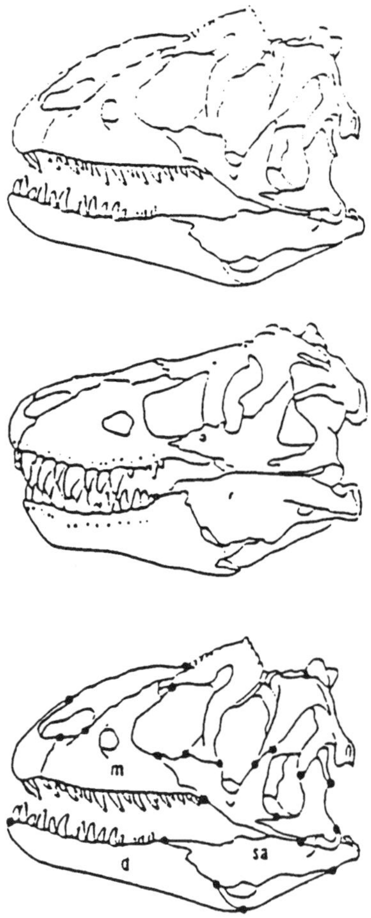

Figure 14 provides landmark/outline diagrams (left) and polygonal figure/outline figure diagrams (right) for *Allosaurus* (top) and *Tyrannosaurus* (bottom). LSTRA and RFTRA analyses, presented as in Figures 6 and 7, are given in Figures 15 and 16, respectively. The fit appears to be similar to that seen for the *Diodon - Mola* comparison in that large vector changes are seen throughout. Note two major differences in the specimens: First, a more robust posterior of the lower jaw in *Tyrannosaurus* (dashed lines) is evident from the superimposed outline figures for that element. Second, there is a higher degree of dorso-ventral vaulting in *Allosaurus*, indicated by antero-ventral vector directions for the landmarks in the nasal and premaxilla regions, and in the posteriorly directed vectors in the jugal and quadratojugal regions. The RFTRA comparison suggests that differences were concentrated in the posterior of the cranium and the top of the snout. The LSTRA fit concentrated

Figure 13. Carnosaur crania used as an example of RFTRA and LSTRA. Top and middle: lateral views of *Allosaurus* and *Tyrannosaurus*. Bottom: 25 landmarks shown on *Allosaurus*.

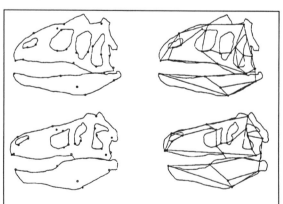

Figure 14. H-points and outlines (left) and polygonal and outline figures (right). Top: *Allosaurus*. Bottom: *Tyrannosaurus*.

the fit on the top of the skull in general. Only additional analyses in other views and using different landmarks will shed further light on which model is best.

The additional strength of the graphical approach possible with this type of Procrustes analysis is shown by the superimposed outlines in Figure 17. The trends follow the same lines as those described, but accentuate some of the areas where landmarks were difficult to identify. The robustness of the tyrannosaurid jaw comes out particularly clear in the comparison.

The Analysis of Outlines

With conventional Procrustes approaches, the anal-

yses of outlines can be made following at least three general approaches. The first is part of the standard procedure used when applying LSTRA and RFTRA or related methods within a software package that allows outlines to be included in the final graphics. After landmarks have been superimposed using the appropriate algorithm, digitized outlines and other figures of interest are superimposed using the same transformation coefficients. Graphics showing the outlines and other figures allow the researcher to inspect these superimpositions and note where interesting differences and similarities occur. For many applications, this will provide important and sufficient insight.

The second approach takes the standard

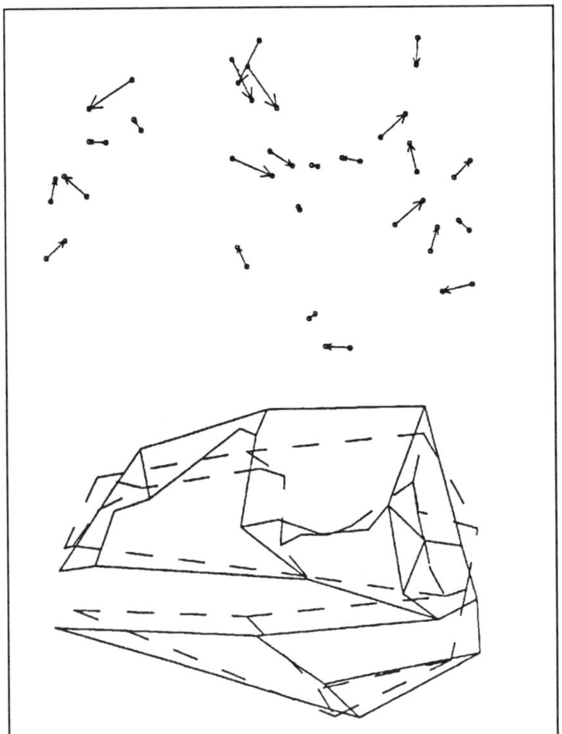

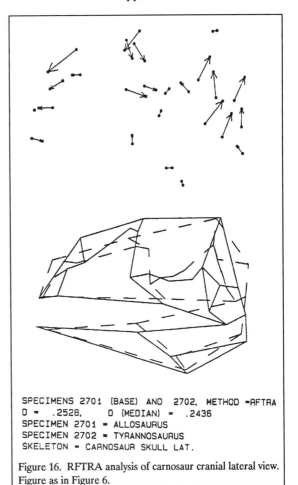

SPECIMENS 2701 (BASE) AND 2702, METHOD =LSTRA
D = .2157,   D (MEDIAN) = .2123
SPECIMEN 2701 = ALLOSAURUS
SPECIMEN 2702 = TYRANNOSAURUS
SKELETON = CARNOSAUR SKULL LAT.

Figure 15. LSTRA analysis of carnosaur cranial lateral views. Figure as in Figure 6.

SPECIMENS 2701 (BASE) AND 2702, METHOD =RFTRA
D = .2528,   D (MEDIAN) = .2436
SPECIMEN 2701 = ALLOSAURUS
SPECIMEN 2702 = TYRANNOSAURUS
SKELETON = CARNOSAUR SKULL LAT.

Figure 16. RFTRA analysis of carnosaur cranial lateral view. Figure as in Figure 6.

analysis a step further. After the landmarks and outlines have been superimposed following standard procedures, differences in the outlines can be quantified further by calculating distances between the two outlines at selected intervals. These intervals can be chosen for equally spaced radii from the center of form (see, for example, Ehrlich and Weinberg, 1970) or from a single landmark, as was done by Benson (1967) and Kaesler and Waters (1972), or they can be equally spaced along the perimeter of the outline (see, for example, Lohmann, 1983; Rohlf, 1986). An average of the squared differences supplies a second distance coefficient that can be used in much the same way as the conventional distance coefficients discussed earlier. The differences can be plotted in bivariate space for comparison between two or among a number of specimens. These analyses can be made following the conventional Procrustes approach, by projecting all specimens onto a base specimen, or by using the generalized methods discussed by Gower (1975), Goodall and Bose (1987), or Rohlf (this volume).

The third approach is more typical of standard outline methods. There are no landmarks used in the analyses, but instead, points found at selected intervals along the outlines, using the same methods discussed above, are used as pseudolandmarks. The fit itself is made by rotating one outline over the other and calculating a distance coefficient at each step. The rotation is performed by "homologizing" the first point in the base specimen with successive points on the outline of the second specimen. This is done because there is no basis for assuming that the researcher will start digitizing the outlines at the exact same point for both specimens, or that the researcher necessarily would be able to recognize equal or "homologous" starting points. Consequently, the first comparison is made by "homologizing" point 1 in the base specimen with point 1 in the other. The next comparison is made by "homologizing" point 1 in the base specimen with point 2 of the other, with the last point in the base lining up with point 1 of the second specimen. This procedure continues

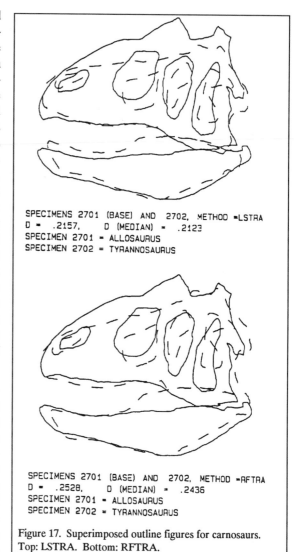

SPECIMENS 2701 (BASE) AND 2702, METHOD =LSTRA
D = .2157,   D (MEDIAN) = .2123
SPECIMEN 2701 = ALLOSAURUS
SPECIMEN 2702 = TYRANNOSAURUS

SPECIMENS 2701 (BASE) AND 2702, METHOD =RFTRA
D = .2528,   D (MEDIAN) = .2436
SPECIMEN 2701 = ALLOSAURUS
SPECIMEN 2702 = TYRANNOSAURUS

Figure 17. Superimposed outline figures for carnosaurs. Top: LSTRA. Bottom: RFTRA.

until point 1 of the base has been compared with all points of the other, providing a series of distance coefficients, and making the optimal fit by determining the smallest distance value. This value is taken as the resulting distance coefficient for the comparison, and a series of these values can be used in the same way as those produced by more standard analyses. As with the second approach, the differences between matched points can be plotted in bivariate space, and more than two specimens can be compared simultaneously. Further, this optimal fit can be used with other, more standard, outline approaches (see Lohmann, 1983, for an example). The least-squares fit is the one more typically used, but optimizing the fit using the RFTRA algorithm can provide very interesting results.

Conclusions

The conventional Procrustes methods discussed herein exhibit a wide range of strengths and weaknesses. They are unusual among morphometric methods in that they concentrate on the deformation of one form relative to another. As such, they share the approach of D'Arcy Thompson's original form-transformation grids and finite-element/biorthogonal approaches. Procrustes methods have limitations for population analyses but provide unexcelled insight into questions of function and the morphological shift from one form to another. These methods have an advantage in that they do not distort the data and provide useful graphics using the data points selected. As such, they are relatively easy to use within any evolutionary framework. They have their limitations in their assumption of isometry and occasionally misleading patterns if the original deformation does not fit the model being considered. However, use of conventional Procrustean methods is usually enlightening in some way.

Within this approach, robust methods have their share of strengths and weaknesses. Past examples of LSTRA and RFTRA comparisons have stressed the strengths of the robust approach, again demonstrated here using the starfish. If the deformation is localized, the robust approach is the best procedure currently available for its analysis. Robust methods work best when comparing very similar forms (see Siegel and Benson, 1982; Benson, et al. 1982).

The LSTRA approach also has its usefulness and is the approach of choice when comparing groups of specimens and when deformation is general. LSTRA provides results that tend to be more directly complementary to those of other landmark methods. RFTRA results tend to be divergent, especially when the deformation is generally distributed.

When the major question of interest is in the comparison of populations, generalized methods should be used, although conventional methods can be useful in some cases as complementary analyses. Within a directly systematic and phylogenetic framework, generalized methods should be more generally applicable, although conventional approaches are more useful for determining functional interpretations and the development of individual phylogenetic characters.

Acknowledgments

There are a number of individuals I would like to thank for assistance provided during the tenure of this research. My past discussions with Richard Benson and Andrew Siegel helped me to develop as a morphometrician and to appreciate the usefulness of conventional Procrustean methods. Joan Richtsmeier and Subhash Lele, both of The Johns Hopkins University, have provided much interesting discussion over the past year which helped me consider many problems not previously considered. I would also like to thank Fred Bookstein, Jim Rohlf, Les Marcus, Richard Reyment and the rest of the attendees at the Ann Arbor symposium for significant discussion and feedback. Jenny Clark provided the starfish illustration, and Linda Deck added considerable moral support throughout.

References

Benson, R. H. 1967. Muscle scar patterns of Pleistocene (Kansan) ostracodes. Pp. 211-241 *in* Essays in paleontology and stratigraphy (Teichert, C. and E. Yochelson, eds.). U. Kansas Press, Lawrence.

Benson, R. H. 1976. The evolution of the ostracode *Costa* analyzed by "Theta-Rho Difference". Abh. Verh. Naturwiss. Ver. Hamburg, (NF) 18/19(Suppl.): 127-139.

Benson, R. H. 1982a. Deformation, Da Vinci's concept of form, and the analysis of events in evolutionary history. Pp. 241-277 *in* Palaeontology, essential of historical geology (Gallitelli, E.M., ed.). S.T.E.M. Mucchi, Modena, Italy.

Benson, R. H. 1982b. Comparative transformation of shape in a rapidly evolving series of structural morphotypes of the ostracode Bradleya. Pp. 147-164 *in* Fossil and recent Ostracodes (Bate, R. H., E. Robinson and L. M. Sheppard, eds.) Ellis Horwood, Chichester, U.K.

Benson, R. H. 1983. Biomechanical stability and sudden change in the evolution of the deep-sea ostracode Poseidonamicus. Paleobiology, 9(4): 398-413.

Benson, R. H., R. E. Chapman and A. F. Siegel. 1982. On the measurement of morphology and its change. Paleobiology, 8(4):328-339.

Bookstein, F. L. 1977. Orthogenesis of the hominids: an exploration using biorthogonal grids. Science, 197:901-904.

Bookstein, F. L. 1978. The Measurement of biological shape and shape change. Springer-Verlag, New York, 191 pp.

Bookstein, F. L. 1986. Size and shape spaces for landmark data in two dimensions. Stat. Sci., 1(2):181-242.

Bookstein, F. L., B. Chernoff, R. Elder, J. Humphries, G. Smith and R. Strauss. 1985. Morphometrics in Evolutionary biology. The Academy of Natural Science of Philadelphia, Spec. Pub. No. 15, 277 pp.

Chapman, R. E. In Press. Shape analysis in the study of dinosaur morphology. in Dinosaur systematics (Currie, P. J. and K. Carpenter, eds.). Cambridge University Press, Cambridge.

Chapman, R. E. and M. K. Brett-Surman. In Press. Morphometric observations of hadrosaurid ornithopods. in systematics (Currie, P. J. and K. Carpenter, eds.). Cambridge University Press,

Ehrlich, R and B. Weinberg. 1970. An exact method for characterization of grain shape. J. Sed. Petrol., 40:205-212.

Goodall, C. R. and P. B. Green. 1986. Quantitative analysis of surface growth. Bot. Gaz., 147(1):1-15.

Goodall, C. R. and A. Bose. 1987. Models and Procrustes methods for the analysis of shape differences. Proceedings, 19th Symposium on the Interface between Computer Science and Statistics. 7 pp.

Gower, J. C. 1971. Statistical methods of comparing different multivariate analyses of the same data. Pp. 138-149 *in* Mathematics in the archaeological and historical sciences (Hodson, F. R., D. G. Kendall and P. Tautu, eds.). Edinburgh University Press, Edinburgh.

Gower, J. C. 1975. Generalized Procrustes analysis. Psychometrika, 40(1):33-51.

Huffman, T., R. A. Christopher and J. E. Hazel. 1978. Orthogonal mapping: a computer program for quantifying shape differences. Comp. Geosc., 4:121-130.

Kaesler, R. L. and J. A. Waters. 1972. Fourier analysis of the ostracode margin. Bull. Geol. Soc. Amer., 83:1169-1178.

Lohmann, G. P. 1983. Eigenshape analysis of microfossils: a general morphometric procedure for describing changes in shape. Math. Geol., 15(6):659-672.

Moss, M. L., H. M. Pucciarelli, L. Moss-Salentijn, R. Skalak, A. Bose, C. Goodall, K. Sen, B. Morgan and M. Winick. 1987. Effects of pre-weaning undernutrition on 21 day-old male rat skull form as described by the finite element method. Gegenbaurs Morphol. Jb., Leipzig, 133(6):837-868.

Neff, N. A. and L. F. Marcus. 1980. A survey of multivariate methods for systematics. Privately Published and A.M.N.H., New York, 243 pp.

Rohlf, F. J. and D. Slice. 1990. Extensions of the Procrustes methods for the optimal superimposition of landmarks. Syst. Zool., 39:40-59.

Siegel, A. F. 1982. Geometric data analysis: an interactive graphics program for shape comparison. Pp. 103-122 *in* Modern data analysis

(Launer, R. L. and A. F. Siegel, eds.). Academic Press, New York.

Siegel, A. F. and R. H. Benson. 1982. A robust comparison of biological shapes. Biometrics, 38(2):341-350.

Sneath, P. H. A. 1967. Trend-surface analysis of transformation grids. J. Zool., London, 151(1): 65-122.

Thompson, D. W. 1942. On growth and form, a new edition. Cambridge Univ. Press, Cambridge, 1116 pp.

Tobler, W. R. 1977. Bidimensional regression. A computer program. Published by the Author, Geography Dept., U. Calif, Santa Barbara, 72 pp.

Tobler, W. R. 1978. Comparison of plane forms. Geogr. Anal., 10(2):154-162.

Chapter 13

Resolving Factors of Landmark Deformation: Miocene Globorotalia, DSDP Site 593

R. Elena Tabachnick[1] and Fred L. Bookstein[2]

[1] *Department of Geological Science and Museum of Paleontology*
[2] *Center for Human Growth and Development*
University of Michigan, Ann Arbor, Michigan 48109

Abstract

Shape coordinates of landmarks taken on Miocene *Globorotalia* from DSDP site 593 were analyzed using two approaches. A principal components analysis produces vectors of correlated landmark displacement with respect to an arbitrary baseline. Separation of the uniform part of this shape variation and analysis of the nonuniform variation by the method of relative warps produces a mixed suite of geometrical and statistical components of shape change. This type of analysis expands the interpretation of shape variation in several ways. First, it allows us to examine the geometric types of shape variation resulting from change in known morphological parameters. Variation of θ, the angle of increment of an equiangular spiral, results in both uniform and nonuniform landmark transformation, while variation of r, the expansion rate, appears in the nonuniform part only. Second, this type of analysis facilitates identification of novel components of morphological variation. For these foraminifera, the tilt of chambers relative to the coiling axis is more strongly correlated with deformation of landmarks in apertural view than was either of the spiral parameters t or r. Third, in light of this separation of landmark displacements into uniform and nonuniform parts, we suggest that predefined morphological features might be interpreted to interact in three ways: responding coherently as a single morphological feature would react to a single process; responding by correlated change reflecting changes due to separate processes that happen to be correlated for the population under study; and responding by independent changes of the separate features. The aperture of the *Globorotalia* test reacts to some extent as a single feature with the spiral parameters and to some extent independently.

Introduction

Planktonic foraminifera are an interesting subject for the study of patterns of morphological diversity inasmuch as the patterns of morphological variation exhibited by these organisms are often richly complicated. Phylogenetic hypotheses are usually erected on the basis of intermediate forms present between species (for example, see Banner and Lowery, 1985), and single samples of planktonic foraminifera often show continuous variation between named species (see, for example, Stainforth et al., 1975; Tabachnick, 1981; Tabachnick and Bookstein, 1990).

There is no guarantee that any variable defined in advance of an analysis, merely because it appears to the researcher to be variable in the population, will vary in accordance with an underlying factor or component of shape change; in fact, it is unlikely to do so. In order to discern the patterns of variation present in populations of foraminifera, it is necessary to analyze that variation by methods sensitive to principal dimensions of shape variation. These "principal dimensions" are of two sorts, geometric and statistical. Our purpose in this study is to demonstrate a landmark-based method of discerning principal dimensions of shape variation in samples of Miocene planktonic foraminifera.

Analysis of sets of homologous points by their shape coordinates results in a description of shape variation in terms of geometric patterns distributed over the form (Bookstein et al., 1985; Bookstein, 1991). Extraction and display of the uniform part of the shape variation, together with the analysis of the nonuniform part of the shape variation by the method of relative warps (Bookstein, this volume), allows greatly increased precision in both the description of that variation and the exploration of the structure of populations within the morphologic spaces so specified. In the present study, vector diagrams of statistical components of the uniform and nonuniform parts of shape variation are generated separately, and then considered jointly, for spirally coiled foraminiferal tests. Those results are compared with the results of a principal components analysis that produces vectors of correlated displacement at the landmarks.

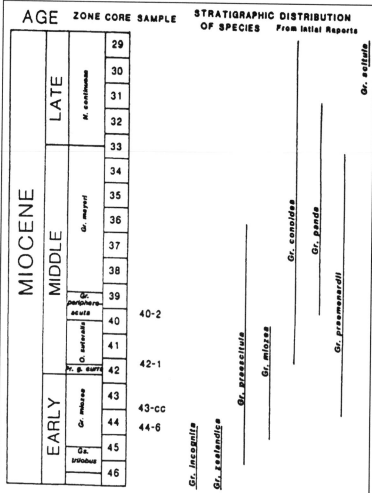

Figure 1. Stratigraphic section of Site 593 showing the position of the samples used in this study: 44-6, 43-cc, 42-1, 40-2. Range chart of named species is taken from the Initial Reports of Dea Sea Drilling Project, Leg 90 (Jenkins and Srinivasan, 1985).

Inspection of statistical components of the uniform and nonuniform parts of shape variation both provides insight into the meaning of change in familiar parameters of the abstract geometry of spirals and allows us to synthesize patterns of morphological change involving landmarks from putatively distinct morphological features. We apply these methods to individuals from Deep Sea

Drilling Project (DSDP) Site 593, which is located between New Zealand and Australia in the southwest Pacific (40° S latitude, 167° E longitude) and to seven named morphologies, presumably representing a single evolutionary lineage, that dominated populations of temperate planktonic foraminifera during the Early to Middle Miocene (Kennett and Srinivasan, 1983).

Analysis

Analysis of Components of Shape Variation

We have examined Early to Middle Miocene *Globorotalia* from the southwest Pacific using four samples from DSDP Site 593 (Figure 1). From each DSDP sample, twenty to thirty individuals were randomly chosen from all *Globorotalia* larger than 0.125 mm in diameter with complete, nonkummerform final chambers. Figure 1 also shows the ranges of named forms as presented in the Initial Reports of DSDP Leg 90 (Jenkins and Srinivasan, 1985).

For a study to support inference about the structure of morphological variation and about the relationship between current taxonomy and this structure of variation, individuals included in the study must have been drawn from an "experimental taxon" imposing a minimum of preconceptions about that structure on the study samples (Scott, 1966). Accordingly, we have chosen the individuals included in the study from all individuals of *Globorotalia* present in the samples without reference to the species-level designations that can be applied to some of those morphologies (as outlined in Tabachnick and Bookstein, 1990).

These *Globorotalia* take the form of an arrangement of chambers along an equiangular spiral. There are three major aspects of their morphological variation: the parameters of the spiral geometry, the shape of the chambers, and the shape of the aperture. Most characters used in traditional qualitative taxonomy can be re-expressed in terms of one or more of these aspects (Stainforth et al., 1975). The geometry of the first

and third of these can be described to a great extent by landmark points (Bookstein et al., 1985) that simplify their comparison. Figure 2 shows the landmarks in the apertural and spiral views of the foraminiferal test that are used in this study.

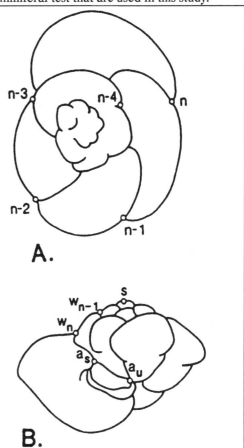

Figure 2. Landmark data collected on individuals of *Globorotalia*. Landmarks used in this study are shown with small circles. A) Spiral view. n: intersection of ultimate and penultimate chambers; n-1 to n-4: intersection of each chamber with the chamber preceding, in sequence. B) Apertural view. s: tip of spiral axis; w_n: intersection of ultimate and penultimate whorls; w_{n-1}: intersection of penultimate and antepenultimate whorls; a_s: spiral end of aperture; a_u: umbilical end of aperture.

The landmarks were selected to incorporate variations in spiral parameters (Raup, 1966) in both the spiral and apertural views. After designation of two landmark points as the baseline, two-point registration of the landmarks results in shape coordinates for each of the other landmarks (Bookstein et al., 1985; Bookstein, 1986, 1991). Joint variation of these shape coordinates expresses shape variation without any reference to size. When the statistics of these reconfigurations are reported properly, they express changes in spiral parameters as well as any other effective factors of variation in a manner nearly independent of the landmarks chosen to be the baseline points.

Multivariate analysis of shape coordinates results in diagrams of vectors of change at each landmark that are correlated from landmark to landmark. These vectors specify directions of displacement in shape space which can then be reinterpreted to find "common names" for the underlying shape processes, acting over the form, that may have resulted in the displacement. In this paper we compare the results of the principal components analysis of shape coordinates (Tabachnick and Bookstein, 1990) to separate analyses of the uniform and nonuniform parts of shape coordinate changes. The results of the first analysis are slightly baseline-dependent; the results of the later pair of analyses are baseline-independent.

The generation of shape coordinates by two-point registration has already discarded dilations, translations and rotations of the configuration. A uniform transformation changes squares into parallelograms; it is uniform over all the landmarks. A uniform transformation displaces each shape coordinate by a multiple of one single vector; the multiple is the landmark's distance from the base-

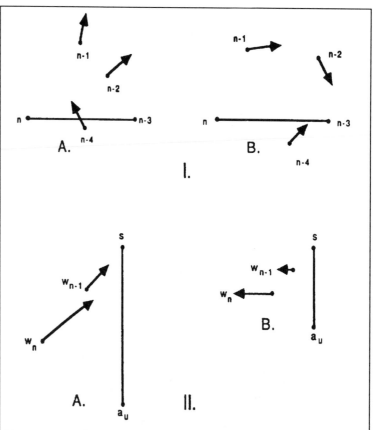

Figure 3. Vector diagrams produced by change, relative to the baseline shown, in each of three spiral parameters in the vicinity of the mean form. I. Spiral view: A) Change in angle of increment, θ: B) Change in spiral view expansion rate, r. II: Apertural view: A) Change in translation rate, t; B) Change in apertural view expansion rate, r.

line. This part of any observed shape change (for example, the difference of any single form from the population mean form) may be estimated as a factor score by the formula given in Bookstein (this volume), and its two statistical components are obvious in a scatter plot of this factor estimate. We call this vector the (two-dimensional) uniform factor.

What remains of variation and covariation of shape coordinates after the uniform part has been projected out is the nonuniform part of the variation. We extract statistical components of this part by the method of relative warps (Bookstein, 1988, 1989). These nonuniform components emerge in descending order of the ratio of sample variation to inverse geometric scale. The unit of this inverse scale is actually the energy it takes to move two landmarks "up" and two "down" on an idealized metal plate. The rationale for using bending energy in this way is presented in Bookstein (this volume). We will interpret the patterns of shape variation generated by these two methods and compare both to unpartitioned principal components of shape coordinates computed without distinguishing the uniform and nonuniform parts of transformations.

In this paper, the phrase "principal components" will refer to the results of the latter analysis. Statistical components of the uniform part of the variation are ordinary principal components of the two-dimensional uniform factor; they will be called the first and second uniform statistical components. Components of variation of the nonuniform part with respect to bending energy will be called relative warps.

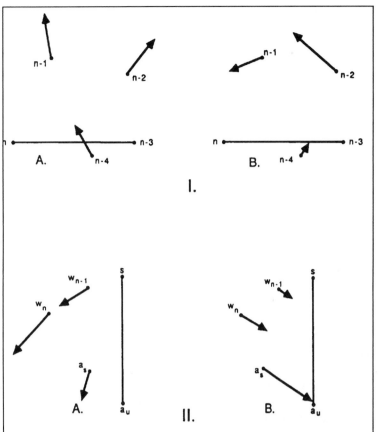

Figure 4. Vector diagrams of the principal components of shape coordinates. I. Spiral landmark shape coordinates: A) PC 1 (change in θ); B) PC 2 (change in r). II Apertural landmark shape coordinates: A) P1a (reflects change in t and aperture tilt); B) PC 2a (change in aperture width).

The first two principal components of shape variation for landmarks in spiral view are shown as vectors at each shape coordinate in Figures 3 IA and IB. These can be interpreted as corresponding to variation in two spiral parameters: θ, the angle at which new chambers are added, and r, the expansion rate (Tabachnick and Bookstein, 1990). The vector diagrams that model these spiral parameters are presented in Figure 4. Figures 3 IIA and IIB present the vector diagrams for the first two principal components of shape variation for landmarks in apertural view. The first principal component combines change in the spiral parameters r and t

(translation down the spiral axis) with a change in chamber tilt relative to the spiral axis. The second principal component of shape variation in apertural view landmarks expresses changes in apertural width (Tabachnick and Bookstein, 1990). Table 1 presents the principal components of the three pairs of shape coordinates in these two views.

A scatter-plot of the uniform factor for the landmarks in spiral view (Figure 5A) is essentially one-dimensional, implying that a single uniform statistical component underlies this variation. This component, explaining 93% of the uniform factor, is a projection of each landmark (weighted by its mean distance to the baseline) on the direction (.70, .71).

Table 1. Principal components of three morphologic spaces, DSDP sample data. A. Spiral view morphospace (see Figure 4), baseline points are n and n-3. B. Apertural view of morphospace (see Figure 4), baseline points are s and a_u. Principal components are computed using correlation matrices.

		A.				B.	
		PC1	PC2			PC1	PC2
% of trace :		49%	21%	% of trace :		52%	21%
Shape Coordinates				Shape Coordinates			
n-1	x	-.06	-.85	w_n	x	.45	.13
	y	.51	-.07		y	-.48	.14
n-2	x	.43	-.33	w_{n-1}	x	.50	.13
	y	.45	.35		y	-.43	.30
n-4	x	-.28	.21	a_s	x	.34	.52
	y	.52	.05		y	-.14	.77

A scatter-plot of the first two relative warps (Figure 5B) suggests that the nonuniform part of the variation in these samples is at least two-dimensional. The higher relative warps explain trivial amounts of sample variation, so we present only the first two (Table 2A). The ratios of sample variance to bending energy for these warps are 0.72 and 0.16. Figure 6 diagrams the uniform factor and the first two relative warps as they apply to the landmarks of the spiral view. The fraction of the total shape variation encompassed by each of these patterns can be approximated (in a slightly base-line-dependent way) as the fraction of the total shape coordinate variance "explained" by a regression of the spiral shape coordinates separately on the factor in question. The uniform factor "explains" 63% of total variance of the three spiral view landmarks. The first relative warp "explains" 68% of the total variance of these landmarks, and 17% of the total variance is "explained" by the second relative warp.

The correlation between the first uniform statistical component and the first relative warp is high ($r = -0.96$). Figure 7A illustrates vectors of shape change that result from summing the first uniform statistical component and the first relative

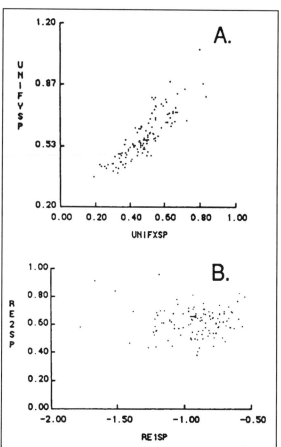

Figure 5. Scatter-plots of individuals in the shape subspaces for the spiral view. A) Uniform factor (UNIFXSP: x dimension; UNIFYSP: y dimension). B) Nonuniform variation (RE1SP: first relative warp; RE2SP: second relative warp).

warp with weights proportional to the fraction of total shape variation accounted for by each. The pattern that results is nearly the same as the model for change in θ presented in Figure 4. That is, variation in the spiral parameter θ is partly uniform and partly nonuniform. Figure 7B illustrates the change in this summed "component" as exemplified by individuals drawn from the study samples. By contrast, the second relative warp closely models change in r. The r-like vector diagram extracted by the principal component analysis (Figure 4) deviated from the pattern of change consequent upon change in r in that the vector at landmark n-4 was

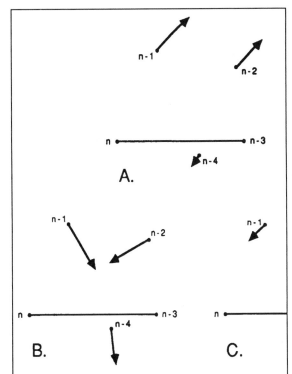

Figure 6. Vector diagrams for the spiral view. A) First uniform statistical component (93% of linear variation, 63% of total variation of the shape coordinates). B) First relative warp (68% of total variation). C) Second relative warp (17% of total variation).

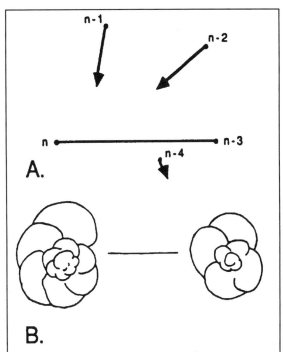

Figure 7. Combined uniform and nonuniform components of landmarks in spiral view. A) Vector diagram: result vectors are the summed first uniform statistical component and first relative warp at each landmark weighted by the proportion of total shape covariance explained. B) Sample individuals with extreme values for the first uniform statistical component.

rotated by 90°. Once the uniform part of shape variation is removed, this discrepancy disappears. The net shape coordinate variance explained by the summed first uniform statistical component and first relative warp is much greater than that explained by the second relative warp; variation of θ far outweighs variation of r as an explanation for covariation of the landmarks in spiral view.

For landmarks taken in the apertural view, scatterplots of the uniform factor and the relative warps suggests that there are two dimensions to both (Figure 8). The first uniform statistical component is estimated as the summed projection of the landmarks, each weighted by its mean distance to the baseline, upon the vector (-.68, .73); the second uniform statistical component is at 90° to the first. The first two relative warps are as in

Table 2B. Figure 9 illustrates the uniform statistical components and the relative warps as vector diagrams. The first uniform statistical component accounts for 50% of the total shape coordinate variance of these landmarks, and the second accounts for 22%. The first relative warp accounts for 43% of the total shape variance, and the second for 19%. (As for the landmarks in spiral view, these fractions of variance overlap.) The ratio of sample variance to bending energy is 0.65 for the first relative warp, 0.21 for the second, and negligible for the others.

Although the uniform and nonuniform parts of shape space are orthogonal in Procrustes distance (the sum of squared distances by which the landmarks have moved, in the coordinate system for which that sum is minimized), they usually

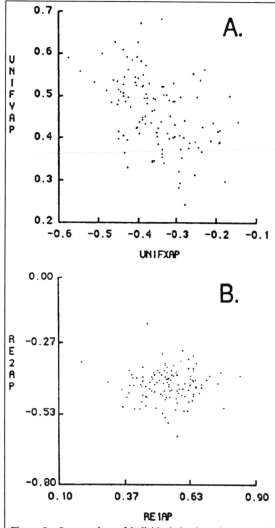

Table 2. Relative eigenvectors of the nonuniform part of shape variation. A. Spiral view landmarks: baseline points are n and n-3. B. Apertural view landmarks: baseline points are s and a_u. Relative eigenvalues represent the ratio of variance to bending energy.

	A.				B.		
		RE 1	RE 2			RE 1	RE 2
ratio of var. to bending energy :		0.72	0.16	ratio of var. to bending energy :		0.62	0.23
Shape Coordinate				Shape Coordinate			
n-1	x	.38	-.51	w_n	x	.09	.22
	y	-.63	-.46		y	-.37	.19
n-2	x	-.54	-.40	w_{n-1}	x	.13	.20
	y	-.30	.71		y	-.20	.05
n-4	x	.03	-.23	a_s	x	.49	-.36
	y	-.48	.10		y	.41	.56

Figure 8. Scatter-plots of individuals in the subspaces of the aperture view. A) Uniform factor (UNIFXAP: x dimension; UNIFYAP: y dimension). B) Nonuniform variation (RE1AP: first relative warp; RE2AP: second relative warp).

exhibit some correlation. For the landmarks in apertural view, the correlation between the first uniform statistical component and the first relative warp is 0.66; between the first uniform statistical component and second relative warp, -0.49. These correlations are unremarkable; they reflect the statistically arbitrary nature of the projection of shape variation into these two incommensurate subspaces. The first principal component represents the pooled effect of correlated uniform and nonuniform parts, a combination of change in spiral parameters at the whorl landmarks and tilt at the aperture landmark; this should be considered to embody two separate, but correlated, components of morphological variation.

The first relative warp expresses a correlated change in spiral parameters and apertural width; the second primarily reflects changes in aperture tilt, associated with small changes in the whorl landmarks that seem unrelated to spiral parameters. What the second uniform statistical component represents also seems unrelated to spiral parameters. The second principal component does not resemble any of the separate uniform components or relative warps in that it associates change in apertural width with a certain direction of change in whorl landmarks.

One identifiable morphological change that could produce the transformation seen here as the first uniform statistical component is a change in the chamber tilt relative to the coiling axis. This would jointly alter translation away from the spire and apertural position relative to the coiling axis, as shown in Figure 10.

Individual Variation at Site 593

Tabachnick and Bookstein (1990) demonstrate, using the principal components, that the *Globorotalia* in these samples are continuously, and roughly elliptically, distributed in both the spiral and apertural landmark spaces. This result is unchanged when the shape variation is partitioned into its uniform and nonuniform parts. The distributions of the individuals from site 593 are shown in Figure 11A for the spiral landmarks and Figure 11B for the apertural landmarks. There are no significant differences among the sample means in the space of the first uniform statistical component and second relative warp for the spiral view. As in the space of the principal components of the apertural view, sample 42-1 is significantly different from all other samples in the space of the first uniform statistical component and the first relative warp of the apertural view. Sample 42-1 lies at the extreme high end of the distributions for both components. There are no significant differences among the means of the other samples.

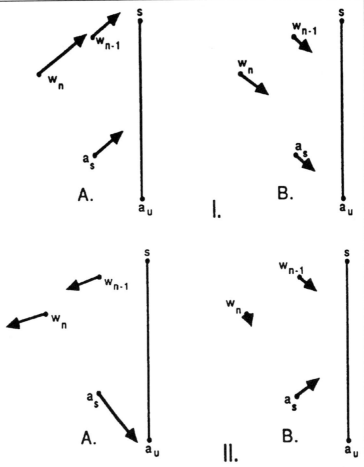

Figure 9. Vector diagrams for the apertural view. I. Uniform factor: A) First uniform statistical component (70% of uniform variation, 50% of total variation of the shape coordinates). B) Second uniform statistical component (30% of uniform variation, 22% of total variation). II. Nonuniform part: A) First relative warp (43% of total variation). B) Second relative warp (19% of total variation).

We have applied this analysis to individuals representing seven named morphologies taken from Kennett and Srinivasan (1983). With one exception, the named specimens fall within the regions occupied by the scatters from site 593 (Figure 12). This individual, labelled *G. archeomenardii*, lies slightly outside the region occupied by the scatters in the apertural landmark morphospace; the morphotype represented by this specimen is apparently not present in the DSDP samples.

Discussion

Interpretation of morphological variation is enhanced by decomposing a generalized system of landmark variation into uniform and nonuniform geometric and statistical components. The particular advantage for interpretation depends on the landmarks and the biologic process whose effect their locations record. For these landmarks, one named aspect of morphological variation, "angle of increment," seems to combine uniform and nonuniform aspects of landmark transformation. Other named aspects of morphological variation can be seen to be primarily nonuniform (e.g., "expansion rate") or primarily uniform (e.g., "chamber tilt"). Principal components analysis of the shape coordinates of the spiral view suggested that θ and r are also statistically orthogonal components of morphological variation for the *Globorotalia*. Yet that result was not completely clear; there were some unexplained deviations between the vector diagrams of the principal components and of the parameters they almost model. Examination of the separate uniform and nonuniform parts and the correlations between them provided clearer interpretations of spiral parameter variation.

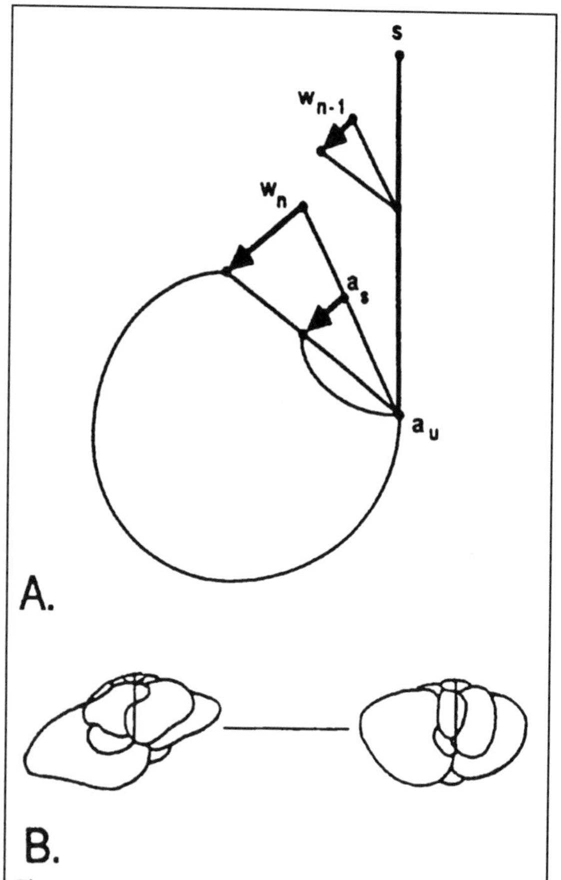

Figure 10. A linear transformation which could result from a change in chamber tilt at a constant translation rate. A) Vector diagram: the position of angular offset for each whorl point is based on the location of the apertural end of the coiling axis as each whorl is created. B) Variation in the first uniform statistical component of landmarks in apertural view as exemplified by sample individuals with extreme values for that component.

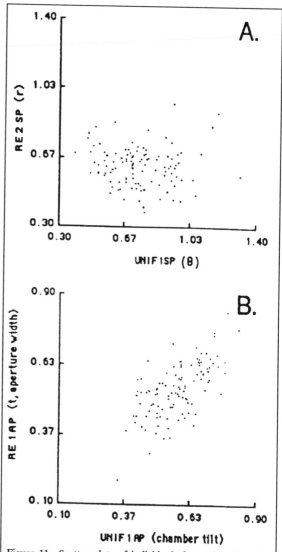

Figure 11. Scatter plots of individuals from Site 593. A) Spiral view variation (UNIF1SP: first uniform statistical component, θ; RE2SP: second relative warp, r). B) Apertural view variation (UNIF1AP: first uniform statistical component, chamber tilt; RE1AP: first relative warp, t and aperture width).

The landmarks in spiral view are homologous among themselves in their location on the biologic geometry of the form. This "serial homology" (Sattler, 1984) is a different sense of homology than is usually applied to landmarks. Each landmark is located where the end of the outer edge of a chamber intersects the wall of the previous one, as projected into two dimensions. The first landmark is located between the ultimate and penultimate chambers, the second between the penultimate and antepenultimate chambers, and so on. Because of this serial homology on the spiral, we would not expect uniform transformations to have any particular biologic meaning.

The apertural view landmarks, on the other hand, combine two features of the form, spiral parameters and aperture shape, which may react independently to the biologic processes that cause morphologic variation. By examining covariation among the landmarks, we can ask whether such features should be considered distinct parts of the morphology. Separation of landmark displacements into uniform and nonuniform factors suggests that these morphologic parts could be interpreted to interact in three ways: responding coherently as a single morphologic feature would react to a single process; responding by correlated change reflecting changes due to separate processes that happen to be correlated for these populations; and responding by independent change of the separate features.

The two principal components of the landmarks in apertural view suggest that displacement of these landmarks is dominated by covariation among the effects of spiral parameters acting on the whorl landmarks (w_n and w_{n-1}) and variation in either aperture width or tilt acting on the aperture landmark (a_s). Analysis of uniform and nonuniform parts of the same landmark variation allowed identification of a prominent uniform component that reflects tilt of the chambers relative to the coiling axis. The aperture and whorl landmarks respond coherently to this single process of uniform displacement. Identification of a prominent nonuniform component for variation of the land-

marks in apertural view suggests that the morphologic "parts" represented by these landmarks also encode biometrically separate features; the aperture exhibits variation that is not merely a result of morphologic variation also affecting the whorl landmarks. Definitive conclusions on this point would require observation of whether actual changes in the aperture result in predicted changes in whorl characters. Our results suggest that this is an interesting question for further investigation.

What we see from the variety of uniform and nonuniform components is that to some extent the aperture and whorl landmarks respond as a single feature to a process of uniform displacement. Changes in aperture are also correlated with changes in spiral parameters in a way which may or may not express some shared underlying process, but which does not suggest that they are a single feature. The feature we call "aperture" might be partitioned into two parts: variation in the chamber as a whole from which the "aperture" cannot be distinguished, and variation of the aperture that is independent of the rest of the test.

Differences in tilt of the whole chamber relative to the coiling axis are more important in explaining variation of landmarks in apertural view than is either change in spiral parameters or variation in the aperture as a separate morphologic feature. Most qualitative taxonomy does not explicitly mention this parameter of morphologic variation; what is discussed instead is the "compression" of the test, a measure of the distance from the tip of the spire to the umbilical opening in apertural view. It is common to orient the test of *Globorotalia* in apertural view so that the plane through the longest extension of the chambers is horizontal in the viewing plane. As a result, the test appears "compressed" as the chambers tilt relative to the axis of coiling (an effect identical to tilting the axis of coiling) since this tilt decreases the height from spire tip to umbilicus in that orientation. By explicitly examining landmark transformation relative to the coiling axis, we have been able to discern that this chamber tilt (or tilt of the coiling

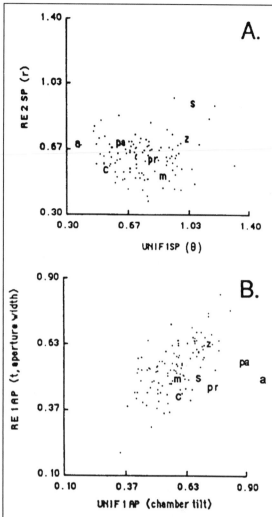

Figure 12. Distribution of the named species exemplars of Kennett and Srinivasan (1983) in each of the morphospaces. z: *G. zealandica*; pr: *G. praescitula*; m: *G. miozea*; s: *G. scitula*; a: *G. archeomenardii*; pa: *G.panda*; c: *G. challengeri*. A) Space of landmarks in spiral view. B) Space of landmarks in apertural view.

tion for the landmarks taken in spiral view, but do vary to some extent with stratigraphic position for the landmarks taken in apertural view. Named morphologies are scattered across the occupied spaces with continuous variation between them, suggesting that, insofar as the variation we have measured validly encompasses the morphology, these "species" names are merely labels for areas of continuously occupied morphologic space. As discussed in Tabachnick and Bookstein (1990), variation in occupation of morphologic space from one stratigraphic sample to another could be used to resolve phylogenetic relationships; yet the species names, while useful as designations of areas within continuous distributions, do not seem to correspond to discrete, gap-bounded parts of the variation. Stratigraphic changes in this lineage seem to involve changes in the shape and position of occupied morphospace without the appearance of gaps in variation. This fits neither a model of strict anagenesis nor one of strict cladogenesis.

Conclusion

The decomposition of variation in landmark configurations into their uniform and nonuniform parts can enhance our ability to interpret shape variation. For Miocene *Globorotalia* this increase in interpretability is fourfold. First, we may examine the geometric details of landmark displacement resulting from variation in known parameters of abstract geometry. We found that θ, the angle of increment of an equiangular spiral, has both a uniform and a nonuniform part, but that r, the expansion rate, is primarily a nonuniform phenomenon. Second, we are able to assess the relative importance of parameters of abstract geometry in explaining morphologic variation. In this study we found that change in the angle of increment explains much more of the variation of landmarks in spiral view than does change in expansion rate. Third, separation of the landmark variation into its uniform and nonuniform parts has allowed us to identify a previously obscured statistical component of morphologic variation and to demonstrate that change in this component

axis), a previously obscured component of variation, accounts for much of the variation of landmarks in apertural view.

The uniform and nonuniform subspaces of shape variation are continuously occupied in both the spiral and apertural view. The distributions of our samples do not change with stratigraphic posi-

accounts for a large fraction of total variation of the landmarks taken in apertural view. Fourth, covariation of suites of landmarks can be used to examine the independence of parts of the morphology. We found that the aperture and whorl landmarks react to some extent as a single feature and to some extent as independent parts.

This increasingly detailed description of shape variation still supports the conclusions of Tabachnick and Bookstein (1990). The relationship of exemplars of named morphospecies to distributions of individuals in morphologic space strongly suggests that the names do not label discrete species. The distributions of individuals in morphologic space are continuous and roughly elliptical, and there is stratigraphic variation between samples in these distributions.

Acknowledgments

The interpretation of uniform variation as reflecting morphologic coherence was stimulated by discussions with Miriam Zelditch on developmental integration. G. Smith, J. Kitchell and D. Eernisse offered editorial suggestions on the manuscript. The research was supported in part by NSF Grant BSR-8708563 to J. Kitchell and N. MacLeod, and NIH Grant GM 37251 to F. L. Bookstein.

References

Banner, F. T. and F. D. M. Lowery. 1985. The stratigraphical record of planktonic foraminifera and its evolutionary implications. Spec. Papers in Paleo., 33:117-130.

Bookstein, F. L. 1986. Size and shape spaces for landmark data in two dimensions. (With discussion and rejoinder). Stat. Sci., 1:181-242.

Bookstein, F. L. 1989. Comment on "A survey of the statistical theory of shape," by David G. Kendall. Stat. Sci., 3:99-105.

Bookstein, F. L. 1991. Morphometric tools for landmark data. Book manuscript, accepted for publication, Cambridge Univ. Press.

Bookstein, F. L., B. Chernoff, R. Elder, J. Humphries, G. Smith and R. Strauss. 1985. Morphometrics in evolutionary biology: the geometry of size and shape change, with examples from fishes. The Academy of Natural Science Philadelphia, Spec. Publ. No. 15, 277 pp.

Jenkins, D. G. and M. S. Srinivasan. 1985. Cenozoic planktonic foraminifers from the Equator to the Subantartic of the Southwest Pacific. in Initial reports of the deep sea drilling project, 90 (J. H. Blakeslee ed.). U.S. Gov. Print. Off., Washington, D.C.

Kennett, J. P. and M. S. Srinivasan. 1983. Neogene planktonic foraminifera: a phylogenetic atlas. Hutchinson Ross Publ. Co., Stroudsburg, PA.

Raup, D. M. 1966. Geometric analysis of shell coiling: general problems. J. Paleontol., 40:1178-1190.

Sattler, R. 1985. Homology -- a continuing challenge. Sys. Bot., 9:382-394.

Scott, G. H. 1966. Description of an experimental class in the *Globigerinidae* (foraminifera), 1. New Zealand J. Geol. and Geophys., 9:513-540.

Stainforth, R. M., J. L. Lamb, H. Luterbacher, J. H. Beard and R. M. Jeffords. 1975. Cenozoic planktonic foraminiferal zonation and characteristics of index forms. Univ. Kansas Paleo. Contrib. Art., 62:1-425.

Tabachnick, R. E. 1981. Morphologic variation in Miocene *Globorotalia* bulloides, D'Orbigny, New Port Bay, California. Unpubl. M.S. Thesis, Univ. California, Los Angeles.

Tabachnick, R. E. and F. L. Bookstein. 1990. The structure of individual variation in Miocene *Globorotalia*, DSDP Site 593. Evolution, 44:416-434.

Chapter 14

Morphometrics and Evolutionary Inference:
A Case Study Involving Ontogenetic and
Developmental Aspects of Foraminiferal Evolution

Norman MacLeod[1,2] *and Jennifer A. Kitchell*[1]

[1] *Department of Geological Sciences and Museum of
Paleontology, University of Michigan, Ann Arbor, MI 48109*
[2] *Present Address: Department of Geological and
Geophysical Sciences, Princeton University
Princeton, NJ 08544*

Abstract

In order to illustrate the use of a predominately landmark-based approach to morphometric analysis, we examine covariant aspects of intraspecific morphologic variation in the planktic foraminiferal species *Subbotina linaperta* by comparing samples prior to and following a well-documented environmental disturbance. These investigations document patterns of size and shape variation in the last three growth stages of shell or test formation, including the terminal ontogenetic phase of sexual maturity. Results indicate that a statistically significant reduction in test size (approximately 30%) occurred between the pre-disturbance sample and the post-disturbance sample. This agrees with previous estimates of the size difference between these samples based on more traditional multivariate size indices. Previous analyses have also shown this size difference to result from a change in the size/frequency distribution of adult individuals and not through the differential preservation of juvenile forms. Post-disturbance mean landmark configura-

tions in shape space are very similar to mean landmark configurations characteristic of smaller-sized pre-disturbance individuals suggesting that changes in developmental patterns may have been responsible for the observed differences in test size and shape. Over a series of size classes, both uniform and non-uniform aspects of shape variation were estimated and used to identify the predominant modes of geometric variation. With respect to a uniform deformation of the shape space, both samples were found to be characterized by a radial expansion/compression and left-to-right/right-to-left lateral translation of the non-constrained shape coordinates, though fundamental differences exist between the samples in the way the predominant axes of variation are oriented relative to these two deformational modes. Non-uniform within-group shape variation primarily involved an asymmetric radial expansion of landmark locations that may be descriptively attributed to a special case of a quadratic deformational mode while between-group shape variation appears to be brought about primarily by a tendency toward less inter-chamber shape differentiation to be exhibited by post-disturbance populations.

Finally, we note that despite the power of morphometrics to efficiently summarize vast amounts of geometric data and provide strategies for the testing of alternative evolutionary hypotheses, solely descriptive models of morphological change are insufficient to the task of explaining how evolutionary processes might have brought such changes about. This limitation derives from the historically contingent nature of evolutionary processes and the fact that morphometric analyses per se maintain an indifferent position with respect to questions of phylogenetic relationship. A primary challenge to the future development of evolutionary morphometrics lies in the ability of its practitioners to employ tools of descriptive and exploratory geometry within a research context that explicitly differentiates between alternative hypotheses of phylogenetic ancestry, and to derive a series of expected patterns of morphological transformation that are based on an understanding of developmental as well as phylogenetic processes.

Introduction

Bookstein (1988) has recently described morphometrics as the study "not of form per se but of covariances with form". This statement alludes to the central problem of developing evolutionary theories, namely that evolution is a contingent process far more constrained by phylogenetic history than by mechanical or geometric optima. Thus, descriptive models of shape change are necessary but not sufficient to the task of explaining morphological change.

In this study of intraspecific variation, we contrast variations in form prior to and following a profound environmental disturbance, as a means of quantifying covariant morphological aspects of a species' response. Within this context, morphometric analyses have been used to provide an efficient means of describing intraspecific changes in form by examining shifts i) across an environmental disturbance, and ii) during different stages of the organism's ontogeny. Although we are unable to posit specific developmental pathways controlling the potential for change in this fossil form, we have made our measurements on a series of separate growth stages within each specimen in order to gather information that can be directly related to the biological process of development. Consequently, these data show strong correspondence between the factors of morphological change that can be defined mathematically, via landmark-based morphometric analysis, and a series of *a priori* models of growth processes and changes in the developmental program.

In an earlier investigation (MacLeod et al., submitted), we examined both geographic and temporal aspects of dwarfing within middle and late Eocene populations of the planktic foraminiferal species *Subbotina linaperta* Finlay. That study was initiated by Keller's (1986; Keller et al., 1987) report of a qualitative size reduction in this species occurring at DSDP Site 612 that was coincident with the first occurrence of microtektites belonging to the late Eocene North American tektite layer at this locality. Presence of the tektite layer in these late Eocene sediments provides unambiguous geochemical and mineralogical evidence for a geologically sudden and widespread environmental disturbance in the form of a terrestrial impact event involving an extraterrestrial object (Barnes and Barnes, 1973; Glass, 1982; Shaw and Wasserburg, 1982). While reductions in planktic foraminiferal test size have been associated with other types of environmental fluctuations and with planktic extinction events (e.g., Berggren 1965, Malmgren and Kennett 1972, Bé et al. 1973, Hecht 1974, Blow 1979, Smit 1982, Erez 1983, Keller et al. 1987), dwarfing in foraminifera has received relatively little study in terms of the quantitative description of patterns of variation found in the dwarfed populations.

As an alternative to the more temporally and geographically comprehensive dataset analyzed in the earlier investigation, we now focus exclusively on two samples from DSDP Site 612: one located in middle Eocene sediments stratigraphically just below the first occurrence of microtektite material, and one located in the overlying late Eocene sediments containing impact ejecta. As opposed to a traditional multivariate morphometric analysis of distances between relocatable landmarks (as was carried out in parts of the earlier study), here we present the results of analyses of variation in the relative positions of the individual landmark coordinates themselves. Aside from providing an illustration of use of some of the newer landmark-based data analytic techniques discussed in the Workshop, the application of different methods of analysis to similar data provides a check on the consistency and methodological robustness of conclusions derived from the earlier investigation.

Materials and Methods

To compare patterns of variation in size and shape in these *S. linaperta* populations, two samples were obtained from DSDP Site 612: the first (sec. 21-5:132-140 cm) representing populations existing prior to the occurrence of tektite debris within the core (herein referred to as the pre-disturbance sample), and the second (sec. 21-5:114-118 cm) representing forms whose occurrence coincides with that of the tektite debris (herein referred to as the post-disturbance sample). A total of 110 individuals were measured from the pre-disturbance interval and 167 individuals from the post-disturbance interval. Since these specimens constitute all individuals of this species found in a random split of the processed residue $> 150\ \mu$, all size classes present in the original fauna (averaged over a relatively short interval of time) are likely to be represented in the samples in their correct relative proportions.

For each specimen, test size and shape were quantified by locating the (x,y) coordinate positions of 11 landmark points (Figure 1). This set of landmarks was chosen because the measured points lie at the ends of the major and minor axes of the three elliptical chambers that comprise the final whorl, thereby insuring some measure of consistency in the sampling of morphometric variation over the entire

test as well as maximizing the ability to detect developmentally significant patterns of variation in the phenotype. However, it must also be recognized that while each of these landmarks is unambiguously relocatable, several are identified as extremal locations and therefore cannot be regarded as having any necessary biological correspondence between individuals.

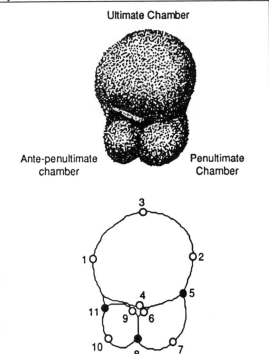

Figure 1. Location of the landmark points used to quantify morphometric variation in *S.linaperta*, shown in umbilical view. Filled circles represent homologous point locations, hollow circles represent positions corresponding to local extrema.

For each specimen, test size was estimated using Bookstein's (1986) centroid size index (S), which is the sum of all squared distances from each landmark to their joint centroid. This size measure has the desirable property of being uncorrelated with all possible shape variables under a model of randomly distributed digitizing error of constant variance at each landmark location. In order to improve the interpretability of the resulting size estimate, however, all sizes are reported as the square root of centroid size ($\sqrt{S}$) so that the units of the size measure and the interlandmark distances used to estimate size correspond. Although use of the centroid size index may not be appropriate in some cases (e.g., if the outline of the shape does not contain the centroid or if a number of the line segments spanning the distance between the centroid and individual landmarks cross the boundary of the organismal outline), it appears to be a convenient multivariate size measure for many globigeriniform planktic foraminifera.

Once individual size had been quantified, each specimen's set of 11 landmark (x,y) coordinates were converted to "shape coordinates" via the two-point registration technique with the line segment between landmarks 1 and 2 serving as the baseline. Selection of the base line coordinates is wholly arbitrary in terms of the scaling calculations with no systematic errors resulting from alternative baseline specifications (Bookstein, 1986). However, there do exist differences in the ease with which results of particular analyses can be interpreted, which depend, to some extent, on baseline placement. In the present study, coordinates 1 and 2 were chosen for the baseline to take advantage of a relatively long shell dimension for size standardization as well as to avoid any unnecessary complications that might arise from using a baseline that spans developmentally distinct units of test structure (e.g., the individual chambers). These shape coordinates were then used to determine average landmark configurations for each of the two samples and, within each sample, to examine the nature of shape variation among a series of eight size classes.

Multiple regression analyses of centroid size ($\sqrt{S}$) on the individual shape coordinates were used to determine whether or not any linear patterns of size/shape variation (allometry) were present in the landmark data. Utilization of the centroid size index for allometric analyses circumvents much of the ambiguity that is inherent in many traditional forms of allometric regression analyses owing to the essentially ad hoc nature of the definitions of "size"

and "shape" employed by these methods (see Bookstein, 1989a). However, since centroid size (both S and $\sqrt{S}$) is statistically uncorrelated with any shape space for homologously located landmark coordinates, the relationship between centroid size and the size-standardized location of the landmark points can be assessed by traditional linear regression techniques with confidence that the resulting statistics will be free from artifactual bias (Bookstein, 1986).

Finally, uniform and non-uniform aspects of shape variation were studied as deformations of a 2-dimensional thin plate spline passing over the locations of the mean shape coordinates for each sample and whose undulations are proportional to the direction and amount of variability in each coordinate location across all measured specimens. In this context, the uniform aspects of shape deformation can be envisioned as accounting for the observed variability in shape coordinate location via a tilting of the thin plate spline with projection of the resulting changes onto a common plane. During this tilting procedure, the baseline remains fixed. Thus, the magnitude of the resulting displacement vectors increases with increasing distance from the baseline and the orientation of the displacement vectors on opposite sides of the baseline changes by 180°. Alternatively, non-uniform aspects of shape variation can be envisioned as a decomposition of the bends in the thin plate spline into a series of hierarchically arranged warps that express progressively more general (less regional) aspects of the shape deformation. When interpreting the results of these uniform and non-uniform aspects of shape deformation it is important to remember that they are conceptually independent ways of representing the variability of landmark location in the size-normalized shape space. Consequently, both uniform and non-uniform representations of shape deformation exhaustively account for all shape variability within each sample. Algorithms for calculation of the uniform and non-uniform (e.g., principal warps and relative eigenvectors) aspects of shape variability are given in Bookstein (1989b, this volume).

Results

Test size distributions for pre-disturbance and post-disturbance samples are given in Table 1. Mean centroid size for the pre-disturbance sample is 10.67 μm while the post-disturbance sample mean centroid size is 7.54 μm. This constitutes a statistically significant (via 2-sample t-test; $p < .05$) size decrease across the tektite datum of slightly less than 30 percent, a result that is in good agreement with previous estimates of size variation between these two samples based on a multivariate analysis of selected interlandmark distances in all three dimensions (MacLeod et al., submitted). In addition, prior SEM observations of test microstructure revealed clear indications that a layer of gametogenic calcite has been deposited on the surface of tests comprising the smaller, post-disturbance sample, thus confirming that the observed size variation is not the result of mechanical sorting or the differential preservation of juvenile individuals, but represents a change in the size-frequency distribution of adult organisms.

Mean shape coordinate configurations for the eight size classes shown in Table 1 are presented in Figure 2. Although several obvious changes in shape in both the pre-disturbance and post-disturbance samples are apparent, the overall nature of observed shape changes is best revealed by superimposing the size class specific landmark configurations upon one another, as in Figure 3. From this figure, it is clear that (holding distance 1-2 constant) as size increased, pre-disturbance specimens tended to develop chambers that were progressively more elongate in the radial direction and that this radial elongation or widening of the ultimate, penultimate and ante-penultimate chambers does not appear to have taken place proportionately through all size classes. While grossly similar variations in shape coordinate location also seem to characterize the post-disturbance sample, the absolute range of variation in landmark location is notably smaller owing, no doubt, to the smaller range of size variation.

		Pre-Disturbance Sample		Post-Disturbance Sample	
Size Class	Centroid Size Index	Freq	% Freq	Freq	% Freq
1	5.54 - 6.90	5	4.50	60	35.93
2	6.90 - 8.26	9	8.11	65	38.92
3	8.26 - 9.63	21	18.92	28	16.77
4	9.63 - 10.99	26	23.88	14	8.38
5	10.99 - 12.36	26	23.88	0	0.00
6	12.36 - 13.72	15	13.51	0	0.00
7	13.72 - 15.08	4	3.60	0	0.00
8	15.08 - 16.45	4	3.60	0	0.00

Table 1. Test size distributions.

*Centroid size index = $(\sqrt{S})$ (see text for discussion)

Overall mean landmark shape configurations for pre- and post-disturbance samples are compared in Figure 4 which shows that observed differences in mean test shape can be attributed to a lesser degree of shape differentiation between chambers in the post-disturbance sample. There is also a suggestion that the ultimate chamber in the pre-disturbance sample may be slightly more elongate tangent to the spiral curve of chamber placement than is the case in the post-disturbance sample, though it is not evident from this diagram whether this is the result of a change in the relative size of the ultimate chamber or a change in the nature of the trochospiral coil (or some combination of the two). Finally, there appears to be a difference in the relative placement of the umbilicus (the central triangle) in the two samples that may indicate that the smaller-sized post-disturbance *S. linaperta* tests exhibit a tighter trochospiral coil.

While it is clear from Figure 3 that some pre-disturbance size classes exhibit mean shapes that closely approximate the overall mean shape of the post-disturbance sample, this relationship must be quantified to be of use within a systematic investigation. One way this quantification may be brought about is by assessing the degree of correspondence between a set of interlandmark distances that completely and redundantly locate the positions of individual landmarks relative to their neighbors. Strauss and Bookstein (1982) have devised a protocol for constructing such a network of inter-landmark distances (termed truss analysis).

For the set of 11 landmarks used to quantify morphologic varia-tion in this study, Figure 5 shows one possible truss network of 21 inter-landmark distances. As suggested by Strauss and Bookstein (1982), the root sum of squares index provides a convenient measure of shape simi-larity:

$$RRS = \sum_{n=1}^{21} [(d_{obs} - d_{exp}) / d_{obs}]^2$$

where: d_{exp} = distance between landmarks in post-disturbance mean shape, d_{obs} = distance between corresponding landmarks in mean shape of a pre-disturbance size class.

Based on the root sum of squares similarity index, both pre-disturbance size classes 1 and 3 can be identified as exhibiting marked similarities with the shape of the post-disturbance *S. linaperta* popula-tions (Figure 5).

Can the patterns of size/shape variation exhibited by the pre-disturbance and post-disturbance *S. linaperta* landmark datasets be inter-preted as manifestations of an allo-metric pattern of growth? *F*-ratios for multiple regressions of $\sqrt{s}$ on the shape coordinates of the 9 non-constrained landmarks are 1.413 and 1.732 for the pre-disturbance and post-disturbance datasets respec-tively (d.f. regression = 18; d.f. deviation = 91 [pre-disturbance], 148 [post-disturbance]). These

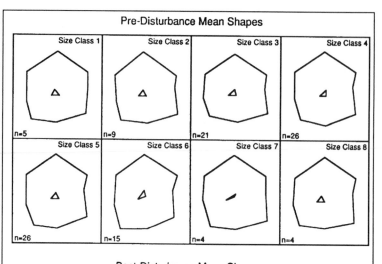

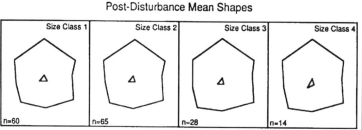

Figure 2. Mean landmark configurations in shape space for the eight size classes of *S. linaperta* in thepre-disturbance sample and the first four corresponding size classes in the post-disturbance sample (see Table1).

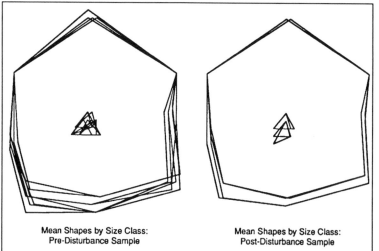

Figure 3. Superimposed mean landmark configurations in shape space for the eight size classes of the pre-disturbance sample and the four size classes of the post-disturbance sample.

values are both non-significant ($p <$ .05) indicating that no linear change in test shape relative to test size is apparent in either sample. Of course, this does not mean that there is no change in overall test shape with increasing test size, only that what changes are there fail to correspond to a strictly linear model of allometric size/shape change in any statistically significant manner.

For the purpose of characterizing non-allometric modes of shape variation, one can consider growth gradients to be an aspect of either uniform or non-uniform shape change whose effects may vary depending on location across the form. In this way, "shape" can be thought of as a series of tilts or bends that represent deviations of the landmarks from strictly linear types of variation. Through the method of thin plate spline interpolations of the shape space (Bookstein, 1989b; this volume), it is convenient to decompose this variation into a hierarchy of "localizability" whose major constituents can, in turn, be related to a series of *a priori* defined geometric models of shape deformation.

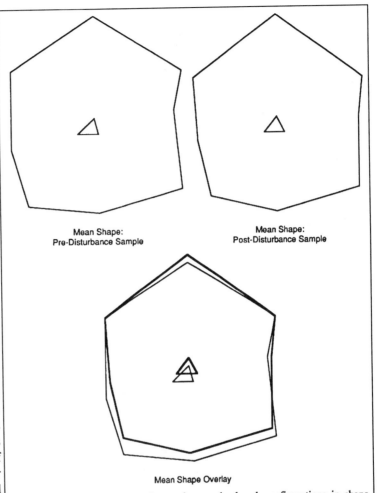

Figure 4. Separate and superimposed mean landmark configurations in shape space for pre-disturbance and post-disturbance samples. Post-disturbance mean shape is indicated by thick lines in the mean shape overlay.

Analysis of the uniform aspects of shape change involves an eigenanalysis of the uniform shape components as defined by Bookstein (this volume). These calculations result in the extraction of two eigenvectors whose relative lengths are proportional to the amount of shape variation accounted for by each axis and whose orientation represents the direction of tilt in the thin plate spline which, when the resultant positions of the landmarks are projected onto a 2-dimensional plane that contains the mean shape, appears as a uniform shearing of the mean shape with respect to the fixed baseline. Results of this analysis are summarized in Figure 6, which reveals several marked differences in this uniform aspect of shape variation to exist between the pre- and post-disturbance samples. In general, the uniform shape space of the post-disturbance sample appears to be much more cleanly partitioned into axes describing shell expansion/compression (uniform factor 1) and differences in the lateral translation of shape coordinates (uniform factor 2) that may reflect a tightening or loosening of the trochospiral coiling mode. Each of these components of variation is also present in the uniform shape space of the pre-disturbance sample, but here they are arranged in such a way as to always link test expansion/compression and lateral translation along the major axes of shape variability. Similarities between these two samples do exist, however, in the relative amount of shape variability accounted for by each of the two uniform shape deformational axes.

Principal warps of the "bending-energy" matrix for pre-disturbance and post-disturbance samples are given in Table 2. These represent mutually orthogonal non-uniform aspects of the observed shape variation ordered in terms of their localizability. Principal warps with high bending energies, as measured by their respective eigenvalues (e.g., 28.523 and 25.248 for pre-disturbance and post-disturbance samples, respectively), exhibit a strong contrast between very high and very low loadings with relatively high loadings occurring at landmarks that are grouped closely together in such a way as to define a very localized region of the form (i.e., landmarks 4, 5, and 6, which together define the umbilical area of the test). These first few principal warps represent patterns of variation occurring between landmarks in these localized regions that are quite pronounced, thus reflecting the presence of substantial bends in the thin plate spline at these locations.

On the other hand, principal warps of relatively low bending energy (e.g., 0.628 and 0.715 for pre-disturbance and post-disturbance samples respectively) exhibit a markedly lower degree of contrast between relatively high and relatively low loadings on the landmarks themselves reflecting the fact that these warps represent non-uniform aspects of morphologic variation that have a broader effect on the form as a whole. In both samples, the last three principal warps (warps 9, 10 and 11), representing the most global aspects of non-uniform shape variability, had no bending energy (eigenvalue = 0.0), and so describe uniform tilts of the thin plate spline that have already been ana-

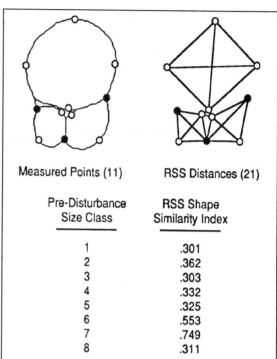

Measured Points (11)	RSS Distances (21)

Pre-Disturbance Size Class	RSS Shape Similarity Index
1	.301
2	.362
3	.303
4	.332
5	.325
6	.553
7	.749
8	.311

Figure 5. Upper figures: positions of the 11 landmarks and 21 interlandmark distances used to measure shape similarity between the post-disturbance mean shape and the mean shapes of eight pre-disturbance size classes. Lower table: Root-sum-of-squares shape similarity indices for the comparison of post-disturbance mean shape with mean shapes for each of the eight pre-disturbance size classes (see text for discussion).

lyzed (by a slightly different method) in the preceding section.

Diagrams of the two most global non-uniform principal warps for pre-disturbance and post-disturbance datasets are shown in Figure 7. From these diagrams it is evident that the patterns of landmark displacement represented by these non-uniform but relatively global principal warps are similar for both samples. The left-to-right pattern of landmark displacements represented by principal warp 8 corresponds to a form of purely inhomogeneous shape change: the "square-to-kite" transformation diagrammed in Figure 8. This type of deformation carries no biological information in terms of energy-normalized variance and is best regarded as a type of analytical artifact possibly resulting from non-correspondence (due to non-homology) of large numbers of the landmark locations between forms and the coincidental approximation of landmarks 1, 2 and 3 to a right triangle.

Interpretation of the second most global principal warp (Figure 7) is more straightforward, however, and in both samples represents a markedly non-uniform radial expansion of landmark locations that is more pronounced in the region of the penultimate and ante-penultimate chambers than it is in the region of the ultimate chamber (though the absolute magnitude of shape coordinate variability is dependent on baseline placement, the relative variability in the location of non-baseline shape coordinates is independent of

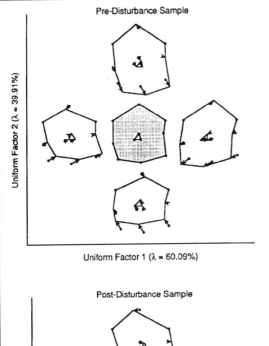

Pre-Disturbance Sample

Uniform Factor 2 (λ = 39.91%)

Uniform Factor 1 (λ = 60.09%)

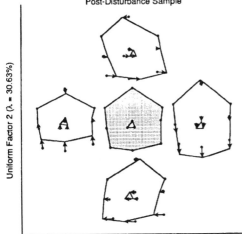

Post-Disturbance Sample

Uniform Factor 2 (λ = 30.63%)

Uniform Factor 1 (λ = 69.37%)

Figure 6. Graphical summary of the two uniform modes of shape deformation for pre-disturbance and post-disturbance landmark data. Stippled figures in the middle of each plot represents mean configurations of the landmark shape coordinates. Dots represent mean shape coordinate configurations for each diagram and arrows represent directions and relative magnitudes of the uniform shape deformations. Polygons drawn at the tips of the arrows show the effect of the uniform deformation on the mean shape. λ values indicate how much relative shape variability is represented by each uniform deformational axis.

Table 2. Principal Warps.								
Pre-Disturbance Sample								
Land-mark Points	1	2	3	4	5	6	7	8
1	.032	.037	-.105	.192	-.461	-.468	.354	.060
2	.038	-.035	-.498	-.246	-.244	-.126	-.509	.118
3	.008	.024	.134	-.005	.302	.468	.185	.431
4	-.539	-.592	-.072	.123	.009	.189	.054	-.454
5	-.061	.129	.728	.270	-.094	-.248	-.232	.071
6	.796	-.137	-.021	.083	.021	.209	.065	-.448
7	-.013	-.017	-.236	.066	.409	-.326	.501	.108
8	-.030	-.055	.189	-.390	-.527	.344	.199	.273
9	-.261	.766	-.152	-.104	.074	.179	.038	-.416
10	.000	.023	-.173	.571	.102	.135	-.408	.324
11	.031	-.143	.207	-.561	.409	-.336	.249	.002
Eigen-value	28.523	13.931	6.124	4.337	2.636	1.063	.946	.628
Post-Disturbance Sample								
Land-mark Points	1	2	3	4	5	6	7	8
1	.021	.041	-.149	.215	-.441	-.485	.339	-.112
2	.041	-.016	-.455	-.298	-.234	-.416	-.516	-.098
3	.011	.021	.137	.005	.264	.516	.171	-.385
4	-.475	-.639	-.111	.110	.022	.150	.671	.470
5	.073	.094	.681	.382	-.055	-.277	-.200	.054
6	.809	-.058	-.038	.089	.003	.175	.085	.450
7	-.012	-.011	-.248	-.016	.388	-.261	.525	-.253
8	-.045	-.055	.230	-.304	-.584	.368	.125	-.255
9	-.329	.746	-.118	-.102	.064	.149	.048	.429
10	-.001	.013	-.233	.505	.195	.132	-.454	-.301
11	.053	-.133	.304	-.585	.379	-.323	-.195	.002
Eigen-value	25.248	17.360	6.488	4.944	3.001	1.089	1.054	.715

this arbitrary decision). The type of deformational mode represented by principal warp 7 has been identified (Bookstein, in press) as a special case of the quadratic transformation and is diagrammed for a simple shape in Figure 8.

Another way of summarizing information on non-uniform aspects of shape variability is through the calculation and analysis of relative eigenvectors. Relative eigenvector analysis produces combinations of shape coordinates that partition observed variances in landmark position into a series of mutually orthogonal components that account for progressively smaller aspects of variation in landmark location weighted by the warping energy (the eigenvalues of Table 2, see Bookstein 1989b). Consequently, relative eigenvectors are re-expressions of the principal warps ordered in such a way as to rank the deformations in terms of the greatest variance represented by the warps relative to the geometric scale over which these deformations take place.

For both pre-disturbance and post-disturbance samples, two relative eigenvectors account for approximately nine-tenths of the observed relative bending energy, and the orientations of these two relative eigenvectors for each of the non-constrained landmark locations are shown in Figure 9. Interpretation of this figure is clarified by comparing it with Figure 3. The relative eigenvectors are the principal directions of each non-constrained landmark's distribution in shape space. Similarly, comparing Figures 7 and 9, it can be seen that the first relative eigenvector is closely aligned with the second principal warp and represents the non-uniform inflation (or deflation) of three chambers comprising the final whorl along with a slight translation of the umbilical area. The second relative eigenvector is a bit more difficult to characterize, but appears, once again, to be related to relative differences in the tightness of the ultimate whorl's trochospiral coil.

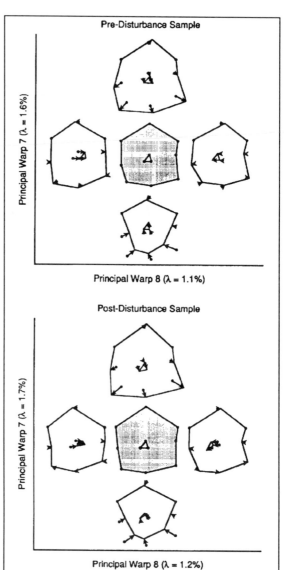

Figure 7. Graphical summary of the two most global principal warps (non-uniform modes of shape deformation) for pre-disturbance and post-disturbance landmark data. Stippled figures in the middle of each plot represent mean configurations of landmark shape coordinates. Dots represent mean shape coordinate configurations for each diagram, and arrows represent directions and relative magnitudes of the non-uniform shape deformations. Polygons drawn at the tips of the arrows show the effect of the uniform deformation on the mean shape. λ values indicate how much relative shape variability is represented by each non-uniform deformational axis.

Discussion

The data analyzed in this study were drawn from a larger database that incorporated measurements of morphologic variation in *S. linaperta* at 15 middle and late Eocene intervals at DSDP Site 612 and (partially) analyzed using a set of more traditional multivariate morphometric techniques (MacLeod et al., submitted). This prior analysis employed distances between landmark points located on the test surface in such a way as to summarize the major aspects of geometric variability in all 3 dimensions as the primary morphologic descriptors rather than the landmark points themselves. There-fore, comparison of the results of these two differ-ent approaches to morphometric analysis should indicate areas of commonality and distinction, as well as providing an external check on the validity of the conclusions derived from the earlier investi-gation. However, the results of this study are both conceptually and computationally independent of those that derive from the earlier investigation and should be evaluated in and of themselves (see Results).

With respect to the analyses of test size, the centroid size index seemed to yield consistently lower estimates of the relative difference between individual specimens than did the PC-1 score (the size estimate used in the previous study), perhaps suggesting that centroid size is a more conservative size estimator. Nevertheless, centroid size should be highly correlated with a number of multivariate size indices. For the *S. linaperta* data, depending on which set of interlandmark distances were chosen for analysis, correlations between centroid size and score on the first principal component of the covariance matrix of log-transformed interlandmark distances were found to be as high as 0.96.

In the context of any analysis that seeks to evaluate changes in the pattern of ontogenetic development, the careful separation of organismal size from organismal shape is of the utmost impor-tance, especially for the recognition of the various types of heterochrony (Alberch et al. 1979; Tissot, 1988a,b). Though the distinctions between the various types of heterochronous variation in terms of patterns of variability in body size and body shape are clear, the relevant comparisons between ancestor and descendent phenotypes are often diffi-cult to make with certainty because of the widely divergent and oftentimes non-comparable estimates of "size" and "shape" that are currently employed in the field of morphometrics (see Bookstein, 1989a). The centroid size index and the simple, yet power-ful, shape coordinate transformation can be used to bring much needed standardization to the practice of estimating size and shape which can, in turn, be easily accommodated within the conceptual frame-work of developmental heterochrony.

In the case of the comparison of pre- and post-disturbance *S. linaperta* populations, the exis-tence of variations in the mean test shape of differ-ent size classes is easily documented by direct

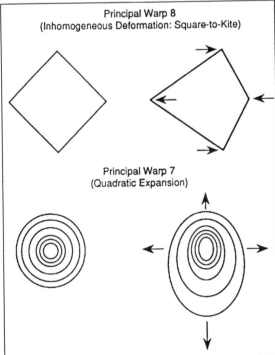

Figure 8. Dominant modes of shape variation (expressed as a deformation) identified by the principal warp analysis (eigenanalysis of the bending energy matrix). See text for discussion.

comparison of shape coordinate plots, as is the overall similarity in the shape of the post-disturbance populations with respect to the smaller size classes of the pre-disturbance sample. These results strongly imply that adult post-disturbance individuals exhibit test sizes and shapes that have been juvenilized, relative to their pre-disturbance, middle Eocene ancestors, a developmental phenomenon known as progenetic dwarfing (Gould, 1977; Alberch et al., 1979; McNamara, 1986). Though this mode of developmental heterochrony was also recognized in the prior analysis of the interlandmark distance data, the method illustrated in this paper represents a computationally independent (and much more efficient) way of achieving the same result.

Analysis of allometry using the landmark-based multiple regression of centroid size ($\sqrt{S}$) on the non-baseline shape coordinates failed to detect any statistically significant pattern of linear allometry in either of these two samples. Despite the fact that previous multivariate analysis did detect a small, but statistically significant, allometric trend in the pooled data from all 15 different middle and late Eocene intervals at this locality, this trend is very subtle and, given the relatively small sample sizes used in this analysis, it is not surprising to see that the landmark-based technique did not reveal its presence (although it should be noted that the F-ratio for the post-disturbance sample is quite close to significance at the 5% level).

However, given these comparabilities with more traditional morphometric methods, the real significance of the landmark-based morphometric methods lies in the fact that, through their use, important patterns of morphometric variability other than size quantification, gross shape similarity, and allometry can be studied in a systematic manner. While allometry is most often employed to analyze the global scaling of size and shape variability, it must be acknowledged that oftentimes highly localized patterns of morphologic variation within a form can be as informative (perhaps even more informative) than strictly global patterns for

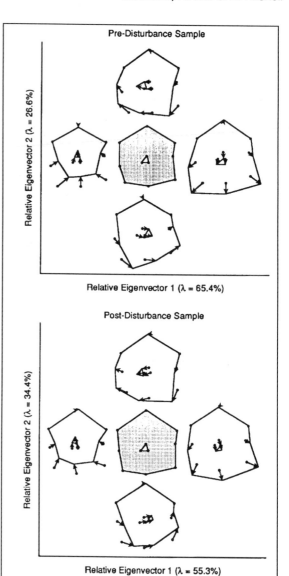

Figure 9. Graphical summary of the two dominant relative eigenvectors (non-uniform modes of shape deformation weighted by the relative scale over which the deformations are taking place) for pre-disturbance and post-disturbance landmark data. Stippled figures in the middle of each plot represent mean configurations of the landmark shape coordinates. Dots represent mean shape coordinate configurations for each diagram, and arrows represent directions and relative magnitudes of the non-uniform shape deformations. Polygons drawn at the tips of the arrows show the effect of the non-uniform deformation on the mean shape. λ values indicate how much relative shape variability is represented by each non-uniform deformational axis.

many types of systematic investigations. For example, from the *S. linaperta* data it is clear from the comparison of the mean positions of non-baseline shape coordinates, as well as from the results of the uniform and non-uniform shape analyses, that a large amount of shape variability is connected to the relative location of the umbilical area (points 4, 6 and 9). This result was totally unexpected and had been completely overlooked in the previous study owing to the fact that no distance measurements between the landmark points that defined the umbilical area were collected even though these same points were used to determine interlandmark distances representing the semi-major axes for each of the three chambers.

Another instance of the importance of regional patterns of variation to the interpretation of morphometric variability can be illustrated by a close look at the morphometrics of the ultimate chamber in the *S. linaperta* data. A number of fossil and modern planktic foraminiferal taxa often exhibit an ultimate chamber that is reduced in size relative to the preceding penultimate chamber, and individuals have been collectively referred to as "kummerform" morphotypes of their respective species (Berger, 1969). In the only quantitative analysis of the kummerform phenomenon to date, Olsson (1971, 1973) has shown that reductions in the apparent size of successive chambers in trochospirally coiled planktic foraminiferal tests can begin prior to ultimate chamber formation and that kummerform morphotypes seem to appear only after a more-or-less fixed number of chamber whorls have been added to the test (e.g., 1000° of whorl in the modern species *Globogerina pachyderma*). Also, in rare instances, kummerform whorl reduction may proceed in such a way as to be maximally expressed in either the penultimate or ante-penultimate chambers with the ultimate chamber exhibiting a slightly larger size.

The kummerform phenomenon can be the result of i) an actual decrease in the absolute size of the ultimate chamber or chamber series, ii) a change in the nature of the coiling parameters for the ultimate chamber or chamber series resulting in an apparent change in chamber size when the test is viewed in certain orientations (e.g., plan view of the spiral side), or iii) some combination of the two. Unfortunately, Olsson's method of test orientation and measurement does not allow for the effective separation of these two alternative sources of morphological variation. On the other hand, Olsson's plots of the apparent sizes of successive chambers for 21 kummerform morphotypes (Olsson, 1971: Figure 9) do show that, owing to the fact that the kummerform phenomenon usually affects a relatively small part of the overall test, this aspect of morphologic variation can easily be missed in the application of many traditional morphometric techniques (i.e., univariate or multivariate allometry) that focus on global linear trends. It would seem that principal warp and relative eigenvector analysis would be ideally suited to the study of such non-linear morphologic phenomena.

In the *S. linaperta* data, it is tempting to consider a kummerform interpretation for principal warp 7 and relative eigenvector 1 which indicate that a relatively larger amount of shape variability is present at landmarks 7 (penultimate chamber) and 10 (ante-penultimate chamber) than at landmark 3 (ultimate chamber), all of which are roughly equivalent in terms of overall chamber geometry. Taking one of the trends represented by principal warp 7 and relative eigenvector 1 to its logical extreme, there is no doubt that kummerform phenotypes would result. However, the operative question in these data is whether or not the actual scale (rather than the relative scale) of the landmark point locations is such that the term kummerform, in its traditional sense, can be applied to any of these phenotypes.

In order to evaluate this question, simple calculations of centroid size and aspect ratios for the ultimate, penultimate, and ante-penultimate chambers for each individual in both pre-disturbance and post-disturbance samples were made to determine i) whether or not the apparent size of the ultimate chamber decreased relative to

Table 3. Centroid size and aspect ratios.

Pre-Disturbance Sample

		Ultimate Chamber	Penultinate Chamber	Ante-penultimate Chamber
Centroid Size[*] :	Mean	5.230	4.418	3.900
	Std. Dev.	1.043	.930	.863
% of Ultimate Chamber :	Mean	-	84.5	74.6
	Std. Dev.	-	9.0	9.5
Aspect Ratio :	Mean	.677	.775	.781
	Std. Dev	.060	.080	.084

Post-Disturbance Sample

		Ultimate Chamber	Penultimate Chamber	Ante-penultimate Chamber
Centroid Size[*] :	Mean	3.906	3.066	2.666
	Std. Dev.	.668	.560	.468
% of Ultimate Chamber :	Mean	-	78.5	68.3
	Std. Dev.	-	7.9	8.3
Aspect Ratio :	Mean	.656	.743	.774
	Std. Dev.	.070	.078	.104

[*] Centroid Size = $\sqrt{S}$

the penultimate or ante-penultimate chambers, and ii) whether or not the apparent decrease in shape variability at landmark 7 corresponds to change in ultimate chamber shape relative to the shapes of the preceding two chambers. Results are presented in Table 3. Compared with the ultimate chamber, the relative sizes of the penultimate and ante-penultimate chambers are actually larger in the pre-disturbance samples, indicating that there may be less interchamber differentiation in the pre-disturbance middle Eocene populations. On the other hand, the overall ellipticity of the ultimate chamber in both samples is much greater than the ellipticity of either the penultimate or ante-penultimate chambers, no doubt reflecting the fact that only a portion of these chambers is exposed. But, comparison of the aspect ratios for the ultimate chamber in each sample shows that during the studied interval, late Eocene post-disturbance populations may have exhibited a slightly more elliptical ultimate chamber than their middle Eocene pre-disturbance ancestors. From these data, it is clear that principal warp 7 and relative eigenvector 1 are describing the greater ellipticity of the ultimate chamber in both samples that arises from a combination of an apparent radial compression of the chamber itself about a line tangent to the trochospiral coil, along with the pattern of overlap of the ultimate (and penultimate) chamber onto the penultimate (and ante-penultimate) chamber that results in the external expression of these chambers being much closer to a sphere than it actually is.

In short, on the basis of the morphometric data, the predominant non-linear trend in the *S. linaperta* skeletal phenotype cannot be interpreted as a manifestation of a kummerform morphotype. Indeed, given the definition of this species, it is difficult to believe that a kummerform individual would be identified as such and not placed into a wholly different taxon. Needless to say, no kummerform *S. linaperta* have ever been reported and it is our impression that kummerform morphotypes as a whole are much rarer in the Paleogene planktic foraminiferal fauna than they are in the Neogene. Nevertheless, the kummerform phenomenon represents but one of many different types of non-linear morphologic variations in planktic foraminifera that could be effectively studied within the general framework of an exclusively landmark-based morphometrics.

Conclusions

The term "evolutionary inference" refers both to causal explanations of observed patterns of evolution and to phylogenetic relationships within and between groups. Morphometrics provides a quantitative means of redescribing biological shapes and their deformations in the economical language of mathematics. In this analysis of populations of *S. linaperta* separated by an environmental disturbance, patterns of variation have been summarized through the exclusive use of landmark-based techniques. Results obtained indicate that while the size class distribution characterizing the post-disturbance population (as measured by the centroid size index) is markedly truncated, the mean test shape exhibited by the smaller, post-disturbance sample is very similar to mean shapes of pre-disturbance specimens of a roughly corresponding size class. This strongly corroborates our previous suggestion that heterochronic variation in the developmental programs of individuals comprising these two populations has occurred.

Aside from this marked reduction in the size of sexually mature adults between the sampled intervals that has been accompanied by a corresponding change in test shape, landmark-based shape analyses have revealed patterns of morphological similarity and difference that were not clearly delineated in our prior investigation. Analyses of uniform aspects of morphologic variation indicate that, while similar modes of shape variation (uniform patterns of test expansion/contraction versus the uniform right-to-left/left-to-right translation of non-baseline landmarks) are present within both samples, the mixture of these two patterns of variation before and after the interval of environmental disturbance is quite different. Overall patterns of non-uniform morphometric variation among measured landmarks can be attributed to at least two modes of geometric deformation: a "square-to-kite" transformation indicative of purely inhomogeneous shape transformation that probably results from a degree of non-correspondence in landmark locations between specimens, and a slightly asymmetric radial expansion that may be characterized as the geometric manifestation of a form of quadratic expansion. When these non-uniform deformational modes are weighted by their geometric scale, the quadratic inflation of the last three chambers of the ultimate whorl dominates the pattern of shape variation by accounting for over 50% of the observed non-uniform shape variability in each sample. Also, a subordinate mode of shape deformation was identified that appears to be related to the tightness of the trochospiral coil. Lastly, although these results,

derived from a 2-dimensional sampling of the foraminiferal test morphology, appear to confirm those obtained using a more traditional multivariate morphometric approach to analyze interlandmark distance data collected in all three dimensions, this should not be interpreted as indicating that we regard 2-dimensional data collection strategies as sufficient for most foraminiferal morphometric investigations; only that dimensional inter-relationships for this particular species appear to be exceptionally well-behaved.

Blackstone (1987a,b) has argued that primarily descriptive approaches to the study of size/shape changes must be tied to causal models of potential size/shape change, and that developmental processes should not be ignored since evolutionary processes may have been expressed through changes in the pattern of ontogenetic development. Certainly, routine determinations of size and shape statistics, devoid of any understanding of the implications that alternative process-oriented hypotheses hold for the interpretation of observed patterns, would be inappropriate. Although morphometrics provides an economical description of form and its transformation, one must be wary of confusing a geometric series of form transformations that either minimizes or maximizes some abstract mathematical property of the data with a form series generated by biologically causal processes.

In addition to these matters, several aspects of morphometric analysis need further consideration, particularly in dealing with characterization and evaluation of inter-specific patterns of variation. One is that the general method of morphometrics reduces, for the purpose of analysis, homologous parts of organisms to points. Although homology of parts may be established, this is not necessarily equivalent to having established homology of points. Also, in some methods, the landmarks are fixed for the purpose of analysis. More biologically meaningful methods allow deformation of the position of landmarks as well (see Bookstein 1989b for an example). Finally, though synonymizing relative degrees of geometric deformation with indices of evolutionary dissimilarity cannot be supported as a general assertion, the economy of description afforded by morphometrics and its potential to summarize biologically and evolutionarily relevant patterns nevertheless provide powerful tools for the study of the evolutionary process.

Acknowledgments

We would like to thank Gerta Keller to first calling our attention to the dwarfing of *S. linaperta* and supplying the samples used to quantify morphometric variation over the study interval, Fred Bookstein for patiently explaining his methods of morphometric analysis to us, Mark Johnson for discussions pertaining to the uniform aspect of shape deformation as well as the graphic summarization of the results of uniform and non-uniform deformational analyses, and an anonymous reviewer for carefully reading the manuscript and making several helpful (and occasionally provocative) comments, especially with regard to the kummerform question. Principal warp and relative eigenvector analyses were performed using Bookstein's FACTOR program; all other data analytic and graphical reconstruction procedures were carried out using programs developed by N. M. These programs and the original landmark data are available upon written request. The landmark data used in this study were collected with the Biosonics Optical Pattern Recognition System (OPRS). This study was supported by NSF grant BSR-8708563 to N. M. and J. K.

References

Alberch, P., S. J. Gould, G. F. Oster and D. B. Wake. 1979. Size and shape in ontogeny and phylogeny. Paleobiology, 5:296-317.

Barnes, V. E. and M. A. Barnes. 1973. Tektites. Benchmark papers in geology, Dowden. Hutchinson and Ross, Stroudsburb, PA, 445 pp.

Bé, A. W. H., S. M. Harrison and L. Lott. 1973. *Orbulina universa* d'Orbigny in the Indian Ocean. Micropaleontology, 19(2):150-92.

Berger, W. H. 1969. Kummerform foraminifera as clues to oceanic environments. Bull. Am. Assoc. Petrol. Geol. [abs.], 53:706.

Berggren, W. A. 1965. Paleocene - A micropaleontologist's point of view. Bull. Am. Assoc. Petrol. Geol., 44:1473-1484.

Blackstone, N. 1987a. Allometry and relative growth: pattern and process in evolutionary studies. Syst. Zool., 36(1):76-8.

Blackstone, N. 1987b. Size and time. Syst. Zool., 36(2):211-15.

Blow, W. H. 1979. The Cainozoic *Globigerinida*, 3 vols., E. J. Brill, Leiden, 1413 pp.

Bookstein, F. L. 1986. Size and shape spaces for landmark data in two dimensions. Stat. Sci., 1(2):181-242.

Bookstein, F. L. 1988. Morphometric tools for landmark data: geometry and biology (distributed in ms at the NSF-Univ. of Michigan Morphometrics Workshop, 1988).

Bookstein, F. L. 1989a. "Size and shape": a comment on semantics. Syst. Zool., 38(2):173-80.

Bookstein, F. L. 1989b. Principal warps: thin plate splines and the decomposition of deformations. IEEE Trans. Pattern Anal. Mach. Intel., 11(6):567-85.

Erez, J. 1983. Calcification rates, photosynthesis and light in planktonic foraminifera. Pp. 307-12 *in* Biomineralization and biological metal accumulation, biological and geological perspectives (Westbroek, P. and E. W. de Jong, eds.). Reidel Publishing Co., Dordrecht, Holland, 783 pp.

Glass, B. P. 1982. Introduction to planetary geology. Cambridge University Press, Cambridge.

Gould, S. J. 1977. Ontogeny and phylogeny. Harvard University Press, Cambridge.

Hecht, A. D. 1974. Intraspecific variation in recent populations of *Globigerinoides ruber* and *Globigerinoides trilobus* and their application to paleoenvironmental analysis. J. Paleontol., 48 (6):1217-1234.

Keller, G., S. L. D'Hondt, C. J. Orth, J. S. Gilmore, P. Q. Oliver, E. M. Shoemaker, and E. Molina.

1987. Late Eocene impact microspherules: stratigraphy, age and geochemistry. Meteoritics, 22(1): 25-60.

MacLeod, N., J. A. Kitchell and G. Keller. Evolutionary opportunity via progenesis: shifts in developmental timing associated with a Late Eocene impact event. (submitted to Marine Micropaleontology).

MacNamara, K. J. 1986. A guide to the nomenclature of heterochrony. J. Paleontol., 60:4-13.

Malmgren, B. A. and J. P. Kennett. 1972. Biometric analysis of phenotypic variation: *Globigerina pachyderma* (Ehrenberg) in the South Pacific Ocean. Micropaleontology, 18(2):241-48.

Olsson, R. K. 1971. The logarithmic spire in planktonic foraminifera: its use in taxonomy, evolution and paleoecology. Trans. Gulf Coast Assoc. Geol. Soc., 21:419-32.

Olsson, R. K. 1973. What is a kummerform planktonic foraminifer. J. Paleontol., 47(2): 327-29.

Shaw, H. F. and G. J. Wasserburg. 1982. Age and provenance of the target materials for tekities and possible impactites as inferred from Sm-Nd and Rb-Sr systematics. Earth Planet. Sc. Lett., 60:155-77.

Smit, J. 1982. Extinction and evolution of planktonic foraminifera after a major impact at the Cretaceous/Tertiary boundary. Geol. Soc. Amer. Spec. Paper, 190:329-352.

Strauss, R. E. and F. L. Bookstein. 1982. The truss: body form reconstructions in morphometrics. Syst. Zool., 31(2):113-35.

Tissot, B. N. 1988a. Geographic variation and heterochrony in two species of cowries (genus *Cypraea*). Evolution, 42(1):103-17.

Tissot, B. N. 1988b. Multivariate analysis. Pp. 35-51 *in* Heterochrony in evolution: a multidisciplinary approach (McKinney, M.L. ed.). Plenum, New York, 432 pp.

Chapter 15

Morphometrics and the Systematics
of Marine Plant Limpets
(Mollusca: Patellogastropoda)

David R. Lindberg

Museum of Paleontology, University of California
Berkeley, CA 94720

Abstract

Marine plant limpets have distinctive shell apertures with parallel lateral margins because apertural width is constrained by the width of the host plant's leaf. This study was conducted to determine if morphometric studies can be useful in determining morphological boundaries between extinct marine plant limpet species. Morphological patterns present in the living species, *Lottia alveus* and *Tectura depicta*, are contrasted with patterns present in four nominal species from the Eocene of the Paris Basin, France; *Patelloida arenarius*, *Patelloida elongata*, *Patelloida concavus* and *Patelloida pyramidale*. Comparison is made of data sets obtained with vernier calipers and the video acquisition system MorphoSys, and between analytical procedures, including principal component analysis, sheared-principal component analysis, principal component analysis using the Burnaby method, generalized least-squares analysis, and a modification of F. Bookstein's uniform versus non-uniform shape change analysis. Different analyses produce different taxa groupings and the use of as many different procedures as possible to explore the data appears useful. When the results of different analyses are viewed in light of biological and ecological knowledge of extant species some results are supported while others can be rejected. In this study the grouping present in the principal component analysis (*P. arenarius*, *P. pyramidale* and *P. concavus/P. elongata*) was rejected and the grouping suggested in the least-squares and uniform/non-uniform shape analyses (*P. concavus*, *P. elongata* and *P. arenarius/P. pyramidale*) was supported.

Introduction

Patellogastropods or "limpets" are gastropod molluscs with cap-shaped or conical shells. They are found in most oceans of the world and live on firm substrata such as rocks, other invertebrates, and marine algae and angiosperms. This paper concerns limpets that live or are thought to have lived on marine angiosperms.

Limpets are improbable candidates for morphometric studies because their shells are simple cones with little morphological relief. Species distinctions are most often made on soft-part criteria, and morphological boundaries between species may become ambiguous due to phenotypic plasticity (Lindberg, 1988). However, morphometric studies may be of value in describing morphological variation within extant species, and in those taxa where morphological variation is

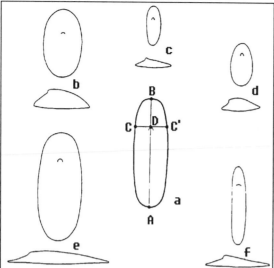

Figure 1. Marine plant limpet landmarks and species used in this study. a) Landmarks and distance measurements. b) *Lottia alveus* (Conrad) Holocene extinction, Northwestern Atlantic; extant Northeastern Pacific. c) *Patelloida arenarius* (Deshayes), d) *Patelloida concavus* (Deshayes), e) *Patelloida elongata* (Lamarck), f)*Patelloida pyramidale* (Cossmann). c - f) Eocene, Paris Basin, France. The distance measure for variable 1 is AB, 2 - CC', 3 - BD, 4 - BC, 5 - BC', 6 - AC, 7 - AC', 8 - CD, and 9 - C'D. The Uniform shape change is AB:CC', and the non-uniform shape change is AD:BD.

constrained, typically by limitations of the substratum to which the limpets are attached. Moreover, for many fossil species gross shell morphology is all that remains, and morphometric studies may provide the only means by which these taxa can be incorporated into a phylogenetic framework.

Extant marine plant limpets are readily recognized by their distinctive shell shape which typically features parallel lateral shell margins and elevated anterior and posterior regions of the aperture (Figure 1). This results because apertural width cannot exceed the width of the plant's leaf. Because of this constraint, fossil limpets showing this morphology are assumed to have lived on marine angiosperms. Many extant marine plants show ecophenotypic plasticity (den Hartog, 1970), and often the range of apertural variation present within a marine plant limpet species is determined

by the variation present within the host plant species (Lindberg, 1982). Therefore, ecophenotypic variation in the plant host may produce ambiguous limpet species boundaries defined strictly on shell-shape criteria. For example, in the extant species *Tectura depicta* (Hinds) plots of width on length suggest three distinct taxa while plots of length on height suggest only a single taxon (Figure 2). It is also possible that a single species of marine plant limpet could occur on several different marine plant species thereby producing similar non-overlapping variation in apertural morphology.

In this study, I examine morphometric patterns present in an extant species to assess the amount of intraspecific variation, and then compare these patterns to those present in fossil taxa. Because the workshop made available numerous data acquisition techniques and analytical procedures, I have compared techniques and procedures using the same specimen data set. Data acquisition comparisons are made between vernier caliper and a video system, and comparisons of analytical procedures include three flavors of principal component analyses, generalized least-squares analysis, and a modification of F. Bookstein's uniform versus non-uniform shape change analysis.

Materials and Methods

Taxa

Marine plant limpets have evolved in at least three clades. In the family Lottiidae, extant taxa are found in the genera *Patelloida*, *Proscutum*, *Lottia*, and *Tectura* (Figure 3). Furthermore, an extensive radiation of marine plant limpets, unequaled in the Holocene, occurred during the Eocene in the Paris Basin of France. Eighteen nominal species of marine plant limpets have been described from this fauna (Cossmann & Pissarro, 1913).

This study focuses on four of the eighteen nominal species that have been described from the Eocene of the Paris Basin, *Patelloida arenarius* (Deshayes), *P. concavus* (Deshayes), *P. elongata* (Lamarck), and *P. pyramidale* (Cossmann). The

Holocene species *Lottia alveus* (Conrad) was also included in this study to estimate intraspecific variation present in a marine plant species.

Data Acquisition

Five landmarks were used (Figure 1), and the *X, Y* coordinates of these landmarks were determined by the MorphoSys video data acquisition system. Only apex position (D in Figure 1) is homologous between specimens; the remaining four landmarks are positional or geometrical landmarks and their homology between specimens is not known. In addition, nine distance measurements (Figure 1) between these landmarks were gathered with both vernier calipers (to the nearest 0.01 mm.) and with the MorphoSys system. Sixteen damaged specimens measured in the caliper study were not used with the MorphoSys data system because the higher resolution of the video system made locating landmarks too arbitrary.

Morphometrics

Standard principal component analyses (PCA, sheared-PCA, and PCA using the Burnaby method) were calculated from the distance data using the SHEAR and BURNABY programs provided by L. Marcus. These programs report the results of a standard PCA before attempting to remove size from the data.

Landmark coordinate data were analyzed with J. Rohlf and Slice's program GRF using a least-squares fit analysis with a consensus form (Gower's generalized Procrustes method) and F. Bookstein's program PROJECT. The least-squares fit method scales the different specimens, and then rotates and translates the landmarks of each speci-

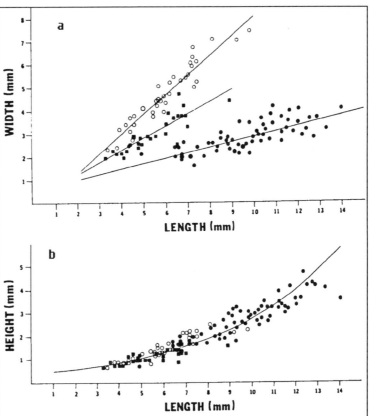

Figure 2. Scatter plot and fitted regression lines of length on a) width and b) height for three ecophenotypes of the marine plant limpet *Tectura depicta*. R^2 values: a) o = .5955, ■ = .8113, • = .8586; b) = .8662. Reprinted with permission from Bull. So. Calif. Acad. Sci., 81(2), 1982.

men until they superimpose as well as possible on the consensus form of the entire data set. Bookstein's PROJECT program uses two of the landmarks for the construction of a baseline along which the comparisons of specimens are scaled and mean shifts of the remaining landmarks calculated. In the marine plant limpet data the high degree of symmetry of the few remaining landmarks (only 2-3 points were available after the baseline was designated) produced results that were difficult to interpret. However, F. Bookstein (pers. comm.) pointed out that the triangles formed by the landmarks could be analyzed using only two ratios: (1) AB:CC', and (2) AD:BD (Figure 1). The first ratio contains information about the uniform shape change (spatially constant deformations), while the second ratio contains information about the non-uniform shape changes (Bookstein et al., 1985: Appendix A.4). These two ratios were calculated from the distance measures for each specimen.

Results

Data Acquisition

Figures 4a and b plot non-uniform shape change versus uniform shape change for four collection lots of fossil marine plant limpets from the Eocene of the Paris Basin. The data presented in Figure 4a were measured with calipers and the data presented in Figure 4b were gathered with the MorphoSys system. Both figures show similar distributions of points. *Patelloida arenarius*, *P. elongata*, and *P. pyramidale* group to the left of the figure, while *P. concavus* is dominant on the right; *P. pyramidale* has the lowest uniform shape ratio. Most of the difference between the two figures is found along the non-uniform shape change axis, and probably results from the difficulty in measuring apex position with calipers (see Figure 1). The data gathered with the MorphoSys system produced more segregated groups and similar mean values.

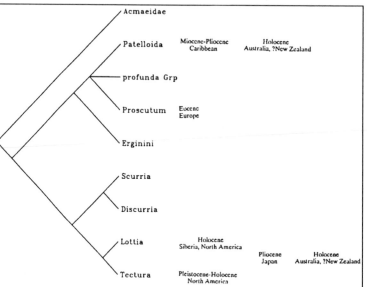

Figure 3. Cladogram of the Acmaeacea with temporal and spatial occurrences of known marine plant limpet taxa.

Morphometrics

The three different PCAs produced identical results; there was no significant difference between the coefficients for the eigenvectors of the individual PCs irrespective of the type of analysis (Figure 5). The redundancy in the truss network (Bookstein et al., 1985) is readily apparent for variable pairs 5, 4 and 7, 6 (Figures 1 and 5). However, the slight differences between the coefficients for the redundant apex width measurements suggests a slight

Table 1. Comparisons of results of principal component analyses, *Lottia alveus*. Eigenvalues of total log-transformed covariance matrix.

Eigen-vectors	Eigenvalues PCA	Eigenvalues Burnaby	% Total Variance PCA	% Total Variance Burnaby	Cumulative % Total Variance PCA	Cumulative % Total Variance Burnaby
1	0.042	0.020	91.736	88.178	91.736	88.178
2	0.003	0.002	6.569	9.249	98.305	97.428
3	0.000	0.000	.891	1.307	99.196	98.734
4	0.000	0.000	.777	1.222	99.973	99.956
5	0.000	0.000	.022	.034	99.995	99.990
6	0.000	0.000	.004	.008	100.000	99.999
7	0.000	0.000	.001	.001	100.000	100.000
8	0.000	0.000	.000	.000	100.000	100.000
9	0.000	0.000	.000	.000	100.000	100.000

Table 2. Eigenvectors of total log-transformed covariance matrix.

Varia-bles	PC2 PCA	PC2 Sheared	PC2 Burnaby	PC3 PCA	PC3 Sheared	PC3 Burnaby
1	.16247	.15939	.16645	.32879	.29837	.31783
2	.14014	.15632	.15104	-.35826	-.35269	-.36971
3	-.58284	-.57991	-.57415	.31323	.28566	.29774
4	-.20377	-.19575	-.20129	.20478	.20249	.19071
5	-.30668	-.29878	-.29577	-.15447	-.18966	-.16781
6	.47216	.47039	.47673	.25552	.24029	.24619
7	.43304	.43049	.44048	.16396	.13860	.15482
8	.26030	.27845	.26235	-.00993	.03337	-.02211
9	.02156	.03602	.04137	-.71230	-.74370	-.72302

asymmetry in apex position.

Because the three PCAs produced identical coefficients for the individual PC eigenvectors only the results of the standard PCA are figured and discussed here. In all three analyses, the first principal component appears to be a size component: all the coefficients are positive and range between 0.40 and 0.26 for the *L. alveus* data set, and between 0.47 and 0.29 for the Paris Basin data set. These values were reduced to between -0.02 and 0.02 when size was removed with the Burnaby method. The remaining principal component eigenvectors were composed of both positive and negative scores and represent mostly shape [shearing and the Burnaby method produced little change to the stan-

dard principal components (Tables 1-4)]. Principal components 2 and 3 accounted for only 7.5% of the initial variance in the *L. alveus* data set and 19.3% of the initial variance in the Paris Basin data set (Tables 3-4). However, principal components 2 and 3 accounted for 90% of the remaining variance after size was removed in the *L. alveus* data set and 92% of the remaining variance in the Paris Basin data set (Table 3). For *L. alveus* there was no discernible pattern or segregation among either Atlantic or Pacific populations (Figure 6). In contrast, the Paris Basin Eocene taxa showed three distinct groups, (1) *P. arenarius*, (2) *P. pyramidale*, and (3) *P. concavus* and *P. elongata* (Figure 7). In

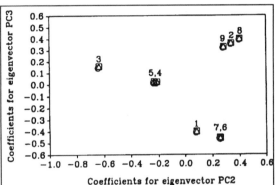

Figure 5. Scatter plot of coefficients for the eigenvectors of PC2 and PC3 of the nine distance measures used in the principal component analyses. o = standard principal component analysis, □ = sheared principal component analysis, and Δ = Burnaby method. See Figure 1 for legend to variables.

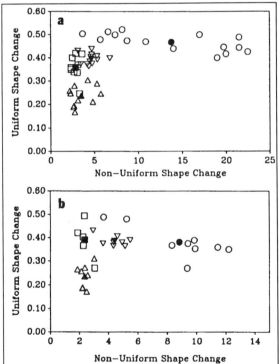

Figure 4. Scatter plot of uniform versus non-uniform shape change values for four nominal species of marine plant limpets from the Eocene of the Paris Basin. a) Data gathered with vernier calipers to the nearest 0.01 mm. b) Data gathered with the MorphoSys video imaging system. Open symbols: □ = *Patelloida arenarius*, o = *P.concavus*, ∇ = *P. elongata*, Δ = *P. pyramidale*. Solid symbols = means.

Table 3. Paris Basin Marine Plant Limpets: Eigenvalues of total log-transformed covariance matrix.

Eigen-vectors	Eigenvalues		% Total Variance		Cumulative % Total Variance	
	PCA	Burnaby	PCA	Burnaby	PCA	Burnaby
1	0.372	0.143	80.308	90.209	80.308	90.209
2	0.073	0.012	15.735	7.322	96.043	97.531
3	0.017	0.003	3.611	1.695	99.654	99.226
4	0.001	0.001	.263	.596	99.917	99.822
5	0.000	0.000	.062	.134	99.979	99.955
6	0.000	0.000	.017	.036	99.996	99.991
7	0.000	0.000	.004	.009	100.000	100.000
8	0.000	0.000	.000	.000	100.000	100.000
9	0.000	0.000	.000	.000	100.000	100.000

this analysis, separation along the second principal component contrasts apex position (negative) with the three width measurements (positive) (Table 4). Separation of taxa along the third principal component was slight and contrasts three length measurements (negative) with the three width measurements (positive) (Table 4).

The least-squares fit analysis (using the GRF program and methods described in Rohlf and Slice, 1990) of the *L. alveus* data shows only slight intraspecific variation among the populations (Figure 8). However, the analysis of the Paris Basin Eocene taxa suggests several distinct groupings. The separation of these groups is primarily found in

apex position and shell width (Figure 9). For apex position there appear to be three distinct groups relative to the consensus shape, (1) *P. arenarius* and *P. pyramidale*, (2) *P. concavus*, and (3) *P. elongata*. Separation is linear along the anterior-posterior axis of the shell. Shell width landmarks, relative to the consensus shape, divide the taxa into two groups, (1) *P. arenarius*, *P. concavus*, *P. elongata*, and (2) *P. pyramidale*. The horizontal separation of these groupings at the width landmarks is produced by the changing apex positions of the taxa; vertical separation results from changes in shell width relative to length.

The analysis of uniform and non-uniform shape changes produces groupings similar to those of the least-squares fit analyses. For *L. alveus* there are no discernible differences between the populations (Figure 10). For the Paris Basin taxa, four distinct groupings are present. However, the four taxa are not randomly scattered relative to the axes. Instead, three taxa, *P. arenarius*, *P. concavus*, and *P. elongata*, have similar mean uniform shape values and different non-uniform shape values, while *P. arenarius* and *P. pyramidale* have similar mean non-uniform shape change values but different uniform shape change values (Figure 4b).

Morphometrics

The differences between the analyses using

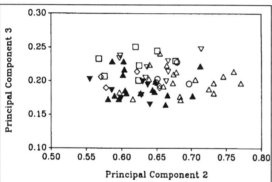

Figure 6. Scatter plot of second and third principal component scores from analysis of seven populations of *Lottia alveus*. Open symbols = Atlantic populations: o = ANSP 39044, □ = ANSP 40962, △ = ANSP 39046, ▽ = ANSP 39047, ◇ = ANSP 66767. Solid symbols = Pacific populations ▲ = SBMNH 8940, ▼ = SBMNH 6258. See Locality Register section for specific localities.

Table 4. Paris Basin Marine Plant Limpets: Eigenvectors of total log-transformed covariance matrix.

Varia-bles	PC2			PC3		
	PCA	Sheared	Burnaby	PCA	Sheared	Burnady
1	.07814	.08268	.08278	-.41076	-.40212	-.40188
2	.33693	.34135	.34107	.35162	.36007	.35952
3	-.64188	-.63491	-.63497	.14324	.15662	.15644
4	-.21676	-.21133	-.21096	.01078	.02116	.02185
5	-.23776	-.23232	-.23225	.01323	.02364	.02376
6	.25394	.25819	.25831	-.46728	-.45923	-.45893
7	.24928	.25352	.25360	-.46334	-.45530	-.45509
8	.39902	.40341	.40335	.38918	.39757	.39745
9	.28109	.28554	.28503	.31992	.32843	.32745

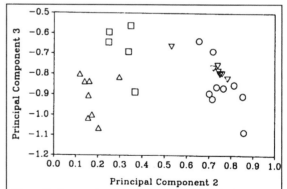

Figure 7. Scatter plot of second and third principal components scores from analysis of four nominal species of marine plant limpet from the Eocene of the Paris Basin. See Figure 4 for legend to symbols.

distance data gathered with calipers versus a video system were surprising. It was suspected that the video system would provide more accurate measurements than the caliper readings, but the apparent large error in estimating apex position (BD in Figure 1) with calipers was unsuspected. The large error associated with measuring apex position with calipers probably results because there are no edges on which to place the calipers. Instead it is necessary to estimate the distance between the apex and anterior margin parallel with the aperture (Figure 1).

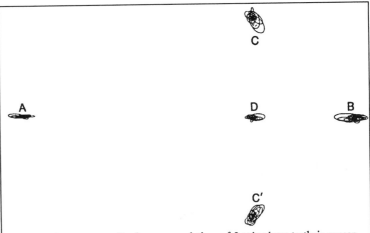

Figure 8. Least-squares fit of seven populations of *Lottia alveus* to their consensus form. A - C' = landmarks from Figure 1.

   The exclusion of sixteen specimens that were measured in the caliper study from the video data acquisition was necessary because geometric landmarks on these specimens were chipped and missing. It was possible to locate and measure these specimens with the calipers because the 2 mm. width of the caliper blades smoothes over smaller irregularities along the aperture. However, magnified under the video system apertural irregularities are readily apparent. Moreover, the width of the measuring point, a single pixel, is only 10 μm wide (adjusted for the magnification used for these specimens with the MorphoSys system), 200 times smaller than the caliper blade. Combined, these two factors make recognition and placement of landmarks on slightly damaged specimens much more difficult with the video system and necessitates more rigorous criteria. Thus, while the use of a video system greatly speeds up data acquisition and provides greater accuracy, the higher resolution and smaller measuring point may reduce the number of specimens available for the analysis, especially when fossils are used.

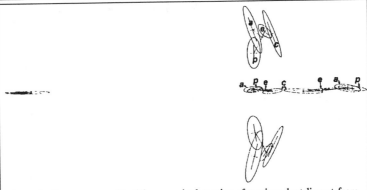

Figure 9. Least-squares fit of four nominal species of marine plant limpet from the Eocene of the Paris Basin. a) = *Patelloida arenarius*, c) = *P. concavus*, e) = *P. elongata*, p) = *P. pyramidale*. See Figure 8 for reference to landmarks.

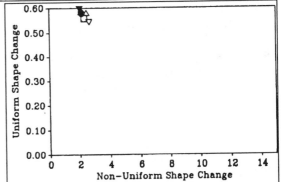

Figure 10. Scatter plot of mean uniform versus non-uniform shape change values for seven populations of *Lottia alveus*. Open symbols = Atlantic populations, solid symbols = Pacific populations. See Figure 6 and Locality Register section for specific localities.

The special procedures used to remove size from the PCA (sheared-PCA, the Burnaby method) had little affect on the marine plant limpet data sets. Most of the variation explained by principal component 1 is size-related and can be ignored. Shape information is mostly contained in principal component 2, with a little more present in principal component 3.

All three of the morphometric analyses (PCA, least-squares, uniform/non-uniform) show little resolution among the populations of *L. alveus* (Figures 6, 8 and 10). However, the different morphometric analyses do not provide similar groupings of the nominal species from the Paris Basin. The PCA distinguishes three groups, *P. arenarius*, *P. pyramidale*, and the combined *P. concavus* and *P. elongata*. These groupings result mainly from differences in shell width and apex position. *Patelloida arenarius* and *P. pyramidale* have more central apices, are narrower in width and are slightly asymmetric (the apex is closer to the left margin) than *P. concavus* or *P. elongata*. *Patelloida concavus* and *P. elongata* differ in apex position, but this difference is not strong enough to produce a separation in the PCA.

Differences in apex position produce the largest and most striking distinction between three of the four nominal species in both the least-squares and in the uniform/non-uniform shape change analyses. These analyses clearly separate *P. concavus* and *P. elongata*, while grouping *P. arenarius* and *P. pyramidale* (Figures 4b and 9). The apertural width differences also appear in both of these latter analyses, but do not appear to be greater or overwhelm apex position differences as in the PCA. However, which of these groupings best represents real, albeit extinct, species?

Extant marine angiosperm limpet species are probably ephemeral. The planktonic larvae settle on young, newly emerging leaves and grow as the leaf enlarges. Seasonal storms and leaf senescence destroys the habitat and new generations of leaves and limpets replace the previous ones.

Three of the four nominal species show similar proportional length/width growth parameters (represented by the uniform shape change variable), although maximum width differs among them (Figures 1 and 4b). This is not the case in *P. pyramidale*. In this nominal species length increases are greater relative to width increases; compared to the other taxa, *P. pyramidale* appears to reach maximum width earlier and subsequent growth is only in length (Figure 11a).

Apex position, which is not affected by the constraints of a marine plant habitat as is aperture shape (and is described by the non-uniform shape change variable), suggests that *P. pyramidale* is identical to *P. arenarius* (Figures 9 and 4b). Geographic and stratigraphic evidence also associates these two taxa. Of the four nominal species studied here only *P. pyramidale* and *P. arenarius* are found at the same Lower Eocene locality; the other two species are known from Upper Eocene localities (Table 5 and Locality Register).

These data suggest that *Patelloida pyramidale* and *P. arenarius* may be two forms of a single species. Bivariate scatter plots of width on

Table 5. Marine plant limpets in the Paris Basin, France.

Species	n	Mer	Lao	Hero	Gob	Fay	Bou	Gri	Par	Bor	Dam	Hou	Her	Auv	Val	Cro	Bor	Ver	Gue	Bra	Sel
P. arenarius	4	X	X	X	X																
P. concavus	2					X	X														
P. elongata	14							X	X	X	X	X	X	X	X	X	X	X	X	X	X
P. pyramidale	1			X																	
Total number of species reported		1	1	2	1	4	1	8	8	5	3	2	2	1	1	1	1	1	2	1	1

length (Figure 11a) and height on length (Figure 11b) show morphological patterns similar to those found in the extant species *T. depicta* (Figure 2). Specimens of *P. arenarius* achieve greater widths than *P. pyramidale* at similar lengths (Figure 11a). However, there is no difference in height versus length relationships between the two species (Figure 11b). In *T. depicta* this pattern results because of ecophenotypic variation in the host plant *Zostera marina* (Lindberg, 1982). When the plant lives intertidally its growth is stunted and its leaves are narrow. Limpets living on this form reflect the plant's narrow leaves in their apertural morphology. Subtidal plants have wider leaves, and, likewise, the limpets become wider at smaller lengths; intergrades between the forms of the plant are rare (den Hartog, 1970). Alternatively, there may have been two species of plants in the Paris Basin Lower Eocene with different width leaves and the taxa *P. arenarius* and *P. pyramidale* represent a single species living on two distinct plants.

Morphometric analyses of marine plant limpet species provide insights into questions of animal/plant interactions, molluscan growth parameters and limpet systematics. These analyses make it possible to include fossil data which are often excluded from ecological and more recent systematic studies. Different analyses produce different patterns and the use of as many different procedures as possible to explore the data appears useful. When the results of different morphological analyses are coupled with biological and ecological knowledge of extant species, some morphological results are supported while others can be rejected. In this study the taxa groupings present in the PCA were rejected and those present in the least-squares and uniform/non-uniform shape analyses supported.

Acknowledgments

I thank K. Warheit, D. Eernisse, and J. Rohlf for their criticism of the manuscript; C. Meacham and T. Duncan for making MorphoSys available to me for data acquisition; F. Bookstein, J. Humphries, L.

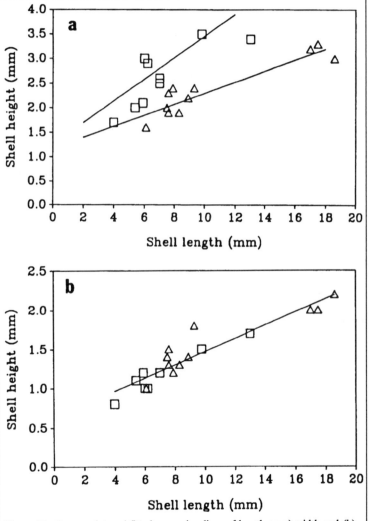

Figure 11. Scatter plot and fitted regression lines of length on a) width and (b) height for *Patelloida arenarius* (□) and *P. pyramidale* (Δ). R^2 values: (a) □ = .6494, Δ = .8566; b) = .8742.

Marcus, and J. Rohlf for providing compiled code of their morphological analyses programs; G. Davis, F. Hochberg, and P. Scott for the loan of specimens; and M. Taylor for preparing Figure 1. F. Bookstein, D. Eernisse, N. MacLeod, L. Marcus, and T. Pearce provided thoughtful discussions and assistance during the workshop. I also wish to thank the workshop organizers, J. Kitchell, W. Fink, and F. Bookstein, as well as D. Schindel, for their efforts and the opportunity to participate.

References

Bookstein, F., B. Chernoff, R. Elder, J. Humphries, G. Smith and R. Strauss. 1985. Morphometrics in evolutionary biology. The Academy of Natural Science of Philadelphia, Spec. Publ. No. 15, 277 pp.

Cossmann, M. and G. Pissarro. 1910-13. Iconographie complete des coquilles fossiles de l'Eocene des environs de Paris, vol. 2, Scaphopodes, gastropodes, brachiopodes, cephalopodes and supplement. 345 pp., 65 plts. de Hartog, C. 1970. The seagrasses of the world. Verh. K. Ned. Akad. Wet. Afd. Natuurk. D. Tweede Reeks 59:1-275.

Lindberg, D. R. 1982. A multivariate study of morphological variation of the limpet *Notoacmea depicta* (Hinds) and its synonyms *Notoacmea gabatella* (Berry) and *Notoacmea lepisma* (Berry) (Gastropoda: Acmaeidae). Bull. Southern Calif. Acad. Sci. 81(2):87- 96.

Lindberg, D. R. 1988. The Patellogastropoda. Malacol. Rev., Suppl. 4:35-63.

Rohlf, F. J. and D. Slice. 1990. Extensions of the Procrustes method for the optimal superimposition of landmarks. Syst. Zool., 39:40-59.

Appendix: Locality Register

Abbreviations used in the text are as follows: ANSP - Academy of Natural Sciences, Philadelphia, PA; SBMNH - Santa Barbara Museum of Natural History, Santa Barbara, CA; UCMP - Museum of Paleontology, University of California, Berkeley, CA. Paris Basin localities in Table 2: Mer - Mercin; Lao - Laon; Hero - Herouval; Gob - St. Gobain; Fay - Le Fayel; Bou - Boucouvilliers; Gri - Grignon; Par - Parnes; Bor - Boursault; Dam - Damery; Hou - Houdan; Her - Hermonville; Auv - Auvers; Val - Valmondois; Cro - Crouy; Ver - Swr; Gue - Le Guepelle.-Angleterre; Bra - Bracklesham; Sel - Selsey.

ANSP 39044	MASSACHUSETTS	Holocene
ANSP 39046	MAINE	Holocene
ANSP 39047	MASSACHUSETTS: Crescent Beach	Holocene
ANSP 40962	MAINE: Islesboro	Holocene
ANSP 66767	?	Holocene
SBMNH 6258	ALASKA: Ketchikan	Holocene
SBMNH 8940	ALASKA: Grant Island	Holocene
UCMP B-5357	FRANCE: Oise; Le Fayel	Bartonian (Upper Eocene), Sables Moyens Fm
UCMP B-5360	FRANCE: Oise; Herouval	Ypresian (Lower Eocene)
UCMP B-5365	FRANCE: Seine et Oise; LeGuepelle	Bartonian (Upper Eocene), Sables Moyens Fm

Chapter 16

Comparative Ontogeny of Cranial Shape in Salamanders Using Resistant Fit Theta Rho Analysis

Stephen M. Reilly

School of Biological Sciences
University of California, Irvine, CA 92717

Abstract

Ontogenetic and evolutionary cranial shape change among three transforming salamanders is studied using Resistant Fit Theta Rho Analysis (RFTRA; see Chapter 12 in this volume). RFTRA is used to describe ontogenetic changes in shape shared by different species, to identify heterochrony, and to demonstrate phyletic differences in early larval, metamorphic and post-metamorphic cranial shape. Metamorphic shape change is greater than differences between species, showing how ontogeny confuses phylogenetic analysis in salamanders. Major gross metamorphic shape changes involve the shortening of the frontal arcade in ambystomatids and increased length of the maxillary bone in newts. Species comparisons reveal that 1) snout width, 2) development of the maxillary bone, 3) design of the neurocranium, and 4) RFTRA morphometric "distance" may be useful phylogenetic characters. Utility of this method is discussed in regard to morphological evolution in salamanders where, the difficulty of comparing ontogenies among taxa is a major hindrance to understanding the phylogeny of the group. The RFTRA method produced results nearly identical to those of generalized resistant fit (GRF), PROJECT, thin plate spline (TPS) and truss analysis.

Introduction

Reconstructing the phylogeny of salamander families (Order Caudata) has been difficult for several reasons, but especially because of the widespread occurrence of paedomorphosis in the group (Duellman and Trueb, 1986). All salamander families seem to share aspects of larval ontogeny but differ in the amount of morphological metamorphosis exhibited by the terminal morphotypes. Thus, the extent of metamorphosis in different taxa confounds the evaluation of characters used in reconstructing evolutionary relationships. Comparative studies are needed to determine what aspects of ontogeny and metamorphosis are shared among salamander taxa, and what features are due to phyletic change.

In this paper, a preliminary attempt to compare ontogenetic and phyletic differences among salamanders is carried out through an analysis of dorsal skull shape in three species of transforming salamanders. Although differences in dorsal skull shape are visibly obvious in salamanders, ontogenetic and phyletic differences have not been studied except as illustrated in cranial ossification sequence studies (Bonebrake and Brandon, 1971; Altig, 1965; Reilly, 1986). There are no quantitative data on skull shape ontogeny in sala-manders.

The importance of defining "shape" change in ontogeny and phylogeny has been formalized by Alberch et al. (1979). These authors, however, restricted the definition of "shape" to non-dimensional parameters of the relative proportions (length, area, volume) of structures. "Shape" in this paper is defined by a constellation of landmark points measured from homologous locations on each skull. Comparison of the two-dimensional shapes defined by homologous points can reveal similarities and differences in form that indicate both ontogenetically shared characters and phyletic differences between species. Using the Resistant Fit Theta Rho Analysis (RFTRA), regions of shape similarity and areas of relative deformation are identified, and a generalized measure of morphometric distance between forms is computed. Interpretations of shape changes based on RFTRA are nearly identical to those based on other shape comparison methods discussed elsewhere in this volume (see discussion for a comparison). The goal of this study is to use this technique to compare and contrast ontogenetic and phyletic differences in cranial shape in several species of transforming salamanders and, thus, to test the utility of the method for a broader phylogenetic analysis of the Order Caudata.

Materials and Methods

RFTRA Methods

Shape analyses and graphics were done with a PC Connection AT computer and Hewlett-Packard laser plotter using the RFTRA package of programs for Resistant Fit Theta Rho Analysis written by Ralph E. Chapman, Scientific Computing, Automatic Data Processing, the National Museum of Natural History, Washington, D.C. (included with this volume).

The RFTRA method was chosen for several reasons. (1) RFTRA has been shown to be especially useful in analyzing the comparative morphol-ogy of animal skeletons to identify similarities and allometric shape differences (Benson, 1976, 1982; Olshan et al., 1982). (2) It has been used to evaluate potential phylogenetic characters (Chapman and Brett-Surman, in press; Chapman, in press). (3) The method is suited to the analysis of cranial shape metamorphosis because it produces easily interpretable output in the form of vector plots showing the relative movement of landmarks in the transformation of one shape to another. (4) RFTRA produces distance coefficient that quantifies overall shape difference of two forms. (5) The mathematical bases of resistant fitting methods are easily understood by researchers trying to interpret the results. And (6), the RFTRA results for this dataset are very similar to results of several other shape analysis methods. The results of five different shape analyses of skull metamorphosis in one of the species examined (*Ambystoma tigrinum*) are compared at the end of the discussion section.

The RFTRA method quantitatively matches points of one shape constellation to those of another, eliminating differences of size and orientation, to reveal only those differences caused by deformation of shape. The two objects being compared are scaled such that the median squared interlandmark distance is unity. This concentrates the registration of parts of the form where there is the least or no isometric shape difference and produces a lack of fit where allometric shape differences actually occur (Siegel and Benson, 1982). As long as 50% or more of the landmark points fit closely, the resistant fit method will effectively match areas with coordinated local (small) variation and identify areas of large deformation. The mathematical framework and discussions of these procedures are presented in Siegel and Benson (1982), Benson and Chapman (1982), and Chapters 10 and 12 in this volume.

The landmark constellations for all specimens in each ontogenetic sample were scaled, rotated, and translated using the RFTRA program (least squares option) to superimpose sets of landmark points without changing their shapes. This produced a new set of superimposed point coordinates for the specimens within a sample that were then averaged to produce a single set of skull shape coordinates for each ontogenetic stage. Average sets of coordinates for ontogenetic samples were then compared using pair-wise RFTRA procedures (resistant fit option) which produced vectors indicating relative size-free deformations in distance and direction for each landmark point in each pair-wise shape comparison.

Between two forms, average vectorial distance and direction can be used as a generalized coefficient of deformation or morphometric difference (Benson, 1976, 1983; Benson and Chapman, 1982). The RFTRA program was used to generate a matrix of mean RFTRA distance coefficients. This distance is the average of the squared distances between corresponding landmark points divided by the average distance of each landmark on both superimposed shapes, from the center of the form. RFTRA distances were clustered using the UPGMA method to illustrate the relative morphometric similarity of the cranial shapes. It is assumed in this study that differences in the average distances between pairs of landmarks in ontogenetic comparisons indicate that some developmental shape change has occurred. Likewise, differences in species comparisons of the same ontogenetic stage indicate that evolutionary shape change has occurred.

Specimens and Shape Measurements

Seventy-four specimens from nine ontogenetic samples representing three species of salamanders from two families were analyzed (specimen numbers and localities in the Appendix). These included 1) two samples of *Ambystoma tigrinum* (Ambystomatidae): 11 larvae at an ontogenetic stage just before metamorphosis (late larvae), and 8 recently transformed individuals; 2) three samples of *Ambystoma talpoideum*: 7 early larvae, 7 late larvae, and 5 transformed individuals; and 3) four samples of *Notophthalmus viridescens* (Salamandridae): 6 early larvae, 10 late larvae, 10 recently transformed juveniles (efts), and 10

transformed individuals. The early larval, late larval, and transformed samples of *N. viridescens* and *Ambystoma talpoideum* represent the same relative stages of development previously determined by ossification sequence of bony elements of

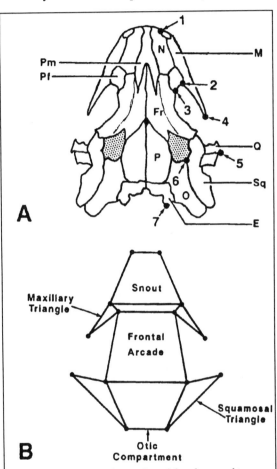

Figure 1. Landmark points and cranial regions used to compare cranial shape in salamanders. A) Newt skull indicating bilateral pairs of homologous points measured from each ontogenetic sample (right half). The landmark pairs are 1) medial point of narial foramen, 2) posterior of maxillary-prefrontal suture, 3) lateral point of frontal-prefrontal suture, 4) posterior tip of maxillary, 5) lateral point of squamosal, 6) anterior point of intersection of squamosal and otic capsule, 7) posteromedial notch of exoccipital. B) Constellation of landmark points taken from the skull shown in A. Morphological regions of the cranial shape discussed in the text are indicated. Abbreviations: E, exoccipital; Fr, Frontal; M, maxillary; N, nasal; O, otic capsule; P, parietal; Pf, prefrontal; Pm, premaxillary; Q, quadrate; Sq, squamosal.

cleared and double stained skulls (Reilly, 1986, 1987). Two stages of metamorphosed newts (efts and adults, both metamorphosed) were included to examine the extent of post-metamorphic change between these forms.

The skull of each specimen was centered in the field of view of a Zeiss SV-8 binocular dissecting microscope and pinned with the top of the skull parallel to the microscope stage. Landmark points were drawn on paper using a camera lucida. Cartesian coordinates for each point were digitized using a Houston Instruments Hipad digitizing pad interfaced with an IBM AT computer. Digitizing measurement error was less than 0.01%. Landmark coordinates were then scaled to millimeters. The 14 landmark points taken from each specimen are illustrated in Figure 1.

To facilitate discussion of regional shape changes, compartments of the cranial shape constellation are defined (Figure 1B) on the basis of functionally important aspects of the cranium. The otic compartment, or foundation of the cranium, defines the relationship between the otic capsules and the foramen magnum. Bridging the otic compartment to the snout is the frontal arcade. The snout region reflects the width of the nasal capsules and anterior neurocranium. The maxillary triangle indicates the displacement of the cheek from the frontal bone, and its lateral side is the length of the jugal process of the maxilla. The free point of the squamosal triangle locates the point of jaw articulation relative to the otic capsule.

Results

Mean shape constellations of the 14 homologous landmarks for each ontogenetic stage are shown in Figure 2. From gross comparison of these shapes qualitative shape differences can be seen. *Notophthalmus viridescens* has a narrower and longer snout relative to the two *Ambystoma* species, while all of the forms have similar otic compartments. The most obvious gross metamorphic shape changes are shortening of the frontal arcade in the ambystomatids, and increased maxillary length in newts.

The RFTRA quantification of the relative vectorial movements of landmarks through ontogeny allows more detailed analysis of ontogenetic shape change within species (Figure 3). The vectors represent actual ontogenetic transforma-

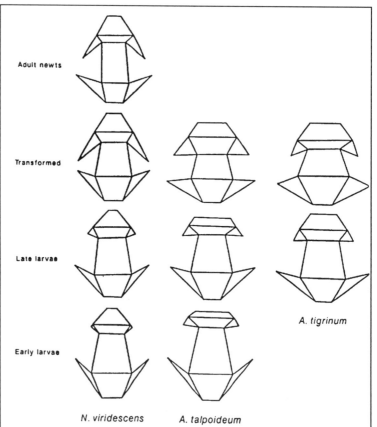

Figure 2. Mean cranial shape configurations of homologous landmark constellations produced by RFTRA for the ontogenetic samples of three species of salamanders. All forms are scaled to the same size and reflect only differences in shape.

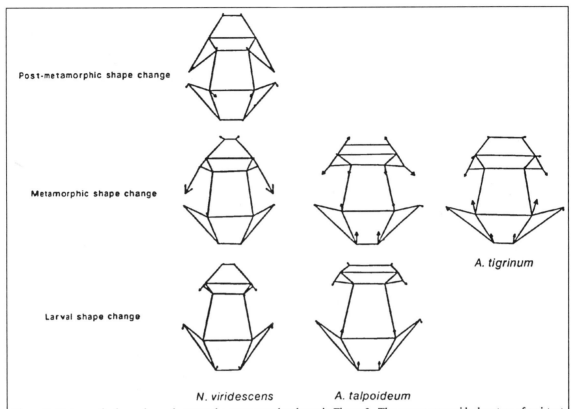

Figure 3. Ontogenetic shape change between the mean samples shown in Figure 2. The arrows are residual vectors of resistant fit methods showing the relative distance and direction of movement of homologous landmarks. Vector deformations describe ontogenetic transformations during larval growth (between early and late larvae), metamorphosis (between late larvae and transformed samples) and postmetamorphic growth for the species indicated.

tions of shape. Through larval development (Figure 3, Larval shape change) *N. viridescens* and *Ambystoma talpoideum* share the following shape changes: 1) shortening of the skull by anterior movement of the otic compartment, 2) enlargement of the maxillary triangle (reflecting increased length of the maxillary bone), and 3) lateral deformation of the squamosal triangle as the point of jaw articulation moves posteriorly relative to the otic compartment. *Ambystoma talpoideum* has greater shortening of the frontal arcade through anterior displacement of the otic compartment (Figure 3).

At metamorphosis all three species undergo lengthening of the maxillary bones and shortening of the frontal arcade but differ in the extent to which these changes occur (Figure 3, Metamorphic change). During transformation there is greater posterior deformation of the maxillary triangle in *N. viridescens* and less shortening of the frontal arcade than in the *Ambystoma* species. In both *Ambystoma* species shortening of the frontal arcade is in part accomplished by anterior movement of the otic compartment, a trend already evident during larval shape change. Metamorphosis in *A. talpoideum* differs from *A. tigrinum* in that the relatively short snout expands posteriorly. Newts differ at metamorphosis by exhibiting anterolateral widening of the snout, while in *A. talpoideum* it becomes narrower and longer.

Shape change between recently transformed efts and older adult newts involves a lengthening and narrowing of the frontal arcade, and a posterolateral shift in the squamosal triangle indicating some shape change after metamorphosis. This increases the distance between the posterior end of the maxillary bone and the point of jaw articulation.

Overall shape difference of the metamorphosed forms of the three species are contrasted in Figure 4. The two species of *Ambystoma* differ in the relative angles of the squamosal and maxillary triangles (Figure 4A). Comparison of all three species (Figure 4B) shows the similarity of the *Ambystoma* spp. relative to the newts, which have a relatively longer, narrower skull (frontal arcade and otic compartment), longer maxillary triangles, and the most anteriorly rotated squamosal triangles.

Resistant fit estimates of the relative morphometric distance among the various skull shapes are shown in Figure 5. The most general pattern seen in the morphometric shape distances is that the larval and transformed shapes are clustered into separate groups indicating that metamorphic shape change is greater than species differences. Within both of these metamorphic groups the newt forms and *Ambystoma* forms were intrinsically similar in shape. The different transformed newt stages were more similar to each other than to the transformed *Ambystoma* spp.

Among the larvae, the early and late newts were the most similar of the comparisons. In contrast to this close ontogenetic shape similarity in newt larvae, late *A. talpoideum* larvae were closer in shape to *Ambystoma tigrinum* larvae of the same stage than they were to their own larvae at an earlier stage.

Discussion

The results of this analysis show the utility of the RFTRA methods for quantifying ontogenetic trajectories in salamander cranial shape (see below for a comparison of RFTRA to other techniques). During larval development prior to metamorphosis, newts and *A. talpoideum* share several modifications of cranial shape, such as shortening of the otic compartment (Figure 3), that indicate similarity of shape ontogeny in the two species. Some of the

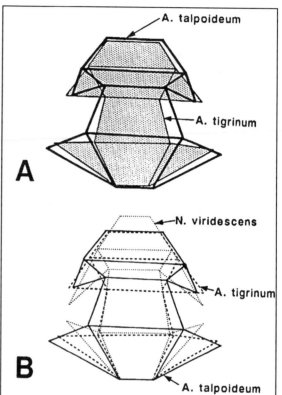

Figure 4. Species comparisons of cranial shape for transformed salamanders using RFTRA. A) Landmark constellation of *Ambystoma talpoideum* (connected by heavy lines) is superimposed on the shaded shape for *A. tigrinum*. B) All three species superimposed. Note the relative rotation of the maxillary and squamosal triangles.

larval shape changes, such as maxillary bone growth, continue, but with a much greater rate of development during metamorphosis in newts than in ambystomatids. Forward movement of the otic compartment and shrinkage of the frontal arcade is a change seen in each species both during larval ontogeny and through metamorphosis, which contributes to the transformed shape differences of the three species. In newts, however, after metamorphosis there is a lengthening of the skull by a reversal of the larval trajectory of reduction in the otic compartment and frontal arcade.

Clustering of RFTRA distances grouped the samples based on overall differences in shape. Larval newts show ontogenetic conservation of shape, while the late larval *A. talpoideum* were more similar to the late *A. tigrinum* than to their own early larvae. This indicates that there is more ontogenetic shape change during larval development in *A. talpoideum* than there is in newts. Perhaps within the *Ambystoma* species there is shape convergence through larval ontogeny of initially different early larval morphologies, which is then followed by species shape divergence during metamorphosis.

Considering the morphological changes that occur during transformation from an aquatic to a terrestrial form (Reilly, 1986, 1987; Lauder and Shaffer, 1988; Reilly and Lauder, 1990; Lauder and Reilly, in press) it is not surprising that morphometric shape distances across metamorphosis were greater than between species (Figure 5). This shows how confusing ontogeny can be when comparing salamanders and reflects why it is difficult to find characters that are not affected by metamorphosis. But the utility of the RFTRA methods for phyletic diagnosis is shown by the consistent separation of newts and ambystomatids within metamorphic stages based on morphometric shape distance.

Comparison of larval shapes of newts and *A. talpoideum* (Figure 2) and the morphometric distance between them (Figure 5) shows that early in their ontogeny there is phyletic divergence in cranial shape, with the latter having much wider snouts. Phyletic differences are also indicated by differences in cranial shape metamorphosis. Newts differ from the others in having much greater development of the maxillary bone at metamorphosis indicating a case of rate heterochrony in maxillary development.

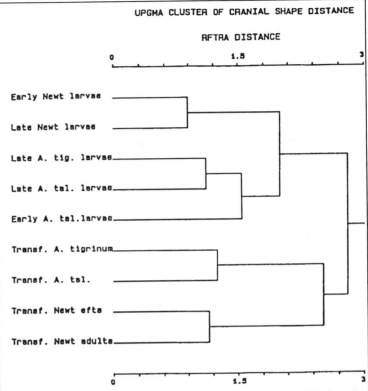

Figure 5. Cluster analysis of RFTRA morphometric distances of cranial shape for nine salamander samples. Note that metamorphic shape differences are greater than taxonomic shape differences but the newts and *Ambystoma* spp. consistently clustered together within ontogenetic stages.

Metamorphosis is very similar within the ambystomatids (Figure 3). Based on the morphometric distance between the *Ambystoma* species they are more similar before metamorphosis than after (Figure 5). After transformation, phyletic difference in the two species can be described in a shape context by a shrinkage and shift in the otic compartment and frontal arcade of *A. talpoideum* (Figure 4B). Such a change rotates the squamosal triangle inward, and the maxillary triangle outward, leading to a convergence of the *A. talpoideum* shape with that of *A. tigrinum*. Thus, patterns of shape differences in transformed crania of the two species might be hypothesized to be due to metamorphic changes in the neurocranium with associated shifts in the peripheral structures.

The longer and narrower skull in newts (Figure 4B) also appears to be due to changes in the frontal arcade and otic compartment. In this case, shape difference can be accounted for by a reversal of the ontogenetic trajectory of these regions which lengthens and narrows the neurocranium and rotates the two triangles inward. As discussed above, the longer maxillary triangle in newts is due to heterochronic development of this bone at metamorphosis. Therefore, changes in the neurocranium and maxillary bone occur both ontogenetically and phyletically in these species. If the pattern of central cranial deformations influencing overall shape was found to be widespread in a broader phyletic comparison, this would provide insights into developmental constraints on design that limit phyletic morphospace available for transformed skulls.

RFTRA is useful in defining ontogenetic changes in shape shared by different species, identifying heterochrony, and demonstrating phyletic differences in early larval, metamorphic and transformed shape. It has great potential for identifying phyletic differences and ontogenetic similarities in shape among families of salamanders. Such data are needed for the study of structural constraints to the evolution of form (*sensu* Lauder, 1981, 1982) and are necessary to test hypotheses relating patterns of organismal design, such as cranial morphology, to functional roles, such as aquatic feeding.

Ideally, however, comparisons of ontogenetic phenomena, should take place with explicit hypotheses of phyletic relationship (Fink, 1982; Lauder, 1982). This is difficult with salamanders because larval ontogeny and metamorphosis are extremely variable. Hypotheses of phyletic relationship are weakened by 1) paedomorphosis, 2) lack of outgroups to identify ancestral character states, and 3) lack of documented patterns of development to which analyses of processes (such as heterochrony) can be applied to determine character polarities. For example, because salamanders share similar ontogenetic trajectories of

maxillary development does this mean they share common ontogeny, common ancestry, or both?

Morphological patterns associated with metamorphosis in different groups must be identified and compared to reveal generalized historical pathways of structural change. Emerson (1988), for example, provides an excellent case study using shape analysis to test historical patterns of structural change in frogs. RFTRA provides a useful method for analyzing differences in shape. The independent analysis of overall skull shape transformation provided valuable insights to interpret data from functional analyses of metamorphosis in the same specimens (see Reilly and Lauder, 1990). RFTRA identified ontogenetic similarities, species differences in skull design, and morphometric distance between forms that could be coded as characters. The usefulness of such characters in phylogenetic analysis will emerge as more comparative data on ontogeny and metamorphosis are gathered, and the congruence of independently derived phylogenetic trees is compared. Whether morphometric analyses of ontogeny in different salamander families can identify new synapomorphic characters to use in phyletic analyses remains to be tested.

Comparison of Shape Analysis Methods

To examine the relative accuracy of the RFTRA method in identifying local and regional shape deformation, metamorphosis of skull shape in *Ambystoma tigrinum* was analyzed using four additional shape analysis methods for comparison (Figure 6). The samples of larval and transformed *Ambystoma tigrinum* skulls were especially useful for this comparison because they are samples taken just before and just after metamorphosis and do not differ in external head-width, external head-length or skull-width (Reilly and Lauder, 1990). Thus, they do not differ in head size and reflect only cranial shape changes that occur at metamorphosis.

All five methods indicate virtually the same metamorphic shape changes (Figure 6). The snout area exhibits small local deformation while the maxillary triangles rotate posteromedially. By far the greatest change illustrated by every method is the shift in the landmarks defining the otic capsule (which expands) and the frontal arcade (which contracts). This brings the rear of the skull forward relative to the snout, resulting in a more posterior point of jaw articulation relative to the entire skull, and decreased skull-length. For this data set, all 5 methods matched areas of local deformation (the snout) and revealed the same large metamorphic deformations in the posterior aspect of the skull. Thus, the RFTRA method fits well with this application for shape analysis. Although all 5 methods independently produce similar results and interpretations regarding shape differences, no one method alone provided an exhaustive analyis of the data. Each of the programs (RFTRA, GRF, TPS) has specific advantages and thus, these three programs complement each other to provide a full range of exploratory analytical approaches for landmark data.

The GRF program is the most versatile and easiest to use. This program provides options for least squares and resistant procrustes fitting to a base specimen (as does the RFTRA program) and generalized least squares and resistant procrustes fitting (to the average form). It is very useful in graphic visualization of variance around landmark points for single or groups of sqpecimens. In addition, the GRF program has the option of adding affine (uniform) fitting and computes tensor strain crosses indicating the direction and

magnitude of affine transformations for landmarks (fit to either a reference or average form).

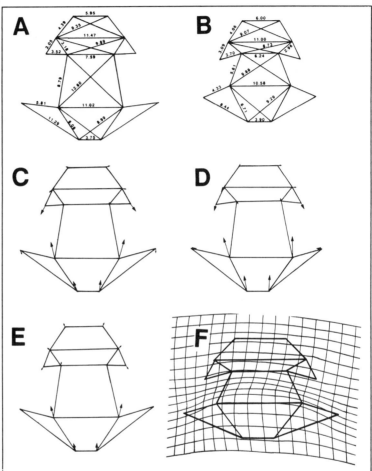

Figure 6. Comparison of five shape analysis methods in describing cranial shape metamorphosis in *Ambystoma tigrinum*. A, B) Truss analysis (Strauss and Bookstein, 1982) of larval A) and transformed B) truss configurations for cranial shape reconstructed using mean truss distances (in mm) between landmark coordinates. Shape changes at metamorphosis are identified by comparing A and B. These forms represent the average shapes input into the remaining analyses as coordinate data. C, D, E) Larval skull configurations with computed residual vectors, indicating movements of larval landmarks during metamorphosis, from the output of C) RFTRA, D) Generalized Resistant Fit (GRF program, Rohlf and Slice, 1990), and E) PROJECT program (Bookstein). Note that the vector directions and magnitudes for individual landmarks are essentially identical for these three methods. F) Thin Plate Spline (TPS program, Rohlf) illustrating the transformed skull configuration (as in B) with grid deformation for skull metamorphosis (grid has an overall horizontal deformation due to program output). Note the expansion of the otic capsule and contraction of the frontal arcade indicated by all five methods.

The RFTRA program has the advantage of providing an overall "morphometric distance" measure between shapes and a clustering option that can be used to visualize overall shape differences in a series of shapes. Even though there are statistical and mathematical problems with this metric (Bookstein, 1991), it provides very useful information about the magnitudes of overall shape differences in a sample of forms. In addition, vector trends for individual landmarks (for example in ontogenetic series) can be easily displayed and plotted using the RFTRA program.

The TPSPLINE program performs pairwise shape comparisons in terms of: 1) the complete fit (translational + rotational + affine (uniform) + non-uniform (warping) components) of one specimen on to another, 2) the affine (uniform) fit alone, 3) the non-uniform fit (warp) alone, and 4) the decomposition of the non-uniform fit into its partial warps (see Chapter 11). The partial warps are ordered either on the basis of landmark contributions to the eigenvectors indicating local (larger eigenvectors) or broad scale (lower eigenvectors) deformations, or on the basis of the contribution of each partial warp to the total bending energy of the fit. Because an understanding of the contribution of these components to shape transformation is now emerging, this method (thin-plate spline) provides a detailed description of shape differences between two landmark shape constellations.

Acknowledgments

This study was facilitated by instruction and software provided by the National Science Foundation-sponsored Workshop on Morphometrics in Systematic Biology, held at the University of Michigan, May 16-28, 1988. I thank Fred Bookstein, Ralph Chapman, Sharon Emerson, Bill Fink, George Lauder, Jim Rohlf, Dennis Slice, Rich Strauss, Peter Wainwright and an anonymous reviewer for helpful discussions and comments on the manuscript. This research was supported by NSF grants DCB 8710210 and BSR 8520305 to George Lauder.

References

Alberch, P., S. J. Gould, G. F. Oster, and D. B. Wake. 1979. Size and shape in ontogeny and phylogeny. Paleobiology, 5:296-317.

Altig, R. G. 1965. Observations on the ontogeny of the osseous skeleton of *Siren intermedia* Le Conte. Masters Thesis, Southern Illinois University, 62 pp.

Benson, R. H. 1976. The evolution of the ostracode *Costa* analyzed by "Theta Rho difference". Abh. Verh. naturwiss. Ver. Hamburg 18/19 (Suppl.), 127-139.

Benson, R. H. 1982. Comparative transformation of shape in a rapidly evolving series of structural morphotypes of the ostracode *Bradleya*. Pp. 147-164. in Fossil and recent Ostracodes (Bate, R. H., E. Robinson and L. M. Shepard, eds.). Ellis Horwood, Chichester, U.K.

Benson, R. H. 1983. Biomechanical stability and sudden change in the evolution of the deep-sea ostracode *Poseidonamicus*. Paleobiology, 9:398-413.

Benson, R. H., and R. E. Chapman. 1982. On the measurement of morphology and its change. Paleobiology, 8:328-339.

Bonebrake, J. E., and R. A. Brandon. 1971. Ontogeny of cranial ossification in the small-mouthed salamander, *Ambystoma texanum*. J. Morphol., 133:189-204.

Bookstein, F. L. 1991. Morphometrics tools for landmark data. Accepted for publication, Cambridge University Press, Cambridge.

Chapman R. E., and M. K. Brett-Surman. In Press. Morphological observations on hadrosaurid ornithopods. *in* Dinosaur systematics (Currie, P. J., and K. Carpenter, eds.). Cambridge Univ. Press, Cambridge.

Chapman, R. E. In press. Shape analysis in the study of dinosaur morphology. *in* Dinosaur systematics (Currie, P. J., and K. Carpenter, eds.). Cambridge Univ. Press, Cambridge.

Duellman, W. E. and L. Trueb. 1986. Biology of amphibians. McGraw Hill, New York.

Emerson, S. B. 1988. Testing for historical patterns of change: a case study with frog pectoral girdles. Paleobiology, 14:174-186.

Fink, W. L. 1982. The conceptual relationship between ontogeny and phylogeny. Paleobiology, 8:254-264.

Lauder, G. V. 1981. Form and function: structural analysis in evolutionary morphology. Paleobiology, 7:430-442.

Lauder, G. V. 1982. Historical biology and the problem of design. J. theor. Biol., 97:57-67.

Lauder, G. V. and H. B. Shaffer. 1988. The ontogeny of functional design in tiger salamanders (*Ambystoma tigrinum*): are motor patterns conserved during major morphological transformations? J. Morphol., 197:249-268.

Lauder, G.V., and S. M. Reilly. In Press. Metamorphosis of the feeding mechanism in tiger salamanders. J. Zool. Lond.

Olsham A. F., A. F. Seigel, and D. R. Swindler. 1982. Robust and least-squares orthogonal mapping: methods for the study of cephalofacial form and growth. Am. J. Phys. Anthropol., 59:131-137.

Reilly, S. M. 1986. Ontogeny of cranial ossification in the eastern newt, *Notophthalmus viridescens* (Caudata: Salamandridae), and its relationship to metamorphosis and neoteny. J. Morphol., 188:215-326.

Reilly, S. M. 1987. Ontogeny of the hyobranchial apparatus in the salamanders *Ambystoma talpoideum* (Ambystomatidae) and *Notophthalmus viridescens* (Salamandridae): the ecological morphology of two neotenic strategies. J. Morphol., 191:205-214.

Reilly, S. M. and G. V. Lauder. 1990. Metamorphosis of cranial design in tiger salamanders: a morphometric analysis of ontogenetic change. J. Morphol., 204:121-137.

Rohlf, F. J. and D. Slice. 1990. Extensions of the Procrustes methods for the optimal superimposition of landmarks. Syst. Zool., 39:40-59.

Siegel, A. F. and R. H. Benson. 1982. A robust comparison of biological shape. Biometrics, 38:341-350.

Strauss, R. E. and F. L. Bookstein. 1982. The truss: body form reconstruction in morphometrics. Syst. Zool., 31:113-135.

Appendix: Specimens Examined

All specimens used in this study are from the Museum of Natural History, University of Kansas, Lawrence, KS.

Ambystoma tigrinum: Colorado Springs, El Paso County, Colorado. Late larvae (*N*=11), Mean snout-vent length (SVL) = 78.8 mm: KU89119-122, 89124, 89128, 89135, 89140-1, 89144-5. Transformed individuals (*N*=8), SVL = 91.1 mm: KU89091, 89096, 89102, 89107-11.

Ambystoma talpoideum: Flamingo Bay, Savannah River Ecology Laboratory, Aiken County, South Carolina. Early Larvae (*N*=7), Stage III (Reilly, 1987), SVL = 25.0 mm: KU204692, 204694-6, 204698-9, 204701. Late Larvae (*N*=7), Stage VI (Reilly, 1987), SVL = 39.1 mm: KU204693, 204697, 204700, 204712, 204714-16. Transformed individuals (*N*=5), 1 yr. old transformed Adults (Reilly, 1987), SVL = 39.8 mm: KU204722-26.

Notophthalmus viridescens: McGuires Pond, 9.6 km South of Carbondale, Jackson County, Illinois. Early larvae (*N*=6), Stage III (Reilly, 1986), SVL = 16.5 mm: KU203908. Late larvae (*N*=10), Stage IV (Reilly, 1986), SVL = 22.7 mm: KU203912. Transformed emergent efts (*N*=10), Efts (Reilly, 1986), SVL = 24.2 mm: KU203916-7, 203922, 203929, 203932-4 203942-4. Transformed newts (*N*=10), Adults (Reilly, 1986), SVL = 46.6 mm: KU203974-83.

Part IV

The Problem of Homology

This Part was originally to be entitled "Morphometrics in the Systematic Context." But of the three goals of the Workshop reviewed in the Introduction to Part I—teaching data acquisition, demonstrating advanced morphometric Analyses exploring the conceptual tie between morphometrics and systematics—the third proved most stubbornly resistant. This was not owing to any reticence on the part of the student participants. In question periods, end-of-day reviews, and bull sessions, they endlessly raised questions about this tie, questions that none of the instructors seemed to be able to answer satisfactorily. While the techniques of data acquisition (Part II) and the various hybrid analyses (Part III) are capable of dealing with outline or landmark data regardless of its provenance (fossils, living specimens, biomedical images, cartoons), the basic tools of systematics— character states and their changes—do not appear very well aligned with the morphometric synthesis we have built here. The mismatch may owe to the new morphometric emphasis on covariances, to the intentional destruction of the idea of any unitary phenetic distance, or to the treatment of a landmark configuration as a single "character" in a space of dimension too high to suit notions of gap-coding

or polarity. We shall not explore these philosophical-methodological perplexities here. However, one conceptual crux of the morphometrics-systematics tie was anticipated in the original Workshop plan: the possible difficulties in the course of using the same word, "homology," with two different meanings, one from evolutionary theory and one from geometry.

In Chapter 17, expanded from a faculty lecture, Gerry Smith reviews the classic systematic construct of homology, arguing how difficult it will be to retrieve the discrete logic of changes in homologous characters from the modern morphometric toolkit of deformations and covariances. In Chapter 18, Spafford Ackerly, a student participant at the Workshop, presents us a bold experiment in extending the notion of landmarks so as to represent something "more homologous," in this case, the record of a parametric growth process. In Chapter 19 Annika Sanfilippo and William Riedel show that while the analysis of the coordinates of seven landmarks was sufficient to differentiate different morphotypes, additional, difficult to quantify, information from pore patterns, outline shape, and appendages is needed in order to differentiate traditional species

Chapter 17

Homology in Morphometrics and Phylogenetics

Gerald R. Smith

Museum of Zoology, University of Michigan
Ann Arbor, Michigan 48109

Abstract

Homology is neither an empirical nor a conceptual problem, but a relation enabling a two-part method for classifying the descent and modification of characters. The two parts are operational and taxic.

Location of homologous characters in terms of landmark topology is a necessary operation prior to comparative anatomical and phylogenetic analyses. Operational homology is a similarity relation defined by positional correspondences among internal and external landmarks. In evolutionary biology, the relation assumes the existence of unknown transformation series of character states which descended with modification and branching from an unknown common ancestor. The analytical construct is a phylogenetic hierarchy of taxa in which the observed character states are derived expressions of ancestral characters. Operational homology cannot remain independent of the evolutionary concept of homology (similarity due to common ancestry) because at some point observed similarities must be put into historical context with a phylogenetic test that discriminates homology from homoplasy (similarity due to reversal, parallelism, or convergence). Speculation about evolutionary transformations of a character may be logically circular until corroborated by congruence with the state distributions of other characters in a cladistic hypothesis (Hennig, 1966).

Taxic homologies are corroborated synapomorphies, i.e., shared, derived character-state identities that diagnose monophyletic groups. Such homologies are correctly named by reference to the character (and usually its state), and the monophyletic group it diagnoses (e.g., the cap-like tarsals of the Dinosauria). The homologous states of a character have a polarized general-to-particular hierarchical relation, which corresponds, except for homoplasy, to the cladistic tree hierarchy for the taxa.

Morphometrics serves to define the relations among landmarks that identify operationally homologous characters and to quantify shape differences across ontogenetic and taxic transformation series. Morphometrics plays an important post-cladistic role in the analysis of trends and responses to evolutionary causes and constraints.

Introduction

This review of the relationship between morphometrics, cladistics, and homology is motivated by claims to exclusivity by advocates of operational homology and taxic homology. I will argue that neither of these approaches, nor a developmental approach, can stand alone. Operational homology is traditionally associated with morphometrics and phenetics; taxic homology is the core of cladistics. Morphometrics is the quantitative, comparative description of shapes of organisms as measured among sets of landmarks on homologous, and therefore comparable, anatomical units. Cladistics is a method that infers the sequence of branching lineages from hierarchical arrangements of derived character-state identities–taxic homologies. Phylogenetics is the analysis of amount and trend in evolutionary changes through time, as measured along transformations among homologous character states.

Phylogenetic information, based on homologous characters from comparative anatomy, molecular genetics, behavior, etc., provides a context for framing or testing hypotheses about evolution of species or clades. Morphometrics may contribute to evolutionary biology by quantifying comparative information about the states in transformation series of homologous structures and by ordinating trajectories of ontogenetic or taxic transformations in morphospace. These steps may provide characters for cladistic analyses or describe evolutionary trends in phylogenetic analyses.

A "character" is a variable used in cladistic estimation or other systematic studies. Morphometric characters are defined by several (not fewer than two) anatomical landmarks. Selection of landmarks, characters, and variables depends on the concept of homology to justify comparability among attributes of different species. **Homology**, defined as similarity due to descent from a common ancestor, is the methodological basis for studies of character-state transformations and diagnosis of monophyletic groups. The definition of homology in terms of descent from a common ancestor means that comparable structures in related organisms are taken to be complementary evolutionary representations of the "identical" ancestral structure (Simpson, 1961). Two current concepts of homology have developed around the two seemingly separate issues involved: (1) recognition by anatomical **similarity** and (2) definition in terms of **descent** with modification.

1) Homologous anatomical structures are operationally identified by similar material and shapes, as indicated by internal landmarks, and similar position, as indicated by spatial relations to external landmarks, at appropriate stages of development (Remane, 1956, 1971; Sneath and Sokal,

1973). The topological relations are of the anatomical sort, "anterior to," "dorsal to," and "distal to" (Jardine, 1967). For example, a homologous pair of bones, nerves, amino acid sequences, or DNA segments is recognized by correspondence of compositional, spatial, and ontogenetic relations among the landmarks in and around each complementary member of the pair. Two sets of structures that maximally satisfy these internal and external landmark relations in appropriate ontogenetic stages of different individuals are operationally said to be homologous (Boyden, 1947; Remane, 1956, 1971; Sneath and Sokal, 1963; Withers, 1964; Inglis, 1966; Fitch, 1966; Key, 1967; Jardine, 1967, 1969; Jardine and Jardine, 1967, 1969; Sokal and Sneath, 1973:78-82; Bookstein et al., 1985). Features of different organisms hypothesized to be homologous in this sense are acknowledged as such by application of the same anatomical name (Owen, 1848), e.g., the fifth gill arch of Elasmobranchii; the protochonch of Gastropoda, the alpha hemoglobin sequence of hominoids. Quantitative aspects of these homologies usually vary continuously, for example when the transformations are tracked vertically through the hierarchy.

2) The evolutionary concept defines homology in terms of the descent of complementary structural identities from their hypothetical complement in a common ancestor (Simpson, 1961). In its recently refined form (Patterson, 1982), this concept creates a special role for the relationship between a character identity and the group of species in which it is found: A homolog is a synapomorphy that helps diagnose a monophyletic group, i.e., a group that is restricted to all of the descendants of a common ancestor. In this concept, the monophyly of the group implies the homology of the synapomorphic character state (Bock, 1969). Analytically, this synapomorphic character state is evidence for the monophyly of the group (Wiley, 1975). Such a "taxic" homology is named with a qualifier indicating the monophyletic taxon it diagnoses, e.g., the feathers of Aves, the spinnerets of Arachnida. A taxic homology bears

an all-or-none relation to other character states within a cladistic hypothesis.

Conflict between advocates of the operational and cladistic schools is not new. In 1961, Simpson observed that usage seemed to demand two different terms, but that if "the argument were only about which concept is to bear the name 'homology', it would be quite useless." He favored using "homology" for the evolutionary concept, suggesting "morphological correspondence" (Woodger, 1945; see also Ghiselin, 1969) to indicate the identity relation that is methodologically independent of evolution. The vocabulary has changed because of the growth of morphometrics and cladistics, but "homology" retains the dualism it has had since Darwin provided a causal mechanism for Owen's idea and substituted the common ancestor for Owen's archetype. The dualism consists of (1) concern to find a conceptual basis and method for identifying comparable characters independent of the evolutionary conclusions sought, and (2) analysis of character identities in the framework of descent with modification. In the spirit of Inglis (1966), I will advocate the necessity of iterative interaction between operational and taxic homology. Operational homology is necessary because cladistic methods cannot begin with characters chosen because they are known to be descended from a common ancestor (it is not possible and if it were, it would be circular). Taxic homology is necessary because only a phylogenetic test can distinguish homologies from homoplasies– "similarities due not to common ancestry but to independent acquisition of similar characters" (Haas and Simpson, 1946).

Operational Homology

The operational concept of homology– correspondence of landmark positions from form to form (Sneath and Sokal, 1973)–refers to designation and definition of characters whose states are being compared in a systematic study. This use of the homology relation is near that which Owen (1848) called general homology–correspondence of parts in relation to an archetypical body plan.

Several problems will be discussed to clarify operational homology: some practical procedures by which homology can by identified (e.g., Nieuwenhuys and Bodenheimer; 1966, Jardine; 1969; Bookstein et al., 1985), difficulties of definition (Bookstein et al., 1985), and the nature of the circularity in the operational concept of homology (Inglis, 1966; Rieppel, 1980).

Historical background. Simpson (1961) discussed a broad range of practical homology criteria, emphasizing "minuteness of resemblance and multiplicity of similarities" of anatomical structure, as well as criteria for recognizing non-homologous parallelisms and convergences. Simpson's discussion of contemporary statistical methods for evaluating similarity was unenthusiastic, but he suggested that, conclusions on affinities (which means largely on homologies) are stronger the more the characters involved." Simpson's criterion for improved inference applies to the number of landmarks in operational homology (and the number of characters in taxic homology, see below).

D'Arcy Thompson (1942) elaborated the geometrical foundation of the operational concept of homology, based on his "naturalist's" concept: "invariant relation of position." Woodger (1945) is credited with developing the modern logical framework for the concept. N. Jardine made it more operational. In Jardine's (1967, 1969) system, like D'Arcy Thompson's, homology is a function specifying the geometrical transformation of one form to another: it is an optimized level of correspondence of position among landmarks. A sufficient matching correspondence relative to external landmarks is a necessary condition for the recognition of homology. [Internal landmark correspondence may be adequate for structures that are sufficiently unique due to the complexity of landmarks, e.g. fossil mammalian teeth. But internal similarity is not a necessary condition (Bock, 1963) if structural changes are being studied across higher taxa (Sattler, 1984), e.g., the transitions between fish jaw bones and mammalian ear bones.]

Difficulties of definition and the role of development. Difficulties arise when homologous structures have different ontogenies (Kluge, 1985, 1988; de Queroz, 1985; Roth, 1988), or when different structures have evolved to occupy the same position relative to external landmarks (Van Valen, 1982), or when iterative homologs confuse character ancestry (Roth, 1984; Wagner, 1989). For example, as a character, the mammalian upper premolar 4 of a taxon must have a number of recognizable correspondences of internal and external landmark locations to justify its comparison to other fourth premolars in a systematic study. We wish not to mistakenly compare a premolar of one taxon to a molar of a related group. This problem is complicated by similarity of internal landmarks of some premolars and molars due to their serial homology. At some point in the evolutionary history of mammals, premolars and molars had a common ancestry; at a later point, premolars and molars differentiated into daughter lineages.

Van Valen (1982) defined homology as continuity of information. In a detailed history of a specific case of the above example, he demonstrated that landmark similarity is not always an unambiguous criterion for homology. Ontogenetic changes, such as growth, movement, regression, repetition, addition, deletion, and changes in genetic and embryonic origin of parts can create missing or iterative homologies that confound comparison (Van Der Klaauw, 1966; Ghiselin, 1976; Roth, 1984, 1988; Kluge, 1985). Wagner (1989) described possible relationships between iterative homologs in radial symmetry. But tracking the continuity of information still requires identification of potentially homologous characters by landmark correspondence (quantitative or not).

Bookstein et al. (1985, Figure 5.1.1) illustrated several kinds of homology relations that require additional information for resolution of morphological correspondence. Structures with accretionary growth may exhibit material as well as anatomical landmark relations resulting from growth along "point paths" (Skalak et al., 1982;

Bookstein et al., 1985, Figure 5.1.1a). The paths are a source of homologous position information. Material landmarks enable visualization of the process of growth and its meaning for homologous landmarks and homologous development. Landmark fusion and division (Bookstein et al., 1985, Figure 5.1.1b,c) in ontogeny and evolution create ambiguity regarding new or lost space in the structure. Shape change through truncation of development (Bookstein et al., 1985, Figure 5.1.1d) also renders landmarks ambiguous and requires that we look to developmental processes for homologous pathways (Alberch et al., 1979; Roth, 1984, 1988).

Missing landmarks create problems for the operational comparison of potentially related forms (Jardine, 1969), but there are operational methods for facilitating comparisons. Changed or missing internal landmarks can be imputed from external landmarks, using a displacement vector model or an elastic mapping model, as illustrated by Bookstein et al. (1985, Figure 5.1.2). Accuracy of the inferred landmark position depends upon the extent to which the morphometric model mimics growth of tissues. (Missing terminal additions, due either to plesiomorphy or truncation, are a different problem, see below.)

van der Klaauw (1966) provided a classification of different kinds of modifications of homology in development (see also Alberch et al., 1979). Fitch (1970) and Patterson (1987) reviewed the comparable sources of variation in data from molecular biology. Wagner (1989) suggested three criteria–conservatism, individuality, and uniqueness–as three biological properties expected of homology on the basis of developmental considerations. The theoretical importance of developmental information to recognition of homology has been emphasized by Owen (1848), Gould (1977), Nelson (1978), and Roth (1984, 1988).

Even if developmental information is included in the definition of homologous characters (e.g., Wagner, 1989), the discernment of its alternate states in different lineages ultimately depends on correspondence of landmark sets, which must be determined empirically as matching similarities across topographic comparisons (Jardine, 1969). Jardine (1969) explored alternative definitions of homology in terms of topology, structure, and ontogenetic status, and concluded that topographic position was more basic because of changes in the migration and orientation of parts during development.

It is clear that homology must be located by landmarks in ontogenetic time as well as anatomical space if we are to gain information about the role of development in evolutionary processes. This can be seen most clearly when ontogenetic changes involve serial replacement of homologous characters or states, e.g., when teeth or exoskeletons go through replacement cycles. The most subtle and informative examples of ontogenetic homologs will involve discovery of the role of heterochrony in evolution (Alberch et al., 1979; Bookstein et al., 1985).

In an insightful review of the developmental basis for character conservatism, individuality, and uniqueness, Wagner (1989) suggested that there are problems with the concept of homology. But the difficulties cited pertain to the attempt to derive process information from a concept with no content extrinsic to the theory of evolution. The historical existence of descent with modification is assumed by evolutionary homology, therefore its role can be no more than the classification of character-state identities. It is a part of a method for describing the character-by-character chronicle, given a hierarchy. Using homology as part of cladistic methodology (see below) has conceptual utility, but no theoretical or predictive content. The difficulty is breaking out of the circularity of operational homology. Wagner's (1989) excellent discussion of development contributes to understanding the processes of transformation of homologous characters, but less to the methodological problem of homology.

Operational tests. Attempts to test homology operationally involve landmark consistency among

the groups constituting the larger taxonomic unit. Riedl (1978) and Patterson (1981) discuss probabilistic contexts for considering homolog identity, based on consistency of included homologous structures among given groups. Riedl considers the consistency of a homologous structure to be a function of the number of homologous parts it comprises. He also considers the uniformity of its occurrence in the groups that possess it, as well as its absence in the groups that lack it. Knowledge of groups is an evident requirement.

Patterson's function calculates the probability P that h homologs will specify the same taxon by chance as:

$$P = (S! \, (N\text{-}S)! \, / \, N! \,)^{h\text{-}1}$$

where N is the total number of groups available for sampling (i.e., the number of groups involved in the disagreement) and S is the sample of groups displaying the homology. The calculated probabilities that h homologies represent S out of N groups by chance ranges from 1/3 to astronomical values for more or less unconvincing homologs, leading Patterson to abandon the method (if not the principle) and evaluate the homologies on the basis of subjective considerations of taxic distribution and anatomy (Patterson, 1982:42; see also Ax, 1987:168).

Patterson (1982, 1987) suggested additional tests for rejection or acceptance of homology, independent of evolutionary theory. These include similarity, congruence, conjunction (two different homologous structures must not both be present in the same individual, within constraints imposed by symmetry relations and iteration), endoparasitism, etc. Of these, only similarity and congruence can play a major role–in hypothesized operational homology and cladistic tests, respectively.

Circularity in operational homology. Potential circularity arises from lack of satisfactory internal tests of two assumptions underlying decisions about operational homology: (1) that the study set of OTUs consists of a related group of organisms, and (2) that the characters are not parallel, reversed, or convergent homoplasies. Operational methods by themselves are insufficient for unambiguous resolution of homology because of the inadequacy of non-cladistic tests. This is most easily seen in the following kinds of difficulties: reversed or lost characters, parts that changed through amplification of serial homologies (Patterson, 1987), or structures that became spatially reoriented during ontogeny and phylogeny (Jardine, 1969). Bookstein et al. (1985) admitted, "The resolution of conflicts in homology, if resolution exists, lies mostly outside the arena of morphometrics." Roth (1984), after discussing the possibility that independently evolved developmental pathways might result in serial and (or) parallel homologs, stated, "For good biological as well as methodological reasons, it may be impossible to distinguish homology from parallelism within a population, or even between species." These authors are speaking of the difficulty of resolving homology from homoplasy, using only information from landmark matching and developmental similarity. At this (precladistic) stage of analysis, resolution requires external testing to relate homologous states to the historical taxonomic hierarchy upon which they are conditional (Bock, 1973, 1977; Ghiselin, 1969).

Taxic Homology

Referring to homology as "synapomorphy" defines it as a derived character state shared by the members of a monophyletic group (Patterson, 1982). Discovery and validation of a shared, derived character state as a homology is the objective of a cladistic procedure that establishes monophyletic groups and identifies the derived character states that diagnose them. Not all synapomorphies diagnose monophyletic groups. Confidence in the homology of a synapomorphy is established by its congruence with the distribution of the states of other characters over a cladogram. The homologies of special interest are those that uniquely diagnose well-supported monophyletic groups.

This sense of homology, called **taxic homology** by Eldredge (1979b), corresponds to the special similarity function that characterizes cladistic but not phenetic classifications. A derived character state is homologous with respect to a specific branch of a cladistic tree; the homologous state diagnoses the monophyletic group that descended from the stem on which it originated. The common ancestor replaces the hypothetical archetype (Rosen, 1973:500; Patterson, 1987:4). In this special sense, a character state is not homologous with less-derived states of the same character below it on the tree or with more-derived states of the same character higher in the tree.

But all states of the character must be generally homologous in their hierarchic relations throughout the tree to justify the comparisons: Hennig (1966:93) regarded homologous characters as "transformation stages of the same original character." This **transformational homology**, as it is called by Eldredge (1979a), is established by evidence that the observed representations of the character descended from one source through a hierarchic path indicated by the cladistic tree. Unlike taxic homology, it is a vertical, not an instantaneous, relation in the tree.

The homology of each synapomorphy is ultimately tested by its congruence with other characters (Hennig, 1966:112). Character states whose similarities are not concordant with the tree are homoplasies (Haas and Simpson, 1948) or "convergent homologies" and "parallel homologies" (Ghiselin, 1976). I would add a third category, "homoplasies that are introgressively transferred homologies."

Circularity in taxic homology. Can taxic homology stand alone, without an operational concept for identification of comparable characters? The possibilities might be illustrated by examples of taxic homologs that would not ordinarily be recognized by operational methods. For example, characters that are not operationally recognizable as homologs because they are dissimilar in composition and development may be indicated by a cladistic tree to be homologous. The horns of the Bovidae and the bone antlers of the Cervidae may be homologous as frontal ornaments at the level of the Pecora of the Artiodactyla, although operationally one finds different composition and different development, with little more than frontal position to suggest homology. But homology is suggested by congruence with other characters.

Non-operationally "similar" synapomorphic attributes in different anatomical positions provide a more extreme example, and we can begin to see the potential for circularity. Operational criteria would not ordinarily suggest that the black-and-white color patterns of spotted and striped skunks are homologous, but the congruence with morphology (and scent) leads to such a hypothesis. These synapomorphies are congruent; they encourage one to search for additional synapomorphies satisfying the congruence criterion.

The emergence of congruence as a prime criterion does not suggest a sufficient, non-circular methodology for the discovery of homology. The red color on the sides of the rainbow trout is homologous at some level with the orange color on the lower jaw of the plesiomorphically similar cutthroat trout. "The search for synapomorphies" could well begin with a plesiomorphic similarity, corroborated with such a "homology," and supplemented by other characters "tested" by congruence, bypassing landmark correspondence and operational homology. Given the available morphological and biochemical variability, additional congruent homoplasies could be found and listed, ad hoc, if not ad infinitum.

A test independent of congruence is clearly necessary as a check on the possibility of beginning with a plesiomorphic similarity and selectively adding homoplasies that agree with it. Since an initial step in cladistics is the assembly of a collection of comparable characters, this is an optimal time for application of morphometrics of landmark data to operationally posit homology. Initial selection of characters with a method independent of phylogenetic considerations avoids the possibility of adding evidence sorted by prior expectations. (But the taxa chosen for comparison are not so indepen-

dent.) The other necessary defense against circularity is the use of numerical phylogenetic methods based on a suitable criterion for tree selection, such as the parsimony methods of Farris (1983; Kluge and Farris, 1969) or the maximum likelihood methods of Felsenstein (1982). These also provide safeguards against biased chains of evidence.

Homology, synapomorphy, and paraphyly. Are homology and synapomorphy redundant terms (Patterson, 1982)? This suggestion has been made by Patterson, but it is weakened to the extent that some synapomorphies do not diagnose monophyletic groups and some unambiguous homologies may characterize paraphyletic assemblages. An occasion arises when an apomorphy is shared by all but a derivative part of a group–for example, where parallel, reversed, lost, or replaced states contribute to the designation of a monophyletic subgroup, leaving the homologous state incompletely represented in the larger group. The lepidotrichia of bony fishes, a paraphyletic group, stand as an example because of the loss or replacement of lepidotrichia in tetrapods, a derived group, as pointed out by Patterson (1982) in presenting the opposite point of view. Other cases exist wherever character states are introgressively transferred between species that are not sister groups. Although the logical and methodological relationship between homology and synapomorphy is fundamental to the homology concept and to cladistic methods, not all synapomorphies are taxic homologs, and the distinction justifies retention of the concepts of transformational as well as taxic homologies in hierarchical relation to each other (Ax, 1987).

In a comprehensive summary of the view that homologies are synapomorphies, Patterson (1982:33) argued that symplesiomorphy and synapomorphy are terms for homologs that stand in hierarchic relation to one another. This is a fundamental point, but, contrary to Patterson, I take this to imply that hierarchic relations of characters validate transformational as well as taxic homologies (see above). Patterson (1982) has criticized attempts to document character transformations, especially those utilizing fossils, on the grounds that they are founded upon paraphyletic groups, which are unnatural (unreal) inventions. If the groups are unnatural, much of what we think we know about evolution is invention of questionable reality (Patterson, 1982). Further, if paraphyletic groups are by definition without homologies, homology has nothing to do with evolution: "If phylogenies have to say something about evolution, then it is evident that homologies can have no role in them" (Patterson, 1982:58).

There are several important points in Patterson's exploration. First is the observation that paraphyletic groups have dominated classifications since their beginnings, and when we restructure systematics to monophyletic form, reflecting the historical chronicle of evolutionary divergence, our inferences about processes can only be improved (O'Hara, 1988). In addition, there is the problem that historical processes of transformation are unknowable because of our dependence on inferences based on extinct and paraphyletic (i.e., unnatural) groups. I think this problem is misspecified, however. Inference based on an assemblage made paraphyletic by misclassification of a monophyletic group is no more misleading than one made false by our ignorance of extinct group members. Conclusions confidently based on groups erroneously thought to be monophyletic will be most damaging of all. Documentation of anagenic transformations through (technically paraphyletic) lineages within monophyletic groups can generate and test important historical hypotheses (Roth, 1988). Ironically, tests of evolutionary hypotheses are most informative, as well as safest, when based on groups at the species level, where the processes occur, where morphometric methods are least in need of corroboration, and where evolutionary units are most likely to be paraphyletic. The natural order in a cladistic hierarchy is necessarily consistent with most evolutionary processes, as understood, as well as the historical narrative implicit in phylogenetic trees (Riedl, 1979; O'Hara, 1988). The results of past cross-testing of these bodies of theory and data justify some optimism for continuing. The correction of the paraphyly problem is desirable but trivial; we should strive toward a monophyletic classification for groups above the species level. Granted that, for many or even most groups, the true cladistic chronicle may be unknown and unknowable, nevertheless, molecular data and the fossil record provide independent consistency tests by which cladistic estimates can be improved.

Patterson (1982:58) seemed to be reaching still deeper: his rejection of the use of transformational homology—the transitions in states from ancestors to descendants over a cladogram—was accompanied by denial of evolutionary content in homology, presumably to avoid circularity in the study of evolutionary events and processes. His argument persuades me that we cannot regard homology as a concept with predictive content about the process of evolution. Homology cannot provide independent confirmation or challenge to any evolutionary hypothesis because the concept is merely a restatement of descent with modification applied to characters rather than taxa. But the definition "similarity due to common ancestry" is essential as a relational statement that facilitates the operational and taxic methods for inferring character-by-character chronicles of evolution. Assuming that the concept of homology is burdened with the circularities discussed above, we can also turn to empirical and conceptual aspects of geology and molecular biology for the theoretical tensions (Laudan, 1977) necessary to develop tests of hypotheses. Transformational homologies and homoplasies will play decisive roles in investigation of predictions from genetic, ontogenetic, ecological, stratigraphic, paleontological, and paleoecological studies of evolution in the context of cladistic hypotheses (Eldredge, 1979b).

The Relationship between Morphometrics and Cladistics

Morphometrics can contribute identification of characters and quantification of character states for cladistics, but cannot contribute directly to cladistic inference. Morphometric character analysis is either precladistic or postcladistic. Precladistic character analysis contributes to the definition of homologous characters and the quantitative description of character states for cladistic analysis. Character states can be quantified by morphometric methods and, when necessary, coded for treatment by a tree-forming algorithm (see Wagner, 1980; Kluge and Farris, 1969). We can define the cladistic character and character state in relation to the concepts of homology and landmarks as follows:

A "character" may be strictly defined as an among-taxon set of putatively homologous structures whose correspondence is recognized by (1) similarity of internal and external landmark positions on the organisms at appropriate stages of their developmental histories, and (2) identity due to inheritance from a common ancestor (as initially indicated by a previous, more inclusive phylogenetic analysis). However loosely the investigator satisfies these two criteria, they are necessary to establish general homology prior to carrying out a comparative and phylogenetic study. For example, an investigator studying Gardiner's (1982) group, "Homiothermia," may assume that numerous parallel aspects of bird and mammal homiothermy are homologous and therefore comparable characters. The investigator compiles characters from more amniote taxa, based on previous broader studies, and cladistically tests the monophyly of the Homiothermia as well as the taxic homology of all of the characters (see Gauthier et al., 1988). (The problem of redundant information content in genetically or functionally covarying characters is treated by Felsenstein, 1988.)

A "character state" is the quantified or coded condition of a varying character (see Colless, 1967; and Hull, 1968). Character states at one level of analysis often become characters at a finer analysis, as the character hierarchy parallels the taxic hierarchy. In morphometrics, the quantitative value or state of a character might be, for example:

1) a distance between landmarks;

2) ratio shape-ratios of distances measured over a structure defined by landmarks;

3) a measure of size, e.g., centroid size: the sum of squares of all of the interlandmark distances (Bookstein, 1991).

Shape characters are rarely independent of size, i.e., are usually allometric, and therefore should be analyzed as ontogenetic trajectories (Creighton and Strauss, 1985; Kluge and Strauss, 1985). An ontogenetic character state might be the allometric trajectory describing shape difference between juvenile and adult structure (Strauss and Fuiman, 1985). Diverse morphometric methods, such as principal component and factor analysis, biorthogonal analysis, medial axis data (Bookstein et al., 1985) or Fourier analysis, (e.g., Ferson et al., 1985), can be used to define ontogenetic trajectories that can be given character-state values. In general, these are less rigorously specified than Bookstein's (1991) shape variables based on transformations of triangles. Analytical triangles, digitized over the form or its parts (e.g., bones or sclerites), can yield at least one character per triangle; these may prove to be the most subtle and information-rich morphometric products for cladistic use. Each triangle can provide a within-group allometric trajectory, the values of which form a hyperbolic metric (Bookstein, 1991).

The states of morphometric characters usually vary continuously and may require some interpretive coding for use in most algorithms used in phylogenetic reconstruction. Homogeneous subset coding and generalized gap coding are solutions to this problem provided by Archie (1985) and Goldman (1988). In principle, states of continuously varying characters need not be treated as coded homology-identities. Most current parsimony algorithms do not take continuously varying characters as input, but this constraint is not inherent in the parsimony method. In principle, multistate characters may be subjected to Transformation Series Analysis (Mickevich, 1982) to optimize the distribution and order of their state transformations over the tree, although an algorithm is not yet available. The most unconventional aspect of the use of continuous characters in estimating synapomorphies would be the use of continuously varying states as putative estimators of homology, with certain states becoming homology identities only after becoming cladistic characters in the sense of Mickevich (1982). Here, morphometric characters would be similar to molecular sequence characters in displaying relative homology. Morphological characters are less likely to provide satisfactory taxonomic distances, presumably because their tempo and mode of evolution is less clock-like (Fitch and Atchley, 1987; Patterson, 1987).

In cladistics, the character-state value must be accompanied by information about direction of evolution as well as similarity of character states. Morphometric methods currently do not distinguish plesiomorphic and synapomorphic information. Inclusion of outgroup taxa in the analysis, and designation of the direction of the character states to be away from outgroup values, may place the morphometric data in the directional context necessary for cladistic analysis. Optimization of the distribution of character states is accomplished by a function based, for example, on a parsimony criterion (Farris, 1983). Quantified ontogenetic information may also be used to indicate polarity and sequence among multiple character states (Creighton and Strauss, 1985; Strauss and Fuiman, 1985).

When a terminal ontogenetic character state is missing, morphometric and ontogenetic data are incapable of discriminating between loss due to truncation of development and absence due to plesiomorphy, except in the context of a cladistic hypothesis (Fink, 1982). The first step in solving this dilemma is assignment of a character-state code value by reference to outgroups (the larger hierarchy), when possible, or entering an agnostic code value into the data matrix when necessary. The next step is a cladistic analysis, which will provide a homology decision based on congruence with other characters (Hennig, 1966). In numerical phylogenetic analysis, all of the characters are used; they

are not sorted by the systematist's prior evaluation of "good" and "bad" characters from other contexts. Good taxic homologies arise from numerical parsimony analysis, although these analyses almost always show reversed and potentially parallel characters among the taxic homologies—i.e., non-unique homologies little different from homoplasies play a role in diagnosis of monophyletic groups. However, prior selection of taxic homologies by their fit to "known" groups, with the discard of characters known to show some homoplasy, is not different from other methods of subjective prejudging. It is potentially circular.

Phylogenetics. Postcladistic character analysis is the use of morphometric methods to describe character variation and the direction and amount of evolution. This is usually approached by measuring the changes in size and shape differences among related lineages. Cladistic methods ignore random-walk fluctuations and amounts of evolution and are unable to control for secondary factors and effects of uncontrolled variables that affect processes driving groups of characters. One may use morphometrics and cladistics together to estimate the sequence of lineage branching and the patterns of responses among characters. Character correlations may be summarized and analyzed as consequences of genetic, developmental, functional, and ecological factors (Bookstein et al., 1985). Together, the cladistic tree and the morphometrically derived factors provide information about the relative timing, amounts, and directions of evolution. When the morphometric factors are examined in the context of ontogenetic or ecological hypotheses, we have a framework for investigating possible forces and constraints influencing evolutionary change.

Not only is a cladistic framework necessary for such an analysis (Fink, 1982), but the study must utilize a different data set to avoid methodological circularity (Ghiselin, 1966; Hull, 1967). For example, in a test of a hypothesis about the evolution of shape change through heterochrony, the shape/size trajectories would not be appropriate cladistic characters. Biochemical data are sometimes available for estimating branching sequences, or at least branch distances. Distances may be useful indicators of taxonomic distance wherever biochemical divergence is consistent and monotonic (Sibley and Ahlquist, 1987). Morphological evolution is still an objective of such studies, however, and morphological data will continue to be necessary for cladistic tests. Once a cladistic frame of reference is established, morphometrically quantified data may be analyzed by regression, ANOVA, or factor analysis, etc., to provide information about how the morphological differences relate to ecological processes (Bookstein et al., 1985).

Summary

Homology is not a concept with theoretical predictive content about processes, but part of a method for classifying character-state identities in terms of their transformations in a hierarchy of monophyletic groups. (Homology of characters is not a conceptual problem unless the paradigm is something other than descent with modification, in which case homology is the central mystery to be divined.) The ultimate goal in evolutionary biology is to understand the history of change and the ecological, genetic, and ontogenetic processes by which change occurs. Progress toward this goal may not change the concept of homology, but it should be aided by effective application of methods based on it.

Operational homology is a character correspondence, among taxa, based on the optimal matching of internal and external landmarks on exemplars, samples, or developmental series of OTUs. It is usually a quantified construct within which landmarks, variables, and characters are oriented for comparison in systematic biology. In this context, morphometrics can provide quantitative values for character states.

Taxic homology is an identity relation that diagnoses a monophyletic group. By diagnosing monophyletic groups, taxic homologs provide evidence for hypothetical ancestral states and a historical context for tests of evolutionary hypotheses. The validity of a taxic homology is tested against homoplasy by congruence with other characters over a hierarchical tree.

Operational and taxic homology are often discussed as concepts that exclude each other's validity. But each definition refers to a necessary step in the process by which homologous characters are chosen, tested, and analyzed in studies of evolution. The circularity in each half of the dual concept is avoided by appropriate use of the other.

A transformational homology is the sequence of modified states of a character over the inferred course of descent from a common ancestor through a cladistic tree hypothesis. It is the basis for study of genetic, ontogenetic, and ecological processes in the context of phylogenetics.

Acknowledgments

Fred Bookstein, Catherine Badgley, William Fink, Mark Johnston, and Mark Wilkinson read the manuscript and made helpful suggestions.

References

Alberch, P., S. J. Gould, G. F. Oster, D. B. Wake. 1979. Size and shape in ontogeny and phylogeny. Paleobiology, 5(3):296-317.

Archie, J. W. 1985. Methods for coding variable morphological features for numerical taxonomic analysis. Syst. Zool., 34:326-345.

Ax, P. 1987. The phylogenetic system. John Wiley & Sons, Chichester, 340 pp.

Bookstein, F. L. 1991. Morphometric tools for landmark data. Book manuscript, accepted for publication, Cambridge University Press..

Bookstein, F. L., B. Chernoff, R. Elder, J. Humphries, G. Smith, R. Strauss. 1985. Morphometrics in evolutionary biology. The Academy of Natural Science of Philadelphia, Spec. Publ. No. 15, 277 p.

Boyden, A. 1947. Homology and analogy. Am. Midl. Nat., 37:648-669.

Colless, D. H. 1967. An examination of certain concepts in phenetic taxonomy. Syst. Zool., 16:6-27.

Creighton, G. K., and R. Strauss. 1985. Comparative patterns of growth and development in cricetine rodents and the evolution of ontogeny. Evolution, 40:94-106.

Eldredge, N. 1979a. Alternative approaches to evolutionary theory. Bull. Carnegie Mus, Nat. Hist., 13:7-19.

Eldredge, N. 1979b. Cladism and common sense. Pp. 165-198 in Phylogenetic analysis and paleontology (Cracraft, J. and N. Eldredge, eds.). Columbia University Press, New York, 233 pp.

Farris, S. J. 1983. The logical basis of phylogenetic inference. Pp. 7-36 in Advances in cladistics (Platnick, N. I. and V.A. Funk, eds.). Vol. 2, Proc. 2nd Ann. Mtg. Hennig Soc. Columbia Univ. Press, New York, 218 pp.

Felsenstein, J. 1982. Numerical methods for inferring evolutionary trees. Quart. Rev. Biol., 57: 127-141.

Felsenstein, J. 1988. Phylogenies and quantitative characters. Ann. Rev. Ecol. Syst., 19:445-471.

Ferson, S., F. J. Rohlf, R. K. Koehn. 1985. Measuring shape variation of two-dimensional outlines. Syst. Zool., 34:59-68.

Fink, W. L. 1982. The conceptual relationship between ontogeny and phylogeny. Paleobiology, 8(3):254-264.

Fitch, W. M. 1966. An improved method for testing for evolutionary homology. J. Mol. Biol., 16:9-16.

Fitch, W. M. and W. R. Atchley. 1987. Divergence in inbred strains of mice: a comparison of three different types of data. Pp 203-216 in Molecules and morphology in evolution (Patterson, C. ed.). Cambridge Univ. Press, Cambridge, 229 pp.

Gardiner, B. 1982. Tetrapod classification. Zool. J. Linnean Soc., 74:207-232.

Gauthier, J., A. G. Kluge, and T. Rowe. 1988. Amniote phylogeny and the importance of fossils. Cladistics, 4:105-209.

Ghiselin, M. T. 1969. The distinction between similarity and homology. Syst. Zool. 18(1):148-149.

Ghiselin, M. T. 1976. The nomenclature of correspondence: a new look at "homology" and "analogy". Pp. 129-142 in Evolution, brain and behavior (Masterton, R. B., W. Hodos, and H. Jerison, eds.). Lawrence Erlbaum, Hillsdale, New Jersey, 276 pp.

Goldman, N. 1988. Methods for discrete coding of morphological characters for numerical analysis. Cladistics, 4:59-71.

Gould, S. J. 1977. Ontogeny and phylogeny. Harvard Univ. Press, Cambridge Massachusetts, 501 pp.

Haas, O., and G. G. Simpson. 1946. Analysis of some phylogenetic terms with attempts at redefinition. Proc. American Phil. Soc., 90:319-349.

Henning, W. 1966. Phylogenetic systematics. Univ. Illinois Press, 263 pp.

Hull, D. L. 1968. The operational imperative: Sense and nonsense in operationism. Syst. Zool., 17:438-457.

Inglis, W. G. 1966. The observational basis of homology. Syst. Zool., 15:219-228.

Jardine, N. 1967. The concept of homology in biology. Brit. J. Phil. Sci., 18:125-139.

Jardine, N. 1969. The observational and theoretical components of homology: a study based on the morphology of the dermal skull-roofs of rhipidistian fishes. Biol. J. Linn. Soc., 1:327-361.

Jardine, N., and C. J. Jardine. 1967. Numerical homology. Nature, 216:301-302.

Jardine, N., and C. J. Jardine. 1969. Is there a concept of homology common to several sciences? Class. Soc. Bull., 2(1):12-18.

Key, K. H. L. 1967. Operational homology. Syst. Zool., 16:275-276.

van der Klaauw, C. J. 1966. Introduction to the philosophic backgrounds and prospects of the supraspecific comparative anatomy of conservative characters in the adult stages of conservative elements of Vertebrata with an enumeration of many examples. Verh. Kon. Ned. Akad. Wetensch., Afd. Natuurk., Tweede Sect., 57:1-196.

Kluge, A. G. 1985. Ontogeny and phylogenetic systematics. Cladistics, 1:13-27.

Kluge, A. G. 1988. The characterization of ontogeny. Pp. 57-81 in Ontogeny and systematics (C. J. Humphries, ed.). Columbia Univ. Press, New York, 236 pp.

Kluge, A. G., and J. S. Farris. 1969. Quantitative phyletics and evolution of anurans. Syst. Zool., 18(1):1-32.

Kluge, A. G., and R. E. Strauss. 1985. Ontogeny and systematics. Ann. Rev. Ecol. Syst., 16:247-268.

Laudan, L. 1977. Progress and its problems: Towards a theory of scientific growth. University of California Press, Berkeley, 257 pp.

Mickevich, M. F. 1982. Transformation series analysis. Syst. Zool., 31(4):461-478.

Nelson, G. 1978. Ontogeny, phylogeny, paleontology, and the biogenetic law. Syst. Zool., 27:324-345.

Nieuwenhuys, R., and T. S. Bodenheimer. 1966. The diencephalon of the primitive bony fish Polypterus, in the light of the problem of homology. J. Morphol., 118:415-450.

O'Hara, R.J. 1988. Homage to Clio, or, toward an historical philosophy for evolutionary biology. Syst. Zool., 37(2):142-155.

Owen, R. 1848. Report on the archetype and homologies of the vertebrate skeleton. Rep. 16th Mtg. British Assoc. Adv. Sci., 169-340.

Patterson, C. 1982. Morphological characters and homology. Pp. 21-74 in Problems of phylogenetic reconstruction (Joysey, K. A. and A. E. Friday, eds.). The Systematics Association Special vol. 21, Academic Press, London, New York, 442 pp.

Patterson, C. 1987. Introduction. Pp. 1-22 in Molecules and morphology in evolution (C. Patterson, ed.). Cambridge University Press, Cambridge and New York, 229 pp.

de Queiroz. K. 1985. The ontogenetic method for determining character polarity and its relevance to phylogenetic systematics. Syst. Zool., 34:280-299.

Remane, A. 1956. Die Grundlagen des naturlichen Systems der vergleichenden Anatomie und der Phylogenetic. Theoretische Morphologie und systematik. (Second Ed.) Geest und Portig, Leipzig. Also 1971. (Reprint of 1st ed.) Koeltz, Konigstein-Tanaus, 364 pp.

Riedl, R. 1978. Order in living organisms. Wiley, New York, 313 pp.

Rieppel, O. 1980. Homology, a deductive concept? Z. Zool. Syst. Evolut.-forsch., 18:315-319.

Rosen, D. E. 1973. Interrelationships of higher euteleostean fishes. Pp. 397-513, in Interrelationships of fishes (Greenwood, P. H., R. S. Miles, and C. Patterson, eds.). Suppl. 1, Zool. J. Linnean Soc. Vol. 53.

Roth, V. L. 1984. On homology. Biol. J. Linnean Soc., 22:13-29.

Roth, V. L. 1988. The biological basis of homology. Pp. 1-26 in Ontogeny and systematics (C. J. Humphries, ed.). Columbia Univ. Press, New York, 236 pp.

Sattler, R. 1984. Homology–A continuing challenge. Syst. Bot., 9(4):382-394.

Sibley, C. G., and J. E. Ahlquist. 1987. Avian phylogeny reconstructed from comparisons of the genetic material, DNA. Pp. 95-121 in Molecules and morphology in evolution: conflict or compromises (Patterson, C., ed.). Cambridge Univ. Press, Cambridge, 229 pp..

Simpson, G. G. 1961. Principles of animal taxonomy. Columbia Univ. Press, New York, 247 pp.

Skalak, R., D. Dasgupta, M. Moss, E. Otten, P. Dullemeiher, H. Vilmann. 1982. Analytical description of growth. J. Theoretical Biol., 94:555-577.

Sokal, R. R., and P. H. A. Sneath. 1963. Principles of numerical taxonomy. W.H. Freeman and Co., San Francisco, 359 pp.

Sneath, P. H. A., and R. R. Sokal. 1973. Numerical taxonomy. W.H. Freeman and Co., San Francisco, 573 pp.

Strauss, R. E., and L. A. Fuiman. 1985. Quantitative comparisons of body form and allometry in larval and adult Pacific sculpins (Teleostei: Cottidae). Can. J. Zool., 63:1582-89.

Thompson, D. W. 1942. On growth and form. Cambridge Univ. Press, Cambridge, 793 pp.

Van Valen, L. 1982. Homology and causes. J. Morphology, 173:305-312.

Wagner, G. P. 1989. The origin of morphological characters and the biological basis of homology. Evolution, 43:1157-1171.

Wagner, W. H. Jr. 1980. Origin and philosophy of the groundplan divergence method of cladistics. Syst. Bot., 5: 173-193.

Wiley, E. O. 1981. Phylogenetics: The theory and practice of phylogenetic systematics. Wiley-Interscience, New York.

Withers, R. F. J. 1964. Morphological correspondence and the concept of homology. Pp. 378-394, *in* Form and strategy in science (Gregg, J. R. and F. T. C. Harris, eds.). Studies dedicated to Joseph Henry Woodger on the occasion of his seventieth birthday. Reidel Publ. Co., Ordrecht, The Netherlands, 476 pp.

Woodger, J. H. 1945. On biological transformations. Pp. 94-120 *in* Essays on growth and form presented to D'Arcy Wentworth Thompson (Clark, W. E. LeGros and P. B. Medawar). Clarendon Press, Oxford, 408 pp.

Chapter 18

Using Growth Functions to Identify Homologous Landmarks on Mollusc Shells

Spafford C. Ackerly

Department of Geological Sciences,
Cornell University, Ithaca, NY 14853

Introduction

Many features of organisms are potentially useful in phylogenetic analyses, including structural arrangements of the skeleton, behavioral repertories, and developmental and transformational features (Roth, 1984). If based on a continuity of genetic information from one generation to the next, any feature may become homologous (Roth, 1988). In molluscs one of the most important features of the skeleton is the ontogenetic record of growth preserved in the shell. The shell records information about rates and directions of shell growth through time. This paper examines how information about growth rates and growth directions might be useful in the identification of homologous landmarks on mollusc shells. I will refer to rates and directions of shell growth as programmatic features of molluscan morphology.

Growth Functions

The gastropod shell provides a model illustration of molluscan shell growth patterns. The shell aperture revolves around the coiling axis expanding at rate W, and translates along the coiling axis at a rate T (W and T are defined by Raup, 1966). The aperture is a snapshot in time of the shape of the marginal mantle tissues and the coiling parameters W and T define the aperture's expansion rate and its path.

While mathematically compact, the spiral model can be criticized on the grounds that the coiling parameters have no obvious biological significance. This analysis attempts to convert the spiral model into a more biologically meaningful growth function. The growth function specifies the magnitude of the growth vector at any given point on the margin and the rate of divergence of adjacent growth vectors.

Growth functions will be illustrated by a computer simulation. The shell forms in Figure 1 were constructed with identical coiling parameters: $W = 3.5$, $T = 2.0$. The aperture shapes are also identical (elliptical, with ellipticity = 0.8) but the orientations of the apertures' long axes vary in 45° increments. In each shell form the aperture touches the coiling axis. Mathematically, the shell forms in Figure 1 are closely related; however, biologically, the forms exhibit significant differences in both the rates and directions of growth, expressed in terms of position on the aperture margin.

The heart-shaped graphs in Figure 1 represent the magnitude of the growth vector at the aperture margin. For example, in aperture A the maximum growth rate occurs at (or near) the point on the aperture most distal from the coiling axis; the location is marked with a circle. The minimum growth rate occurs at a point on the "upper left"

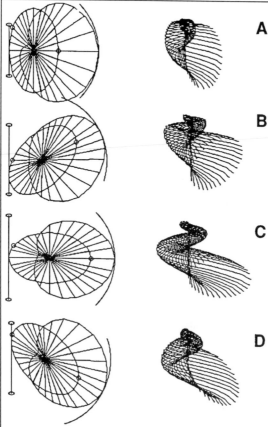

Figure 1. Simulated shell forms (right) generated by the coiling parameters (after Raup, 1966): W = 3.5, T = 2, and D = 0, and with elliptical apertures (e = 0.8) at different orientations to the coiling axis (45° increments). The left column shows the position of the coiling axis, the aperture, and a polar plot of the magnitudes of the growth vector at different positions around the aperture.

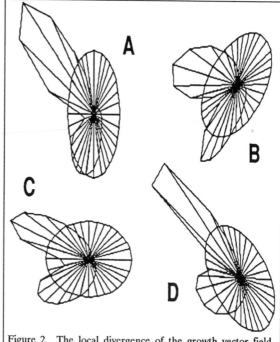

Figure 2. The local divergence of the growth vector field around the edge of the aperture. The divergence field is bimodal with peaks occurring in the vicinity of the coiling axis. A, B, C, and D refer to the shell forms in Figure 1.

sector of the aperture, also marked with a small circle. The asymmetrical heart-shaped graph represents the distribution of growth rates around the shell margin.

The polar graphs in Figure 2 represent the *divergence rate* of adjacent growth vectors along the aperture margin. The divergence rate is expressed in radians per unit distance along the aperture margin, $\Delta\theta / \Delta a$, where $\Delta\theta$ is the angle between two adjacent growth vectors and Δa is the arc distance between them, measured along the aperture edge. The divergence field shows a bimodal distribution on the aperture, with high rates of divergence generally occurring just "above" and "below" the point of adherence of the aperture to the coiling axis.

To a first approximation, the surface area of shell secretion at any point along the aperture margin is proportional to the product of the divergence function and the growth magnitude function squared. The surface area function also shows a bimodal distribution along the aperture (not shown as a figure).

A third component of the growth field is closely related to the aperture's shape (which is held fixed in this example) and is represented by the curvature of the growth vector field. Thus, the growth vectors sweep around the shell, following the aperture's edge. The form of the curvature function can be traced directly to the shape of the embryonic shell aperture, and hence to embryogenic processes of shell formation (Raven, 1966).

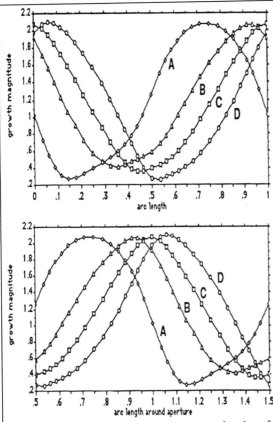

Figure 3. Magnitude of the growth vector as a function of arc position around the shell margin for the four shell forms in Figure 1. Arc lengths of 0 and 1 correspond to points on the long axis of the ellipse most distal from the coiling axis. The two graphs show the same data but one is shifted by one half the arc length (90°) to show the peaks of the curve while the other shows the troughs. A, B, C, and D refer to the shell forms in Figure 1.

A Programmatic Criterion for Identifying Landmarks

In the preceding analysis, the growth program was defined by the magnitude and divergence of growth vectors along the aperture margin. Using this approach, it is possible to locate growth maxima and minima on the margin and to use these positions as "programmatic" landmarks. The points of growth maxima and minima are biologically relevant because they reflect processes of mantle secretion.

In Figure 1, the growth magnitude was illustrated with a polar graph, and points of maxima and minima with circles on the aperture margin. Figure 3 shows a different representation of these data, where the aperture margin is "unrolled" onto a straight line, the x-axis, and the growth magnitude represented on the y-axis. The zero point on the aperture arbitrarily coincides with the position of the aperture's long axis. The bottom graph shows the same data but shifted by half the aperture's length. Figure 3 permits a comparison of the growth functions in the different shell forms, by inspection of the points of growth maxima and minima, and of inflection points showing the maximum rate of change of the growth function.

Figure 4 also shows the four apertures, redrawn with large circles indicating maxima and minima, and small circles indicating inflection points. These "programmatic" landmarks are the basis for four simple trusses. Similar trusses could be constructed for the divergence field, showing points of maximum and minimum divergence of growth vectors along the aperture margin. The four apertures are unrolled as straight lines at the bottom of Figure 4, and registered to one of the extrema.

Transformation Analysis

In the preceding analysis, components of the growth vector field (i.e., the magnitude and divergence) were mapped onto the shell aperture using polar graphs (Figures 1 and 2). The aperture shape in each shell is identical, and the distribution of growth vectors around the aperture determines the final shell form. Components of the growth field are biologically meaningful because they reflect net rates of cell growth and division along the aperture margin.

If phylogenetically related, then the shell forms in Figure 1 represent modifications of a common growth program. In other words, the vector field along the aperture margin was deformed (i. e., modified) in order to produce the observed diversity of shell forms. Representing the deformation of a vector field is a complex problem. Consider, for example, the magnitude of the growth vector and its distribution around the shell margin, as represented by the heart-shaped graphs in Figure 1. If genetically related, then each heart-shaped distribution is a modification of a common growth program. The modifications are apparently complex because the shape of each growth function is slightly different (Figure 3). On the other hand, the transformation series is hypothetical. Are biological transformations constrained to simpler pathways, for example, by correlated changes between the growth program and aperture shape?

Because the aperture shape in the four shells is identical, the variation in shell form can be related to a deformation of the function that determines the magnitude (and also the direction) of shell growth at any given position on the aperture. Thus, we see that the maximum and minimum growth points shift both relative to one another, and relative to the long axis of the aperture. These shifts, or deformations, are represented in Figure 4 by a series of trusses, where the vertices represent the extrema and the inflection points on the growth function. The differences in shell form are readily visible as deformations of the trusses representing the growth data.

A second method of representing these data is by breaking the aperture edge and rolling it out onto a straight line (bottom, Figure 4). Homologous landmarks on the growth function are registered to one another, and differences between the growth functions can be explained in terms of (1) a uniform deformation of the function, accounting for the relative shifts in the position of growth maxima and minima, and (2) a non-uniform deformation of the function, accounting for shifts in the positions of the growth inflection points.

The preceding discussion was based on an arbitrary choice of reference points on the growth function curve (i.e., extrema and inflection points). In principle, we would seek to describe the deformation between one growth curve and the next in terms of a continuous deformation function, but the

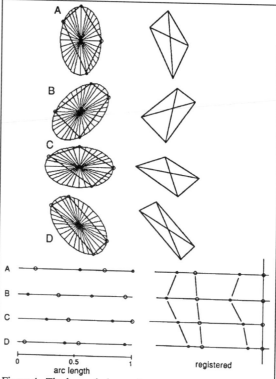

Figure 4. The large circles on the apertures represent points of maximum and minimum growth, and the small circles represent the position of inflection points on the growth function curve (see Figure 2). Lines connecting the extrema and inflection points form a truss for each shell form. Differences in geometry between the shell forms can be expressed, to a first order, as deformations in the trusses. In the bottom figure, the apertures have been "broken" and rolled out onto a straight line, thus showing the relative positions of the points on the aperture as linear functions. The relative shifts of extrema and inflection points reflect the transformations of the growth magnitude field between shell forms.

techniques for such an analysis are not well developed. In addition we must describe the deformation of the entire vector field, not just its scalar component as illustrated above.

Discussion

The mollusc shell is an ontogenetic record of growth by skeletal accretion. The shell is a developmental pathway, recording the trajectories of cell lineages. The developmental path is preserved more or less in plants, turtle shells, or fish scales. In contrast, the path is difficult to establish in structures such as skulls because the physical connections from one growth stage to another are frequently resorbed. In skulls, one can specify a sequence of structural arrangements, but not the precise path between them.

The fundamental differences in the nature of form in different phyla require different styles of morphological analysis. In vertebrates, for example, considerable attention is given to the configuration of structural elements of the skeleton, such as the bones of the hand. Morphometric analyses focus on spatial deformations of coordinate data (e.g., Bookstein et al., 1985). In molluscs, on the other hand, studies of shell form diversity tend to emphasize the ontogenetic component of the organism. The shell is a collection of growth vectors, and morphometric analyses focus on the geometry of the vector field. The vector field represents dynamic growth processes and hence reflects transformational, as opposed to structural, features of the organism (Sattler, 1984).

Traditionally, the growth field in molluscs has been defined in terms of the logarithmic spiral model. The coiling parameters define a growth vector field, and the shape of the aperture, and its position in the field, define the resulting shell form (see Bayer, 1978). Mathematically, the aperture's shape is decoupled from its path, since differently shaped apertures may follow the same growth trajectory field (e.g., Raup, 1966). However, biologically, the growth "field" is confined to the roughly one-dimensional edge of the aperture, and the coding of the growth program that produces a given shell form must be specified in relation to a particular configuration of cells at the shell margin. The growth functions derived in this paper are biologically relevant insofar as they represent patterns of cell growth, division, and divergence. The growth functions provide a biological basis for defining homologous landmarks that reflect dynamic growth processes.

Acknowledgments

I would like to thank the following individuals for critical discussions and for their willingness to share thoughts and ideas: David Ackerly, Neil Blackstone, Fred Bookstein, John Carr, Dan Fisher, Julian Humphries, David Lindberg, Amy McCune, Jim Rohlf, David Schindel, Rachel Tabachnick, and Peter Wimberger. This research was supported in part by an NSF grant to participate in the University of Michigan's Morphometrics Workshop, and by an NSF grant to Cornell University (EAR86-07469).

References

Bayer, U. 1978. Morphogenetic programs, instabilities, and evolution - a theoretical study. Neues Jahrbuch für Geologie und Paläontologie, Abhandlungen 156:226-261.

Bookstein, F. L., B. Chernoff, R. L. Elder, J. M. Humphries Jr., G. R. Smith, and R.E. Strauss. 1985. Morphometrics in evolutionary biology: the geometry of size and shape change with examples from fishes. The Academy of Natural Sciences of Philadelphia, Special Publication No. 15, 277 pp.

Raup, D. M. 1966. Geometric analysis of shell coiling: general problems. J. Paleontol., 40:1178-1190.

Raven, C. P. 1966. Morphogenesis: the analysis of molluscan development. Pergamon, Oxford. 2nd Edition, 365 pp.

Roth, V. L. 1984. On homology. Biol. J. Linnean Soc., 22:13-29.

Roth, V. L. 1988. The biological basis of homology. in Ontogeny and systematics (C. J. Humphries, ed.). Columbia Univ. Press., New York, pp 1-26.

Sattler, R. 1984. Homology - a continuing challenge. Syst. Bot., 9:382-394.

Appendix: Growth Functions

Following Raup (1966) we define the position of a point on the shell by its radial distance from the coiling axis r, its angular distance around the coiling axis θ, and its translational distance along the coiling axis z, by

$$r = r_{0i} \, e^{\cot \alpha \, \theta}$$

and

$$z = z_{0i} \, e^{\cot \alpha \, \theta} + r_c \, T(e^{\cot \alpha \, \theta} - 1)$$

where α and T are spiral constants, r_{0i} and z_{0i} are the radial and translational positions of the ith point along the shell margin at the position $\theta = 0$, and r_c is the radial position of the aperture's centroid at $\theta = 0$.

Differentiating the preceding functions gives the growth vector

$$\mathbf{G} = \frac{dr}{d\theta} r + \frac{dc}{d\theta} c + \frac{dz}{d\theta} z,$$

where the terms represent the angular rates of growth in the radial, circumferential, and translational directions respectively, and where

$$\frac{dr}{d\theta} = r_{0i} \cot \alpha \, e^{\cot \alpha \theta},$$

$$\frac{dc}{d\theta} = r_{0i} \, e^{\cot \alpha \theta},$$

and

$$\frac{dz}{dh} = \cot \alpha \, e^{\cot \alpha \theta} \, (z_{0i} + r_c T).$$

The magnitude of the growth vector is

$$\mathbf{G} = \sqrt{ \left[\frac{dr}{d\theta} \right]^2 + \left[\frac{dc}{d\theta} \right]^2 + \left[\frac{dz}{d\theta} \right]^2 },$$

and the local divergence of the vector field is $Q\nabla G$ The local divergence is estimated by a finite equation:

$$Q\nabla G = \frac{\Delta\delta}{\Delta a},$$

where $\Delta\delta$ is the angle between two adjacent growth vectors and Δa is the arc distance between them.

Chapter 19

Morphometric Analysis of Evolving Eocene Podocyrtis (Radiolaria) Morphotypes Using Shape Coordinates

Annika Sanfilippo and William R. Riedel

Scripps Institution of Oceanography
University of California at San Diego
La Jolla, California 92093

Abstract

Species of the Eocene radiolarian genus *Podocyrtis* from the Caribbean (Deep Sea Drilling Project Site 29) have been described using five pairs of shape coordinates in lateral view. The assemblages are from five stratigraphic levels. Principal components and resistant-fit theta-rho analyses (RFTRA) were effective in distinguishing morphotypes based on landmark data. Principal components analysis indicated that the thoracic and lumbar coordinates contribute most to the variation. Cluster analysis of the principal component scores, RFTRA and principal components analysis produced groupings of specimens similar to those used in conventional taxonomy, but with the advantage of quantifying similarities and differences.

The stratigraphic distribution of the sampled morphotypes shows the development, through time, of two or three distinct populations with continuous variation within each morphocline. New morphologies can be explained by cladogenesis and anagenesis. However, the traditional species used for stratigraphic interpretations cannot be recognized by using seven landmarks alone. Additional information from pore patterns, outline shape, and appendages are needed to differentiate adequately the various morphotypes.

Introduction

Geologic history reveals few fossil groups with as complete a record as polycystine radiolarians. Ranging throughout Phanerozoic time, these protozoans were apparently as diverse in the Paleozoic as they are now. They have left behind a detailed evolutionary record that makes them potentially one of the most important marine microfossil groups.

Radiolarians are exclusively marine and are found in all oceans. They are unicellular, with an internal skeleton made of amorphous silica. Polycystine radiolarians are classified principally on the basis of their skeletal shape and symmetry into two major groups: Spumellaria and Nassellaria. Spumellarians are commonly spherical, discoidal or ellipsoidal, with radial spines extending from the surface of the shell and often connecting internal shells. Nassellarians are characterized by axial, often conical symmetry, although various modifications can be seen in the multitude of individual forms.

In the nassellarians, an apical spine is termed a "horn," while prominent basal spines are termed "feet." At level G in Figure 1, where a collar stricture separates the first and second segments, there is an internal spicular structure clearly homologous in all the diverse nassellarian forms. The cephalic shell wall associated with this internal spicule is the basis for distinguishing families. Further subdivision depends generally on structures peculiar to individual family-level taxa. For example, pore size, shape and arrangement are commonly significant at the species and genus levels, but may also characterize family-level groups.

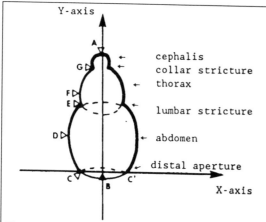

Figure 1. Generalized nassellarian sketch indicating the landmarks used in this study.

It is usually considered that the stratigraphical succession in radiolarians provides evidence for gradual change of one species into another. This evolutionary process is referred to as phyletic speciation or anagenesis. This continuum is arbitrarily divided into a number of chronospecies (Mayr, 1942). Since chronospecies are arbitrarily defined, some paleontologists (e.g., Eldredge and Gould, 1972) would argue that anagenesis does not produce new species and therefore call it phyletic evolution or micro-evolution (Eldredge and Cracraft, 1980).

The term "lineage" has been used to describe an evolving succession of two or more species. Divergences are called speciations.

Reconstruction of a phylogeny is based on the search for characters shared by two or more organisms. The species is usually considered the only non-arbitrary unit in spite of many different definitions and limits, being limited in nature only to interbreeding members. Supraspecific categories such as genera, families, orders etc., are groupings chosen by the systematist to organize the diversity of organisms on the basis of relationships.

To document the evolution of one species into another it is necessary to have a good fossil record. In recent years oceanographic expeditions and the Deep Sea Drilling Project have collected long, more or less continuous sequences of highly radiolarian sediments representing most of Cenozoic time from several biogeographic regions suitable for biostratigraphic and paleontological studies.

Radiolarian evolution during the Cenozoic is illustrated by many lineages among the nassellarians (Sanfilippo et al., 1985). One such lineage—the Eocene *Podocyrtis* lineage, in the family Pterocorythidae — is used in this morphometric study. These phyletic lineages, each regarded as comprising a genus or subgenus, are based on one or two characters e.g., the long-term gradual morphologic change observed in the subgenus *Lampterium* (Figure 2) as a decrease in the thoracic size and increase in the size (decrease in the number) of the abdominal pores. Speciations are observed in a few genus-level taxa and we treat the two resulting branches as different subgenera.

Traditional Taxonomy

The Eocene radiolarian genus *Podocyrtis* offers an interesting group for the study of morphological diversity over approximately 13 million years. The morphotypes display complicated patterns of variation.

This lineage exhibits continuous morphological change through time (Figure 2). Any sample contains a range of variability, with some specimens resembling the commonest form in the older samples, and other specimens resembling the forms in younger samples. It is the "center of gravity" of the morphology that changes unidirectionally. The *Podocyrtis* lineage has previously been described (Riedel and Sanfilippo, 1970; Sanfilippo and Riedel, 1973; Nigrini, 1974; Sanfilippo et al., 1985) from a large number of tropical Deep Sea Drilling and Ocean Drilling Project sites. The early *Podocyrtis papalis* gives rise to three lineages in the Eocene. Within each lineage we recognize several stratigraphically useful anagenetic species. This genus has been chosen for the analysis of changes in shape using shape coordinates (Bookstein, 1991) to explore the changes in variation of five populations as morphospecies appear and disappear over the time interval studied.

Figure 2 illustrates the lineage under investigation. Levels sampled from DSDP Site 29 are indicated in the left column. Species marked with an open arrow are included in this study. The range chart is a composite, based on information from a number of widely scattered localities, and therefore some of the species represented in it are absent at Site 29. To the nominate subgenus *Podocyrtis* are assigned members of the anagenetic lineage leading from *P. papalis* via *P. diamesa* and *P. phyxis* to *P. ampla*. Included in this study are *P. diamesa* morphotypes which are clearly transitional between *P. papalis* and *P. ampla*. They differ from *P. papalis* in their larger size and the presence of a lumbar stricture, and from *P. ampla* in their general form, being spindle-shaped rather than conical.

The subgenus *Lampterium* comprises those forms that developed from *P. papalis*, beginning with *P. aphorma* and evolving via *P. sinuosa*, *P. mitra*, and *P. chalara* into *P. goetheana*.

The forms *P.* (*L.*) *fasciolata* and *P.* (*L.*) *trachodes* are closely related to members of the subgenus *Lampterium* but their placement in this anagenetic lineage has not been fully ascertained.

Nigrini (1974, p. 1069) considered *P. fasciolata* a geographic variant (Indian Ocean) of *P. ampla*. However, *P. fasciolata* was later found (Sanfilippo and Riedel, 1974) to co-occur with *P. ampla* in the Indian Ocean as well as in many other localities. Because its characteristics correspond better with those of *P. sinuosa* it was transferred to the subgenus *Lampterium* to indicate this relationship. At the time of its acme, *P. fasciolata* far outnumbered the co-occurring *P. sinuosa*.

P. (*L.*) *trachodes*, which is similar to *P.* (*L.*) *mitra* in its general shape, is distinguished from the latter by its rough abdomen. It is considered a branch from the *Lampterium* lineage.

The anagenetic *Lampterium* lineage shows a decrease in thoracic volume (second segment) and an increase in the abdominal volume (third segment) over time.

Conventional descriptions and diagrams (Figure 2) are unable to accommodate all the variation and all the intermediate forms included in each species at each level in the continuum. They illustrate only the general form at a single level during the range of that particular species. The data necessary to describe the variation cannot be illustrated this way. Techniques developed for morphometric studies provide a uniform way of describing the variation and transformations involved in the evolution of these forms.

Morphometrics

Materials

This study utilized five Middle Eocene sediment samples (each approx. 50 cc) from the Caribbean (DSDP Site 29, at Lat. 14° 47.15' N, Long. 69° 19.38' W: samples 29B-8-5, 91-93 cm; 29-9-6, 72-74 cm; 29-12-2, 87-89 cm; 29-16-5, 14-16 cm; 29B-10-2, 29-31 cm). For each sample, thirty to fifty specimens comprised of *Podocyrtis* morphotypes (Figure 2) were randomly chosen from prepared strewn slides (sieved at 63 microns) mounted in Canada balsam. Each slide contains approximately 30,000 radiolarians, of which 1-5% belong to the genus *Podocyrtis*.

A total of 195 specimens were used in this analysis. To avoid problems associated with tilting, only specimens with their longitudinal axis parallel or nearly parallel to the glass slide were chosen. Each stratigraphic level was analyzed separately. To avoid preconceptions about the species content, all the individuals were lumped into one experimental taxon without reference to species level designations. Thus the phenetic patterns of variability and the relationship between and within previously designated species within this genus can be evaluated.

Measurement Techniques

Members of *Podocyrtis* are basically conical, consisting of the horn and three segments (cephalis, thorax and abdomen). Each specimen was digitized using a V150 Videometric System developed by American Innovision of San Diego, California. A program was written to rotate the coordinates to a specific orientation so that the longitudinal axis, chosen to represent the baseline for the homologous landmarks, is placed along the Y-axis of the coordinate system, with landmark B at the origin (Figure 1). The final output is written to a file as seven pairs of (*x,y*) coordinates. Only one side of each specimen is digitized. "B" is a derived landmark — the point where the longitudinal axis intersects the line connecting the opposite sides of the distal aperture. Morphological variation is traditionally expressed in width and length measurements of the segments in addition to a description of outline shape, presence and nature of horn and feet (Figure 2). A photographic record was made of each individual.

Landmarks

The shape of a group of three or more landmarks can be described and analyzed by selecting two landmarks as a "baseline" fixed in position and scale. The effect of changing the choice of baseline does not alter the original form (Bookstein, 1984a; 1984b). After standardization of the distance between one particular pair of landmarks, these two landmarks are at a constant separation (Bookstein,

1986; 1991). In our case the baseline is set to 100 by dividing each set of landmark coordinates by the AB distance and multiplying by 100. Movements in relation to the baseline induce correlations among distinct shape coordinates (Bookstein, 1986). Each landmark is designated by an (*x,y*) coordinate pair (shape coordinates) relative to the baseline. Multivariate analysis of shape coordinates results in an extraction of the main factors describing the morphological variations within a sample, which can be translated into variables useful in taxonomic work.

The seven landmarks (Figure 1) in the lateral view of *Podocyrtis* morphotypes describe a morphological space that can be captured by five pairs of shape coordinates. The landmarks can be described as follows: A is the top of the cephalis, G is the change of contour at the collar stricture marked internally by the internal spicular structure, F is at the widest level of the thorax, E is the change of contour at the lumbar stricture marked by an internal shelf in the skeleton, D is at the widest level of the abdomen, C is at a level where the shell wall changes from porous to hyaline. These landmarks are digitized using the Videometric V150 software. The complete data set is given in the Appendix and on a supplied disk. Software rotates the baseline AB to the vertical, and all the other landmarks are digitized in relation to this line. Scaling is performed in a separate operation using LOTUS 123. Point B of the baseline is a derived landmark, chosen so that the line from A to B approximately corresponds to the longitudinal axis around which the form is considered symmetrical. Landmarks F and E describe the increase or decrease in relative thoracic size for each species. Differences in the position of landmark D in relation to E and C indicate differences in the shape of the third segment from inverted conical (*papalis*), via inflated (*sinuosa*), to conical (*ampla*). Points F and D represent the widest points of the thorax and the abdomen respectively. Repeated collection of these coordinates indicated a greater error in locating them than the coordinates at the other points

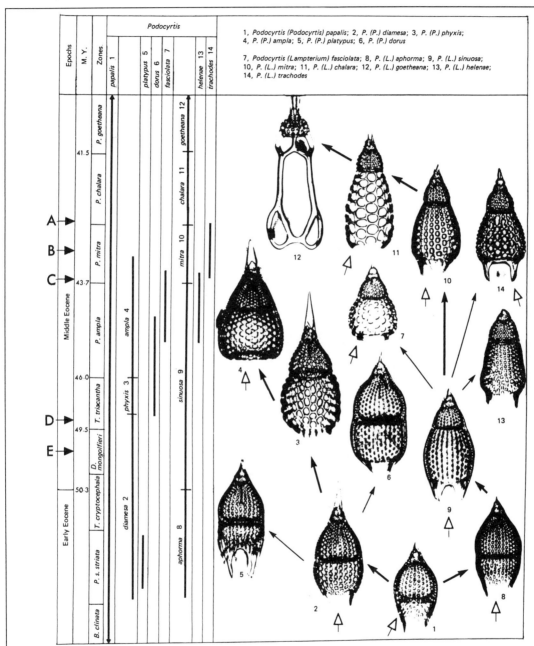

Figure 2. Evolution of the Eocene *Podocyrtis* lineage (modified from Sanfilippo et al., 1985; magnification 180X). Sample levels used in this study are indicated in the far left column: A, DSDP 29B-8-5, 91-93 cm; B, DSDP 29-9-6, 72-74 cm; C, DSDP 29-12-2, 87-89 cm; D, DSDP 29-16-5, 14-16 cm; E, 29B10-2, 29-13 cm. Species used in this study are marked by open arrows in the main body of the figure. Species present at each level can be deduced from their time-ranges indicated by heavy vertical lines.

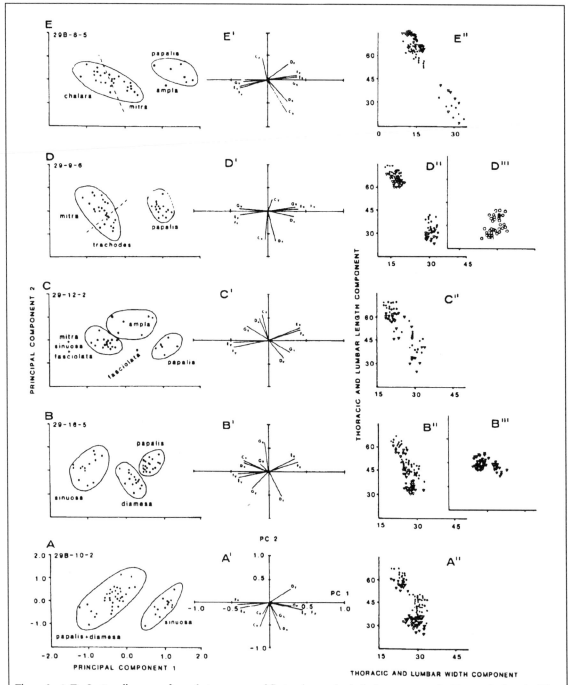

Figure 3. A-E: Scatter diagrams of morphotype scores of first and second principal components for Eocene *Podocyrtis*. The scores show definite clustering into identifiable stratigraphical clusters. For explanation, refer to text "Principal component analysis." For original data see Appendix.

because they occur on the gentle curve of the out-line of the two segments. The C_y and $B_{x,y}$ coordinates for the distal aperture and the derived midpoint, respectively, should equal zero since the line BC is placed along the x-axis with B at the origin. However, rounding errors in the calculations sometimes cause these numbers to have values slightly different from zero.

Results of NTSYS-pc Principal Components and Cluster Analysis

We have used NTSYS-pc version 1.40 (Rohlf, 1988) to do a principal components analysis using the correlations among five pairs of (x,y) shape coordinates. The steps were: 1) standardize each data matrix by columns; 2) compute a matrix of correlations among the variables (columns); 3) extract five eigenvectors from the correlation matrix; 4) project and plot the standardized data onto the eigenvectors; and 5) make scatterplots of the specimens. In addition, scores for five principal components were computed, plotted and clustered to reveal possible groupings.

Principal component analysis

Results of the principal components analysis are reported below in stratigraphical order starting with the lowermost (oldest) level. The three first principal components explain 80-90% of the variance in each of the five samples. The first principal component accounts for up to 66% of the total variance, and corresponds to change in shape of the thoracic segment. The second principal component, which accounts for 13-22% of the total variance, represents mainly variation in the width element of the abdomen and the distal aperture. As can be seen in Figure 3A-E, morphologic as well as stratigraphic differences can be described adequately by the first and second principal components. Table 1 gives the loadings on the first and second principal components for each of the ten variables in each of the five samples, and the eigenvalues in percent (%) explaining the variance expressed by the first two principal components.

The panel of Figure 3 are as follows. A'-E': Plot of eigenvector A"-E": Information relating to the measured variables having the greatest influence on PCA 1 and PCA 2. Thoracic width and length components are plotted as dots, and lumbar width and length components as triangles. B''': Using the same scale as the data plot at this level, information (solid triangles) on the abdominal width component (X-axis) and its distance from the aperture (Y-axis). Insert D''' gives the same information (open circles) at the DSDP 29-9-6 sample level.

Sample DSDP 29B-10-2

The coordinates indicative of the thoracic morphospace ($E_{x,y}$; $F_{x,y}$) load very highly on the first principal component (Table 1). The scatter diagram (Figure 3A) of the two first principal component scores show two distinct clusters (*papalis + diamesa* and *sinuosa*). One contains morphotypes dominated by a large thorax, traditionally identified as *P. papalis* and *P. diamesa*, while the other cluster, traditionally identified as *P. sinuosa*, is made up of members with a small thorax. There is an inverse relationship between the width of the thorax (F_x) and its distance from the distal aperture (F_y). This is illustrated in the accompanying data plot (Figure 3A") which contains additional information relating to the measured variables. The x-axis gives widths and the y-axis lengths of the thoracic (dots) and the lumbar (triangles) coordinates. The length component is the distance from the point at which the width component is digitized to the distal aperture. Since variation in the size of the cephalis is small, changes in position of the lumbar stricture along the longitudinal axis describe changes in the thoracic morphospace. This indicates the importance of the relative size of the thorax to the total size of the morphotype in separating two distinct groups within the genus *Podocyrtis*. Increases in the distance to the widest point on the thorax (F_x), as well as the lumbar (E_x) stricture, from the longitudinal axis correspond to an increase in the volume of this segment. The influence, magnitude, and direction of these components on the principal component scores can readily be seen in the accompanying plot of the eigenvector coefficients (Figure 3A'). Figures 3A and A' together are used as a biplot (Krzanowski, 1988, p. 128).

Sample DSDP 29-16-5

Three more or less distinct groups of specimens are represented in the scatter diagrams (Figure 3B - *sinuosa*, *diamesa*, *papalis*). The appearance of the third cluster (*diamesa*) is indicative of a speciation before or at this level (observed as a gap in the existing variation) at least as far as the data in the seven landmarks chosen for this study are concerned. Again, the shape coordinates influencing the thoracic morphospace load highly on the first PC, as do the abdominal (D_x) and apertural (C_x) widths (Figure 3B'). Along PC 1 in Figure 3B, there is a trend between members of the clusters labeled *diamesa* and *papalis*, indicating a gradation in the degree of inflation of the abdomen. Members of the *sinuosa* cluster are distinct on the basis of this character, as can be seen in the data plots from this sample (Figure 3B'''). In Figure 3B, the *papalis*-group actually includes two *diamesa* specimens because their thoracic and lumbar components are more similar to those of the *papalis*-group. The *diamesa*-group includes one *papalis* member. It plots lower than the other *papalis* members on the second principal component. One reworked *sinuosa* is more similar to the *diamesa* group than to its own group.

The eigenvector plot (Figure 3B') shows the redundancy of the two pairs of thoracic shape coordinates and the influence of the width component at the collar stricture (G_x). The nature of the collar stricture in this genus (as well as the family) makes this an unreliable measurement since the orientation of the specimen influences this shape coordinate.

Sample DSDP 29-12-2

The first PC corresponds to change in shape of the thoracic segment. Scatter diagrams (Figure 3C) of the first two principal component scores indicate three clusters (*sinuosa*+*mitra*+*fasciolata*, *ampla* and *papalis*). One specimen, which plots further to the right of the *sinuosa*+*mitra*+*fasciolata*- cluster than expected, is a representative of a stratigraphically earlier level (not represented here), possibly reworked, at which these morphotypes have a larger thorax. Using characters other than the ones employed here (alignment of pores, absence of costae, presence of a thickened peristome), this specimen would be identified as *P. fasciolata*. The vectors seen in the eigenvector plot (Figure 3C') indicate that the thoracic and lumbar *x* and *y* coordinates have greatest influence, placing a specimen along PC-axis 1, and that the E and F vectors, in pairs, have the same magnitude and direction for *x* and *y*. Abdominal and apertural widths are important for determining positions along the second principal component axis.

Sample DSDP 29-9-6

The first principal component represents variation in the shape of the thorax due to increase or decrease of its morphological space as compared to the total morphological space. Scatter diagrams of

Table 1. Loadings on the first and second principal components for each of the ten variables in each of the five samples. Eigenvalues in percent explain the variance expressed by the first two principal components.

Sample Variable	29B-8-5 PC 1	PC 2	29-9-6 PC 1	PC 2	29-12-2 PC 1	PC 2	29-16-5 PC 1	PC 2	29B-10-2 PC 1	PC 2
1	0.352	-0.000	0.333	0.037	0.285	-0.228	-0.043	0.589	0.124	-0.317
2	-0.337	-0.038	-0.364	0.048	-0.285	0.341	-0.240	-0.327	0.313	-0.055
3	0.379	0.096	0.399	0.035	0.452	0.257	0.394	0.201	-0.432	-0.046
4	-0.373	-0.190	-0.393	-0.064	-0.452	-0.111	-0.420	-0.039	0.434	-0.136
5	0.380	0.071	0.398	0.027	0.423	0.288	0.354	0.272	-0.410	-0.130
6	-0.374	-0.165	-0.395	-0.075	-0.457	-0.051	-0.421	-0.109	0.444	-0.153
7	0.270	-0.461	0.099	-0.671	-0.099	0.490	-0.368	0.217	0.155	-0.571
8	0.257	0.328	0.339	-0.129	0.189	-0.310	0.183	-0.508	0.304	0.293
9	0.234	-0.619	-0.038	-0.677	-0.069	0.576	-0.348	0.265	-0.140	-0.586
10	-0.051	0.466	0.042	0.244	-	-	-0.125	0.207	-0.099	-0.285
Eigenvalue(%)	66.43	12.52	61.45	18.02	47.86	20.67	51.86	19.27	46.96	21.50

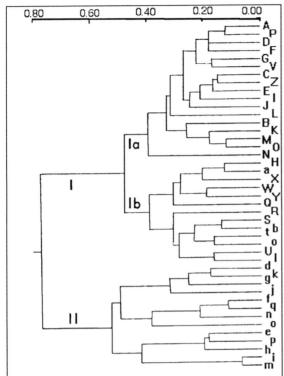

Figure 4. UPGMA clustering of individual scores of PC 1 - PC 5 reveals two distinct clusters of Eocene *Podocyrtis* in DSDP sample 29-16-5. Cluster I forms the largest cluster corresponding to "large thorax" morphotypes, cluster Ia referable to *papalis* and cluster Ib referable to *P. diamesa*. Cluster II includes "small thorax" morphotypes represented by *P. sinuosa*. (A-P *papalis*, Q-c *diamesa*, d-q *sinuosa*). $r_c = 0.862$.

the first two principal component scores again show discontinuous variation represented by two clusters (*mitra+trachodes* and *papalis*, Figure 3D). The third cluster (*ampla*) seen in the level stratigraphically below is no longer distinguishable. Sampling at closer intervals and further examination of this assemblage is necessary in order to verify whether the "cluster" represents the results of a branching event, of extinction, or of absorption into one of the two clusters at this level. The second principal component, which explains 18% of the total variance, is not effective in separating any clusters within the assemblage. As can be seen in the plot (Figure 3D'), the second principal component is

influenced by vectors C_x and D_x, both of which show continuous variation within a small range in the data set and therefore do not separate any groupings. Figure 3D' indicates that the vectors most useful in separating groupings along the first principal component are $E_{x,y}$, $F_{x,y}$, $G_{x,y}$ and D_y. It can also be noted that $E_{x,y}$ and $F_{x,y}$ are of the same magnitude. One of them is sufficient for distinguishing the groupings. However, within the elongated *mitra+trachodes* cluster, although continuous variation exists in the abdominal (D_x) and apertural (C_x) width components, this PC separates the narrow *P. mitra* morphotypes from the wider *P. trachodes* morphotype (Figure 3D).

The data plots (Figure 3D", D'"), showing thoracic and lumbar width components relative to their respective distances from the aperture, clearly indicate the separation of two groups based on the inverse relationship between these components. The data reflecting the abdominal maximum width and distance of this point from the aperture play a smaller role in distinguishing the groupings.

Sample DSDP 29B-8-5

The first PC corresponds to high positive loadings on the distance from the longitudinal axis of the maximum thoracic (F_x) and lumbar (E_x) widths and high negative loadings on the corresponding Y-coordinates. The scatter diagram (Figure 3E) of the two first principal component scores shows two distinct clusters (*chalara+mitra* and *papalis*). The *chalara+mitra* cluster comprises a large number of morphotypes displaying continuous variation in the thoracic components without apparent gaps. In identifying the species within this cluster it becomes apparent that, although there is continuous variation in the thoracic shape, the ancestral forms, (*P. mitra*) have larger scores on PC 1, while the descendants (*P. chalara*) have smaller scores on PC 1. One specimen close to the *papalis* cluster has a somewhat smaller score on PC 2 than the other *P. papalis* members. This specimen is identified as *P. ampla*, reworked from a stratigraphically lower level. From Figure 3E' it can be seen that the apertural $(C_{x,y})$ and abdominal $(D_{x,y})$ vectors have the greatest influence on PC 2. The most interesting vector from the viewpoint of conventional taxonomy is D_y, indicative of the shape of the abdomen. One of the morphological distinctions between *P. ampla* and *P. papalis* is the fact that while the former is conical in general shape the latter is inverted conical. Moreover, the maximum width of *P. ampla* is normally greater than in *P. papalis*, although this is not clearly demonstrated in this sample.

Cluster Analysis

Cluster analysis was carried out using procedures available in NTSYS-pc. A distance matrix was computed and the unweighted pair-group method, arithmetic average (UPGMA) clustering method was used. Results were plotted in the form of phenograms (results from only one of the analyses are shown and discussed here).

Cluster analysis of individual specimen scores (from DSDP sample 29-16-5) using distances and UPGMA clustering of the five principal components reveals two distinct clusters (marked I and II in Figure 4) corresponding to the conventional taxonomic groupings — the "large thorax" group *papalis* (A-P) and *diamesa* (Q-c), and the "small thorax" group *sinuosa* (d-q). Within the *papalis*+*diamesa* group, two clusters, representing the two species, can be seen. Because of differences in the thoracic volume as compared to members of their own groups, two specimens of *diamesa* are displaced within the *papalis* cluster, one *papalis* within the *diamesa* cluster, and one reworked older morphotype of *sinuosa* within the *diamesa* cluster. Although taxonomic differences are adequately described by the first and second principal component, all five principal components were included in this analysis to reveal as many species clusters as possible using all the available data.

Comparison with Results of Resistant-Fit Theta-Rho (RFTRA)-Analysis

The same data set of landmark coordinates used for the principal components analysis was edited into a series of 195 files, one for each specimen, in the format required for RFTRA analysis (Chapman, Chapter 12). For the group of specimens from each of the five assemblages, a matrix of RFTRA distances was determined (using Chapman's R-OUTD program), and each of these matrices was subjected to a UPGMA analysis, using his R-CLUSTR program (Figure 5). For two of the five samples, Least-Squares Theta-Rho-Analysis (LSTRA) distance matrices were also established, for comparison with the clustering results on the corresponding RFTRA matrices. LSTRA distances are of course somewhat less than their RFTRA counterparts, but the clustering results on the two types of matrices were indistinguishable.

Figure 5 shows the dendrograms of RFTRA distance coefficients clustered using UPGMA, for the five investigated assemblages of *Podocyrtis* specimens. In stratigraphic order from oldest to youngest, they are A, DSDP 29B-10-2; B, 29-16-5; C, 29-12-2; D, 29-9-6; E, 29B-8-5. Letters to the left of each dendrogram correspond to the code for each specimen given in the tables constituting the Appendix.

In the oldest sample, DSDP 29B-10-2, there are two groups separated at a distance of 3.5 (Figure 5A), corresponding to the two clusters revealed by cluster analysis of the specimen scores from principal components analysis. In DSDP 29-16-5 there are two distinct groups (marked I and II in Figure 5B) separated at a distance of 3.0, one of them divided into two subgroups (Ia and Ib) at a distance of 1.5. Group II comprises specimens that would be assigned to *P. sinuosa*, and completely matches the similar group resulting from clustering the principal component scores. Subgroup Ia comprises all the specimens of *P. papalis* and one specimen of *P. diamesa*, whereas the corresponding group from clustering the principal component scores excludes one of the specimens of *P. papalis* and includes one additional specimen of *P. diamesa*. Inspection of the data reveals that the two specimens of *P. diamesa* included in this subgroup have wider thoraxes than other specimens of this species in this sample, but this difference is too subtle to be detected by viewing the specimens. Subgroup Ib from both RFTRA and clustering principal component scores is composed of specimens of *P. diamesa* plus one *P. sinuosa*. This specimen represents an older morphotype of *P. sinuosa*, which was discussed with the results of the principal component analysis in the section referring to this sample. It had a larger thorax than later specimens of this species. DSDP 29-12-2 has two groups separated at a distance of 3.9 (Figure 5C). One of these (group I in Figure 5C) is comprised exclusively of *P. papalis* and is identical with a parallel group detected by clustering the principal component scores. Group II is divided into two subgroups at a distance of 2.5. Subgroup IIa contains only *P. ampla*, which corresponds to a cluster from clustering principal component scores as well. Subgroup IIb, again identical in RFTRA and PC clustering, comprises specimens of *P. sinuosa*, *P. fasciolata* and *P. mitra*. There is some indication of these forming subgroups, but the numbers of specimens in each are too small to form the basis for any reliable conclusions.

In DSDP 29-9-6 there are two distinct groups (detected identically by clustering the principal component scores) separated at a distance of about 5.0 (Figure 5D). Group I comprises exclusively specimens of *P. papalis*. Group II comprises specimens of *P. mitra* and *P. trachodes*. In principal components clustering, there are indications of two subgroups which separate the two species, with the exception that the *P. mitra* subgroup contains one specimen of *P. trachodes*; in RFTRA this separation into two species is less clear.

In the youngest sample, DSDP 29B-8-5, RFTRA shows two groups separated at a distance of 5.4 (Figure 5E). Group I comprises *P. papalis* and one specimen of *P. ampla*. This group is similarly separated by clustering the principal component scores, and the single specimen of *P. ampla* is discussed with the results of principal components analysis, in the section referring to this sample. Group II is neatly divided into two subgroups (one comprising *P. chalara* and the other *P. mitra*) at a RFTRA distance of 2.4. Principal component clustering indicates a subgroup comprising late specimens of *P. chalara*, but early specimens of this species are mixed with their ancestral species in the other subgroup.

Conclusions

This study demonstrates the utility of using shape coordinates in describing individual variation in radiolarian assemblages. Principal components analysis points out which variables are the most important in grouping the specimens based on five pairs of shape coordinates. Cluster analysis of individual specimen scores using distances and UPGMA clustering of the five principal components reveal distinct clusters confirming the groupings suggested by principal components analysis. The eigenvector plot comprehensively illustrates the importance of each shape coordinate on the respective principal component score and thus, for plotting purposes, its position along the principal component axis. The scatter diagrams of specimens from the five sampled levels at DSDP Site 29 on the first principal component show two or three distinct clusters corresponding to changes in the shape of the thorax. The appearance of a third cluster in DSDP samples 29-16-5 and 29-12-2 indicates a gap in the variation of thoracic shape — a branching event. Although continuous variation exists in the thoracic shape component of the *mitra+chalara* cluster, the first principal component separates *P. chalara* (the descendant) from *P. mitra* (the ancestor) in the topmost sample. Traditional taxonomy

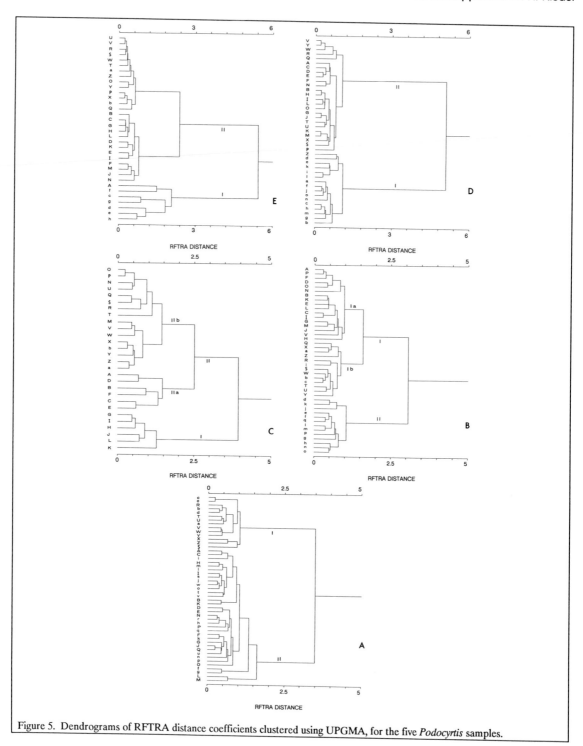

Figure 5. Dendrograms of RFTRA distance coefficients clustered using UPGMA, for the five *Podocyrtis* samples.

separated these species based on characters other than the thoracic ones, namely, the size of the abdominal pores.

The relationship of *P. fasciolata* to the *Lampterium* subgenus (*sinuosa*-cluster in sample 29-12-2) rather than the *Podocyrtis* subgenus (*ampla*-cluster) is borne out by the fact that the first principal component groups morphotypes representing *sinuosa* and *fasciolata* together, and separates *ampla* types.

For DSDP sample 29-9-6, the second principal component represents changes in abdominal (D_x) and distal (C_x) aperture widths. The two species *P. trachodes* and *P. mitra* are within a cluster of continuous variation in this width component. In conventional taxonomy these two species are distinguished by characters other than the ones here included, differences relating to changes of the abdominal outline from smooth to knobby.

Both the RFTRA and clustering of the principal component scores of the shape coordinates scale the data to remove size, but in different ways. The results are very similar, with only minor disagreements.

The eigenvector plots indicate that the two pairs of shape coordinates relating to the thorax ($E_{x,y}$ and $F_{x,y}$) can be reduced to one thoracic shape coordinate, effective in separating two or three clusters. Difficulties in reproducibility of the shape coordinate pair relating to the maximum width of the thorax, discussed in "measurement techniques," speak for discarding these coordinates and for retaining the shape coordinate pair for the lumbar stricture which is less error-prone. Further stratigraphic sampling will be required to resolve whether the placement of the clusters representing morphological space is both morphologic and stratigraphic.

To discriminate the traditional species used for stratigraphic and paleoenvironmental interpretations, one needs more information on each specimen than the seven easily captured landmarks used in this study. The additional characters needed are more difficult to quantify, and a case could be made for continuing their use as qualitative descriptors until the needs of a particular research objective justifies the expenditure of time and effort necessary to express them numerically (e.g., the measurement of rates of morphologic change, quantification of degrees of difference or similarity etc.).

Acknowledgments

We thank Jeannie Westberg-Smith for generous technical advice and programming the V150 system to make the necessary calculations. We express our sincere appreciation to Linda Tway for valuable assistance in all phases of this study and for critical review of the manuscript. Leslie M. Marcus and F. James Rohlf, as editors and colleagues, provided an extensive critique of the entire manuscript, and especially its statistical aspects.

The research was supported in part by NSF Grant OCE87-07417.

References

Bookstein, F. L. 1984a. A statistical method for biological shape change. J. Theor. Biol., 107:475-520.

Bookstein, F. L. 1984b. Tensor biometrics for changes in cranial shape. Ann. Hum. Biol., 11:413-437.

Bookstein, F. L. 1986. Size and shape spaces for landmark data in two dimensions. (With Discussion and Rejoinder). Stat. Sci., 1:181-242.

Bookstein, F. L. 1991. Morphometric tools for landmark data. Cambridge University Press. In press.

Bookstein, F., B. Chernoff, R. Elder, J. Humphries, G. Smith and R. Strauss. 1985. Morphometrics in evolutionary biology. The Academy of Natural Sciences of Philadelphia, Spec. Publ. No. 15, 277 pp.

Elderedge, N. and S. J. Gould. 1972. Punctuated equilibrium: an alternative to phyletic gradualism. Pp. 82-115 *in* Models in paleobiology

(Schopf T. J. M., ed.). Freeman, Cooper and Co., San Fransico, 250 pp.

Eldredge, N. and J. Cracraft. 1980. Phylogenetic patterns and the evolutionary process: method and theory in comparative biology. Columbia Univresity Press, New York, 349 pp.

Krzanowski, W. J. 1988. Principles of multivariate analysis: A user's perspective. Clarendon Press, Oxford, 563 pages.

Mayr, E. 1942. Systematics and the origin of species. Columbia University Press, New York, 334 pp.

Nigrini, C. A. 1974. Cenozoic Radiolaria from the Arabian Sea, DSDP Leg 23. Pp 1051-1121 *in* Initial reports of the deep sea drilling project, Volume 23 (Davies, T. A., B. P. Lyuendyk, et al.). Washington, D. C., U.S. Government Printing Office, 1183 pp.

Riedel, W. R. and A. Sanfilippo. 1970. Radiolaria, Leg 4, Deep Sea Drilling Project. Pp 503-575 *in* Intitial reports of the deep sea drilling project, Volume 4 (Bader, R. G., R. D. Gerard, et al.).

Washington, D. C., U. S. Government Printing Office, 753 pp.

Rohlf, J. F. 1988. NTSYS-pc. Numerical taxonomy and multivariate analysis system (Version 1.40). Exeter Publishing, Ltd. New York.

Sanfilippo, A. and W. R. Riedel. 1973. Cenozoic Radiolaria (exclusive of theoperids, artostrobiids and amphipyndacids) from the Gulf of Mexico, DSDP Leg 10. Pp 475-611 *in* Initial reports of the deep sea drilling project, Volume 10 (Worzel, J. L., W. Bryant, et al.). Washington, D. C., U. S. Government Printing Office, pp. 475-611.

Sanfilippo, A. and W. R. Riedel. 1974. Radiolaria from the west-central Indian Ocean and Gulf of Aden, DSDP Leg 24. Pp 997-1035 *in* Initial reports of the deep sea drilling project (Fisher, R. L., E. T. Bunce, et al.). Washington, D. C., U. S. Government Printing Office, 1183 pp.

Sanfilippo, A., M. J. Westberg-Smith and W. R. Riedel. 1985. Cenozoic Radiolaria. Pp. 631-712 *in* Plankton stratigraphy (Bolli, H. M., K. Perch-Nielsen and J. B. Saunders, eds.). Cambridge University Press, Cambridge, 1032 pp.

Appendix

Data matrices for five sample levels from Deep Sea Drilling Site 29, are listed below in stratigraphic order, youngest to oldest. The rows correspond to species as indicated, and the columns to the X and Y coordinates of the seven landmarks, representing: A cephalis, G collar stricture, F thorax, E lumbar stricture, D abdomen, C distal aperture, and B midpoint of distal diameter (See Figure 1). Code letters preceding species names tie these data to the cluster analysis (Figure 4) and RFTRA diagrams (Figure 5). The data set represents the digitized and rotated landmarks. Scaling has not yet been performed.

Sample DSDP 29B-8-5, 91-93 cm.															
cephalis		collar		thorax		lumbar		abdomen		aperture		midpoint		code	species
A_x	A_y	G_x	G_y	F_x	F_y	E_x	E_y	D_x	D_y	C_x	C_y	B_x	B_y		
0	290	24	246	84	125	85	106	91	37	59	0	0	1	A	*ampla*
0	348	26	306	57	246	52	233	92	52	65	2	0	0	B	*mitra*
0	325	31	280	53	236	55	214	93	43	67	0	0	1	C	*mitra*
0	347	27	318	52	263	51	246	94	41	67	3	1	1	D	*mitra*
0	351	24	322	53	271	51	248	97	55	61	3	1	1	E	*mitra*
0	289	26	247	52	200	52	180	84	22	63	1	0	1	F	*mitra*
0	343	31	301	56	246	54	229	99	44	67	0	0	1	G	*mitra*
0	353	34	305	57	253	54	230	95	40	70	0	0	0	H	*mitra*
0	373	28	335	51	273	53	258	95	52	66	1	0	0	I	*mitra*
0	322	31	283	59	229	58	214	99	29	57	4	0	1	J	*mitra*
0	369	26	333	55	273	56	257	102	31	69	2	1	0	K	*mitra*

A_x	A_y	G_x	G_y	F_x	F_y	E_x	E_y	D_x	D_y	C_x	C_y	B_x	B_y	code	species
0	324	29	287	51	233	54	209	88	45	65	3	0	1	L	*mitra*
0	341	32	301	61	237	61	222	99	29	71	4	0	1	M	*mitra*
0	327	26	292	60	222	59	207	95	40	69	3	0	0	N	*mitra*
0	393	22	361	46	318	46	305	103	65	71	4	0	1	O	*chalara*
0	420	29	381	55	345	50	323	101	35	72	0	1	1	P	*chalara*
0	478	20	444	50	402	51	385	110	40	77	1	1	1	Q	*chalara*
0	392	28	346	54	305	53	286	100	36	73	1	0	1	R	*chalara*
0	381	23	341	51	299	47	278	87	38	66	3	0	1	S	*chalara*
0	375	24	339	52	292	54	279	103	36	80	0	0	0	T	*chalara*
0	383	24	347	52	299	49	286	100	43	75	2	0	1	U	*chalara*
0	412	27	368	57	322	53	305	102	48	70	3	0	1	V	*chalara*
0	410	29	373	54	320	49	308	99	42	70	9	0	1	W	*chalara*
0	389	25	353	47	314	48	301	89	37	69	3	0	1	X	*chalara*
0	395	24	368	54	318	53	300	103	51	73	1	0	0	Y	*chalara*
0	348	27	310	53	276	51	255	100	32	65	6	0	0	Z	*chalara*
0	394	19	350	43	304	41	291	84	38	58	5	0	1	a	*chalara*
0	397	29	355	49	317	42	296	93	33	62	3	0	0	b	*chalara*
0	229	29	194	79	80	73	42	65	27	49	1	0	0	c	*papalis*
0	237	28	199	77	91	77	72	75	59	53	3	1	0	d	*papalis*
0	213	23	179	66	89	65	70	64	59	42	0	0	0	e	*papalis*
0	261	25	228	62	125	64	107	67	63	48	2	0	0	f	*papalis*
0	194	26	156	59	69	55	48	52	32	37	3	0	0	g	*papalis*
0	240	27	205	65	103	66	92	65	71	43	1	0	0	h	*papalis*

Sample DSDP 29-9-6, 72-74 cm.															
cephalis		collar		thorax		lumbar		abdomen		aperture		midpoint		code	species
A_x	A_y	G_x	G_y	F_x	F_y	E_x	E_y	D_x	D_y	C_x	C_y	B_x	B_y		
0	348	30	300	58	244	57	230	106	35	80	1	1	2	A	*mitra*
0	352	29	310	59	240	59	226	100	30	80	1	0	1	B	*mitra*
0	338	33	300	56	238	56	226	91	24	79	1	1	0	C	*mitra*
0	384	25	346	51	278	52	261	93	27	80	1	0	0	D	*mitra*
0	380	28	337	57	277	58	258	102	31	83	2	0	1	E	*mitra*
0	358	31	324	58	263	56	245	91	48	70	2	0	0	F	*mitra*
0	334	22	296	51	232	52	216	91	40	77	1	0	0	G	*mitra*
0	366	37	323	67	252	64	234	103	28	78	2	1	2	H	*mitra*
0	326	29	287	61	219	57	201	90	31	74	0	1	0	I	*mitra*
0	329	23	295	52	225	53	212	91	41	78	3	1	0	J	*mitra*
0	293	26	263	57	201	58	183	90	39	68	3	0	0	K	*mitra*
0	325	25	282	54	219	53	203	84	36	71	6	0	1	L	*mitra*
0	314	23	280	57	211	57	194	92	42	69	2	1	1	M	*mitra*
0	346	24	306	50	251	53	230	90	36	64	2	0	1	N	*mitra*
0	322	28	279	56	212	58	197	91	28	67	3	1	1	O	*mitra*
0	322	26	282	51	208	50	194	81	41	61	0	0	1	P	*mitra*
0	298	25	261	54	206	53	188	99	72	70	0	0	0	Q	*trachodes*
0	296	25	254	51	192	49	176	91	63	72	1	0	1	R	*trachodes*
0	314	29	273	61	210	60	193	107	37	87	3	0	0	S	*trachodes*
0	313	22	279	51	213	51	203	102	43	80	0	1	1	T	*trachodes*
0	318	26	275	51	211	51	202	97	36	83	0	1	1	U	*trachodes*
0	288	23	249	50	191	51	178	93	47	75	1	0	0	V	*trachodes*
0	302	23	267	50	206	49	189	93	57	73	3	0	1	W	*trachodes*
0	304	23	267	55	214	57	195	98	42	71	1	0	1	X	*trachodes*
0	308	27	267	56	206	54	186	102	49	81	0	0	1	Y	*trachodes*
0	278	27	231	83	112	81	71	77	55	56	1	1	0	Z	*papalis*

0	253	27	215	77	85	75	70	74	59	58	1	0	1	a	*papalis*
0	234	27	199	79	70	74	54	72	44	56	4	0	0	b	*papalis*
0	258	24	217	80	81	79	65	76	48	58	1	1	0	c	*papalis*
0	246	27	206	74	102	76	82	78	55	57	0	1	0	d	*papalis*
0	260	30	215	78	95	75	76	71	57	48	2	0	1	e	*papalis*
0	265	30	222	88	89	83	72	83	57	61	1	0	1	f	*papalis*
0	257	26	219	76	82	75	69	73	57	56	2	0	0	g	*papalis*
0	257	23	219	77	82	74	68	73	47	58	3	1	0	h	*papalis*
0	263	22	224	81	98	79	80	78	67	59	4	0	0	i	*papalis*
0	264	30	219	85	88	84	74	83	59	66	3	0	0	j	*papalis*
0	264	30	225	86	106	85	85	81	65	54	2	1	0	k	*papalis*
0	256	25	214	75	96	74	76	74	65	56	1	0	1	l	*papalis*
0	264	28	223	79	91	79	68	74	52	58	1	0	1	m	*papalis*
0	248	28	201	79	78	76	64	73	51	55	1	0	0	n	*papalis*
0	245	33	203	82	84	80	68	75	51	57	2	0	1	o	*papalis*

Sample DSDP 29-12-2, 87-89 cm.

| cephalis | | collar | | thorax | | lumbar | | abdomen | | aperture | | midpoint | | code | species |
A_x	A_y	G_x	G_y	F_x	F_y	E_x	E_y	D_x	D_y	C_x	C_y	B_x	B_y		
0	281	28	243	94	129	96	116	103	31	82	3	0	0	A	*ampla*
0	325	35	280	93	164	94	153	110	48	75	0	1	0	B	*ampla*
0	314	23	281	77	158	76	149	114	36	94	6	0	1	C	*ampla*
0	255	28	217	82	115	83	104	103	36	73	2	1	1	D	*ampla*
0	320	24	275	69	163	70	147	100	40	90	0	0	1	E	*ampla*
0	327	31	290	93	190	96	176	122	71	92	3	0	0	F	*ampla*
0	268	24	229	78	113	78	91	78	71	50	0	0	1	G	*papalis*
0	261	32	220	73	108	75	90	73	72	49	0	0	0	H	*papalis*
0	276	27	238	85	107	84	94	84	78	54	1	0	0	I	*papalis*
0	249	25	210	75	97	70	75	69	61	54	0	0	1	J	*papalis*
0	234	27	200	72	89	73	58	72	46	56	0	1	1	K	*papalis*
0	218	26	176	72	80	70	67	69	53	46	0	1	0	L	*papalis*
0	268	26	226	55	173	55	155	87	40	61	1	0	0	M	*sinuosa*
0	279	28	240	54	188	55	171	95	80	67	0	1	1	N	*sinuosa*
0	252	25	218	51	163	53	147	88	66	61	0	1	1	O	*sinuosa*
0	273	27	237	56	176	60	161	94	71	72	0	0	0	P	*sinuosa*
0	251	22	218	54	163	56	148	84	56	48	0	0	0	Q	*fasciolata*
0	264	25	238	61	186	64	165	107	51	57	0	0	1	R	*fasciolata*
0	247	28	218	57	160	63	148	93	53	49	0	0	0	S	*fasciolata*
0	239	30	194	55	146	55	126	86	60	57	2	0	1	T	*fasciolata*
0	267	27	236	57	184	61	165	96	72	62	2	1	1	U	*fasciolata*
0	267	21	235	53	179	56	163	92	46	59	0	1	1	V	*fasciolata*
0	261	28	225	51	179	50	154	83	43	64	0	0	1	W	*mitra*
0	287	23	246	51	195	54	174	86	25	62	3	1	0	X	*mitra*
0	261	24	223	52	168	52	155	84	23	65	1	0	1	Y	*mitra*
0	296	29	256	56	195	57	181	87	34	69	2	0	0	Z	*mitra*
0	273	28	236	55	183	57	166	85	40	65	0	0	1	a	*mitra*
0	295	25	258	54	205	56	187	92	32	67	1	0	0	b	*mitra*

Sample DSDP 29-16-5, 14-16 cm.

cephalis		collar		thorax		lumbar		abdomen		aperture		midpoint		code	species
A_x	A_y	G_x	G_y	F_x	F_y	E_x	E_y	D_x	D_y	C_x	C_y	B_x	B_y		
0	260	28	217	80	109	76	88	71	72	45	0	1	2	A	*papalis*
0	263	25	228	75	115	72	90	68	74	43	4	0	1	B	*papalis*
0	265	28	226	77	114	75	99	73	85	45	0	0	1	C	*papalis*
0	251	28	213	72	94	70	81	69	70	52	0	0	0	D	*papalis*
0	244	28	204	72	108	67	84	63	71	40	0	0	1	E	*papalis*
0	252	25	209	77	101	72	78	69	69	46	0	1	2	F	*papalis*
0	246	28	203	76	117	73	80	70	68	48	2	1	1	G	*papalis*
0	270	27	238	73	134	70	101	67	91	45	0	0	0	H	*papalis*
0	263	31	222	77	116	77	97	75	82	49	2	0	1	I	*papalis*
0	250	30	206	71	119	70	88	71	77	46	0	0	0	J	*papalis*
0	242	23	204	68	107	66	81	63	67	45	4	0	0	K	*papalis*
0	264	27	230	81	123	77	92	72	75	48	0	0	1	L	*papalis*
0	268	26	224	78	132	70	91	68	76	49	2	0	0	M	*papalis*
0	272	31	229	91	102	91	89	90	78	53	4	1	0	N	*papalis*
0	249	25	208	70	96	69	84	65	71	43	2	1	0	O	*papalis*
0	240	28	199	73	100	68	78	66	67	44	1	0	0	P	*papalis*
0	314	22	266	78	152	79	143	86	111	60	1	1	0	Q	*diamesa*
0	312	28	257	76	157	78	143	85	83	60	1	0	0	R	*diamesa*
0	367	33	310	92	179	94	153	100	115	62	0	1	0	S	*diamesa*
0	320	29	270	80	152	80	138	88	90	55	3	0	0	T	*diamesa*
0	323	34	273	80	162	82	139	89	86	54	0	1	0	U	*diamesa*
0	298	31	248	85	135	86	115	84	78	56	4	0	0	V	*diamesa*
0	346	29	300	96	167	92	148	96	103	63	0	1	0	W	*diamesa*
0	351	32	302	91	175	90	160	93	119	53	0	0	0	X	*diamesa*
0	349	30	307	90	183	89	164	98	108	62	1	0	0	Y	*diamesa*
0	323	31	269	98	155	97	143	99	106	58	4	1	0	Z	*diamesa*
0	327	34	284	89	166	87	144	94	119	61	1	0	0	a	*diamesa*
0	330	29	282	89	165	88	147	93	102	58	0	0	0	b	*diamesa*
0	325	26	276	88	162	86	145	95	94	59	2	0	1	c	*diamesa*
0	223	28	192	57	134	57	116	76	54	47	0	1	1	d	*sinuosa*
0	270	25	233	57	172	62	155	90	79	52	1	1	0	e	*sinuosa*
0	248	27	213	57	154	61	144	81	65	55	5	1	2	f	*sinuosa*
0	242	30	200	55	146	59	133	79	68	57	3	1	0	g	*sinuosa*
0	258	25	224	51	166	53	157	89	74	58	0	1	0	h	*sinuosa*
0	263	30	222	64	137	67	127	77	75	50	3	1	0	i	*sinuosa*
0	228	27	194	56	130	60	118	80	51	59	0	1	1	j	*sinuosa*
0	230	31	196	55	147	59	120	83	57	48	3	1	0	k	*sinuosa*
0	234	21	204	57	145	60	135	90	61	59	0	1	1	l	*sinuosa*
0	239	24	210	56	146	59	137	85	58	59	1	1	0	m	*sinuosa*
0	264	29	226	55	172	58	156	88	66	65	4	0	1	n	*sinuosa*
0	275	31	242	61	184	63	169	94	77	61	10	1	1	o	*sinuosa*
0	258	22	228	57	156	58	148	87	81	58	1	1	1	p	*sinuosa*
0	260	28	226	60	162	59	149	90	70	54	4	1	1	q	*sinuosa*

Sample DSDP 29B-10-2, 29-31 cm.

cephalis		collar		thorax		lumbar		abdomen		aperture		midpoint		code	species
A_x	A_y	G_x	G_y	F_x	F_y	E_x	E_y	D_x	D_y	C_x	C_y	B_x	B_y		
0	272	27	234	89	128	87	94	88	69	68	2	0	0	A	papalis
0	270	25	231	86	115	86	95	88	71	68	1	1	0	B	papalis
0	294	31	258	93	138	89	97	90	66	65	2	0	1	C	papalis
0	256	31	215	86	103	84	74	83	51	63	0	0	1	D	papalis
0	254	26	208	81	99	78	69	80	52	64	5	0	0	E	papalis
0	298	24	249	85	135	81	89	80	69	65	3	1	1	F	papalis
0	296	29	253	91	128	85	86	83	73	66	6	0	0	G	papalis
0	295	26	251	80	148	77	101	73	82	53	1	0	1	H	papalis
0	281	31	233	85	132	74	93	74	72	56	2	0	0	I	papalis
0	282	28	239	81	119	81	89	81	67	57	2	0	1	J	papalis
0	274	28	240	83	115	84	97	84	77	59	3	0	1	K	papalis
0	238	24	189	81	75	79	60	75	52	56	2	0	0	L	papalis
0	240	22	195	75	86	72	62	70	48	55	6	1	0	M	papalis
0	289	29	241	87	110	87	87	84	68	66	3	1	1	N	papalis
0	288	28	238	80	115	75	93	72	74	47	1	1	1	O	papalis
0	281	30	238	85	103	84	82	80	69	54	3	1	1	P	papalis
0	280	25	238	84	118	81	87	78	68	54	0	0	1	Q	papalis
0	263	30	232	63	165	67	154	84	75	56	0	0	1	R	sinuosa
0	260	30	230	64	164	64	154	86	90	58	2	0	0	S	sinuosa
0	255	23	222	59	164	60	142	84	58	56	5	1	0	T	sinuosa
0	253	31	222	61	163	60	142	80	62	58	7	0	0	U	sinuosa
0	284	28	243	61	189	60	169	87	72	63	2	0	0	V	sinuosa
0	288	33	250	65	194	65	169	85	82	64	4	0	0	W	sinuosa
0	244	30	209	64	147	65	130	84	63	49	1	0	1	X	sinuosa
0	264	25	228	53	176	53	159	76	75	53	1	1	1	Y	sinuosa
0	232	28	194	56	136	55	123	79	65	55	3	0	1	Z	sinuosa
0	260	25	226	59	159	58	143	80	64	49	2	0	0	a	sinuosa
0	262	32	228	61	164	61	153	84	67	53	4	1	0	b	sinuosa
0	240	33	199	61	141	60	132	81	49	64	1	0	1	c	sinuosa
0	262	29	228	59	166	60	152	85	71	58	1	1	0	d	sinuosa
0	268	30	237	66	171	68	159	87	89	57	0	0	1	e	sinuosa
0	264	26	226	88	113	83	72	81	43	69	4	0	1	f	diamesa
0	261	29	220	88	122	85	75	80	41	71	8	0	1	g	diamesa
0	292	24	245	88	118	86	89	85	66	71	2	0	0	h	diamesa
0	290	35	244	92	143	90	103	90	60	70	2	0	0	i	diamesa
0	304	33	259	93	143	91	104	90	80	66	1	0	1	j	diamesa
0	296	28	253	86	138	86	93	84	74	68	3	1	1	k	diamesa
0	298	29	260	87	147	81	110	82	85	55	3	0	0	l	diamesa
0	292	33	251	85	145	75	99	73	83	57	3	1	0	m	diamesa
0	282	29	230	85	131	79	89	80	61	57	3	0	1	n	diamesa
0	280	30	235	83	136	80	92	81	68	58	2	0	0	o	diamesa
0	306	25	260	88	130	83	107	87	69	64	4	1	0	p	diamesa
0	288	36	236	89	114	88	92	87	74	64	1	0	0	q	diamesa
0	288	31	244	89	114	84	91	87	69	65	3	0	0	r	diamesa
0	300	29	249	90	140	86	96	83	76	56	1	1	0	s	diamesa
0	282	34	241	81	129	79	98	79	73	58	1	1	0	t	diamesa
0	277	28	228	84	117	77	85	77	63	58	4	0	1	u	diamesa
0	306	33	252	89	138	86	109	87	83	64	0	0	1	v	diamesa
0	300	30	256	92	150	88	106	87	81	66	3	0	1	w	diamesa

Combined References From All Chapters

Abe, K., R. Reyment, F. L. Bookstein, A. Honigstein, and O. Hermalin. 1988. Microevolutionary changes in *Veenia fawwarensis* (Ostracoda, Crustacea) from the Cretaceous (Santonian) of Israel. Hist. Biol., 1:303-322.

Ahmavaara, Y. 1957. On the unified factor theory of mind. Ann. Acad. Sci. Fenn., Ser. B, 106, Helsinki, 176 pp.

Aitchison, J. 1986. The statistical analysis of compositional data. Monographs on statistics and applied probability. Chapman and Hall, London, 416 pp.

Alberch, P., S. J. Gould, G. F. Oster and D. B. Wake. 1979. Size and shape in ontogeny and phylogeny. Paleobiology, 5:296-317.

Altig, R. G. 1965. Observations on the ontogeny of the osseous skeleton of *Siren intermedia* Le Conte. Masters Thesis, Southern Illinois University, 62 pp.

Anderson, T. W. and R. R. Bahadur. 1962. Two sample comparisons of dispersion matrices for alternatives of immediate specificity. Ann. Math. Stat., 33:420-431.

Archie, J. W. 1985. Methods for coding variable morphological features for numerical taxonomic analysis. Syst. Zool., 34:326-345.

Ashkar, G. P. and J. W. Modestino. 1978. The contour extraction problem with biomedical applications. Computer Graphics Image Processing, 7:331-355.

Atchley, W. R., C. T. Gaskins and D. Anderson. 1976. Statistical properties of ratios. I. Empirical results. Syst. Zool., 25(2):563-583.

Atkinson, L. V. and P. J. Harley. 1983. An introduction to numerical methods with Pascal. Addison-Wesley, Reading, Mass., 300 pp.

Aubrey, D. G. 1979. Seasonal patterns of onshore/offshore sediment movement. J. Geophys. Res., 84:6347-6354.

Aubrey, D. G. and K. O. Emery. 1986a. Relative sea levels of Japan from tide-gauge records. Geol. Soc. Amer. Bull., 97:194-205.

Aubrey, D. G. and K. O. Emery. 1986b. Relative sea levels of Japan from tide-gauge records: Reply. Geol. Soc. Amer. Bull., 97:1282.

Ax, P. 1987. The phylogenetic system. John Wiley & Sons, Chichester, 340 pp.

Ballard, D. H. and L. M. Brown. 1982. Machine vision. Prentice-Hall, New York. 523 pp.

Banner, F. T. and F. D. M. Lowery. 1985. The stratigraphical record of planktonic foraminifera and its evolutionary implications. Spec. Papers in Paleo., 33:117-130.

Bargmann, R. E. 1970. Interpretation and use of a generalized discriminant function. Pp. 35-60 in Essays in probability and statistics (Bose, R. C., I. M. Chakravarti, P. C. Mahalanobis, C. R. Rao and K. J. C. Smith, eds). The University of North Carolina Press, Chapel Hill.

Barnes, V. E. and M. A. Barnes. 1973. Tektites. Benchmark papers in geology, Dowden. Hutchinson and Ross, Stroudsburb, PA, 445 pp.

Barnett, V. and Lewis, T. 1978. Outliers in statistical data. Wiley and Sons: Chichester, 365 pp.

Barnsley, M. F., V. Ervin, D. Hardin, and J. Lancaster. 1986. Solution of an inverse problem for fractals and other sets. Proc. Natl. Acad. Sci. USA, 83:1975-1977.

Baxes, G. A. 1984. Digital image processing: a practical primer. Prentice-Hall, Englewood Cliffs, 182 pp.

Bayer, U. 1978. Morphogenetic programs, instabilities, and evolution - a theoretical study. Neues Jahrbuch für Geologie und Paläontologie, Abhandlungen 156:226-261.

Bé, A. W. H., S. M. Harrison and L. Lott. 1973. *Orbulina universa* d'Orbigny in the Indian Ocean. Micropaleontology, 19(2):150-92.

364

Bennack, D. E. 1988. The lock and key hypothesis of mechanical reproductive isolation: Variation in genitalic fit and its influence on spermatiphore transfere in the grasshopper *Barytettix humphreysii*. Ph.D. dissertation, Michigan State University, East Lansing, Michigan. 148 pp.

Benson, R. H. 1967. Muscle scar patterns of Pleistocene (Kansan) ostracodes. Pp. 211-241 *in* Essays in paleontology and stratigraphy (Teichert, C. and E. Yochelson, eds.). U. Kansas Press, Lawrence.

Benson, R. H. 1976. The evolution of the ostracode *Costa* analyzed by "Theta-Rho Difference". Abh. Verh. Naturwiss. Ver. Hamburg, (NF) 18/19(Suppl.): 127-139.

Benson, R. H. 1982a. Deformation, Da Vinci's concept of form, and the analysis of events in evolutionary history. Pp. 241-277 *in* Palaeontology, essential of historical geology (Gallitelli, E.M., ed.). S.T.E.M. Mucchi, Modena, Italy.

Benson, R. H. 1982b. Comparative transformation of shape in a rapidly evolving series of structural morphotypes of the ostracode Bradleya. Pp. 147-164 *in* Fossil and recent Ostracodes (Bate, R. H., E. Robinson and L. M. Sheppard, eds.) Ellis Horwood, Chichester, U.K.

Benson, R. H. 1983. Biomechanical stability and sudden change in the evolution of the deep-sea ostracode Poseidonamicus. Paleobiology, 9(4): 398-413.

Benson, R. H., and R. E. Chapman. 1982. On the measurement of morphology and its change. Paleobiology, 8:328-339.

Bentler, P. M. 1985. Theory and implementation of EQS: a structural equations program. BMDP Statistical Software, Inc., Los Angeles. (standalone program manual for structural relations— a little easier to use than LISREL, but does not include multiple samples).

Benzecri, J. P. 1973. L'analyse des correspondances. Dunod: Paris, 619 pp.

Berger, W. H. 1969. Kummerform foraminifera as clues to oceanic environments. Bull. Am. Assoc. Petrol. Geol. [abs.], 53:706.

Berggren, W. A. 1965. Paleocene - A micropaleontologist's point of view. Bull. Am. Assoc. Petrol. Geol., 44:1473-1484.

Bezier, P. E. 1970. Emploi des machines a commande numérique. Masson:Paris.

Blackith, R. E. and R. Reyment. 1971 Multivariate morphometrics. Academic Press: London, 71 pp.

Blackstone, N. 1987a. Allometry and relative growth: pattern and process in evolutionary studies. Syst. Zool., 36(1):76-8.

Blackstone, N. 1987b. Size and time. Syst. Zool., 36(2):211-15.

Blondel, J., F. Vuilleumier, L. F. Marcus and E. Terouanne. 1984. Is there ecomorphological convergence among Mediterranean bird communities of Chile, California, and France. *in* Evolutionary biology, 18:141-213 (M. K. Hecht, B. Wallace and G. T. Prance, eds.). Plenum Publishing, New York.

Blow, W. H. 1979. The Cainozoic *Globigerinida*, 3 vols., E. J. Brill, Leiden, 1413 pp.

Blum, H. 1967. A transformation for extracting new descriptors of shape. Pp. 362-380 *in* Models for the perception of speech and visual form (W. Whaten-Dunn, ed.). MIT Press, Cambridge, MA, 470 pp.

Blum, H. 1973. Biological shape and visual science (part I). J. Theor. Biol., 38:205-287.

Blum, H. and R. N. Nagel. 1978. Shape description using weighted symmetric axis features. Patt. Recogn., 10:167-180.

Bonebrake, J. E., and R. A. Brandon. 1971. Ontogeny of cranial ossification in the small-mouthed salamander, *Ambystoma texanum*. J. Morphol., 133:189-204.

Bookstein, F. L. 1977. Orthogenesis of the hominids: an exploration using biorthogonal grids. Science, 197:901-904.

Bookstein, F. L. 1978. The Measurement of biological shape and shape change. Lecture Notes in Biomathematics, v. 24. Berlin: Springer, 191 pp.

Bookstein, F. L. 1979. The line skeleton. Comp. Graph. Image Proc., 11:123-137.

Bookstein, F. L. 1981. Looking at mandibular growth: some new geometric methods. Pp. 83-103 in Craniofacial biology. Univ. Michigan Center for Human Growth and Development. (D. S. Carlson, ed.) Ann Arbor.

Bookstein, F. L. 1982. Foundations of morphometrics. Ann. Rev. Ecol. Syst., 13:451-470.

Bookstein, F. L. 1984a. A statistical method for biological shape change. J. Theor. Biol., 107:475-520.

Bookstein, F. L. 1984b. Tensor biometrics for changes in cranial shape. Ann. Human Biol., 11:413-437.

Bookstein, F. L. 1985a. A geometric foundation for the study of left ventricular motion: some tensor considerations. Pp. 65-83 in Digital Cardiac Imaging (A. J. Buda and E. J. Delp, eds.). Martinus Nijhoff, The Hague.

Bookstein, F. L. 1985b. Transformations of quadrilaterals, tensor fields, and morphogenesis. Pp. 221-265 in Mathematical essays on growth and the emergence of form (Antonelli, P. L., ed.). University of Alberta Press.

Bookstein, F. L. 1986. Size and shape spaces for landmark data in two dimensions. (With discussion and rejoinder). Stat. Sci., 1:181-242.

Bookstein, F. L. 1987. Describing a craniofacial anomaly: Finite elements and the biometrics of landmark location. Am. J. Phys. Anthropol., 74:495-509.

Bookstein, F. L. 1989a. Comment on D. G. Kendall, "A survey of the statistical theory of shape." Statistical Science, 4:99-105.

Bookstein, F. L. 1989b. Principal warps: Thin-plate splines and the decomposition of deformations. I.E.E.E. Transactions on Pattern Analysis and Machine Intelligence, 11:567-585.

Bookstein, F. L. 1989c. "Size and shape": a comment on semantics. Systematic Zool., 38:173-180.

Bookstein, F. L. 1990. Four metrics for image variation. in Proceedings of the XI international conference on information processing in medical imaging (Ortendahl, D. and J. Llacer, eds.). Alan R. Liss, Inc, New York.

Bookstein, F. L. 1991. Morphometric tools for landmark data. Book manuscript, accepted for publication, Cambridge University Press.

Bookstein, F. L., B. Chernoff, R. Elder, J. Humphries, G. Smith, and R. Strauss. 1982. A comment on the uses of Fourier analysis in systematics. Systematic Zool., 31:85-92.

Bookstein, F. L., B. Chernoff, R. L Elder, J. M. Humphries, Jr., G. R. Smith, and R. E. Strauss. 1985. Morphometrics in evolutionary biology. The Academy of Natural Science Philadelphia. Special Publ. No. 15, 277 pp.

Bookstein, F. L. and R. A. Reyment. 1989. Micro-evolution in Miocene Brizalina (Foraminifera) studied by canonical variate analysis and analysis of landmarks. Bull. Math. Biol., 51:657-679.

Bookstein, F. L., and P. Sampson. 1987. Statistical models for geometric components of shape change. Proceedings of the section on statistical graphics, 1987 Annual Meeting of the American Statistical Association, pp. 18-27.

Bookstein, F. L. and P. D. Sampson. 1990. Statistical models for geometric components of shape change. Communications in Statistics: theory and methods, in press.

Boyd, E. M. 1980. Origins of the study of human growth. University of Oregon Health Sciences Center.

Boyden, A. 1947. Homology and analogy. Am. Midl. Nat., 37:648-669.

Brice, C. R. and C. L. Fennema. 1970. Scene analysis using regions. Artificial Intell., 1: 205-226.

Campbell, N. A. 1979. Canonical variate analysis: Some practical aspects. Ph.D. Thesis, Imperial College, London.

Campbell, N. A. 1980a. Robust procedures in multivariate analysis. I. Robust covariance estimation. Appl. Statist., 29:231-237.

Campbell, N. A. 1980b. Shrunken estimators in discriminant and canonical variate analysis. Appl. Statist., 29: 5-14.

Campbell, N. A. 1984. Some aspects of allocation and discrimination. Pp. 177-192 in Multivariate statistical methods in physical anthropology (van Vark, G. N. and W. W. Howells, eds.). Reidel, Dordrecht, 433 pp.

Campbell, N. A. and W. R. Atchley. 1981. The geometry of canonical variate analysis. Syst. Zool., 30:268-280.

Campbell, N. A. and R. A. Reyment. 1978. Discriminant analysis of a Cretaceous foraminifer using shrunken estimators. Math. Geol., 10: 347-359.

Campbell, N. A. and R. A. Reyment. 1980. Robust multivariate procedures applied to the interpretation of atypical individuals of a Cretaceous foraminifer. Cret. Res., 1:207-221.

Casasent, D., J. Pauly, and D. Fetterly. 1981. Infrared ship classification using a new moment pattern recognition concept. SPIE, 302:126-133.

Castleman, R. R. 1979. Digital image processing. Prentice-Hall, Englewood Cliffs, New Jersey, 429 pp.

Cattell, R. B. and D. K. Khanna. 1977. Principles and procedures for unique rotation in factor analysis. Chapter 9 in Enslein, K. and A. Ralston (eds.) Mathematical methods for digital computers, vol. III. Wiley-Interscience:New York.

Chapman, R. E. In Press. Shape analysis in the study of dinosaur morphology. in Dinosaur systematics (Currie, P. J. and K. Carpenter, eds.). Cambridge University Press, Cambridge.

Chapman, R. E. and M. K. Brett-Surman. In Press. Morphometric observations of hadrosaurid ornithopods. in systematics (Currie, P. J. and K. Carpenter, eds.). Cambridge University Press,

Chatterjee, S. 1984. Variance estimation in factor analysis: an application of the bootstrap. Brit. J. Math. Stat. Psychol., 37:252-262.

Cheung, E. and J. Vlcek. 1986. Fractal analysis of leaf shapes. Can. J. For. Res., 16:124-127.

Cifelli, R. 1965. Planktonic foraminifera from the western North Atlantic. Smithsonian Miscellaneous Collections, 148(4):36pp.

Cock, A. G. 1966. Genetical aspects of metrical growth and form in animals. Q. Rev. Biol., 41:131-190.

Cohn, T. and I. Cantrall. 1974. Variation and speciation in the grasshoppers of the Conalcaeini (Orthoptera: Acrididae: Melanoplinae): The lowland forms of western Mexico, the genus Barytettix. Memoirs, San Diego Society of Natural History, Number 6, 131 pp.

Colless, D. H. 1967. An examination of certain concepts in phenetic taxonomy. Syst. Zool., 16:6-27.

Cooley, W. S. and P. R. Lohnes. 1971. Multivariate data analysis. Wiley and Sons, New York, 364 pp.

Cooley, J. W. and J. W. Tukey. 1965. An algorithm for the machine computation of complex Fourier series. Math. Comput., 19:297-301.

Cossmann, M. and G. Pissarro. 1910-13. Iconographie complete des coquilles fossiles de l'Eocene des environs de Paris, vol. 2, Scaphopodes, gastropodes, brachiopodes, cephalopodes and supplement. 345 pp., 65 plts. de Hartog, C. 1970. The seagrasses of the world. Verh. K. Ned. Akad. Wet. Afd. Natuurk. D. Tweede Reeks 59:1-275.

Creighton, G. K., and R. Strauss. 1985. Comparative patterns of growth and development in cricetine rodents and the evolution of ontogeny. Evolution, 40:94-106.

Davis, R. E. 1976. Predictability of sea surface temperature and sea level pressure anomalies over the North Pacific Ocean. J. Phys. Oceanogr., 6(3):249-266.

de Boor, C. 1978. A practical guide to splines. Springer-Verlag, New York, 392 pp.

de Queiroz. K. 1985. The ontogenetic method for determining character polarity and its relevance to phylogenetic systematics. Syst. Zool., 34:280-299.

Dempster, A. P. 1969. Elements of continuous multivariate analysis. Addison-Wesley, Reading, Mass.

Diaconis, P. 1985. Theories of data analysis: from magical thinking through classical analysis. Pp. 1-36 in Understanding robust and exploratory data analysis (Hoaglin, D. C., F. Mosteller and J. W. Tukey, eds.). John Wiley & Sons, New York.

Diaconis, P. and B. Efron. 1983. Computer-intensive methods in statistics. Sc. Amer., 248: 96-108. (semi-popular article on the bootstrap).

Didier, R. 1962. Note sur l'os penien de quelques rongeurs de l'Amerique du Sud. Mammalia, 26:408-430.

Dillon, W. R. and M. Goldstein. 1984. Multivariate analysis. Methods and applications. John Wiley & Sons, New York.

Dixon, W. J. (Ed.) 1983. BMDP Statistical Software, 1983 revised printing. University of California Press, Berkeley.

Dixon, W. J. and Massey, F. J. 1969. Introduction to statistical analysis, third edition. McGraw Hill Book Company, New York.

Dudani, S. A., K. J. Breeding, and R. B. McGhee. 1977. Aircraft identification by moment invariants. IEEE Trans. Computers, C26:39-46.

Duellman, W. E. and L. Trueb. 1986. Biology of amphibians. McGraw Hill, New York.

Dunn, L. C. 1922. The effect of inbreeding on the bones of the fowl. Bulletin 152. Storrs Agricultural Experiment Station, pp.1-112.

Dyer, C. R. and A. Rosenfeld. 1979. Thinning algorithms for gray-scale pictures. IEEE Trans. Pattern Anal. Mach. Intell., PAMI-1:88-89.

Eastment, H. T. and W. J. Krzanowski. 1982. Cross-validatory choice of the number of components from a principal component analysis. Technometrics, 24:73-77.

Eckart, C. and G. Young. 1936. The approximation of one matrix by another of lower rank. Psychometrika, 1:211-218.

Efrom, B. 1979. Bootstrap methods: another look at the jacknife. Ann. Statist., 7: 1-26.

Efrom, B. 1982. The jacknife, the bootstrap and other resampling plans. SIAM Monograph 38. Soc. for Indust. and Applied Math., Philadelphia. (small monograph on the subject).

Efrom, B. and G. Gong. 1983. A leisurely look at the bootstrap, the jacknife and cross-validation. Amer. Statist., 37: 36-48.

Ehrlich, R., P. J. Brown, J. M. Yarus, and R. Przygocki. 1980. The origin of shape frequency distributions and the relationship between size and shape. J. Sedim. Petrol., 50(2):475-484.

Ehrlich, R and B. Weinberg. 1970. An exact method for characterization of grain shape. J. Sed. Petrol., 40:205-212.

Eldredge, N. 1979a. Alternative approaches to evolutionary theory. Bull. Carnegie Mus, Nat. Hist., 13:7-19.

Eldredge, N. 1979b. Cladism and common sense. Pp. 165-198 in Phylogenetic analysis and paleontology (Cracraft, J. and N. Eldredge, eds.). Columbia University Press, New York, 233 pp.

Eldredge, N. and J. Cracraft. 1980. Phylogenetic patterns and the evolutionary process: method and theory in comparative biology. Columbia Univresity Press, New York, 349 pp.

Elderedge, N. and S. J. Gould. 1972. Punctuated equilibrium: an alternative to phyletic gradualism. Pp. 82-115 in Models in paleobiology (Schopf T. J. M., ed.). Freeman, Cooper and Co., San Fransisco, 250 pp.

Emerson, S. B. 1988. Testing for historical patterns of change: a case study with frog pectoral girdles. Paleobiology, 14:174-186.

Engel, H. 1986. A least-squares method for estimation of Bezier curves and surfaces and its applicability to multivariate analysis. Math. Biosci., 79:155-170.

Erez, J. 1983. Calcification rates, photosynthesis and light in planktonic foraminifera. Pp. 307-12 *in* Biomineralization and biological metal accumulation, biological and geological perspectives (Westbroek, P. and E. W. de Jong, eds.). Reidel Publishing Co., Dordrecht, Holland, 783 pp.

Evans, D. G., P. N. Schweitzer, and M. S. Hanna, 1985. Parametric cubic splines and geological shape descriptions. Math. Geol., 17:611-624.

Eves, H. 1972. Analytic geometry. in Standard mathematical tables (Selby, S. M., ed.). Chemical Rubber Co., Cleveland, 20th edition, 353 pp.

Farris, S. J. 1983. The logical basis of phylogenetic inference. Pp. 7-36 *in* Advances in cladistics (Platnick, N. I. and V.A. Funk, eds.). Vol. 2, Proc. 2nd Ann. Mtg. Hennig Soc. Columbia Univ. Press, New York, 218 pp.

Feinberg, S. E. 1977. The analysis of cross-classified categorical data. The MIT Press, Cambridge, Mass.

Felsenstein, J. 1982. Numerical methods for inferring evolutionary trees. Quart. Rev. Biol., 57: 127-141.

Felsenstein, J. 1988. Phylogenies and quantitative characters. Ann. Rev. Ecol. Syst., 19:445-471.

Ferson, S., F. J. Rohlf, and R. K. Koehn. 1985. Measuring shape variation of two-dimensional outlines. Syst. Zool., 34:59-68.

Fink, W. L. 1982. The conceptual relationship between ontogeny and phylogeny. Paleobiology, 8:254-264.

Fink, W. L. 1987. Video digitizer: a system for systematic biologists. Curator, 30(1):63-72.

Fisher, R. A. 1936. The use of multiple measurements in taxonomical problems. Ann. Eugenics London, 10:422-429.

Fitch, W. M. 1966. An improved method for testing for evolutionary homology. J. Mol. Biol., 16:9-16.

Fitch, W. M. and W. R. Atchley. 1987. Divergence in inbred strains of mice: a comparison of three different types of data. Pp. 203-216 *in*

Molecules and morphology in evolution (Patterson, C. ed.). Cambridge Univ. Press, Cambridge, 229 pp.

Flury, B. 1984. Common principal components in k groups. J. Amer. Statist. Assoc., 79:892-898.

Flury, B. and G. Constantine. 1985. The FG diagonalization algorithm. Algorithm AS 211, Appl. Stat., 34:177-183.

Flury, B. and H. Riedwyl. 1988. Multivariate statistics: a practical approach. Chapman and Hall, London.

Frei, W. 1977. Image enhancement by histogram hyperbolization. Comp. Graphics and Image Processing, 6:286-294.

Friedman, J. H., and W. Stuetzle. 1982. Projection pursuit methods for data analysis. Pp. 123-147 *in* Modern data analysis (Launer, R. L, and A. F. Siegel, eds.). Academic Press, New York.

Full, W. E. and R. Ehrlich. 1986. Fundamental problems associated with "eigenshape analysis" and similar "factor" analysis procedures. Math. Geol., 18:451-463.

Gabriel, K. R. 1968. The biplot graphical display of matrices with application to principal component analysis. Biometrika, 58:453-467.

Gabriel, K. R. 1971. The biplot graphical display of matrices with applicatin to principal componenet analysis. Biometrika, 58:453-467.

Gardiner, B. 1982. Tetrapod classification. Zool. J. Linnean Soc., 74:207-232.

Gasson, P. C. 1983. Geometry of spatial forms. Ellis Horwood Limited, Chichester, 601 pp.

Gauthier, J., A. G. Kluge, and T. Rowe. 1988. Amniote phylogeny and the importance of fossils. Cladistics, 4:105-209.

Ghiselin, M. T. 1969. The distinction between similarity and homology. Syst. Zool. 18(1):148-149.

Ghiselin, M. T. 1976. The nomenclature of correspondence: a new look at "homology" and "analogy". Pp. 129-142 *in* Evolution, brain and behavior (Masterton, R. B., W. Hodos, and H.

Jerison, eds.). Lawrence Erlbaum, Hillsdale, New Jersey, 276 pp.

Gibson, A. R., A. J. Baker and A. Moeed. 1984. Morphometric variation in introduced populations of the common Myna (Acridotheres tristis): an application of the jackknife to principal component analysis. Syst. Zool., 33(4): 408-421.

Glass, B. P. 1982. Introduction to planetary geology. Cambridge University Press, Cambridge.

Gnanadesikan, R. 1977. Methods for statistical data analysis of multivariate observations. Wiley and Sons, New York, 311 pp.

Goldman, N. 1988. Methods for discrete coding of morphological characters for numerical analysis. Cladistics, 4:59-71.

Golub, G. H. and C. Reinsch. 1970. Singular value decomposition and least squares solutions. Num. Math., 14:403-420.

Goodall, C. R. 1989. WLS estimators and tests for shape differences in landmark data. J. Roy. Statist. Soc., submitted for publication.

Goodall, C. 1990. Procrustes methods in the statistical analysis of shape. J. Royal Statistical Soc., Ser. B, *in press*.

Goodall, C. R. and A. Bose. 1987. Procrustes techniques for the analysis of shape and shape change. Pp. 86-92 in R. Heiberger (ed.) Computer science and statistics: proceedings of the 19th symposium on the interface. Alexandria, Virginia: Amer. Stat. Assn.

Goodall, C. R. and P. B. Green. 1986. Quantitative analysis of surface growth. Bot. Gaz., 147(1):1-15.

Gordon, A. D. 1981. Classification. Monographs on applied probability and statistics. Chapman and Hall, London, 193 pp.

Gould, S. J. 1977. Ontogeny and phylogeny. Harvard Univ. Press, Cambridge Massachusetts, 501 pp.

Gower, J. C. 1966a. Some distance properties of latent root and vector methods used in multivariate analysis. Biometrika 53(3&4):325-338.

Gower, J. C. 1966b. A Q-technique for the calculation of canonical variates. Biometrika, 53(3 & 4):588-590.

Gower, J. C. 1971. Statistical methods of comparing different multivariate analyses of the same data. Pp. 138-149 *in* Mathematics in the archaeological and historical sciences (Hodson, F. R., D. G. Kendall and P. Tautu, eds.). Edinburgh University Press, Edinburgh.

Gower, J. C. 1975. Generalized Procrustes analysis. Psychometrika, 40(1):33-51.

Gower, J. 1984. Multivariate analysis: ordination, multidimensional scaling and allied topics. Pp. 727-781 in E. Lloyd (ed.). Handbook of applicable mathematics, Volume VI: Statistics. Wiley:New York.

Gower, J. C. 1987. Introduction to ordination techniques. Pp. 2-64 *in* Developments in Numerical ecology, (Legendre, P. and L. Legendre, eds.). NATO A 51 series, G 14, Springer Verlag, New York.

Gower, J. C. and P. G. and N. Digby. 1984. Some recent advances in multivariate analysis applied to anthropometry. Pp. 21-36 in Multivariate statistical methods in physical anthropology (van Vark, N. and W. W. Howells, eds.). Reidel: Dordrecht, 433 pp.

Grainger, J. E. 1981. A quantitative analysis of photometric data from aerial photographs for vegetation survey. Vegetatio, 48:71-82

Grayson, B., C. Cutting, F. L. Bookstein, H. Kim, and J. McCarthy. 1988. The three dimensional cephalogram: Theory, technique, and clinical application. Am. J. Orthod., 94:327-337.

Grayson, B., N. Weintraub, F. L. Bookstein, and J. McCarthy. 1985. A comparative cephalometric study of the cranial base in craniofacial syndromes. Clef. Pal. J., 22:75-87.

Greenacre, M. J. 1984. Theory and applications of correspondence analysis. Academic Press, London.

Groch, W. D. 1982. Extraction of line shaped objects from aerial images using a special operator to analyze the profiles of functions.

370

Computer Graphics Image Processing, 18:347-358

Haas, O., and G. G. Simpson. 1946. Analysis of some phylogenetic terms with attempts at redefinition. Proc. American Phil. Soc., 90:319-349.

Hall, E. L. 1979. Computer Image Processing and Recognition Academic Press, New York, 584 pp.

Hampel, F. R., E. N. Ronchetti, P. J. Rousseeuw, and W. A. Stahel. 1986. Robust statistics. Wiley and Sons, New York, 502 pp.

Hand, D. J. 1981. Discrimination and classification. John Wiley & Sons, Chichester.

Harris, R. J. 1975. A primer of multivariate statistics. Academic Press, New York.

Hawkins, D. M. 1974. The detection of errors in multivariate data using principal components. J. Amer. Statist. Assoc., 69:340-344.

Hawkins, D. M. 1980. Identification of outliers. Chapman and Hall, London.

Healy-Williams, R. Ehrlich, and D. F. Williams. 1985. Morphometric and stable isotopic evidence for subpopulations of *Globorotalia truncatulinoides*. J. Foraminiferal Res., 15(4):242-253.

Hecht, A. D. 1974. Intraspecific variation in recent populations of *Globigerinoides ruber* and *Globigerinoides trilobus* and their application to paleoenvironmental analysis. J. Paleontol., 48 (6):1217-1234.

Henning, W. 1966. Phylogenetic systematics. Univ. Illinois Press, 263 pp.

Hoaglin, D. C., F. Mosteller and J. W. Tukey (eds.) 1985. Understanding robust and exploratory data analysis. John Wiley & Sons, New York.

Hocking, R. R. 1983. Developments in linear regression methodology: 1959-1983. Technometrics, 25:219-30.

Horn, B. K. P. 1986. Robot vision. M.I.T. Press, Cambridge, MA, 509 pp.

Howarth, R. J. 1971. An empirical discriminant method applied to sedimentary rock classification from major element geochemistry. Math. Geol., 3:51-60.

Hu, M. K. 1962. Visual pattern recognition by moment invariants. IRE Trans. Information Th., 8:179-187.

Hubbs, C. L. and K. F. Lagler. 1941. Guide to the Great Lakes and tributary waters. Cranbrook Institute of Science Bulletin, 18, 100 pp.

Huffman, T., R. A. Christopher and J. E. Hazel. 1978. Orthogonal mapping: a computer program for quantifying shape differences. Comp. Geosc., 4:121-130.

Hull, D. L. 1968. The operational imperative: Sense and nonsense in operationism. Syst. Zool., 17:438-457.

Hummel, R. 1977. Image enhancement by histogram transformation. Computer Graphics and Image Processing, 6:184-195.

Hurley, J. R. and R. B. Cattell. 1962. The Procrustes program: producing direct rotation to test a hypothesised factor structure. Computers in Behavioral Sci., 7:258-262.

Huxley, J. 1932. Problems of relative growth. Methuen and Co., London.

Inglis, W. G. 1966. The observational basis of homology. Syst. Zool., 15:219-228.

Jacobs, L. J. and H. Claeys. 1987. A digitizing tablet as an efficient and accurate tool in morphometric studies on nematodes. Ann. Soc. r. zool. Belg., 117:15-20.

Jardine, N. 1967. The concept of homology in biology. Brit. J. Phil. Sci., 18:125-139.

Jardine, N. 1969. The observational and theoretical components of homology: a study based on the morphology of the dermal skull-roofs of rhipidistian fishes. Biol. J. Linn. Soc., 1:327-361.

Jardine, N., and C. J. Jardine. 1967. Numerical homology. Nature, 216:301-302.

Jardine, N., and C. J. Jardine. 1969. Is there a concept of homology common to several sciences? Class. Soc. Bull., 2(1):12-18.

Jenkins, D. G. and M. S. Srinivasan. 1985. Cenozoic planktonic foraminifers from the Equator to the Subantartic of the Southwest Pacific. in Initial reports of the deep sea drilling project, 90 (J. H. Blakeslee ed.). U.S. Gov. Print. Off., Washington, D.C.

Johnson, R. A. and D. W. Wichern. 1982. Applied multivariate statistical analysis. Prentice Hall, Englewood Cliffs, New Jersey.

Jolicoeur, P. 1959. Multivariate geographical variation in the wolf, *Canis lupus* L. Evolution, 13:283-299.

Jolicoeur, P. 1963. The multivariate generalization of the allometry equation. Biometrics, 19:497-499.

Jolicoeur, J. and J. E. Mosimann. 1960. Size and shape variation in the Painted Turtle. Growth, 24: 339-354.

Joliffe, I. T. 1986. Principal component analysis. Springer-Verlag, New York.

Jöreskog, K. G., J. E. Klovan and R. A. Reyment. 1976. Geological factor analysis. Elsevier, Amsterdam.

Jöreskog, K. G. and D. Sorbom. 1979. Advances in factor analysis and structural equation models. Abt Books, Cambridge, Mass.

Jöreskog, K. G. and D. Sorbom. 1985. LISREL VI. Analysis of linear structural relationships by maximum likelihood, instrumental variables, and least-squares methods. Scientific Software, Inc., Mooresville, Ind.

Kaesler, R. L. and J. A. Waters. 1972. Fourier analysis of the ostracode margin. Bull. Geol. Soc. Amer., 83:1169-1178.

Katz, M. J. and E. B. George. 1985. Fractals and the analysis of growth paths. Bull. Math. Biol., 47:273-286.

Keller, G., S. L. D'Hondt, C. J. Orth, J. S. Gilmore, P. Q. Oliver, E. M. Shoemaker, and E. Molina. 1987. Late Eocene impact microspherules: stratigraphy, age and geochemistry. Meteoritics, 22(1): 25-60.

Kendall, D. G. 1981. The statistics of shape. Pp. 75-80 in V. Barnett (ed.) Interpreting multivariate data. Wiley: New York, 374 pp.

Kendall, D. G. 1984. Shape-manifolds, Procrustean metrics and complex projective spaces. Bulletin of the London Mathematical Society, 16:81-121.

Kendall, D. G. 1985. Exact distributions for shapes of random triangles in convex sets. Adv. Appl. Prob., 17:308-329.

Kennett, J. P. and M. S. Srinivasan. 1983. Neogene planktonic foraminifera: a phylogenetic atlas. Hutchinson Ross Publ. Co., Stroudsburg, PA.

Key, K. H. L. 1967. Operational homology. Syst. Zool., 16:275-276.

Kirsch, R. A. 1971. Computer determination of the constituent structure of biological images. Computers and Biomedical Res., 4:315-328.

Kluge, A. G. 1985. Ontogeny and phylogenetic systematics. Cladistics, 1:13-27.

Kluge, A. G. 1988. The characterization of ontogeny. Pp. 57-81 *in* Ontogeny and systematics (C. J. Humphries, ed.). Columbia Univ. Press, New York, 236 pp.

Kluge, A. G., and J. S. Farris. 1969. Quantitative phyletics and evolution of anurans. Syst. Zool., 18(1):1-32.

Kluge, A. G., and R. E. Strauss. 1985. Ontogeny and systematics. Ann. Rev. Ecol. Syst., 16:247-268.

Kristof, W. and B. Wingersky. 1971. Generalization of the orthogonal Procrustes rotation procedure to more than two matrices. Proc. 79th Ann. Convention, APA, pp. 89-90.

Kruskal, J. B. and M. Wish. 1978. Multidimensional scaling. No. 11 in Series: Quantitative applications in the social sciences. Sage Publications, Beverly Hills.

Krzanowski, W. J. 1979. Between-group comparison of principal components. J. Amer. Stat. Assoc., 74:703-707.

Krzanowski, W. J. 1987a. Cross-validation in principal component analysis. Biometrics, 43:575-584.

Krzanowski, W. J. 1987b. Selection of variables to preserve multivariate data structure using principal components. J. Roy. Stat. Soc. C, 36:22-33.

Krzanowski, W. J. 1988. Principles of multivariate analysis: a user's perspective. Oxford University Press, Oxford, 563 pp.

Kuhl, F. P. and C. R. Giardina, 1982. Elliptic Fourier features of a closed contour. Computer Graphics and Image Processing, 18:236-258.

Kvalheim, O. 1987. Doctoral thesis in Physical Chemistry presented to the Faculty of Science, University of Bergen, Dec. 4th, 1987.

Laudan, L. 1977. Progress and its problems: Towards a theory of scientific growth. University of California Press, Berkeley, 257 pp.

Lauder, G. V. 1981. Form and function: structural analysis in evolutionary morphology. Paleobiology, 7:430-442.

Lauder, G. V. 1982. Historical biology and the problem of design. J. theor. Biol., 97:57-67.

Lauder, G.V., and S. M. Reilly. In Press. Metamorphosis of the feeding mechanism in tiger salamanders. J. Zool. Lond.

Lauder, G. V. and H. B. Shaffer. 1988. The ontogeny of functional design in tiger salamanders (*Ambystoma tigrinum*): are motor patterns conserved during major morphological transformations? J. Morphol., 197:249-268.

Lee, D. T. 1982. Medial axis transformation of a planar shape. IEEE Trans. Pattern Anal. Mach. Intell., vol. PAMI-4:363-369.

Lindberg, D. R. 1982. A multivariate study of morphological variation of the limpet *Notoacmea depicta* (Hinds) and its synonyms *Notoacmea gabatella* (Berry) and *Notoacmea lepisma* (Berry) (Gastropoda: Acmaeidae). Bull. Southern Calif. Acad. Sci. 81(2):87- 96.

Lindberg, D. R. 1988. The Patellogastropoda. Malacol. Rev., Suppl. 4:35-63.

Little, R. J. A. and D. B. Rubin. 1987. Statistical analysis with missing data. John Wiley & Sons, New York.

Loehlin, J. C. 1987. Latent variable models: an introduction to factor, path and structural analysis. Lawrence Erlbaum Associates, Hillsdale, NJ.

Lohmann, G. P. 1983. Eigenshape analysis of microfossils: A general morphometric procedure for describing changes in shape. Math. Geol., 15(6):659-672.

Lohmann, G. P. and J. J. Carlson. 1981. Oceanographic significance of Pacific Late Miocene calcareous nannoplankton. Mar. Micropaleo., 6:553-579.

Lohmann, G. P. and B. A. Malmgren. 1983. Equatorward migration of *Globorotalia truncatulinoides* ecophenotypes through the Late Pleistocene: Gradual evolution or ocean change? Paleobiology, 9(4):414-421.

Long, C. A. 1985. Intricate sutures as fractal curves. J. Morph., 185:285-295.

Lorenz, E. N. 1959. Empirical orthogonal functions and statistical weather prediction. Report No. 1, Statistical Weather Forcasting Project, MIT.

Machuca, R. and A. L. Gilbert. 1981. Finding edges in noisy scenes. IEEE Trans. Pattern Anal. Mach. Intell., PAMI-3:103-111.

MacLeod, N., J. A. Kitchell and G. Keller. Evolutionary opportunity via progenesis: shifts in developmental timing associated with a Late Eocene impact event. (submitted to Marine Micropaleontology).

MacNamara, K. J. 1986. A guide to the nomenclature of heterochrony. J. Paleontol., 60:4-13.

Maitra, S. 1979. Moment invariants. Proc. of the IEEE., 67:697-699.

Malmgren, B. A., W. A. Berggren, and G. P. Lohmann. 1983. Evidence for punctuated gradualism in the Late Neogene *Globorotalia tumida* lineage of planktonic foraminifera. Paleobiology, 9(4):377-389.

Malmgren, B. A. and J. P. Kennett. 1972. Biometric analysis of phenotypic variation: *Globigerina pachyderma* (Ehrenberg) in the South Pacific Ocean. Micropaleontology, 18(2):241-48.

Marcus, L. F. 1982. A portable, digital, printing caliper. Curator, 25(3):233-226.

Mardia, K. V. and I. L. Dryden. 1989. The statistical analysis of shape data. Biometrika, 76:271-281.

Maronna, R. A. 1976. Robust M-estimators of multivariate location and scatter. Ann. Statist., 4: 51-67.

Mayr, E. 1942. Systematics and the origin of species. Columbia University Press, New York, 334 pp.

Mero, L. and Z. Vassy. 1975. A simplified and fast version of the Hueckel operator for finding optimal edges in pictures. Proc. 4th. Intl. Conf. on Artificial Intelligence. Tbilisi, USSR. 650-655.

Mickevich, M. F. 1982. Transformation series analysis. Syst. Zool., 31(4):461-478.

Mittelbach, G. G. 1984. Predation and resources partitioning in two sunfishes (Centrarchidae). Ecology, 65:499-513.

Mix, A. C., W. F. Ruddiman, and A. McIntyre. 1986. Late Quaternary paleoceanography of the Tropical Atlantic, 1: Spatial variability of annual mean sea-surface temperatures, 0-20,000 years B. P. Paleoceanography, 1(1):43-66.

Moellering, H. and J. N. Raynor. 1981. The harmonic analysis of spatial shapes using dual axis Fourier shape analysis (DAFSA). Geographical Anal., 13:64-77.

Moellering, H. and J. N. Raynor. 1982. The dual axis Fourier shape analysis of closed cartographic forms. The Cartographic J., 19:53-59.

Montanari, U. 1969. Continuous skeletons from digitized images. J. Assn. Comput. Mach., 16:534-549.

Morrison, D. F. 1976. Multivariate statistical methods. Mc-Graw Hill, New York.

Morse, D. R., J. H Lawton, M. M. Dodson, and M. H. Williamson. 1985. Fractal dimension of vegetation and the distribution of arthropod body lengths. Nature, 314:731-733.

Mortenson, M. E. 1985. Geometric modeling. John Wiley & Sons, New York, 763 pp.

Mosier, C. I. 1939. Determining a simple structure when loading for certain tests are known. Psychometrika, 4:149-162.

Mosimann, J. E. 1970. Size allometry: size and shape variables with characterizations of the lognormal and generalized gamma distributions. J. Amer. Stat. Assoc., 65:930-945.

Mosimann, J. E. and F. C. James. 1979. New statistical methods for allometry with application to Florida red-winged blackbirds. Evolution, 33:444-459.

Moss, M. L., H. M. Pucciarelli, L. Moss-Salentijn, R. Skalak, A. Bose, C. Goodall, K. Sen, B. Morgan and M. Winick. 1987. Effects of pre-weaning undernutrition on 21 day-old male rat skull form as described by the finite element method. Gegenbaurs Morphol. Jb., Leipzig, 133(6):837-868.

Mosteller, F. and J. W. Tukey. 1977. Data analysis and regression. A second course in statistics. Addison-Wesley Pub. Co., Reading, Mass.

Mulaik, S. A. 1986. Factor analysis and psychometrika: major developments. Psychometrika, 51: 23-33.

Mulaik, S. A. 1972. The foundations of factor analysis. McGraw-Hill, New York, 452 pp.

Nagao, M. and T. Matsuyama. 1979. Edge preserving smoothing. Computer Graphics Image Processing, 9:394-407.

Neff, N. A. and L. F. Marcus. 1980. A survey of multivariate methods for systematics. Privately Published and A.M.N.H., New York, 243 pp.

Nelson, G. 1978. Ontogeny, phylogeny, paleontology, and the biogenetic law. Syst. Zool., 27:324-345.

Nieuwenhuys, R., and T. S. Bodenheimer. 1966. The diencephalon of the primitive bony fish Polypterus, in the light of the problem of homology. J. Morphol., 118:415-450.

Nigrini, C. A. 1974. Cenozoic Radiolaria from the Arabian Sea, DSDP Leg 23. Pp. 1051-1121 *in* Initial reports of the deep sea drilling project, Volume 23 (Davies, T. A., B. P. Lyuendyk, et

374

al.). Washington, D. C., U.S. Government Printing Office, 1183 pp.

O'Hara, R.J. 1988. Homage to Clio, or, toward an historical philosophy for evolutionary biology. Syst. Zool., 37(2):142-155.

Olsham A. F., A. F. Seigel, and D. R. Swindler. 1982. Robust and least-squares orthogonal mapping: methods for the study of cephalofacial form and growth. Am. J. Phys. Anthropol., 59:131-137.

Olson, E. C. and R. J. Miller. 1958. Morphological integration. Univ. Chicago Press, Chicago.

Olsson, R. K. 1971. The logarithmic spire in planktonic foraminifera: its use in taxonomy, evolution and paleoecology. Trans. Gulf Coast Assoc. Geol. Soc., 21:419-32.

Olsson, R. K. 1973. What is a kummerform planktonic foraminifer. J. Paleontol., 47(2): 327-29.

Owen, R. 1848. Report on the archetype and homologies of the vertebrate skeleton. Rep. 16th Mtg. British Assoc. Adv. Sci., 169-340.

Oxnard, C. E. 1972. Some problems in the comparative assessment of skeletal form. Pp. 1-23 in Symposium on human evolution (M. H. Day, ed.). Taylor and Francis, London.

Oxnard, C. 1973. Form and pattern in human evolution: some mathematical, physical and engineering approaches. The University of Chicago Press, Chicago.

Oxnard, C. 1975. Uniqueness and diversity in human evolution. University of Chicago Press, Chicago.

Papoulis, A. 1965. Probability, random variables, and stochastic variables. McGraw-Hill, New York.

Parnell, J. N. and P. E. Lestrel. 1977. A computer program for comparing irregular two-dimensional forms. Computer Programs in Biomedicine, 7:145-161.

Paton, K. 1979. Line detection by local methods. Computer Graphics Image Processing, 9: 316-332.

Patterson, C. 1982. Morphological characters and homology. Pp. 21-74 in Problems of phylogenetic reconstruction (Joysey, K. A. and A. E. Friday, eds.). The Systematics Association Special vol. 21, Academic Press, London, New York, 442 pp.

Patterson, C. 1987. Introduction. Pp. 1-22 in Molecules and morphology in evolution (C. Patterson, ed.). Cambridge University Press, Cambridge and New York, 229 pp.

Patton, J. L. 1987. Species groups of spiny rats, genus Proechimys (Rodentia:Echimyidae). Fieldiana (Zoology), 39:305-346.

Patton, J. L. and L. H. Emmons. 1985. A review of the genus Isothrix (Rodentia:Echimyidae). Amer. Museum Novitates, 2817:1-14.

Patton, J. L. and M. A. Rogers. 1983. Systematic implications of non-geographic variation in the spiny rat genus Proechimys (Echimyidae). Zeitshrift fur Saugertierkunde, 48:363-370.

Pavlidis, T. 1982. Algorithms for graphics and image processing. Computer Science Press: Rockville, MD, 416 pp.

Pearson, K. 1897. Mathematical contributions to the theory of evolution. Proc. Roy. Soc., 60: 489-498.

Peli, T. and J. S. Lim. 1982. Adaptive filtering for image enhancement. Optical Engineering, 21: 108-112.

Petite-Maire, N. and J. F. Ponge. 1979. Primate cranium morphology through ontogenesis and phylogenesis, factorial analysis of global variation. J. Hum. Evol., 8:233-234.

Pielou, E. C. 1984. The interpretation of ecological data: a primer on classification and ordination. John Wiley & Sons, New York.

Pimentel, R. A. 1979. Morphometrics, the multivariate analysis of biological data. Kendall/ Hunt Publishing Co., Dubuque, Iowa, 276 pp.

Preparata, F. P. and M. I. Shamos. 1985. Computational geometry: An introduction. Springer-Verlag, New York, 390 pp.

Press, W. H., B. P. Flannery, S. A. Teukolsky, and W. T. Vetterling. 1986. Numerical recipes: the art of scientific computing, FORTRAN and Pascal version. Cambridge, New York, 818 pp.

Rao, C. R. 1964. The use and interpretation of principal components analysis in applied research. Sankhya, 26: 329-358.

Rao, C. R. 1966. Covariance adjustment and related problems in multivariate analysis. in Multivariate analysis. (P. R. Krishnaiah, ed.) Academic Press, New York.

Raup, D. M. 1966. Geometric analysis of shell coiling: general problems. J. Paleontol., 40:1178-1190.

Raven, C. P. 1966. Morphogenesis: the analysis of molluscan development. Pergamon, Oxford. 2nd Edition, 365 pp.

Ray, T. S. 1981. Growth and heterophylly in an herbaceous tropical vine, *Syngonium* (Araceae). Ph.D. dissertation, Harvard University.

Ray, T. S. 1983a. *Monstera tenuis*. Pp. 278-280 in Costa Rican natural history (D. Janzen ed.). University of Chicago Press, Chicago, 816 pp.

Ray, T. S. 1983b. *Syngonium triphyllum*. Pp. 333-35 *in* Costa Rican natural history (D. Janzen ed.). University of Chicago Press, Chicago, 816 pp.

Ray, T. S. 1986. Growth correlations within the segment in the Araceae. Amer. J. Bot., 73(7): 993-1001.

Ray, T. S. 1987a. Cyclic heterophylly in *Syngonium* (Araceae). Amer. J. Bot., 74(1): 16-26.

Ray, T. S. 1987b. Leaf types in the Araceae. Amer. J. Bot., 74(9): 1359-1372.

Ray, T. S. 1987c. Diversity of shoot organization in the Araceae. Amer. J. Bot., 74(9): 1373-1387.

Ray, T. S. 1988. Survey of shoot organization in the Araceae. Amer. J. Bot., 75(1): 56-84.

Ray, T. S. Accepted a (in revision). Foraging behavior in climbing Araceae. J. Ecol.

Ray, T. S. Accepted b (in revision). Metamorphosis in the Araceae. Amer. J. Bot.

Ray, T. S., and S. Renner. 1990. Comparative studies on the morphology of the Araceae. A. Engler, 1877. Translation with an introduction, updated nomenclature, and a glossary. Englera 12. 140 pp.

Reddi, S. S. 1981. Radial and angular moment invariants for image identification. EEE Trans. Pattern Anal. Mach. Intell., PAMI-3:240-242.

Reilly, S. M. 1986. Ontogeny of cranial ossification in the eastern newt, *Notophthalmus viridescens* (Caudata: Salamandridae), and its relationship to metamorphosis and neoteny. J. Morphol., 188:215-326.

Reilly, S. M. 1987. Ontogeny of the hyobranchial apparatus in the salamanders *Ambystoma talpoideum* (Ambystomatidae) and *Notophthalmus viridescens* (Salamandridae): the ecological morphology of two neotenic strategies. J. Morphol., 191:205-214.

Reilly, S. M. and G. V. Lauder. 1990. Metamorphosis of cranial design in tiger salamanders: a morphometric analysis of ontogenetic change. J. Morphol., 204:121-137.

Remane, A. 1956. Die Grundlagen des naturlichen Systems der vergleichenden Anatomie und der Phylogenetic. Theoretische Morphologie und systematik. (Second Ed.) Geest und Portig, Leipzig. Also 1971. (Reprint of 1st ed.) Koeltz, Konigstein-Tanaus, 364 pp.

Rempe, U. and E. E. Weber. 1972. An illustration of the principal ideas of MANOVA. Biometrics, 28:235-238.

Rencher, A. C. 1988. On the use of correlations to interpret canonical functions. Biometrika, 75 (2):363-365.

Reyment, R. A. 1969. A multivariate paleontological growth problem. Biometrics, 25:1-8.

Reyment, R. A. 1985. Multivariate morphometrics and analysis of shape. Math. Geol., 17(6):591-609.

Reyment, R. A., Blackith, R. E., Campbell, and N. A. 1984. Multivariate morphometrics. Second Edition, Academic Press, London, 232 pp.

376

Rhoads, J. G. 1984. Improving the sensibility, specificity, and appositeness of morphometric analyses. Pp. 257-259 in Multivariate methods in physical anthropology (van Vark, G. N. and W. W. Howells, eds.). Reidel: Dordrecht, 433 pp.

Riedl, R. 1978. Order in living organisms. Wiley, New York, 313 pp.

Riedel, W. R. and A. Sanfilippo. 1970. Radiolaria, Leg 4, Deep Sea Drilling Project. Pp. 503-575 in Intitial reports of the deep sea drilling project, Volume 4 (Bader, R. G., R. D. Gerard, et al.). Washington, D. C., U. S. Government Printing Office, 753 pp.

Rieppel, O. 1980. Homology, a deductive concept? Z. Zool. Syst. Evolut.-forsch., 18:315-319.

Rindskopf, D. and T. Rose. 1988. Some theory and application of confirmatory second order factor analysis. Multivariate Behavior Research, 23: 51-67.

Riseman, E. M. and M. A. Arbib. 1977. Computational techniques in visual systems, part II: segmenting static scenes. IEEE Computer Soc. Repository, R:77-87

Rock, N. M. S. 1987. Robust: an interactive FORTRAN 77 package for exploratory data analysis using parametric robust and non parametric locations and scale estimates, data transformations, normality tests and outlier assessment. Comp. Geosc., 13: 463-494.

Rohlf, F. J. and S. Ferson. 1983. Image analysis. Pp. 583-599. in Numerical taxonomy. (J. Felsenstein, ed.) Springer-Verlag, Berlin.

Rohlf, F. J. 1972. An empirical comparison of three ordination techniques in numerical taxonomy. Syst. Zool., 21:271-280.

Rohlf, F. J. 1986. The relationships among eigenshape analysis, Fourier analysis, and the analysis of coordinates. Math. Geol., 18:845-854.

Rohlf, J. F. 1988. NTSYS-pc. Numerical taxonomy and multivariate analysis system (Version 1.40). Exeter Publishing, Ltd. New York.

Rohlf, F. J. 1990. The analysis of shape variation using ordinations of fitted functions. In Ordi-

nations in the study of morphology, evolution and systematics of insects: applications and quantitative genetic rationales. (Sorensen, J. T., ed.) Elsevier, Amsterdam. In press.

Rohlf, F. J. and J. Archie, 1984. A comparison of Fourier methods for the description of wing shape in mosquitoes (Diptera: Culicidae). Syst. Zool., 33:302-317.

Rohlf, F. J. and F. L. Bookstein. 1987. A comment on shearing as a method for "size correction". Syst. Zool., 36: 356-367.

Rohlf, F. J. and D. Slice. 1990. Extensions of the Procrustes method for the optimal superimposition of landmarks. Systematic Zool., 39:40-59.

Rosen, D. E. 1973. Interrelationships of higher euteleostean fishes. Pp. 397-513, in Interrelationships of fishes (Greenwood, P. H., R. S. Miles, and C. Patterson, eds.). Suppl. 1, Zool. J. Linnean Soc. Vol. 53.

Rosenfield, A. and A. C. Kak. 1982. Digital picture processing, volume 1. Academic Press, New York, 435 pp.

Roth, V. L. 1984. On homology. Biol. J. Linnean Soc., 22:13-29.

Roth, V. L. 1988. The biological basis of homology. Pp. 1-26 in Ontogeny and systematics (C. J. Humphries, ed.). Columbia Univ. Press, New York, 236 pp.

Samet, H. 1981a. An algorithm for converting rasters to quadtrees. IEEE Trans. Pattern Anal. Mach. Intell., PAMI-3:93-95

Samet, H. 1981b. Computing perimeters of regions in images represented by quadtrees. IEEE Trans. Pattern Anal. Mach. Intell., PAMI-3: 683-687.

Samet, H. 1983. A quadtree medial axis transform. Communications ACM, 26:680-393.

Sampson, P. D. 1982. Fitting conic sections to "very scattered" data: An iterative refinement of the Bookstein algorithm. Comp. Graph. Im. Proc., 18:97-108.

Sanfilippo, A. and W. R. Riedel. 1973. Cenozoic Radiolaria (exclusive of theoperids, artostrobiids and amphipyndacids) from the Gulf of Mexico, DSDP Leg 10. Pp. 475-611 *in* Initial reports of the deep sea drilling project, Volume 10 (Worzel, J. L., W. Bryant, et al.). Washington, D. C., U. S. Government Printing Office, pp. 475-611.

Sanfilippo, A. and W. R. Riedel. 1974. Radiolaria from the west-central Indian Ocean and Gulf of Aden, DSDP Leg 24. Pp. 997-1035 *in* Initial reports of the deep sea drilling project (Fisher, R. L., E. T. Bunce, et al.). Washington, D. C., U. S. Government Printing Office, 1183 pp.

Sanfilippo, A., M. J. Westberg-Smith and W. R. Riedel. 1985. Cenozoic Radiolaria. Pp. 631-712 *in* Plankton stratigraphy (Bolli, H. M., K. Perch-Nielsen and J. B. Saunders, eds.). Cambridge University Press, Cambridge, 1032 pp.

Sattler, R. 1984. Homology - a continuing challenge. Syst. Bot., 9:382-394.

Schweitzer, P. N. 1990. Inference of ecology from ontogeny of microfossils. Ph. D. Thesis, MIT/WH01, WH01-90-04.

Schweitzer, P. N., R. L. Kaesler, and G. P. Lohmann. 1986. Ontogeny and heterochrony in the ostracode Cavellina Coryell from the Lower Permian rocks in Kansas. Paleobiology, 12(3):290-301.

Schweitzer, P. N. and G. P. Lohmann. 1990. Life-history and the evolution of ontogeny in the ostracode genus, *Cyprideis*. Paleobiology, In press.

Scott, G. H. 1966. Description of an experimental class in the *Globigerinidae* (foraminifera), 1. New Zealand J. Geol. and Geophys., 9:513-540.

Scott, G. H. 1980. The value of outline processing in the biometry and systematics of fossils. Paleontology, 23:757-768.

Scott, G. H. 1981. Upper Miocene biostratigraphy: Does *Globorotalia conomiozea* occur in the Messinian? Revista Espanola Micropaleo., 12:489-506.

Scott, D. S. and S. Iyengar. 1986. TID--A translation invariant data structure for storing images. Comm. ACM., 29:418-429.

Seber, G. A. F. 1984. Multivariate observations. Wiley Series in Probability and Mathematical Statistics. Wiley and Sons, New York, 686 pp.

Serra, J. 1982. Image analysis and mathematical morphology. Academic Press, London, 610 pp.

Shaw, G. B. 1979. Local and regional edge detectors: some comparisons. Computer Graphics and Image Processing, 9:135-149

Shaw, H. F. and G. J. Wasserburg. 1982. Age and provenance of the target materials for tekities and possible impactites as inferred from Sm-Nd and Rb-Sr systematics. Earth Planet. Sc. Lett., 60:155-77.

Shea, B. T. 1985. Bivariate and multivariate growth allometry: statistical and biological considerations. J. Zool. Lond. (A) 206:367-390.

Sibley, C. G., and J. E. Ahlquist. 1987. Avian phylogeny reconstructed from comparisons of the genetic material, DNA. Pp. 95-121 *in* Molecules and morphology in evolution: conflict or compromises (Patterson, C., ed.). Cambridge Univ. Press, Cambridge, 229 pp..

Sibson, R. 1978. Studies in robustness of multidimensional scaling: Procrustes statistics. J. Roy. Statist. Soc., B, 40: 234-238.

Siegel, A. F. 1982a. Robust regression using repeated medians. Biometrika, 69:242-244.

Siegel, A. F. 1982b. Geometric data analysis: an interactive graphics program for shape comparisons. Pp, 103-122 in Launer, R. L. and Siegel, A. F. Modern Data Analysis. Academic Press:New York.

Siegel, A. F. and R. H. Benson. 1982. A robust comparison of biological shapes. Biometrics, 38:341-350.

Simpson, G. G. 1961. Principles of animal taxonomy. Columbia Univ. Press, New York, 247 pp.

Sjovold, T. 1975. Some notes on the distribution and certain modifications of Mahalanobis' generalized distance (D2). J. Hum. Evol., 4:549-558.

Skalak, R., D. Dasgupta, M. Moss, E. Otten, P. Dullemeiher, H. Vilmann. 1982. Analytical

378

description of growth. J. Theoretical Biol., 94:555-577.

Small, C. G. 1987. A survey of shape statistics. Pp. 1-9 in Proc. of the section on statistical graphics, 1987 Annual Meeting of the Amer. Statistical Assn.

Small, C. G. 1988. Techniques of shape analysis on sets of points. Internat. Statistical Rev., 56:243-257.

Smit, J. 1982. Extinction and evolution of planktonic foraminifera after a major impact at the Cretaceous/Tertiary boundary. Geol. Soc. Amer. Spec. Paper, 190:329-352.

Smith, D., B. Crespi, and F. L. Bookstein. 1990. Asymmetry and morphological abnormality in the honey bee, *Apis mellifera*: effects of ploidy and hybridization. Evolution, *submitted*.

Sneath, P. H. A. 1967. Trend-surface analysis of transformation grids. J. Zool., London, 151(1): 65-122.

Sneath, P. H. A., and R. R. Sokal. 1973. Numerical taxonomy. W.H. Freeman and Co., San Francisco, 573 pp.

Sokal, R. R. and F. J. Rohlf. 1981. Biometry. The principles and practice of statistics in biological research. W. H. Freeman, San Francisco.

Sokal, R. R., and P. H. A. Sneath. 1963. Principles of numerical taxonomy. W.H. Freeman and Co., San Francisco, 359 pp.

Spath, H. 1974. Spline algorithms for curves and surfaces. Utilitas Mathematica Publishing, Winnipeg, Canada, 198 pp.

Stainforth, R. M., J. L. Lamb, H. Luterbacher, J. H. Beard and R. M. Jeffords. 1975. Cenozoic planktonic foraminiferal zonation and characteristics of index forms. Univ. Kansas Paleo. Contrib. Art., 62:1-425.

Stigler, S. H. 1986. The history of statistics: the measurement of uncertainty to 1900. Harvard University Press, Cambridge, MA.

Strauss, R. E., and F. L. Bookstein. 1982. The truss: body form reconstruction in morphometrics. Syst. Zool., 31:113-135.

Strauss, R. E., and L. A. Fuiman. 1985. Quantitative comparisons of body form and allometry in larval and adult Pacific sculpins (Teleostei: Cottidae). Can. J. Zool., 63:1582-89.

Tabachnick, R. E. 1981. Morphologic variation in Miocene *Globorotalia* bulloides, D'Orbigny, New Port Bay, California. Unpubl. M.S. Thesis, Univ. California, Los Angeles.

Tabachnick, R. E. and F. L. Bookstein. 1990. The structure of individual variation in Miocene *Globorotalia*, DSDP Site 593. Evolution, 44: 416-434.

Tanimoto, S. and T. Pavlidis. 1975. A hierarchical data structure for picture processing. Computer Graphics Image Proc., 4:104-119.

Teissier, G. 1938. Un essai d'analyse factorielle. Les variants sexuels de Maia squinaday.

Ten Berge, J. M. F. and K. Nevels. 1977. A general solution to Mosier's oblique Procrustes problem. Psychometrika, 42:593-600.

ter Braak, C. A. J. 1986. Canonical correspondence analysis: a new eigenvector technique for multivariate direct gradient analysis. Ecology, 67:1167-1179.

Thompson, D. W. 1917. On growth and form. Cambridge:London, 793 pp.

Thompson, D. W. 1942. On growth and form. Cambridge Univ. Press, Cambridge, 793 pp.

Thorpe, R. S. 1988. Multiple group principal component analysis and population differentiation. J. Zool. Lond., 216:37-40.

Tissot, B. N. 1988a. Geographic variation and heterochrony in two species of cowries (genus *Cypraea*). Evolution, 42(1):103-17.

Tissot, B. N. 1988b. Multivariate analysis. Pp. 35-51 *in* Heterochrony in evolution: a multidisciplinary approach (McKinney, M.L. ed.). Plenum, New York, 432 pp.

Tobler, W. R. 1977. Bidimensional regression. A computer program. Published by the Author, Geography Dept., U. Calif, Santa Barbara, 72 pp.

Tobler, W. R. 1978. Comparison of plane forms. Geogr. Anal., 10(2):154-162.

van der Klaauw, C. J. 1966. Introduction to the philosophic backgrounds and prospects of the supraspecific comparative anatomy of conservative characters in the adult stages of conservative elements of Vertebrata with an enumeration of many examples. Verh. Kon. Ned. Akad. Wetensch., Afd. Natuurk., Tweede Sect., 57:1-196.

van Morkhoven, F. C.P. M. 1962. Post-Palaeozoic Ostracoda: Their morphology, taxonomy, and economic use. Elsevier, Amsterdam, 204 pp.

van Valen, L. 1978. The statistics of variation. Evol. Theor., 4:33-43.

Van Valen, L. 1982. Homology and causes. J. Morphology, 173:305-312.

Vlcek, J. and E. Cheng. 1986. Fractal analysis of leaf shape. Can. J. Forestry Res., 16:124-127.

Voss, R. S., L. F. Marcus and P. Escalante. in press. Morphological evolution in muroid rodents. I. Conservative patterns of craniometric covariance and their ontogenetic basis in the neotropical genus Zygodotomys. Syst. Zool.,

Wagner, G. P. 1989. The origin of morphological characters and the biological basis of homology. Evolution, 43:1157-1171.

Wagner, W. H. Jr. 1980. Origin and philosophy of the groundplan divergence method of cladistics. Syst. Bot., 5: 173-193.

Wang, S., A. Y. Wu, and A. Rosenfeld. 1981. Image approximation from gray scale "medial axes." IEEE Trans. Pattern Anal. Mach. Intell, PAMI-3:687-??.

Webber, R. L. and H. Blum. 1979. Angular invariants in developing human mandibles. Science, 206(4419):689-691.

Werdelin, L. 1983. Morphological patterns in the skulls of cats. Biol. J. Linnean Soc., 19:375-391.

Werdelin, L. 1988. Correspondence analysis and the analysis of skull shape and structure. Ossa, 13:181-189.

Werner, E. E. and D. J. Hall. 1976. Niche shifts in sunfishes: Experimental evidence and significance. Science, 191:404-406.

White, R. J. and H. C. Prentice. 1987. Comparison of shape description methods for biological outlines. Pp. 395-402 in Classification and related methods of data analysis. (H. H. Bock, ed.). North-Holland, New York.

Wiley, E. O. 1981. Phylogenetics: The theory and practice of phylogenetic systematics. Wiley-Interscience, New York.

Willig, M. R., R. D. Owen, and R. L. Colbert. 1986. Assessment of morphometric variation in natural populations: the inadequacy of the univariate approach. Syst. Zool., 35(2):195- 203.

Withers, R. F. J. 1964. Morphological correspondence and the concept of homology. Pp. 378-394, in Form and strategy in science (Gregg, J. R. and F. T. C. Harris, eds.). Studies dedicated to Joseph Henry Woodger on the occasion of his seventieth birthday. Reidel Publ. Co., Ordrecht, The Netherlands, 476 pp.

Woodger, J. H. 1945. On biological transformations. Pp. 94-120 in Essays on growth and form presented to D'Arcy Wentworth Thompson (Clark,W. E. LeGros and P. B. Medawar). Clarendon Press, Oxford, 408 pp.

Woods, C. A. 1984. Hystricognath rodents. Pp. 389-446 in Orders and families of recent mammals of the World (S. Anderson and J.K. Jones, Jr., eds.). John Wiley & Sons, New York, 686 pp.

Wright, S. 1921. Correlation and causation. Jour. Agric. Res., 20:557-585.

Wright, S. 1954. The interpretation of multivariate systems. Pp. 11-33 in Statistics and mathematics in biology (O. Kempthorne, T. A. Bancroft, J. W. Gowen and J. L. Lush, eds.). The Iowa State College Press, Ames.

Wright, S. 1968. Evolution and genetics of populations. I. Genetic and biometric foundations. Univ. of Chicago Press, Chicago.

Yin, B. H. and H. Mack. 1981. Target classification algorithms for video and forward looking infrared (FLIR) imagery. SPIE, Infrared Technology for Target Detection and Classification, 302:134-140.

Younker, J. L. and R. Ehrlich. 1977. Fourier biometrics: harmonic amplitudes as multivariate shape descriptors. Syst. Zool., 26:336-342.

Zablotny, J. E. 1988. A non conventional morphometric technique for measuring ontogenetic shape changes of the fifth ceratobranchial in two species of centrarchid fishes. M.S. thesis, Michigan State University, East Lansing, Michigan, 82 pp.

Zahn, C. T., and R. Z. Roskies. 1972. Fourier descriptors for plane closed curves. IEEE Trans. Comp., C21(3): 269-281.

Zelditch, M. L. 1987. Evaluating models of developmental integration in the laboratory rat using confirmatory factor analysis. Syst. Zool., 36(4): 368-380.

Zelditch, M. L. 1988. Ontogenetic variation and patterns of phenotypic integration in the laboratory rat. Evolution, 42(1):28-41.

Zelditch, M. L., R. W. DeBry, and D. O. Straney. 1989. Triangulation measurement schemes in the analysis of size and shape. J. Mamm., 70:571-579.

Zhang, T. Y. and C. Y. Suen. 1984. A fast parallel algorithm for thinning digital patterns. Comm. ACM, 27:236-239.